Springer Collected Works in Mathematics

For further volumes:
http://www.springer.com/series/11104

With Father, 1917

At Nankai Institute of Mathematics, 1986

Shiing-Shen Chern

Selected Papers III

Reprint of the 1989 Edition

Springer

Shiing-Shen Chern
(1911 Jiaxing, China – 2004 Tianjin, China)
The University of Chicago
Chicago, IL
USA

ISSN 2194-9875
ISBN 978-1-4614-4396-4 (Softcover)
978-0-387-96817-9 (Hardcover)
DOI 10.1007/978-1-4614-4397-1
Springer New York Heidelberg Dordrecht London

Library of Congress Control Number: 2012954381

Mathematics Subject Classification: 01A75, 53-02

Printed on acid-free paper

Springer is part of Springer Science+Business Media (www.springer.com)

Contents

*Numbers in brackets refer to the Bibliography of the Publications of S.S. Chern (see pages vii–xiv).

Bibliography of the Publications of S.S. Chern

Note: **Boldface** numbers at the end of each entry denote the volume in which the entry appears.

I. *Books and Monographs*

1. *Topics in Differential Geometry* (mimeographed), Institute for Advanced Study, Princeton (1951), 106 pp. **(IV)**
2. *Differentiable Manifolds* (mimeographed), University of Chicago, Chicago (1953), 166 pp.
3. *Complex Manifolds*
   a. University of Chicago, Chicago (1956), 195 pp.
   b. University of Recife, Recife, Brazil (1959), 181 pp.
   c. Russian translation, Moscow (1961), 239 pp.
4. *Studies in Global Geometry and Analysis* (Editor), Mathematical Association of America (1967), 200 pp.
5. *Complex Manifolds without Potential Theory*, van Nostrand (1968), 92 pp. Second edition, revised, Springer-Verlag (1979) 152 pp.
6. *Minimal Submanifolds in a Riemannian Manifold* (mimeographed), University of Kansas, Lawrence (1968), 55 pp. **(IV)**
7. (with Wei-huan Chen) *Differential Geometry Notes*, in Chinese, Beijing University Press (1983), 321 pp.
8. *Studies in Global Differential Geometry* (Editor), Mathematical Association of America (1988), 350 pp.

II. *Papers*

1932

[1] Pairs of plane curves with points in one-to-one correspondence. *Science Reports Nat. Tsing Hua Univ.* **1** (1932) 145–153. **(II)**

1935

*[2] Triads of rectilinear congruences with generators in correspondence. *Tohoku Math. J.* **40** (1935) 179–188.

[3] Associate quadratic complexes of a rectilinear congruence. *Tohoku Math. J.* **40** (1935) 293–316. **(II)**

[4] Abzählungen für Gewebe. *Abh. Math. Sem. Univ. Hamburg* **11** (1935) 163–170. **(I)**

1936

[5] Eine Invariantentheorie der Dreigewebe aus r-dimensionalen Mannigfaltigkeiten im R_{2r}. *Abh. Math. Sem. Univ. Hamburg* **11** (1936) 333–358. **(I)**

* Does not appear in these volumes.

1937

[6] Sur la géométrie d'une équation différentielle du troisième ordre. *C. R. Acad. Sci. Paris* **204** (1937) 1227–1229. **(II)**

[7] Sur la possibilitè de plonger un espace à connexion projective donné dans un espace projectif. *Bull. Sci. Math.* **61** (1937) 234–243. **(I)**

1938

[8] On projective normal coördinates. *Ann. of Math.* **39** (1938) 165–171. **(II)**

[9] On two affine connections. *J. Univ. Yunnan* **1** (1938) 1–18. **(II)**

1939

[10] Sur la géométrie d'un système d'équations différentielles du second ordre. *Bull. Sci. Math* **63** (1939) 206–212. **(II)**

1940

*[11] The geometry of higher path-spaces. *J. Chin. Math. Soc.* **2** (1940) 247–276.

[12] Sur les invariants intégraux en géométrie. *Science Reports Nat. Tsing Hua Univ.* **4** (1940) 85–95. **(II)**

[13] The geometry of the differential equation $y''' = F(x, y', y'')$. *Science Reports Nat. Tsing Hua Univ.* **4** (1940) 97–111. **(I)**

[14] Sur une généralisation d'une formule de Crofton. *C.R. Acad. Sci. Paris* **210** (1940) 757–758. **(II)**

[15] (with C.T. Yen) Sulla formula principale cinematica dello spazio ad *n* dimensioni. *Boll. Un. Mat. Ital.* **2** (1940) 434–437. **(II)**

*[16] Generalization of a formula of Crofton. *Wuhan Univ. J. Sci.* **7** (1940) 1–16.

1941

[17] Sur les invariants de contact en géométrie projective différentielle. *Acta Pontif. Acad. Sci.* **5** (1941) 123–140. **(II)**

1942

[18] On integral geometry in Klein spaces. *Ann. of Math.* **43** (1942) 178–189. **(I)**

[19] On the invariants of contact of curves in a projective space of *N* dimensions and their geometrical interpretation. *Acad. Sinica Sci. Record* **1** (1942) 11–15. **(I)**

[20] The geometry of isotropic surfaces. *Ann. of Math.* **43** (1942) 545–559. **(II)**

[21] On a Weyl geometry defined from an $(n - 1)$-parameter family of hypersurfaces in a space of *n* dimensions. *Acad. Sinica Sci. Record* **1** (1942) 7–10. **(I)**

1943

[22] On the Euclidean connections in a Finsler space. *Proc. Nat. Acad. Sci. USA*, **29** (1943) 33–37. **(II)**

[23] A generalization of the projective geometry of linear spaces. *Proc. Nat. Acad. Sci. USA*, **29** (1943) 38–43. **(I)**

1944

[24] Laplace transforms of a class of higher dimensional varieties in a projective space of *n* dimensions. *Proc. Nat. Acad. Sci. USA*, **30** (1944) 95–97. **(II)**

[25] A simple intrinsic proof of the Gauss–Bonnet formula for closed Riemannian manifolds. *Ann of Math.* **45** (1944) 747–752. **(I)**

[26] Integral formulas for the characteristic classes of sphere bundles. *Proc. Nat. Acad. Sci. USA* **30** (1944) 269–273. **(II)**

*[27] On a theorem of algebra and its geometrical application. *J. Indian Math. Soc.* **8** (1944) 29–36.

1945

*[28] On Grassmann and differential rings and their relations to the theory of multiple integrals. *Sankhya* **7** (1945) 2–8.

[29] Some new characterizations of the Euclidean sphere. *Duke Math. J.* **12** (1945) 279–290. **(II)**

[30] On the curvature integra in a Riemannian manifold. *Ann. of Math.* **46** (1945) 674–684. **(I)**

*[31] On Riemannian manifolds of four dimensions. *Bull. Amer. Math. Soc.* **51** (1945) 964–971.

1946

[32] Some new viewpoints in the differential geometry in the large. *Bull. Amer. Math. Soc.* **52** (1946) 1–30. **(II)**

[33] Characteristic classes of Hermitian manifolds. *Ann. of Math.* **47** (1946) 85–121. (I)

1947

[34] (with H.C. Wang). Differential geometry in symplectic space I. *Science Report Nat. Tsing Hua Univ.* **4** (1947) 453–477. **(II)**

[35] Sur une classe remarquable de variétés dans l'espace projectif à *N* dimensions. *Science Reports Nat. Tsing Hua Univ.* **4** (1947) 328–336. **(I)**

*[36] On the characteristic classes of Riemannian manifolds. *Proc. Nat. Acad. Sci USA*, **33** (1947) 78–82.

*[37] Note of affinely connected manifolds. *Bull. Amer. Math. Soc.* **53** (1947) 820–823; correction ibid **54** (1948) 985–986.

*[38] On the characteristic ring of a differentiable manifold. *Acad. Sinica. Sci. Record* **2** (1947) 1–5.

1948

[39] On the multiplication in the characteristic ring of a sphere bundle. *Ann. of Math.* **49** (1948) 362–372. **(I)**

[40] Note on projective differential line geometry. *Acad. Sinica Sci. Record* **2** (1948) 137–139. **(II)**

*[41] (with Y.L. Jou) On the orientability of differentiable manifolds. *Science Reports Nat. Tsing Hua Univ.* **5** (1948) 13–17.

[42] Local equivalence and Euclidean connections in Finsler spaces. *Science Reports Nat. Tsing Hua Univ.* **5** (1948) 95–121. **(II)**

1949

[43] (with Y.F. Sun). The imbedding theorem for fibre bundles. *Trans. Amer. Math. Soc* **67** (1949) 286–303. **(II)**

*[44] (with S.T. Hu) Parallelisability of principal fibre bundles. *Trans. Amer. Math. Soc.* **67** (1949) 304–309.

1950

[45] (with E. Spanier). The homology structure of sphere bundles. *Proc. Nat. Acad. Sci. USA*, **36** (1950) 248–255. **(II)**

[46] Differential geometry of fiber bundles. *Proc. Int. Congr. Math.* (1950) **II** 397–411. **(II)**

1951

[47] (with E. Spanier). A theorem on orientable surfaces in four-dimensional space. *Comm. Math. Helv.* **25** (1951) 205–209. **(I)**

BIBLIOGRAPHY OF THE PUBLICATIONS OF S.S. CHERN

1952

[48] On the kinematic formula in the Euclidean space of *N* dimensions. *Amer. J. Math* **74** (1952) 227–236. **(II)**

[49] (with C. Chevalley). Elie Cartan and his mathematical work. *Bull. Amer. Math. Soc.* **58** (1952) 217–250. **(II)**

[50] (with N.H. Kuiper) Some theorems on the isometric imbedding of compact Riemann manifolds in Euclidean space. *Ann. of Math.* **56** (1952) 422–430. **(II)**

1953

[51] On the characteristic classes of complex sphere bundles and algebraic varieties. *Amer. J. of Math.*, **75** (1953) 565–597. **(I)**

*[52] Some formulas in the theory of surfaces. *Boletin de la Sociedad Matematica Mexicana*, **10** (1953) 30–40.

[53] Relations between Riemannian and Hermitian geometries. *Duke Math. J.*, **20** (1953) 575–587. **(II)**

1954

[54] Pseudo-groupes continus infinis *Colloque de Geom. Diff.* Strasbourg (1954) 119–136. **(I)**

[55] (with P. Hartman and A. Wintner) On isothermic coordinates. *Comm. Math. Helv.* **28** (1954) 301–309. **(I)**

1955

[56] La géométrie des sous-variétés d'un espace euclidien à plusieurs dimensions. *l'Ens. Math.*, **40** (1955) 26–46. **(II)**

[57] An elementary proof of the existence of isothermal parameters on a surface. *Proc. Amer. Math. Soc.*, **6** (1955) 771–782. **(II)**

[58] On special *W*-surfaces. *Proc. Amer. Math. Soc.*, **6** (1955) 783–786. **(II)**

[59] On curvature and characteristic classes of a Riemann manifold. *Abh. Math. Sem. Univ. Hamburg* **20** (1955) 117–126. **(II)**

1956

*[60] Topology and differential geometry of complex manifolds. *Bull. Amer. Math. Soc.*, **62** (1956) 102–117.

1957

[61] On a generalization of Kähler geometry. *Lefschetz jubilee volume*. Princeton Univ. Press (1957) 103–121. **(I)**

[62] (with R. Lashof) On the total curvature of immersed manifolds. *Amer. J. of Math.* **79** (1957) 306–318. **(I)**

[63] (with F. Hirzebruch and J-P. Serre) On the index of a fibered manifold. *Proc. Amer. Math. Soc.*, **8** (1957) 587–596. **(I)**

[64] A proof of the uniqueness of Minkowski's problem for convex surfaces. *Amer. J. of Math.*, **79** (1957) 949–950. **(II)**

1958

*[65] Geometry of submanifolds in complex projective space. *Symposium International de Topologia Algebraica* (1958) 87–96.

[66] (with R.K. Lashof) On the total curvature of immersed manifolds, II. *Michigan Math. J.* **5** (1958) 5–12. **(II)**

[67] Differential geometry and integral geometry. *Proc. Int. Congr. Math. Edinburgh* (1958) 441–449. **(II)**

1959

[68] Integral formulas for hypersurfaces in Euclidean space and their applications to uniqueness theorems. *J. of Math. and Mech.* **8** (1959) 947–956. **(I)**

1960

[69] (with J. Hano and C.C. Hsiung) A uniqueness theorem on closed convex hypersurfaces in Euclidean space. *J. of Math. and Mech.* **9** (1960) 85–88. **(I)**

[70] Complex analytic mappings of Riemann surfaces I. *Amer. J. of Math.* **82** (1960) 323–337. **(I)**

[71] The integrated form of the first main theorem for complex analytic mappings in several complex variables. *Ann. of Math.* **71** (1960) 536–551. **(I)**

*[72] Geometrical structures on manifolds. *Amer. Math. Soc. Pub.* (1960) 1–31.

*[73] La géométrie des hypersurfaces dans l'espace euclidean. *Seminaire Bourbaki*, **193** (1959–1960).

*[74] Sur les métriques Riemanniens compatibles avec une reduction du groupe structural. *Séminaire Ehresmann*, January 1960.

1961

[75] Holomorphic mappings of complex manifolds. *L'Ens. Math.* **7** (1961) 179–187. **(I)**

1962

*[76] Geometry of quadratic differential form. *J. of SIAM* **10** (1962) 751–755.

1963

[77] (with C.C. Hsiung) On the isometry of compact submanifolds in Euclidean space. *Math. Annalen* **149** (1963) 278–285. **(II)**

[78] Pseudo-Riemannian geometry and Gauss–Bonnet formula. *Academia Brasileira de Ciencias* **35** (1963) 17–26. **(I)**

1965

[79] Minimal surfaces in an Euclidean space of N dimensions. *Differential and Combinatorial Topology*, Princeton Univ. Press (1965) 187–198. **(I)**

[80] (with R. Bott) Hermitian vector bundles and the equidistribution of the zeroes of their holomorphic sections. *Acta. Math.* **114** (1965) 71–112. **(II)**

[81] On the curvatures of a piece of hypersurface in Euclidean space. *Abh. Math. Sem. Univ. Hamburg* **29** (1965) 77–91. **(III)**

[82] On the differential geometry of a piece of submanifold in Euclidean space. *Proc. of U.S.–Japan Seminar in Diff. Geom.* (1965) 17–21. **(III)**

1966

[83] Geometry of G-structures. *Bull. Amer. Math. Soc.* **72** (1966) 167–219. **(III)**

[84] On the kinematic formula in integral geometry. *J. of Math. and Mech.* **16** (1966) 101–118. **(III)**

*[85] Geometrical structures on manifolds and submanifolds. *Some Recent Advances in Basic Sciences*, Yeshiva Univ. Press (1966) 127–135.

1967

[86] (with R. Osserman) Complete minimal surfaces in Euclidean n-space. *J. de l'Analyse Math.* **19** (1967) 15–34. **(III)**

[87] Einstein hypersurfaces in a Kählerian manifold of constant holomorphic curvature. *J. Diff. Geom.* **1** (1967) 21–31. **(III)**

1968

[88] On holomorphic mappings of Hermitian manifolds of the same dimension. *Proc. Symp. Pure Math.* **11**. Entire Functions and Related Parts of Analysis (1968) 157–170. **(I)**

1969

[89] Simple proofs of two theorems on minimal surfaces. *L'Ens. Math.* **15** (1969) 53–61. **(I)**

1970

[90] (with H. Levine and L. Nirenberg) Intrinsic norms on a complex manifold. *Global analysis*, Princeton Univ. Press (1970) 119–139. **(I)**

[91] (with M. do Carmo and S. Kobayashi) Minimal submanifolds of a sphere with second fundamental form of constant length. *Functional Analysis and Related Fields*, Springer-Verlag (1970) 59–75. **(I)**

[92] (with R. Bott) Some formulas related to complex transgression. *Essays on Topology and Related Topics*, Springer-Verlag, (1970) 48–57. **(I)**

*[93] Holomorphic curves and minimal surfaces. *Carolina Conference Proceedings* (1970) 28 pp.

[94] On minimal spheres in the four–sphere, Studies and Essays Presented to Y. W. Chen, Taiwan, (1970) 137–150. **(I)**

[95] Differential geometry: Its past and its future. *Actes Congrès Intern. Math.* (1970) **1**, 41–53. **(III)**

[96] On the minimal immersions of the two-sphere in a space of constant curvature. *Problems in Analysis*, Princeton Univ. Press, (1970) 27–40. **(III)**

1971

[97] Brief survey of minimal submanifolds. *Differentialgeometrie im Grossen.* W. Klingenberg (ed.), **4** (1971) 43–60. **(III)**

*[98] (with J. Simons) Some cohomology classes in principal fibre bundles and their application to Riemannian geometry. *Proc. Nat. Acad. Sci. USA*, **68** (1971) 791–794.

1972

[99] Holomorphic curves in the plane. *Diff. Geom., in honor of K. Yano*, (1972) 73–94. **(III)**

*[100] Geometry of characteristic classes. *Proc. 13th Biennial Sem. Canadian Math. Congress*, (1972) 1–40. Also pub. in Russian translation.

1973

[101] Meromorphic vector fields and characteristic numbers. *Scripta Math.* **29** (1973) 243–251. **(I)**

*[102] The mathematical works of Wilhelm Blaschke. *Abh. Math. Sem. Univ. Hamburg* **39** (1973) 1–9.

1974

[103] (with. J. Simons) Characteristic forms and geometrical invariants. *Ann. of Math.* **99** (1974) 48–69. **(I)**

[104] (with M. Cowen, A. Vitter III) Frenet frames along holomorphic curves. *Proc. of Conf. on Value Distribution Theory*, Tulane Univ. (1974) 191–203. **(III)**

[105] (with J. Moser) Real hypersurfaces in complex manifolds. *Acta. Math.* **133** (1974) 219–271. **(III)**

1975

[106] (with S.I. Goldberg) On the volume decreasing property of a class of real harmonic mappings. *Amer. J. of Math.* **97** (1975) 133–147. **(III)**

[107] On the projective structure of a real hypersurface in C_{n+1}. *Math. Scand.* **36** (1975) 74–82. **(III)**

1976

[108] (with J. White) Duality properties of characteristic forms. *Inv. Math.* **35** (1976) 285–297. **(III)**

1977

*[109] Circle bundles. *Geometry and topology, III.* Latin Amer. School of Math, Lecture Notes in Math. Springer-Verlag, **597** (1977) 114–131.

*[110] (with P.A. Griffiths) Linearization of webs of codimension one and maximum rank. *Proc. Int. Symp. on Algebraic Geometry, Kyoto* (1977) 85–91.

1978

[111] On projective connections and projective relativity. *Science of Matter*, dedicated to Ta-you Wu, (1978) 225–232. **(III)**

[112] (with P.A. Griffiths) Abel's theorem and webs. *Jber. d. Dt. Math. Verein.* **80** (1978) 13–110. **(III)**

[113] (with P.A. Griffiths) An inequality for the rank of a web and webs of maximum rank. *Annali Sc. Norm. Super.–Pisa, Serie IV*, **5** (1978) 539–557. (III)

[114] Affine minimal hypersurfaces. *Minimal Submanifolds and Geodesics.* Kaigai Publications, Ltd. (1978) 1–14. **(III)**

1979

*[115] Herglotz's work on geometry. *Ges. Schriften Gustav Herglotz*, Göttingen (1979) xx–xxi.

[116] (with C.L. Terng) An analogue of Bäcklund's theorem in affine geometry. *Rocky Mountain J. Math.* **10** (1979) 105–124. **(III)**

[117] From triangles to manifolds. *Amer. Math. Monthly* **86** (1979) 339–349. **(III)**

[118] (with C.K. Peng) Lie groups and KdV equations. *Manuscripta Math.* **28** (1979) 207–217. **(III)**

1980

[119] General relativity and differential geometry. *Some Strangeness in the Proportion: A Centennial Symp. to Celebrate the Achievements of Albert Einstein, Harry Woolf* (ed.), Addison-Wesley Publ. (1980) 271–287. **(III)**

[120] (with W.M. Boothby and S.P. Wang) The mathematical work of H.C. Wang. *Bull. Inst. of Math,* **8** (1980) xiii–xxiv. **(III)**

*[121] Geometry and physics. *Math. Medley*, Singapore, **8** (1980) 1–6.

*[122] (with R. Bryant and P.A. Griffiths) Exterior differential systems. *Proc. of 1980 Beijing DD-Symposium*, (1980) 219–338.

1981

[123] Geometrical interpretation of the sinh-Gordon equation. *Annales Polonici Mathematici* **39** (1981) 63–69. **(IV)**

[124] (with P.A. Griffiths) Corrections and addenda to our paper: "Abel's theorem and webs." *Jber. d. Dt. Math.–Verein.* **83** (1981) 78–83. **(III)**

[125] (with R. Osserman) Remarks on the Riemannian metric of a minimal submanifold. *Geometry Symposium Utrecht 1980*, Springer Lecture Notes **894** (1981) 49–90. **(IV)**

[126] (with J. Wolfson) A simple proof of Frobenius theorem. *Manifolds and Lie Groups, Papers in Honor of Y. Matsushima.* Birkhäuser (1981) 67–69. **(IV)**

[127] (with K. Tenenblat) Foliations on a surface of constant curvature and modified Korteweg–de Vries equations. *J. Diff. Geom.* **16** (1981) 347–349. **(IV)**

[128] (with C.K. Peng) On the Bäcklund transformations of KdV equations and modified KdV equations. *J. of China Univ. of Sci. and Tech.*, **11** (1981) 1–6. **(IV)**

1982

[129] Web geometry. *Proc. Symp. in Pure Math.* **39** (1983) 3–10. **(IV)**
[130] Projective geometry, contact transformations, and *CR*-structures. *Archiv der Math.* **38** (1982) 1–5. **(IV)**

1983

[131] (with J. Wolfson) Minimal surfaces by moving frames. *Amer. J. Math.* **105** (1983) 59–83. **(IV)**
[132] On surfaces of constant mean curvature in a three–dimensional space of constant curvature. *Geometric Dynamics*, Springer Lecture Notes **1007** (1983) 104–108. **(IV)**

1984

[133] Deformation of surfaces preserving principal curvatures, *Differential Geometry and Complex Analysis*, Volume in Memory of H. Rauch, Springer-Verlag (1984) 155–163. **(IV)**

1985

[134] (with R. Hamilton) On Riemannian metrics adapted to three-dimensional contact manifolds. *Arbeitstagung Bonn 1984* Springer Lecture Notes **1111** (1985) 279–308. **(IV)**
[135] (with J. Wolfson) Harmonic maps of S^2 into a complex Grassmann manifold. *Proc. Nat. Acad. Sci.* USA **82** (1985) 2217–2219. **(IV)**
[136] Moving frames, *Soc. Math. de France*, Astérisque, (1985) 67–77. **(IV)**
*[137] Wilhelm Blaschke and web geometry, Wilhelm Blaschke—Gesammelte Werke. **5**, Thales Verlag, (1985) 25–27.
*[138] The mathematical works of Wilhelm Blaschke—an update. Thales Verlag, (1985), 21–23.

1986

[139] (with K. Tenenblat) Pseudospherical surfaces and evolution equations. *Studies in Applied Math.* MIT **74** (1986) 55–83. **(IV)**
[140] On a conformal invariant of three-dimensional manifolds. *Aspects of Mathematics and Its Applications* Elsevier Science Publishers B.V. (1986) 245–252. **(IV)**
*[141] (with P.A. Griffiths) Pfaffian systems in involution. *Proceedings of 1982 Changchun Symposium on Differential Geometry and Differential Equations*, Science Press, China, (1986) 233–256.

1987

[142] (with J. Wolfson) Harmonic maps of the two–sphere into a complex Grassmann manifold II. *Ann. of Math.* **125** (1987) 301–335. **(IV)**
[143] (with T. Cecil) Tautness and Lie Sphere geometry *Math. Annalen*, Volume Dedicated to F. Hirzebruch **278** (1987) 381–399. **(IV)**

1988

[144] Vector bundles with a connection. *Studies in Global Differential Geometry*, MAA, no. 27 (1988), 1–26.

1989

[145] (with T. Cecil) Dupin submanifolds in Lie sphere geometry, to appear in *Differential Geometry and Topology*, Springer Lecture Notes 1989.
[146] Historical remarks on Gauss–Bonnet, to appear in Moser Volume, Academic Press.
[147] An introduction to Dupin submanifolds, to appear in Do Carmo Volume.

陳省身数學
論文選集

On the Curvatures of a Piece of Hypersurface in Euclidean Space

By SHIING-SHEN CHERN[1])

Dedicated to Professor EMANUEL SPERNER on His Sixtieth Birthday

Let $z = z(x, y)$ be a C^2-surface in euclidean three-space, defined over the disk $x^2 + y^2 < R^2$ in the (x, y)-plane. Let H and K denote respectively its mean and Gaussian curvatures. In 1955 HEINZ ([3], cf. Bibliography at the end) proved the following theorem:

If $|H| \geqq c > 0$, then $R \leqq \frac{1}{c}$. If $K \geqq c > 0$, then $R \leqq \frac{1}{\sqrt{c}}$. If $K \leqq -c < 0$, then $R \leqq e\left(\frac{3}{c}\right)^{\frac{1}{2}}$, ($c$ = const. in all cases).

The purpose of this paper is to extend this theorem to a hypersurface in an euclidean space of dimension $m + 1$. (Cf. Theorems 1, 2, 4 below). Having the global problems in mind, we will consider an immersed hypersurface and establish, in so far as possible, the intermediary results in this general setting. In this sense some of our formulations are more general even in the classical case $m = 2$.

1. Algebraic Preliminaries

Let $x : M \to E$ be an immersion of an oriented manifold M of class two and dimension m into an euclidean space E of dimension $m + 1$. We will consider x as a vector-valued function on M. For $x, y \in E$ we denote by (x, y) their scalar product. Let $x e_1, \ldots, e_{m+1}$ be orthonormal frames, such that $x = x(p)$, $p \in M$, and e_{m+1} is the unit normal vector at $x(p)$. Then we have

$$(1)\qquad dx = \sum_A \omega_A e_A, \qquad de_A = \sum_B \omega_{AB} e_B,$$

where

$$(2)\qquad \omega_{AB} + \omega_{BA} = 0, \quad \omega_{m+1} = 0,$$

[1]) This work is done under partial support by grants from the National Science Foundation and Office of Naval Research, USA.

and the ω's satisfy the structure equations

$$\tag{3} \begin{aligned} d\omega_A &= \sum_B \omega_B \wedge \omega_{BA}, \\ d\omega_{AB} &= \sum_C \omega_{AC} \wedge \omega_{CB}. \end{aligned}$$

(Throughout this paper we will agree on the following ranges of indices:

$$\tag{4} \begin{aligned} 1 \leqq A, B, C \leqq m+1 \\ 1 \leqq i, j, k, h \leqq m.) \end{aligned}$$

As is well-known, we have

$$\tag{5} \omega_{i,m+1} = \sum_k h_{ik}\omega_k, \quad h_{ik} = h_{ki}.$$

Let

$$\tag{6} \det(\lambda\delta_{ik} + h_{ik}) = \sum_{0 \leqq s \leqq m} \binom{m}{s} \sigma_s \lambda^{m-s}.$$

Then σ_s is called the sth curvature of $x(M)$. In particular,

$$\tag{7} \sigma_1 = \frac{1}{m} \sum_i h_{ii}$$

is called the mean curvature and

$$\tag{8} \sigma_m = \det(h_{ik})$$

is the GAUSS-KRONECKER curvature. For $m = 2$ they were denoted above by H and K respectively.

Let a_A be a fixed orthonormal frame in E. Then (a_A, x) is a scalar function on M, and is in fact the height function in the direction a_A. We put

$$\tag{9} \begin{aligned} A_{m-h} = \sum \varepsilon_{i_1, \ldots, i_m} d(a_{i_1}, x) \wedge \cdots \wedge d(a_{i_h}, x) \wedge d(a_{i_{h+1}}, e_{m+1}) \\ \wedge \cdots \wedge d(a_{i_m}, e_{m+1}), \end{aligned}$$

where $\varepsilon_{i_1, \ldots, i_m}$ is the KRONECKER symbol which is $+1$ or -1, according as $i_1, \ldots, i_m$ form an even or odd permutation of $1, \ldots, m$, and is otherwise zero, and where the summation is over all the indices $i_1, \ldots, i_m$. Then A_h is a multiple of σ_h according to the following formula:

$$\tag{10} A_h = (-1)^h m!\, t\sigma_h \Phi,$$

where

$$\tag{11} t = (a_{m+1}, e_{m+1})$$

and

$$\Phi = \omega_1 \wedge \cdots \wedge \omega_m \tag{12}$$

is the volume element of M.

We wish to prove the formula (10). Let α_k, β_k be linear differential forms, and u_{ik} be functions. It follows from the definition of a determinant that

$$\begin{aligned} \sum_{i_1,\dots,i_m} \varepsilon_{i_1,\dots,i_m} (\textstyle\sum u_{i_1 k}\alpha_k) \wedge \cdots \wedge (\sum u_{i_m k}\alpha_k) &= m!\det(u_{ik})\alpha_1 \wedge \cdots \wedge \alpha_m \\ &= \det(u_{ik}) \textstyle\sum \varepsilon_{i_1,\dots,i_m} \alpha_{i_1} \wedge \cdots \wedge \alpha_{i_m}. \end{aligned} \tag{13}$$

Replacing α_k by $\alpha_k + \beta_k$ in the formula and equating the terms of degree h in the α's and $m - h$ in the β's, we get the „polarized form" of (13):

$$\begin{aligned} &\textstyle\sum \varepsilon_{i_1,\dots,i_m} (\sum u_{i_1 k}\alpha_k) \wedge \cdots \wedge (\sum u_{i_h k}\alpha_k) \wedge (\sum u_{i_{h+1} k}\beta_k) \wedge \cdots \wedge (\sum u_{i_m k}\beta_k) \\ &\quad = \det(u_{ik}) \textstyle\sum \varepsilon_{i_1,\dots,i_m} \alpha_{i_1} \wedge \cdots \wedge \alpha_{i_h} \wedge \beta_{i_{h+1}} \wedge \cdots \wedge \beta_{i_m}. \end{aligned} \tag{14}$$

To derive (10) from this relation, we put

$$u_{ik} = (a_i, e_k). \tag{15}$$

Then

$$\begin{aligned} d(a_i, x) &= \textstyle\sum (a_i, e_k)\omega_k = \sum u_{ik}\omega_k, \\ d(a_i, e_{m+1}) &= -\textstyle\sum (a_i, e_k)\,\omega_{k,m+1} = -\sum u_{ik}\omega_{k,m+1}. \end{aligned} \tag{16}$$

Observing that $t = \det(u_{ik})$, we get, from (9), (14), and (16):

$$A_h = (-1)^h t \tilde{A}_h, \tag{17}$$

where

$$\tilde{A}_h = \textstyle\sum \varepsilon_{i_1,\dots,i_m} \omega_{i_1} \wedge \cdots \wedge \omega_{i_{m-h}} \wedge \omega_{i_{m-h+1},m+1} \wedge \cdots \wedge \omega_{i_m,m+1}. \tag{18}$$

Now we have

$$\begin{aligned} &\textstyle\sum \varepsilon_{i_1,\dots,i_m} (\lambda\omega_{i_1} + \omega_{i_1,m+1}) \wedge \cdots \wedge (\lambda\omega_{i_m} + \omega_{i_m,m+1}) \\ &\quad = m!\det(\lambda\delta_{ik} + h_{ik})\Phi = \sum_{0 \leqq s \leqq m} \binom{m}{s} \lambda^{m-s} \tilde{A}_s. \end{aligned}$$

Comparing with (6), we get

$$\tilde{A}_s = m!\,\sigma_s \Phi,$$

and (10) follows.

In particular, we have

$$\begin{aligned} A_1 = (m-1)!\sum_i d(a_1, x) \wedge \cdots \wedge d(a_{i-1}, x) \wedge d(a_i, e_{m+1}) \wedge d(a_{i+1}, x) \\ \wedge \cdots \wedge d(a_m, x) = -(m-1)!\,d\alpha, \end{aligned} \tag{19}$$

The left hand side of (3.62) will involve y^α and z^α, but it is not clear that the compatibilities conditions necessary for it to have the form (3.60) will be satisfied. The computation we are about to give will show that these compatibilities are a consequence of the integrability conditions (3.48) and harmonic property in the form (3.54). It will show, moreover, that along the path we may choose $k^\varrho(x(s))$ with given initial value and with $K^\varrho = 0$ in (3.62).

The first step is to compute $\mathrm{d}Z$. For this we use the notations

$$\begin{cases} \Delta = y^\alpha t_i^\alpha \\ D k^\varrho t_i^\varrho = \mathrm{d}k^\varrho t_i^\varrho + k^{\alpha+\beta}(t_i^{\gamma+\beta}\phi_\gamma^\alpha + t_i^{\gamma+\alpha}\phi_\gamma^\beta). \end{cases} \tag{3.63}$$

Recall that repeated Greek indices are summed. In the second equation Dk^ϱ is defined to be the coefficient of t_i^ϱ on the right hand side. With these conventions, the formula is

$$\begin{aligned} \mathrm{d}Z/\mathrm{d}s = \sum_i (&-k^\varrho t_{i,\varrho} + 1/\Delta((Dk^\varrho/\mathrm{d}s)t_i^\varrho) \\ &+ 1/\Delta\{k^{\alpha+\beta}(t_i^\beta t_{i\alpha\gamma}y^\gamma + t_i^\alpha t_{i\beta\gamma}y^\gamma \\ &- 1/\Delta^2 k^\varrho t_i^\varrho(z^\alpha t_i^\alpha + t_{i\alpha\beta}y^\alpha y^\beta)\})Z_i . \end{aligned} \tag{3.64}$$

Proof. By (3.34)

$$\mathrm{d}Z_i = -\pi^i Z_i + Z_i' t_i^\gamma y^\gamma \mathrm{d}s$$

and so

$$\begin{aligned} \sum_i (k^\varrho t_i^\varrho/y^\alpha t_i^\alpha)\mathrm{d}Z_i &= -\sum_i (k^\varrho t_i^\varrho/\Delta)\pi^i Z_i + \sum_i (k^\varrho t_i^\varrho)Z_i'\mathrm{d}s \\ &= -\sum_i (k^\varrho t_i^\varrho/\Delta)\pi^i Z_i - \sum_i (k^\sigma t_{i,\sigma})Z_i\mathrm{d}s \end{aligned}$$

by (3.56). Next, by the definition of z^α

$$\begin{aligned} \mathrm{d}(y^\alpha t_i^\alpha) &= \mathrm{d}y^\alpha t_i^\alpha + y^\alpha(\pi^i t_i^\alpha + t_i^\beta \phi_\alpha^\beta + t_{i\alpha\beta}y^\beta \mathrm{d}s) \\ &= z^\alpha t_i^\alpha + y^\alpha \pi^i t_i^\alpha + t_{i\alpha\beta}y^\beta \mathrm{d}s . \end{aligned}$$

Finally, setting $\varrho = \alpha + \beta$ and using (3.48)

$$\begin{aligned} \mathrm{d}(k^\varrho t_i^\varrho) &= \mathrm{d}k^\varrho t_i^\varrho + k^{\alpha+\beta}(\mathrm{d}t_i^\alpha t_i^\beta + t_i^\alpha \mathrm{d}t_i^\beta) \\ &= Dk^\varrho t_i^\varrho + 2k^\varrho t_i^\varrho \pi^i + k^{\alpha+\beta}(t_i^\beta t_{i\alpha\gamma}y^\gamma + t_i^\alpha t_{i\beta\gamma}y^\gamma)\mathrm{d}s . \end{aligned}$$

Putting everything together

$$\begin{aligned} \mathrm{d}Z/\mathrm{d}s &= \sum_i ((k^\varrho t_i^\varrho)/\Delta)\mathrm{d}Z_i - \sum_i ((k^\varrho t_i^\varrho)/\Delta^2)\mathrm{d}\Delta Z_i \\ &\quad + \sum_i (\mathrm{d}(k^\varrho t_i^\varrho)/\Delta)Z_i \\ &= (3.64) \times \mathrm{d}s \end{aligned}$$

since the four terms containing π^i cancel. Q.E.D.

We have

$$\frac{1}{(m-1)!}\underbrace{dx \wedge \cdots \wedge dx}_{m-1} = \frac{1}{(m-1)!}(\sum \omega_{i_1} e_{i_1}) \wedge \cdots \wedge (\sum \omega_{i_{m-1}} e_{i_{m-1}})$$

$$= \sum_i p_{1,\ldots,i-1,i+1,\ldots,m}(e_1 \wedge \cdots \wedge e_{i-1} \wedge e_{i+1} \wedge \cdots \wedge e_m)\Psi,$$

so that

(29) $$Q^2 = \sum_i p^2_{1,\ldots,i-1,i+1,\ldots,m}$$

Similarly, the coefficient of Ψ in $\frac{1}{(m-1)!}\underbrace{dx' \wedge \cdots \wedge dx'}_{m-1}$ is

$$\frac{1}{(m-1)!}\sum_i p_{i_1,\ldots,i_{m-1}}(e_{i_1} \wedge \cdots \wedge e_{i_{m-1}})$$

$$+ \frac{1}{(m-1)!}\sum_i \{\sum_{1\leqq s\leqq m-1}(-1)^{m+s} v_k v_{i_s} p_{i_1,\ldots,i_{s-1},i_{s+1},\ldots,i_{m-1}k}\}(e_{i_1} \wedge \cdots \wedge e_{i_{m-1}})$$

$$- \frac{t}{(m-2)!}\sum v_k p_{i_1,\ldots,i_{m-2}k}(e_{i_1} \wedge \cdots \wedge e_{i_{m-2}} \wedge e_{m+1})$$

$$= \sum_{i_1<\cdots<i_{m-1}} \{p_{i_1,\ldots,i_{m-1}} + \sum_{1\leqq s\leqq m-1}(-1)^{m+s} v_k v_{i_s} p_{i_1,\ldots,i_{s-1},i_{s+1},\ldots,i_{m-1}k}\}(e_{i_1} \wedge \cdots \wedge e_{i_{m-1}})$$

$$- t \sum_{i_1<\cdots<i_{m-2}} v_k p_{i_1,\ldots,i_{m-2}k}(e_{i_1} \wedge \cdots \wedge e_{i_{m-2}} \wedge e_{m+1}).$$

From this it follows that

$$(m-1)!\,P^2 = \sum \{p_{i_1,\ldots,i_{m-1}} + \sum_{1\leqq s\leqq m-1}(-1)^{m+s} v_k v_{i_s} p_{i_1,\ldots,i_{s-1},i_{s+1},\ldots,i_{m-1}k}\}$$
$$\{p_{i_1,\ldots,i_{m-1}} + \sum_s (-1)^{m+s} v_l v_{i_s} p_{i_1,\ldots,i_{s-1},i_{s+1},\ldots,i_{m-1}l}\}$$
$$+ (m-1)t^2 \sum v_k v_l p_{i_1,\ldots,i_{m-2}k} p_{i_1,\ldots,i_{m-2}l},$$

where the summations are now over all the i's and k, l. This simplifies to:

$$(m-1)!\,P^2 = \sum p^2_{i_1,\ldots,i_{m-1}} - 2(m-1)\sum v_k v_l p_{i_1,\ldots,i_{m-2}k} p_{i_1,\ldots,i_{m-2}l}$$
$$+ (m-1)\sum v_k v_l v^2_{i_{m-1}} p_{i_1,\ldots,i_{m-2}k} p_{i_1,\ldots,i_{m-2}l}$$
$$+ (m-1)t^2 \sum v_k v_l p_{i_1,\ldots,i_{m-2}k} p_{i_1,\ldots,i_{m-2}l}$$
$$= \sum p^2_{i_1,\ldots,i_{m-1}} - (m-1)\sum v_k v_l p_{i_1,\ldots,i_{m-2}k} p_{i_1,\ldots,i_{m-2}l},$$

since $\sum v_i^2 + t^2 = 1$. It follows that

$$P^2 = \sum p^2_{1,\ldots,i-1,i+1,\ldots,m}$$
$$- \sum p^2_{1,\ldots,i-1,i+1,\ldots,m}(v_1^2 + \cdots + v_{i-1}^2 + v_{i+1}^2 + \cdots + v_m^2)$$
$$+ 2\sum_{i<j}(-1)^{i+j} p_{1,\ldots,i-1,i+1,\ldots,m} p_{1,\ldots,j-1,j+1,\ldots,m} v_i v_j$$
$$= t^2 Q^2 + \sum v_i^2 p^2_{1,\ldots,i-1,i+1,\ldots,m}$$
$$+ 2\sum_{i<j}(-1)^{i+j} v_i v_j p_{1,\ldots,i-1,i+1,\ldots,m} p_{1,\ldots,j-1,j+1,\ldots,m} = t^2 Q^2 + R^2,$$

as to be proved.

2. Geometrical Applications

Suppose the absolute value of the mean curvature of M be bounded below by a positive constant, i. e., $|\sigma_1| \geqq c > 0$. By changing the orientation of M if necessary, we can assume $\sigma_1 \geqq c > 0$. Now the integral

$$V_a = \int_M |t| \Phi \tag{30}$$

is the volume of the projection $x'(M)$. If therefore $t \geqq 0$, we will get, from (22), (26), (30).

$$m \subset V_a \leqq m \int_M t \sigma_1 \Phi = \int_{\partial M} \alpha \leqq L_a$$

where L_a is the volume of $x'(\partial M)$. The condition $t \geqq 0$ is fulfilled, when M is defined by the equation

$$z = F(x_1, \ldots, x_m), \quad x_1^2 + \cdots + x_m^2 \leqq R^2, \tag{31}$$

where $x_1, \ldots, x_m, z$ are rectangular coordinates in the space E. In this case, if a is the unit vector $(0, \ldots, 0, 1)$, we have furthermore the relation

$$m V_a = L_a R.$$

It follows that $c R \leqq 1$. We state these results in the theorem:

Theorem 1. *Let M be a compact piece of an oriented hypersurface of dimension m with smooth boundary ∂M, which is immersed in an euclidean space of dimension $m + 1$. Suppose the mean curvature $\sigma_1 \geqq c > 0$. Let a be a fixed unit vector which makes an angle $\leqq \frac{\pi}{2}$ with all the normals of M. Then*

$$m \subset V_a \leqq L_a, \tag{32}$$

where V_a is the volume of the orthogonal projection of M and L_a that of ∂M in the hyperplane perpendicular to a. If M is defined by the equation (31), then $c R \leqq 1$.

Corollary. *Let $z = F(x_1, \ldots, x_m)$ be a hypersurface of constant mean curvature defined for all values of the x's. Then the mean curvature must be zero. For $m = 2, 3$, such a hypersurface must be a hyperplane.*

The last statement follows from the theorem that, for $m = 2, 3$, a minimal hypersurface $z = F(x_1, \ldots, x_m)$ defined for all values of the x's is a hyperplane. This theorem in the case $m = 2$ is the classical theorem of S. Bernstein. The case $m = 3$ was recently established by E. di Georgi (not yet published).

Theorem 1 leads to a similar theorem involving the second curvature σ_2. For, by the definition of σ_i in (6), we derive

$$m^2\sigma_1^2 - m(m-1)\sigma_2 = \sum_{i,k} h_{ik}^2 \geqq 0.$$

We have therefore the theorem:

Theorem 2. *Let M be a compact piece of oriented hypersurface with smooth boundary ∂M, such that $\sigma_2 \geqq b > 0$. Let a be a fixed unit vector which makes an angle $\leqq \frac{\pi}{2}$ with all the normals of M. Then*

$$\sqrt{m(m-1)b}\, V_a \leqq L_a, \tag{33}$$

where V_a, L_a have the same meaning as in Theorem 1.

The corollary to Theorem 1 suggests naturally the problem of studying the non-compact complete hypersurfaces of constant mean curvature. In particular, it would be of interest to know whether a generalization of OSSERMAN's theorem is true, i. e., whether the plane and the circular cylinder are the only complete surfaces of constant mean curvature in the three-space, whose image under the GAUSS mapping omits an open subset on the sphere.

Locally the hypersurfaces of constant mean curvature have a characterization, which may be derived as follows: Taking the exterior derivative of (5) and making use of the structure equations (3), we get

$$\sum_k \left(dh_{ik} - \sum_j h_{ij}\omega_{kj} - \sum_j h_{kj}\omega_{ij}\right) \wedge \omega_k = 0.$$

It follows that the covariant differential of h_{ik} relative to the riemannian metric of M can be written

$$Dh_{ik} = dh_{ik} - \sum_j h_{ij}\omega_{kj} - \sum_j h_{kj}\omega_{ij} = \sum_j h_{ikj}\omega_j, \tag{34}$$

where h_{ikj} is symmetric in all three indices. For a fiexd unit vector a we have

$$d(a, e_{m+1}) = -\sum (a, e_i) h_{ik}\omega_k,$$

$$d\{-(a, e_i)h_{ik}\} = -\sum (a, e_i)h_{ij}\omega_{kj} - (a, e_{m+1})\sum h_{ik}h_{ij}\omega_j - \sum (a, e_i)h_{ikj}\omega_j.$$

If Δ denotes the Laplacian on M, we have

$$\Delta(a, e_{m+1}) = -\sum (a, e_i)h_{ikk} - (a, e_{m+1})m\{\sigma_1^2 - (m-1)\sigma_2\}. \tag{35}$$

This gives the following theorem:

Theorem 3. *If M is a hypersurface of constant mean curvature, the function (a, e_{m+1}), for any fixed vector a, satisfies the differential equation*

$$\Delta(a, e_{m+1}) + m\{\sigma_1^2 - (m-1)\sigma_2\}(a, e_{m+1}) = 0. \tag{36}$$

Conversely, if this equation holds for all a, M is of constant mean curvature.

6*

3. Hypersurfaces with negative second curvature

Unlike σ_1 the second curvature σ_2 does not depend on the orientation of the hypersurface. The method used by HEINZ for the study of the case that $m = 2$ and σ_2 is bounded above by a negative constant can also be generalized and reduced to an integral formula very much similar to (19) and (22). In fact, the formalisms are quite parallel, except that the analysis will naturally be more complicated, for the problem is now strictly non-linear. We suppose $t > 0$ and introduce the differential forms

$$(37)\quad \begin{aligned} B_{m-h} = \sum \varepsilon_{i_1,\ldots,i_m} d(a_{i_1}, x) \wedge \cdots \wedge d(a_{i_h}, x) \\ \wedge d\frac{(a_{i_{h+1}}, e_{m+1})}{t} \wedge \cdots \wedge d\frac{(a_{i_m}, e_{m+1})}{t}. \end{aligned}$$

We put

$$(38)\quad y_i = (a_i, e_{m+1}),$$

so that

$$\begin{pmatrix} (a_i, e_k) & (a_i, e_{m+1}) \\ (a_{m+1}, e_k) & (a_{m+1}, e_{m+1}) \end{pmatrix} = \begin{pmatrix} u_{ik} & y_i \\ v_k & t \end{pmatrix}$$

is a properly orthogonal matrix. Using (16), we get

$$t^2 d\frac{(a_i, e_{m+1})}{t} = -t\sum u_{ik}\omega_{k,m+1} + y_i \sum v_k \omega_{k,m+1}.$$

These, and the expression for $d(a_i, x)$ in (16), can be substituted into (37), after which the u_{ik}, y_i can be eliminated. This will allow us to express B_{m-h} in terms of the curvatures and the components v_i, t of a relative to the frame e_A. The calculation is a little long and we will not need the general formula in this paper. We therefore content ourselves to give the final result, which is the formula

$$(39)\quad t^{2m-2h} B_{m-h} = (-1)^{m-h} m!\, t^{m-h+1} \sigma_{m-h} \Phi + (-1)^{m-h-1}(m-h)\, t^{m-h+1} C_{m-h},$$

where

$$(40)\quad \begin{aligned} C_{m-h} = -h\sum \varepsilon_{i_1,\ldots,i_m} \omega_{i_1} \wedge \cdots \wedge \omega_{i_h} \wedge \omega_{i_{h+1},m+1} \wedge \cdots \wedge \omega_{i_{m-1},m+1} \\ \wedge \omega_{i_1,m+1} v_{i_1} v_{i_m} - \sum \varepsilon_{i_1,\ldots,i_m} \omega_{i_1} \wedge \cdots \wedge \omega_{i_h} \wedge \omega_{i_{h+1},m+1} \wedge \cdots \\ \wedge \omega_{i_m,m+1} v_{i_m}^2. \end{aligned}$$

Some of the notable special cases of this formula, which can be derived directly with much less calculation, are

$$(41)\quad B_m = (-1)^m m! \frac{\sigma_m}{t^{m+1}} \Phi,$$

$$(42)\quad B_1 = -\frac{m!}{t^2}\left(\sum h_{ij} v_i v_j\right)\Phi,$$

$$(43)\quad t^3 B_2 = m!\, t^2 \sigma_2 \Phi + 2(m-2)!\left\{-\sum_i \left(\sum_j h_{ij} v_j\right)^2 + m\sigma_1 \sum_{i,j} h_{ij} v_i v_j\right\}\Phi.$$

It is the last formula that we will utilize in the rest of this paper.

In fact, the definition of B_2 can also be written

$$(44)\quad B_2 = \sum \varepsilon_{i_1,\dots,i_m} d\frac{(a_{i_1}, e_{m+1})}{t} \wedge d\frac{(a_{i_2}, e_{m+1})}{t} \wedge d(a_{i_3}, x) \wedge \cdots \wedge d(a_{i_m}, x),$$

from which it follows that

$$(45)\qquad B_2 = d\beta,$$

where

$$(46)\quad \beta = \sum \varepsilon_{i_1,\dots,i_m} \frac{(a_{i_1}, e_{m+1})}{t}\, d\frac{(a_{i_2}, e_{m+1})}{t} \wedge d(a_{i_3}, x) \wedge \cdots \wedge d(a_{i_m}, x).$$

In order to apply the formula (45) we restrict ourselves to the case that the hypersurface is given in the non-parametric form. In fact, let a_A be the unit coordinate vectors of a rectangular coordinate system in E, and let

$$(47)\qquad (a_i, x) = x_i, \quad (a_{m+1}, x) = z.$$

Suppose the hypersurface be given by the equation (31), and put

$$(48)\qquad p_i = \frac{\partial z}{\partial x_i}, \qquad p_{ik} = \frac{\partial^2 z}{\partial x_i \partial x_k}, \qquad w = \sqrt{1 + \sum p_k^2}.$$

We choose the unit normal vector to be

$$(49)\qquad e_{m+1} = \left(\frac{-p_1}{w}, \dots, \frac{-p_m}{w}, \frac{1}{w}\right),$$

so that

$$(50)\qquad t = \frac{1}{w}, \qquad (a_i, e_{m+1}) = -\frac{p_i}{w}.$$

Then we have

$$(51)\qquad \beta = \sum \varepsilon_{i_1,\dots,i_m} p_{i_1} d p_{i_2} \wedge d x_{i_3} \wedge \cdots \wedge d x_{i_m}.$$

In the space $(x_1, \dots, x_m)$ consider the ball $\sum_r$ defined by

$$(52)\qquad x_1^2 + \cdots + x_m^2 \leq r^2.$$

Let Θ_r be the differential form of degree $m-1$, which is the volume element of $\partial \sum_r$. Then (cf. [2])

$$(53)\qquad \frac{x_i}{r}\Theta_r = \frac{1}{(m-1)!} \sum \varepsilon_{i, i_2, \dots, i_m} d x_{i_2} \wedge \cdots \wedge d x_{i_m},$$

and we find

$$\int_{\partial \Sigma_r} \beta = \frac{(m-2)!}{r} \int_{\partial \Sigma_r} \Big\{\sum_{i \neq k} x_i p_i p_{kk} - \sum_{i \neq k} x_i p_k p_{ik}\Big\}\Theta_r.$$

Now

$$\frac{1}{2}\frac{\partial}{\partial r}\left(\sum p_i^2\right) = \frac{1}{2r}\sum x_k \frac{\partial}{\partial x_k}\left(\sum p_i^2\right) = \frac{1}{r}\sum_{i,k} x_k p_i p_{ik}$$

$$= \frac{1}{r}\Big\{\sum_{i \neq k} x_i p_k p_{ik} - \sum_{i \neq k} x_i p_i p_{kk} + \left(\sum x_i p_i\right)\left(\sum p_{kk}\right)\Big\}.$$

It follows that

$$\int_{\partial \Sigma_r} \beta = \frac{1}{(m-2)!}\int_{\partial \Sigma_r}\left\{-\frac{1}{2}\frac{\partial}{\partial r}\sum p_i^2 + \frac{\partial z}{\partial r}\sum p_{kk}\right\}\Theta_r. \tag{54}$$

The sums inside the braces are

$$\sum p_i^2 = |\mathrm{grad}\, z|^2, \quad \sum p_{kk} = \Delta z \tag{55}$$

where the norm of the gradient and the Laplacian Δz are taken relative to the euclidean metric in the $(x_1, \ldots, x_m)$-space.

In order to transform the integrand in the right-hand side of (54), we face a situation which occurs in the more general context of a riemannian manifold. For possible future applications we will digress for a moment and give a discussion of the general case. We consider a riemannian manifold X of dimension m and in it a scalar function u with no critical point. Let Y_u be the level hypersurfaces of u, defined by equating u to constants. Now suppose z be another scalar function on X, of class C^2 (say). From z are defined its gradient grad z and its Laplacian Δz. The restriction of z to Y_u gives rise also to a gradient, which we will denote by $\mathrm{grad}_Y z$.

To analyze the situation more completely consider in X a family of orthonormal frames e_i, such that e_α, $1 \leqq \alpha \leqq m-1$, are tangent to Y_u, so that e_m is normal to Y_u. Let θ_i be the coframe dual to e_k. Then θ_m is a multiple of du. We *suppose the function u to be such that* $\theta_m = du$. Geometrically this means that u is the arc length of the orthogonal trajectories of Y_u, measured in a proper sense. We can write

$$dz = \sum_\alpha z_\alpha \theta_\alpha + z_m \theta_m \tag{56}$$

Then

$$|\mathrm{grad}\, z|^2 = \sum_i z_i^2, \qquad |\mathrm{grad}_Y z|^2 = \sum_\alpha z_\alpha^2. \tag{57}$$

The function z_m remains invariant under a rotation of the vectors e_α, and is completely determined by the function z. It is usually called the normal derivative, and we will write $z_m = \frac{\partial z}{\partial u}$. We have then the relation

(58) $$|\operatorname{grad} z|^2 = |\operatorname{grad}_Y z|^2 + \left(\frac{\partial z}{\partial u}\right)^2.$$

Let θ_{ij} be the connection forms of the riemannian metric relative to this family of frames. The absence of torsion of the connection is expressed by the relation

(59) $$d\theta_i = \sum_j \theta_j \wedge \theta_{ji},$$

and the covariant differential of z_i can be written

(60) $$dz_i = \sum_j z_j \theta_{ij} + \sum_k z_{ik} \theta_k,$$

with

(61) $$z_{ik} = z_{ki}.$$

By definition we have

(62) $$\Delta z = \sum z_{ii}.$$

Since $d\theta_m = 0$, we can write

(63) $$\theta_{\alpha m} = \sum_\beta A_{\alpha m \beta} \theta_\beta, \quad 1 \leqq \alpha, \beta \leqq m-1,$$

where

(64) $$A_{\alpha m \beta} = A_{\beta m \alpha}.$$

The ordinary quadratic differential form

(65) $$\mathrm{II} = \sum \theta_\alpha \theta_{\alpha m} = \sum A_{\alpha m \beta} \theta_\alpha \theta_\beta$$

is called the second fundamental form of Y_u. It gives rise to a symmetric bilinear form $\mathrm{II}(\xi, \eta)$, where the arguments ξ, η are tangent vectors at the same point of Y_u. In particular, we have

(66) $$\mathrm{II}(\operatorname{grad}_Y z, \operatorname{grad}_Y z) = \sum_{\alpha,\beta} A_{\alpha m \beta} z_\alpha z_\beta.$$

With all these preparations and under the further assumption that Y_u is compact, we have the integral formula

(67) $$\begin{aligned} &\int_{Y_u} \left\{-\frac{1}{2}\frac{\partial}{\partial u}|\operatorname{grad} z|^2 + \frac{\partial z}{\partial u}\Delta z\right\} dV \\ &= \int_{Y_u} \left\{-\frac{\partial}{\partial u}|\operatorname{grad}_Y z|^2 - (Tr\,\mathrm{II})\left(\frac{\partial z}{\partial u}\right)^2 + \mathrm{II}(\operatorname{grad}_Y z, \operatorname{grad}_Y z)\right\} dV, \end{aligned}$$

where dV is the volume element on Y_u. We proceed to prove (67).

This follows from the observation that the $(m-2)$-form

(68) $$\lambda = \sum_\alpha (-1)^{\alpha-1}\theta_1 \wedge \cdots \wedge \theta_{\alpha-1} \wedge (z_\alpha z_m) \wedge \theta_{\alpha+1} \wedge \cdots \wedge \theta_{m-1}$$

is globally defined on Y_u, so that $\int_{Y_u} d\lambda = 0$. By using (59) and (60), we get (on Y_u),

$$d\lambda = \{\sum z_{\alpha\alpha} \cdot z_m + \sum z_\alpha z_{m\alpha} + z_m^2 \sum A_{\alpha m\alpha} - \sum A_{\alpha m\beta} z_\alpha z_\beta\}\, \theta_1 \wedge \cdots \wedge \theta_{m-1},$$

where the product of the θ's is equal to dV. Now the lefthand side of (67) is equal to

$$LHS = \int_{Y_u} \left\{-\frac{1}{2}\frac{\partial}{\partial u}(\sum z_\alpha^2 + z_m^2) + z_m(\sum_i z_{ii})\right\} dV,$$

and the right-hand side of (67) is

$$RHS = \int_{Y_u} \left\{-\frac{\partial}{\partial u}\sum z_\alpha^2 - z_m^2 \sum A_{\alpha m\alpha} + \sum A_{\alpha m\beta} z_\alpha z_\beta\right\} dV.$$

It follows that their difference

$$LHS - RHS = \int_{Y_u} d\lambda = 0.$$

This proves (67).

We now apply (67) to the family of spheres $\partial \sum_r$ in the space $(x_1, \ldots, x_m)$. Then $u = r$ and

(69) $$\theta_{\alpha m} = -\frac{1}{r}\theta_\alpha,$$

or

(69a) $$A_{\alpha m\beta} = -\frac{1}{r}\delta_{\alpha\beta}.$$

From (54) and (67) we get

(70) $$(m-2)! \int_{\partial\Sigma r} \beta = \int_{\partial\Sigma r} \left\{-\frac{\partial}{\partial r}(\sum z_\alpha^2) + \frac{m-1}{r} z_r^2 - \frac{1}{r}\sum z_\alpha^2\right\} \Theta_r.$$

For $x \in \partial \sum_1$, the map $x \to rx$ establishes a diffeomorphism between $\partial \sum_1$ and $\partial \sum_r$. Under this diffeomorphism we have

$$\Theta_r = r^{m-1}\Theta_1.$$

It follows that

(71) $$\begin{aligned} &(m-2)! \int_{\Sigma r} B_2 = (m-2)! \int_{\partial\Sigma r} \beta \\ &= \int_{\partial\Sigma_1} \left\{-\frac{\partial}{\partial r}(r^{m-1}\sum z_\alpha^2) + (m-1) r^{m-2} z_r^2 + (m-2)\, r^{m-2}\sum z_\alpha^2\right\} \Theta_1. \end{aligned}$$

For $m = 2$ this formula is due to S. Bernstein (cf. [1]).

The fact that the last two terms in the last integral are $\geqq 0$ is the basic reason for the following theorem:

Theorem 4. *Let*

$$z = F(x_1, \ldots, x_m), \quad x_1^2 + \cdots + x_m^2 \leqq R^2, \tag{31}$$

be a hypersurface in euclidean $(m+1)$-space, such that there exists a constant $c > 0$ with the properties:

1. *The second curvature $\sigma_2 \leqq -c$;*
2. *For all v_i,*

$$-\sum_i \Big(\sum_j h_{ij} v_j\Big)^2 + m\sigma_1 \sum_{i,j} h_{ij} v_i v_j \leqq -\frac{m(m-1)}{2} c (v_1^2 + \cdots + v_m^2). \tag{72}$$

Then we have

$$R \leqq e\sqrt{3}\frac{1}{\sqrt{c}}, \quad m = 2,$$

$$R \leqq \frac{(m-2)\sqrt{6}}{\sqrt{m(m-1)}\,\cancel{(m-3)!}\left\{1-(\tfrac{1}{2})^{\frac{m}{2}-1}\right\}} \cdot \frac{1}{\sqrt{c}}, \quad m \geqq 3. \tag{73}$$

We remark that, for $m = 2$, the condition 2) is equivalent to 1), as follows from an easy calculation. In this case, therefore, only the first condition is necessary. The factor before $1/\sqrt{c}$ in the right-hand sides of (73) can most likely be improved. The conclusion for $m = 2$ being that of HEINZ, we suppose from now on that $m \geqq 3$. We will need the following:

Lemma. *Let $f(r)$, $0 \leqq r \leqq R$, be a positive-valued C^2-function, which satisfies the conditions*

$$f(r) \geqq a r^m, \quad f''(r) \geqq b r^{-m} f(r)^2, \quad m \geqq 3, \quad f(0) = 0, \tag{74}$$

where a, b are two positive constants. Then

$$R \leqq \frac{\sqrt{b}\,(m-2)}{1-(\tfrac{1}{2})^{\frac{m}{2}-1}} \frac{1}{(ab)^{\frac{1}{2}}}. \tag{75}$$

To prove the lemma notice that the second inequality implies

$$\frac{d}{d\varrho}\left\{f'(\varrho)^2 - \frac{2b}{3r^m}(f(\varrho))^3\right\} \geqq 0, \quad 0 \leqq \varrho \leqq r$$

Since the expression between the braces is $\geqq 0$ for $\varrho = 0$, we get

$$f'(r)^2 \geqq \frac{2b}{3r^m} f(r)^3.$$

The latter can be written

$$-\frac{d}{dr}\left(f(r)^{-\frac{1}{2}}\right) \geqq \left(\frac{b}{6}\right)^{\frac{1}{2}} r^{-m/2}.$$

By integration it follows that, for $0 < R_1 < R_2 < R$,

$$(a R_1^m)^{-\frac{1}{2}} \geqq f(R_1)^{-\frac{1}{2}} \geqq f(R_1)^{-\frac{1}{2}} - f(R_2)^{-\frac{1}{2}} \geqq \left(\frac{b}{6}\right)^{\frac{1}{2}} \frac{2}{m-2}\left(R_1^{-\frac{m}{2}+1} - R_2^{-\frac{m}{2}+1}\right).$$

As $R_2 \to R$, this gives

$$\frac{m-2}{2}\left(\frac{6}{ab}\right)^{\frac{1}{2}} \geqq R_1 - \left(\frac{R_1}{R}\right)^{m/2} R.$$

This inequality is true for all $0 < R_1 < R$. Putting $R = 2R_1$, we get the statement in the Lemma.

To give a proof of Theorem 4 notice that

$$\Phi = w\, dx_1 \wedge \cdots \wedge dx_m = \frac{1}{t} dx_1 \wedge \cdots \wedge dx_m. \tag{76}$$

From (43) we get

$$\frac{1}{m!} t^4 B_2 = \left\{\sigma_2 t^2 + \frac{2}{m(m-1)}\left[-\sum_i\left(\sum_j h_{ij} v_j\right)^2 + m\sigma_1 \sum_{i,j} h_{ij} v_i v_j\right]\right\} dx_1 \wedge \cdots \wedge dx_m.$$

The conditions 1) and 2) of the theorem imply that the expression between the braces is $\leqq -c$, because $t^2 + \sum_i v_i^2 = 1$. It follows from (71) that

$$-\int_{\partial \Sigma_1} \frac{\partial}{\partial r}\left(r^{m-1} \sum z_\alpha^2\right) \Theta_1 \leqq -(m-2)!\, m!\, c \int_{\Sigma_r} w^4 dx_1 \cdots dx_m,$$

which can also be written

$$\frac{d}{dr}\left\{r^{m-1} \int_{\partial \Sigma_1} \sum z_\alpha^2 \Theta_1\right\} \geqq (m-2)!\, m!\, c \int_{\Sigma_r} w^4 dx_1 \cdots dx_m.$$

Notice that the integrand in the first integral is a function of r and the same is true of the integral itself.

Now consider the function

$$f(r) = \int_{\Sigma_r} \left(1 + \sum z_\alpha^2\right) dx_1 \cdots dx_m.$$

We have

$$f(r) \geqq \gamma_m r^m,$$

where γ_m is the volume of a unit ball in m-space. By the Schwarz inequality we have

$$f(r) \leqq (\gamma_m r^m)^{\frac{1}{2}} \left\{\int_{\Sigma_r} \left(1 + \sum z_\alpha^2\right)^2 dx_1 \cdots dx_m\right\}^{\frac{1}{2}} \leqq (\gamma_m r^m)^{\frac{1}{2}} \left\{\int_{\Sigma_r} w^4 dx_1 \cdots dx_m\right\}^{\frac{1}{2}}.$$

But $f(r)$ can be written

$$f(r) = \int_0^r \varrho^{m-1} d\varrho \int_{\partial \Sigma_1} (1 + \sum z_\alpha^2)\, \Theta_1$$

By differentiation we get

$$f''(r) = \frac{d}{dr}\left\{ r^{m-1} \int_{\partial \Sigma_1} (\sum z_\alpha^2)\, \Theta_1 \right\} + m(m-1)\gamma_m r^{m-2}.$$

It follows that

$$f''(r) \geqq \frac{d}{dr}\left\{ r^{m-1} \int_{\partial \Sigma_1} (\sum z_\alpha^2)\, \Theta_1 \right\} \geqq (m-2)!\, m!\, c \int_{\Sigma_r} w^4 dx_1 \cdots dx_m$$

$$\geqq \frac{(m-2)!\, m!\, c}{\gamma_m r^m} f(r)^2.$$

The function $f(r)$ thus satisfies the conditions of the lemma, a direct application of which gives Theorem 4.

Bibliography

[1] S. Bernstein, Sur la généralisation du problème de Dirichlet. Math. Annalen 69 (1910) 82—136.

[2] S. Chern, The integrated form of the first main theorem for complex analytic mappings in several complex variables. Annals of Math. 71 (1960) 536—551.

[3] E. Heinz, Über Flächen mit eineindeutiger Projektion auf eine Ebene, deren Krümmungen durch Ungleichungen eingeschränkt sind. Math. Annalen 129 (1955) 451—454.

Eingegangen am 20. 1. 1965

Reprinted from
Abh. Math. Sem. Univ. Hamburg **29** (1965) 77–91

On the Differential Geometry of a Piece of Submanifold in Euclidean Space

Shiing-shen CHERN[1)]

1. Introduction.

To illustrate the type of problems that will be discussed, we shall mention two known results. One is a theorem of Efimow, which states that if a surface $z = F(x, y)$ in the ordinary euclidean space defined over a square in the (x, y)-plane of side length a has Gaussian curvature $K \leqq -1$, then $a \leqq 14$.

The second is an inequality of Erhard Heinz concerning a minimal surface

$$z = F(x, y), \quad x^2 + y^2 < R^2, \tag{1}$$

defined over a disk of radius R in the (x, y)-plane. Heinz proved that if K_0 denotes the Gaussian curvature at the center of the disk, there is a universal constant c such that

$$-K_0 \leqq \frac{c}{R^2}. \tag{2}$$

In particular, if $K_0 \neq 0$, we have $R \leqq (c/-K_0)^{1/2}$.

Both theorems concern with surfaces satisfying differential equations or inequalities, generally non-linear, and give bounds on the domains over which they are defined. We will state some theorems which generalize these results. Proofs are to be found in the references [1] and [2]. Needless to say, many simple and natural questions in this area remain open.

2. Hypersurfaces.

Let E be the euclidean n-space and $x : M \longrightarrow E$ be an immersed compact submanifold with boundary. This means that M is a proper subset of a submanifold M' of the same dimension such that at a boundary point M behaves like a half-space. Among

1) This work is done under partial support by grants from the National Science Foundation and Office of Naval Research, USA.

17

the simplest cases is the one when M is a hypersurface, i.e., the codimension of M is 1. Then M has, in addition to its first fundamental form I, a second fundamental form II, which is defined up to sign depending on the orientation of M. The expansion

$$\frac{\det(\lambda I + II)}{\det I} = \sum_{0\leq r\leq n-1} \binom{n-1}{r} \lambda^r \sigma_{n-1-r}, \quad \sigma_0 = 1, \tag{3}$$

gives rise to the coefficients $\sigma_1, \cdots, \sigma_{n-1}$, which are the simplest scalar invariants of an oriented hypersurface. The σ's with even indices are independent of the orientation, while those with odd indices change their signs when the orientation of M is reversed. In particular, σ_1 is called the mean curvature and σ_{n-1} the Gauss-Kronecker curvature.

The following theorems give results on a piece of hypersurface whose curvatures satisfy some inequalities.

THEOREM 1. *Let M be a compact hypersurface with boundary ∂M. Suppose $\sigma_1 \geqq c$ (=const.) > 0, and suppose a be a fixed unit vector which makes an angle $\leqq \pi/2$ with all the normals of M. Then*

$$(n-1)cV_a \leqq L_a, \tag{4}$$

where V_a is the volume of the orthogonal projection of M and L_a that of ∂M in the hyperplane perpendicular to a.

If M is defined by the equation

$$x_n = F(x_1, \cdots, x_{n-1}), \quad x_1^2 + \cdots + x_{n-1}^2 < R^2, \tag{5}$$

and $a = (0, \cdots, 0, 1)$, we have

$$(n-1)V_a = L_a R.$$

Hence we get the corollary:

COROLLARY. *If the mean curvature of the hypersurface* (5) *satisfies the inequality $\sigma_1 \geqq c > 0$, then $cR \leqq 1$.*

It follows that if the hypersurface (5) is defined for all values of $x_1, \cdots, x_{n-1}$, and is of constant mean curvature, the mean curvature must be zero. By a classical theorem of S. Bernstein and a recent theorem of E. di Georgi, such a hypersurface must be a hyperplane for $n=3, 4$.

Unlike σ_1 the second curvature σ_2 is defined with its sign, and the inequalities

$$\sigma_2 \geqq c > 0, \quad \sigma_2 \leqq -c < 0$$

have quite different geometrical meanings. The first case is easier, because of the algebraic inequality[1)]

$$\sigma_1^2 - \sigma_2 \geqq 0. \tag{6}$$

Actually a deeper reason is a difference in the "types" of the two inequalities, the first being "elliptic" and the second "hyperbolic". The inequality (6) allows us to derive from Theorem 1 the following theorem, which sharpens a corresponding theorem in [2]:

THEOREM 2. *Let M be a compact hypersurface with boundary ∂M. Supposc $\sigma_2 \geqq c > 0$, and suppose a be a fixed unit vector which makes an angle $\leqq \pi/2$ with all the normals of M. Then*

$$(n-1)c^{1/2}V_a \leqq L_a\,, \tag{7}$$

where V_a and L_a have the same meaning as in Theorem 1.

The analogue of the corollary to Theorem 1 is also valid. The example of a hemisphere of radius R shows that the conclusions

1) To prove the inequality (6) choose a basis so that the matrix of I is the unit matrix and the matrix of II is (h_{ik}), $1 \leqq i, k \leqq n-1$. By definition we have

$$(n-1)\sigma_1 = \sum_i h_{ii}$$

$$(n-1)(n-2)\sigma_2 = \sum_{i,k}(h_{ii}h_{kk} - h_{ik}^2) = (n-1)^2\sigma_1^2 - \sum_{i,k} h_{ik}^2\,.$$

It follows that

$$\begin{aligned}(n-1)(n-2)(\sigma_1^2 - \sigma_2) &= \sum_{i,k} h_{ik}^2 - (n-1)\sigma_1^2 \\ &= \sum_{i \neq k} h_{ik}^2 + \sum_i h_{ii}^2 - \frac{1}{n-1}\Big(\sum_i h_{ii}^2 + 2\sum_{i<k} h_{ii}h_{kk}\Big) \\ &= \sum_{i \neq k} h_{ik}^2 + \frac{1}{n-1}\sum_{i<k}(h_{ii} - h_{kk})^2 \geqq 0\,,\end{aligned}$$

which proves (6).

in Theorems 1 and 2 cannot be improved.

For hypersurfaces satisfying the inequality $\sigma_2 \leqq -c < 0$ it seems that an additional condition is necessary in order to get a conclusion of the same type. We have the theorem:

THEOREM 3. *Consider the hypersurface* (5). *Suppose there exists a constant* $c > 0$ *with the properties*:

1) $\sigma_2 \leqq -c$;

2) *If* h_{ij} *are the coefficients of the second fundamental form relative to an orthonormal frame, the relation*

$$-\sum_i (\sum_j h_{ij} v_j)^2 + (n-1)\sigma_1 \sum_{i,j} h_{ij} v_i v_j \leqq -\frac{1}{2}(n-1)(n-2)c(v_1^2 + \cdots + v_{n-1}^2)\,, \tag{8}$$

is true for all v_i.

Then we have

$$\left\{\begin{array}{ll} R \leqq e\sqrt{3}\dfrac{1}{\sqrt{c}}\,, & n=3\,, \\ R \leqq \dfrac{\alpha(n)}{\sqrt{c}}\,, & n \geqq 4\,,\quad \alpha(n) \textit{ depending only on } n\,. \end{array}\right. \tag{9}$$

It should be remarked that, for $n=3$, the conditions 1) and 2) are equivalent.

An important rôle in the proof of the above theorems is played by certain integral formulas, and the proof of Theorem 3 is considerably more complicated than that of Theorem 1. In the next section we will get similar estimates for a different case by an entirely different method.

3. Minimal surfaces.

Let x: $M \longrightarrow E$ be an oriented minimal (two-dimensional) surface in E. The Gauss mapping maps M into the manifold of oriented planes through a point in E, the latter being isomorphic to a complex hyperquadric in a complex projective space $P_{n-1}(C)$ of dimension $n-1$. M is a minimal surface, if and only if the Gauss mapping is anti-holomorphic. This property makes it possible to apply complex function theory to the study of minimal surfaces in E. We suppose $P_{n-1}(C)$ to be endowed with the standard Kähler metric.

THEOREM 4. *Let M be a geodesic disk of radius s about the center p_0. Suppose its image in $P_{n-1}(C)$ under the Gauss mapping be at a distance $\delta > 0$ from a fixed hyperplane. Then*

$$-K_0 s^2 \leqq 2(n-1)\cot^2\delta \csc^2\delta\{1+2(n-2)\cot^2\delta\}, \tag{10}$$

where K_0 is the Gaussian curvature of M at p_0.

This theorem generalizes the inequality of E. Heinz mentioned in the Introduction. One of its consequences is the theorem that if M is complete and satisfies the above condition, it is a plane.

References

[1] Chern, S.: Minimal surfaces in an euclidean space of n dimensions, "Differential Topology", Morse Jubilee Volume, Princeton Univ. Press. (1965) 177-198.

[2] ———: On the curvatures of a piece of hypersurface in euclidean space, to appear in Hamburger Abhandlungen.

[3] Efimow, N. W.: Untersuchung der eindeutigen Projektion auf einer Fläche negativer Krümmung, Dokl. Akad. SSSR (N. S.), **93** (1953), 609-611.

[4] Heinz, E.: Uber Flächen mit eineindeutiger Projektion auf eine Ebene, deren Krümmungen durch Ungleichungen eingeschränkt sind, Math. Ann., **129** (1955), 451-454.

[5] Osserman, R.: Global properties of minimal surfaces in E^3 and E^n, Ann. of Math., **80** (1964), 340-364.

University of California, Berkeley

Reprinted from
Proc. of U.S.-Japan Seminar in Diff. Geom. (1965) 17–21

THE GEOMETRY OF G-STRUCTURES[1]

BY S. S. CHERN

1. **Introduction.** Differential geometry studies differentiable manifolds and geometric objects or structures on them. It is now customary to distinguish it from differential topology by the presence of a structure in addition to the differentiable structure. What a differential geometric structure is or should be is a matter of taste. At the present state of the field a definition general enough to include all the significant structures will certainly contain many uninteresting ones. Among the attempts at a general definition is the notion of a geometric object initiated by Oswald Veblen [**130**].

Not all the geometrical structures are "equal". It would seem that the riemannian and complex structures, with their contacts with other fields of mathematics and with their richness in results, should occupy a central position in differential geometry. A unifying idea is the notion of a G-structure, which is the modern version of a local equivalence problem first emphasized and exploited in its various special cases by Elie Cartan [**36**], [**41**], [**129**]. Generally we will restrict ourselves in this article to the discussion of problems which fall under the notion of a G-structure.

Two general problems are of importance:

I. *Existence or nonexistence of certain structures on a manifold.*

EXAMPLE 1. A positive definite riemannian structure always exists.

EXAMPLE 2. On a compact manifold M a nowhere zero differentiable vector field exists if and only if the Euler-Poincaré characteristic of M is zero.

EXAMPLE 3. One may ask whether a nonzero vector field exists on M (supposed to be compact and orientable) which is parallel with respect to a riemannian metric. By Hodge's harmonic forms a necessary condition is that the first Betti number b^1 of M is ≥ 1. One can further prove that the second Betti number $b^2 \geq b^1 - 1$ (cf. §5). These conditions are probably not sufficient.

II. *Local and global properties of a given structure.*

EXAMPLE 1. For a riemannian structure this means riemannian geometry.

EXAMPLE 2. If we are only interested in the existence or nonexist-

---

[1] Based on the Colloquium Lectures delivered at East Lansing, Michigan, August 30–September 2, 1960; revised and expanded, October 1965. Work done under partial support of Air Force, National Science Foundation, and Office of Naval Research.

ence of a nowhere zero vector field, it is not necessary to distinguish between contravariant and covariant vector fields, because a riemannian metric will transform one into the other. However, the distinction is essential when we study their properties. Any two nonzero contravariant vector fields are locally equivalent. On the other hand, a covariant vector field is the same as a linear differential form ω. To ω one can associate a pair of integers (k, l), defined to be the largest integers such that

$$(1) \qquad (d\omega)^k \neq 0, \qquad \omega \wedge (d\omega)^l \neq 0 \qquad l = k \text{ or } k-1.$$

These vector fields define differential systems. Of importance is the study of the local and global properties of their integral manifolds.

It is natural to combine I and II and ask for conditions such that a structure exists with given local or global properties. A simple problem of this nature is: What are the conditions that there exists a covariant vector field on a manifold with given values of k, l $(=k$ or $k-1)$ at every point? I do not know the answer even in the following special case: Does there exist on a compact orientable three-dimensional manifold a linear differential form ω such that $\omega \wedge d\omega \neq 0$ everywhere? The tangent bundle of an orientable three-dimensional manifold is a product bundle, but this fact does not seem to help the problem.

Both in the last problem and in the problem of Example 3, I the conditions in question are differential conditions (i.e., involving the partial derivatives of the tensor fields in question), in contrast to Examples 1, 2, I, where the conditions are algebraic in nature. For problems of the latter kind various techniques for their treatment have been developed in fiber bundles (obstructions, characteristic classes, cohomology operations, etc.). For existence and properties of structures satisfying differential conditions much less is known in the way of general methods, except perhaps the conclusions which can be derived by application of harmonic forms. We will consider in this paper principally structures satisfying differential conditions.

When manifolds have certain structures, it is natural to consider their mappings which are in a sense admissible. Examples are isometric mappings of riemannian manifolds and holomorphic mappings of complex manifolds. Studies of such mappings and of problems mentioned above all lead to systems of differential equations or inequalities, generally nonlinear.

This paper will be devoted to a review of some of the important developments in differential geometry in recent years, following the above guideline. With a task of this scope omissions are inevitable and results unmentioned are in no way less important. We will emphasize simple and concrete problems, at the expense of generality.

2. **Riemannian structure.** A riemannian manifold of two dimensions has at every point a scalar invariant, the gaussian curvature. In higher dimensions its generalization is the riemannian or sectional curvature, which is a function of (p, λ), p being a point and λ a two-dimensional plane element through p. Geometrically this is the gaussian curvature at p of the surface through p generated by the geodesics through p and tangent to λ. The knowledge of the sectional curvature for all (p, λ) determines the Riemann-Christoffel tensor and hence in a sense all the local properties of the riemannian structure. The class of riemannian manifolds whose sectional curvature keeps a constant sign is of obvious geometrical interest.

A complete riemannian manifold with sectional curvature $\leqq 0$ has a universal covering space homeomorphic to the euclidean space. This is due to the fact that the geodesics through a point 0 contain no point conjugate to 0. By studying the isometries in the universal covering space one derives properties of the fundamental group of the manifold. An example is the following theorem [**112**]: A compact riemannian manifold with sectional curvature <0 has a fundamental group which cannot be abelian and of which every abelian subgroup is cyclic.

The Euler-Poincaré characteristic of a compact orientable riemannian manifold M of even dimension $n = 2m$ is given by the Gauss-Bonnet formula [**4**], [**39**],

$$\chi(M) = \frac{(-1)^m}{2^{3m}\pi^m m!} \cdot \int_M \Big(\sum_{i,j} \epsilon_{i_1 \cdots i_{2m}} \epsilon_{j_1 \cdots j_{2m}} R_{i_1 i_2 j_1 j_2} \cdots R_{i_{2m-1} i_{2m} j_{2m-1} j_{2m}} \Big) dV \tag{2}$$

where dV is the volume element and R_{ijkl} are the components of the curvature tensor relative to orthonormal frames; $\epsilon_{i_1 \cdots i_{2m}}$ is zero if $i_1, \cdots, i_{2m}$ do not form a permutation of $1, \cdots, 2m$, and is equal to 1 or -1 according as the permutation is even or odd. It does not seem to be an easy problem to draw conclusions on the sign of the integrand from properties of the sign of sectional curvature. As a consequence the following conjecture has not been decided in its full generality: If M has sectional curvature $\geqq 0$, then $\chi(M) \geqq 0$; if M has sectional curvature $\leqq 0$, then $\chi(M) \geqq 0$ or $\leqq 0$, according as $n \equiv 0$ or 2, mod 4. The statement is true for $n = 4$ [**42**] and for the case that M has constant sectional curvature.

Much work has been done recently on complete riemannian manifolds with sectional curvature bounded below by a positive constant. Such a manifold is always compact. This is proved by examining the second variation of arc length and estimating the distance from a point to its first conjugate point on a geodesic. For $n=2$ the manifold is homeomorphic to the sphere. This is not true for higher dimensions. In fact, the complex projective space with the usual elliptic hermitian metric has sectional curvatures R satisfying the inequalities $A/4 \leqq R \leqq A$, A being a positive constant.

Again by considering the second variation of arc length Myers proved the following theorem [**98**]: A complete riemannian manifold with Ricci curvature $\geqq c$ $(=\text{constant}) > 0$ is compact. (The Ricci curvature is a function of (p, X), where X is a vector through p. It is the arithmetic mean of the sectional curvatures at (p, λ_i), $1 \leqq i \leqq n-1$, where λ_i are $n-1$ mutually perpendicular plane elements through X.) It follows that its universal covering manifold is also compact and hence that its fundamental group is finite. From the last result one concludes that the first Betti number of the manifold is zero.

The last conclusion can also be derived by a method of Bochner [**141**]. Bochner's method is an extremely important tool in differential geometry. It applies to the high-dimensional Betti numbers of riemannian manifolds and, more importantly, to the derivation of criteria (in terms of curvature properties) for the vanishing of more general cohomology groups (cf. §8). The idea is to study the measure of a harmonic form of degree r and the curvature properties at a point where the measure attains its maximum. From the absence of any point with such curvature properties one concludes that the harmonic form must be zero and hence that the rth Betti number is zero. For $r=1$ one gets the above corollary to Myers' theorem. For $r>1$ the geometrical significance of Bochner's conditions has not been sufficiently studied (see below).

With the example of the complex projective space in mind Rauch introduced the notion of a pinched riemannian manifold. A riemannian is said to be α-pinched if its sectional curvatures R satisfy the inequalities: $\alpha A \leqq R \leqq A$ for a positive constant A. Rauch proved the following theorem: *A complete simply connected α-pinched riemannian manifold with $\alpha = 0.75$ is homeomorphic to the sphere,* [**114**], [**115**].

By utilizing Bochner's conditions for the vanishing of the second Betti number Berger derived the following theorem [**17**]: *A compact α-pinched riemannian manifold M such that* dim $M=2m$ (*resp.* dim $M=2m+1$) *and* $\alpha > \frac{1}{4}$ (*resp.* $>2m-2/8m-5$) *has its second Betti number equal to zero.*

It was Klingenberg who observed that in the even-dimensional case a lemma of Synge can be utilized to give a simpler proof of Rauch's theorem and improve the result. Synge's lemma says the following: *Let M be a simply connected compact even-dimensional riemannian manifold whose sectional curvature is* >0. *Let g be a closed geodesic of minimum length among the closed geodesics of M. Then there is a family of closed curves converging toward g, which are all shorter than g.*

The combined efforts of Klingenberg and Berger lead to the following theorem [**15**], [**72**]: *Let M be a simply connected riemannian manifold which is* α*-pinched. If* $\alpha > \frac{1}{4}$, *M is homeomorphic to a sphere. If* $\alpha = \frac{1}{4}$ *and M is even-dimensional and not homeomorphic to a sphere, then M is a compact symmetric riemannian manifold of rank* 1 *with its canonical metric.* The proof of this theorem is difficult; the main tools are the Alexandrow-Rauch-Toponogov comparison theorem and the Morse critical point theory [**74**].

The symmetric riemannian manifolds of rank 1 mentioned above are the complex or quaternionic projective spaces or the Cayley plane. According to a theorem of H. C. Wang [**132**], these, together with the spherical and elliptic spaces, are the only compact and connected two-point homogeneous spaces.

Generally it is difficult to decide whether a differentiable manifold can be given a riemannian metric of positive curvature. The answer is not known even for the manifold $S^2 \times S^2$ (S^2 = two-dimensional sphere) and for Milnor's exotic spheres. Besides the examples mentioned above the only known simply connected spaces having a riemannian metric of positive curvature are two examples given by Berger [**16**]: These are homogeneous spaces of dimensions 7 and 13 respectively and are not homeomorphic to the sphere. No topological property is known for a compact simply connected manifold to carry a riemannian metric of positive curvature.

The homeomorphism theorem of Klingenberg and Berger was improved by D. Gromoll [**57**] and E. Calabi to a diffeomorphism theorem: *There exists a number* δ_n *depending on n* ($0 < \delta_n < 1$, $\lim_{n\to\infty} \delta_n = 1$) *such that if a simply connected riemannian manifold M of dimension n is* δ_n*-pinched, then M is diffeomorphic to the n-sphere with its usual differentiable structure. The best value of* δ_n *is not known; current estimates of it tend rapidly to* 1.

These pinching theorems have been extended to complex Kähler manifolds (cf. §5, 7). For instance, Andreotti and Frankel proved that if a compact Kähler manifold of complex dimension 2 has positive sectional curvature, it is complex analytically equivalent to the complex projective plane [52]. More general results have been obtained by doCarmo, S. Kobayashi, Klingenberg, Bishop-Goldberg, etc. [21], [22], [48], [73], [75].

Given a manifold, it is natural to ask for the "simplest" riemannian metric which can be defined on it. For a compact two-dimensional manifold this will be one of constant gaussian curvature; the curvature is positive for the sphere, zero for the torus, and negative for a surface of genus >1. For high dimensions constancy of the sectional curvature will lead to the space forms and it is known that not every manifold has a riemannian metric of constant curvature. A more general class of riemannian metrics consists of the Einstein metrics, which are defined by the condition that the Ricci tensor is a scalar multiple of the fundamental tensor. It is not known whether every simply connected manifold can have an Einstein metric. As an Einstein metric on a three-dimensional manifold is necessarily of constant curvature, the problem has a bearing on the Poincaré hypothesis that a compact simply connected three-dimensional manifold is homeomorphic to the sphere.

A result in this direction is the following theorem of Yamabe [140]: *A compact riemannian manifold of dimension $\geqq 3$ is conformally equivalent to one of constant scalar curvature. It is not known whether the sign of the scalar curvature is a global conformal invariant.*

3. **Connections** [5], [8], [40], [43], [50], [63], [78], [94]. The classical example of a connection is the parallelism of Levi-Civita in riemannian geometry. When the riemannian manifold is a surface in ordinary euclidean space, the parallelism along a curve is obtained by taking the tangent developable surface and rolling it on a plane. Parallelism is at the basis of the notion of curvature, for geometrically curvature measures the dependence of parallelism on the curve. For higher dimensions algebraic concepts enter and there is no extra work to develop the theory in a more general setting.

Let G be a Lie group of dimension r, and let $L(G)$ be its Lie algebra. On a differentiable manifold M we consider exterior differential forms (to be abbreviated as e.d.f.), with values in $L(G)$. A multiplication can be introduced as follows: Let X_i, $1 \leqq i \leqq r$, be a basis in $L(G)$. An $L(G)$-valued e.d.f. can be written

$$\lambda = \sum \lambda^i \otimes X_i, \tag{3}$$

where λ^i are ordinary e.d.f.'s. If

$$\mu = \sum_j \mu^j \otimes X_j \tag{4}$$

is a second $L(G)$-valued form, we define the product

$$[\lambda, \mu] = \sum_{i,j} \lambda^i \wedge \mu^j \otimes [X_i, X_j]. \tag{5}$$

This definition is clearly independent of the choice of basis in $L(G)$. If λ, μ are of degrees l, m respectively, we have

$$[\lambda, \mu] = (-1)^{lm+1}[\mu, \lambda]. \tag{6}$$

Let ω^i, $1 \leqq i \leqq r$, be the dual basis of X_i. By left translations they can be identified with left-invariant linear differential forms on the group manifold G. Then

$$\omega = \sum_i \omega^i \otimes X_i \tag{7}$$

is an $L(G)$-valued left-invariant linear differential form on G, called the Maurer-Cartan form. The Maurer-Cartan equation can be written

$$d\omega = -\tfrac{1}{2}[\omega, \omega]. \tag{8}$$

Consider an inner automorphism of G defined by $s \rightarrow asa^{-1}$, where a, $s \in G$ and a is fixed. This leaves fixed the unit element of G and induces an automorphism on $L(G)$, which we call $\operatorname{ad}(a)$. Clearly $\operatorname{ad}(a)$ has the properties

$$\begin{aligned} \operatorname{ad}(ab) &= \operatorname{ad}(a)\operatorname{ad}(b), \qquad a, b \in G \\ \operatorname{ad}(a)[\lambda, \mu] &= [\operatorname{ad}(a)\lambda, \operatorname{ad}(a)\mu], \end{aligned} \tag{9}$$

where λ, μ are the forms considered above.

A connection is most conveniently defined as a structure on a principal fiber bundle. A simple example of a principal fiber bundle is the space of all orthonormal frames in euclidean space. It is a fiber space over the euclidean space itself, the projection being defined by taking for an orthonormal frame its origin, so that a fiber consists of all orthonormal frames with the same origin. The generalization of this classical concept in geometry and kinematics leads to the method of moving frames of Elie Cartan in differential geometry and to the notion of a principal fiber bundle in modern algebraic topology. Formally the latter is defined by a differentiable mapping $\psi: B \rightarrow M$ and, relative to an open covering $\{U, V, \cdots\}$ of M, a family of "transition functions" $g_{UV}(x) \in G$, $x \in U \cap V$, defined for every pair of members of the covering satisfying $U \cap V \neq \phi$, such that: (1) $\psi^{-1}(U)$ is a product $U \times G$ and has the local coordinates (x, s), $x \in U$, $s \in G$; (2) For $x \in U \cap V$, the local coordinates (x, s), (x, t), s, $t \in G$, relative to U and V respectively, define the same point if and only if $s=g_{UV}(x)t$, the multiplication at the right-hand side being in the sense of group multiplication in G. Thus a fiber $\psi^{-1}(x)$ has the structure of a group manifold defined up to left translations. Right translations have a meaning, not only on a fiber, but on B itself; it is locally defined by $b=(x, s)\rightarrow ba=(x, sa)$.

At a point $b\in B$ we call the tangent space $V(b)$ to the fiber the vertical space. A subspace in the tangent space of B at b, which has only the zero vector in common with $V(b)$ and which, together with $V(b)$, spans the tangent space, is called a horizontal space. A connection in the bundle B is a field of horizontal spaces which is stable under the right translations of B. Instead of the horizontal spaces we can equally well define the connection by their orthogonal spaces in the cotangent spaces. This in turn can be described as an $L(G)$-valued linear differential form $\phi(b)$ in B, such that its restriction to a fiber is the Maurer-Cartan form and such that under right translations of B it satisfies the condition

$$\phi(ba) = \mathrm{ad}(a^{-1})\phi(b). \tag{10}$$

In terms of the local coordinates (x, s) in $\psi^{-1}(U)$, this condition means that $\phi(b)$ is of the form

$$\phi(b) = \omega(s) + \mathrm{ad}(s^{-1})\theta_U(x, dx), \tag{11}$$

where $\theta_U(x, dx)$ is an $L(G)$-valued linear differential form in U. By equating the expressions for $\phi(b)$ in $\psi^{-1}(U\cap V)$, we get

$$\omega(g_{UV}(x)) + \mathrm{ad}(\overset{-1}{g_{UV}})\theta_U(x, dx) = \theta_V(x, dx). \tag{12}$$

This formula gives the relation between θ_U, θ_V in $U\cap V$.

A sectionally smooth curve in B is called a horizontal curve, if it is everywhere tangent to a horizontal space of the connection. To a given point $b\in B$ all the elements $a\in G$ such that b and ba can be joined by a horizontal curve form a group H_b, called the holonomy group at b. It is easy to see that the holonomy groups H_b, $H_{b'}$ at two points b, $b'\in B$ are conjugate to each other in G.

The connection defines an absolute differentiation or covariant differentiation in B, as follows: Let η be an exterior differential form of degree q in B, with values in a vector space E. Its absolute differential $D\eta$ will be of degree $q+1$, also with values in E. To define it we consider an exterior differential form to be an alternating multilinear function of vector fields. If X is a vector field in B, we denote by VX and HX its projections in the vertical and horizontal spaces respectively, so that $X=VX+HX$. Then $D\eta$ is defined by the equation

$$D\eta(X_1, \cdots, X_{q+1}) = d\eta(HX_1, \cdots, HX_{q+1}), \tag{13}$$

where $X_1, \cdots, X_{q+1}$ are $q+1$ vector fields in B.

Using this definition we compute the absolute differential of the connection form ϕ. By (11) we find

$$d\phi + \tfrac{1}{2}[\phi, \phi] = \operatorname{ad}(s^{-1})\{d\theta_U + \tfrac{1}{2}[\theta_U, \theta_U]\}. \tag{14}$$

From this it is seen that the common expression is $D\phi$. Putting $\Phi = D\phi$ and

$$\Theta_U = d\theta_U + \tfrac{1}{2}[\theta_U, \theta_U], \tag{15}$$

we have

$$\Phi = \operatorname{ad}(s^{-1})\Theta_U. \tag{16}$$

The $L(G)$-valued quadratic differential form Φ is called the curvature form. In $U \cap V$ we have

$$\Theta_U = \operatorname{ad}(g_{UV})\Theta_V. \tag{17}$$

In the usual treatment the curvature form is described by the family of forms $\{\Theta_U\}$ with the transformation law (17).

Now let G be a subgroup of a Lie group G'. Every element $a \in G$ defines an automorphism $\operatorname{ad}(a): L(G') \to L(G')$, which leaves $L(G)$ invariant. It induces a linear mapping of the quotient space $L(G')/L(G)$ into itself, which we also denote by $\operatorname{ad}(a)$. The situation occurs that we have a G-connection in a G'-bundle, or, more precisely, that we have a connection in the associated principal bundle with the group G', when the bundle B is considered to have the structural group G. With the covering $\{U, V, \cdots\}$ of M the curvature form of the connection will be given by a family of $L(G')$-valued quadratic differential forms $\{\Theta'_U\}$, satisfying the relation $\Theta'_U = \operatorname{ad}(g_{UV}(x))\Theta'_V$ in $U \cap V$. The projections of these forms into the quotient space $L(G')/L(G)$ constitute the torsion form.

The curvature forms define the local properties of a connection and the holonomy groups H_b its global property. Ambrose and Singer showed how the Lie algebra of H_b can be determined from the curvature forms [5].

EXAMPLE. Consider the frame bundle of a manifold M of dimension n (a frame is an ordered set of n linearly independent tangent vectors with the same origin). In this case $G = \mathrm{GL}(n, R)$ and G can be considered to be the group of all $n \times n$ nonsingular matrices. Its Lie algebra has as underlying vector space that of all $n \times n$ matrices, nonsingular or singular. For $g \in G$, the Maurer-Cartan form is $\omega = g^{-1}dg$, while $\operatorname{ad}(g)X = gXg^{-1}$, $X \in L(G)$, in the sense of matrix multiplication. If u^i, v^k, $1 \leq i, k \leq n$, are local coordinates in U, V respectively, we have

$$g_{UV} = \left(\frac{\partial u^i}{\partial v^k}\right), \qquad \theta_U = \left(\sum_j \Gamma^i_{kj}\, du^j\right). \tag{18}$$

The transformation formula (12) becomes

$$\frac{\partial^2 u^i}{\partial v^k \partial v^j} = \sum_{m,l} \Gamma^i_{ml} \frac{\partial u^m}{\partial v^k} \frac{\partial u^l}{\partial v^j} + \sum_l \frac{\partial u^i}{\partial v^l} \tilde{\Gamma}^l_{kj} \tag{19}$$

where $\tilde{\Gamma}^i_{kj}$ are the coefficients in θ_V. This is the transformation formula for the "components" of an affine connection in the usual form.

Returning to the general case, it follows from (12) and a simple extension argument that a connection always exists in a principal fiber bundle. In the case of the frame bundle of a manifold this implies that the bundle space is topologically parallelisable, i.e., there are n^2+n linear differential forms in the space, which are everywhere linearly independent.

The notion of a connection is at the basis of many fields of differential geometry, such as riemannian, hermitian, as well as projective and conformal differential geometries. Our treatment adapts well both to the local theory and to the study of the homology (with real coefficients) of principal fiber bundles. A close relationship exists between the curvature forms and the characteristic classes of a fiber bundle (cf. §9).

4. *G*-structure [18], [33], [41], [129]. Among the important structures on a manifold is the reduction of the structural group of the tangent bundle, which we explain as follows:

Let T be an n-dimensional real vector space and T^* be its dual space. Denote their pairing by $\langle y, \xi\rangle \in R$, $y \in T$, $\xi \in T^*$. We let $GL(n, R)$ act on T on the left and on T^* on the right, so that the following relation holds:

$$\langle gy, \xi\rangle = \langle y, \xi g\rangle, \qquad g \in GL(n, R). \tag{20}$$

The tangent bundle over M has the local charts (x, y_U), $x \in U$, $y_U \in T$, which are the local coordinates of the tangent vectors relative to U. The local coordinates (x, y_U) and (x, y_V) in $U \cap V$ define the same tangent vector if and only if $y_U = g_{UV}(x) y_V$, where $g_{UV}: U \cap V \to GL(n, R)$. Consider a subgroup G of $GL(n, R)$; we say that the structural group of the tangent bundle is reduced to G, if all $g_{UV}(x) \in G$. Such a reduction will be simply called a G-structure.

EXAMPLE 1. An $O(n)$-structure is nothing else but a riemannian structure.

EXAMPLE 2. For $n=2m$ consider $\mathrm{GL}(m, C)$ as a subgroup of $\mathrm{GL}(n, R)$. Then a $\mathrm{GL}(m, C)$-structure is what is called an almost complex structure.

Many methods are known in the theory of fiber bundles to find necessary conditions for a manifold to have a G-structure (obstruction theory, characteristic classes, cohomology operations, etc.) and we will not discuss them. We will go further to study the properties of a given G-structure. We take a basis in T and call all frames obtained from this basis frame by the transformations of G permissible frames. If M has a G-structure, it makes sense to call a frame of M permissible, if it is so in a local coordinate system. The dual basis of a permissible frame in the cotangent space will be called a permissible coframe. In the associated principal bundle of the G-structure the permissible coframes give rise to n linearly independent linear differential forms, which are globally defined.

The problem of local invariants of a G-structure is essentially the following problem of equivalence of Elie Cartan [33]: Given a set of n linearly independent linear differential forms $\theta_U^i(u, du)$ in the coordinates u^k, and another such set $\theta_V^i(v, dv)$ in the coordinates v^k, $1 \leq i, k \leq n$, and given a Lie group $G \subset \mathrm{GL}(n, R)$. To determine the condition that there exist functions

$$v^i = v^i(u^1, \cdots, u^n), \tag{21}$$

such that the $\theta_V^i(v(u), dv(u))$ differ from the $\theta_U^i(u, du)$ by a transformation of G. The last condition gives rise to an exterior differential system. As a first step to the solution of the problem one considers the differential forms $\phi^i = \sum_j g_j^i \theta_U^j$, $(g_j^i) \in G$, with the coordinates of G as auxiliary variables. In our terminology these are the permissible coframes. We illustrate this by some examples for $n=2$:

EXAMPLE 3. $G=\mathrm{SO}(2)$ and is the group of all matrices of the form

$$\begin{pmatrix} \cos\lambda & -\sin\lambda \\ \sin\lambda & \cos\lambda \end{pmatrix}.$$

With λ as auxiliary variable we have

$$\phi^1 = \cos\lambda\theta^1 + \sin\lambda\theta^2, \qquad \phi^2 = -\sin\lambda\theta^1 + \cos\lambda\theta^2. \tag{22}$$

There is a linear differential form

$$\pi = d\lambda + \text{lin comb of } \theta^1, \theta^2, \tag{23}$$

which is uniquely determined by the conditions

(24) $$d\phi^1 = \pi \wedge \phi^2, \qquad d\phi^2 = -\pi \wedge \phi^1.$$

The exterior derivative of π is of the form

(25) $$d\pi = -K\phi^1 \wedge \phi^2.$$

One sees that K is the gaussian curvature of the riemannian structure. The forms ϕ^1, ϕ^2, π in the associated principal bundle with the group NSO(2) (we denote by NG the nonhomogeneous linear group whose homogeneous part is G) can be interpreted geometrically as defining a connection in the bundle.

EXAMPLE 4. G is the group of all matrices of the form

$$\begin{pmatrix} \lambda & 0 \\ 0 & \lambda \end{pmatrix},$$

$\lambda \neq 0$. We put

(26) $$\phi^1 = \lambda\theta^1, \qquad \phi^2 = \lambda\theta^2.$$

There is a uniquely determined linear differential form

(27) $$\pi = \frac{d\lambda}{\lambda} + \text{lin comb of } \theta^1, \theta^2,$$

satisfying the conditions

(28) $$d\phi^1 = \pi \wedge \phi^1, \qquad d\phi^2 = \pi \wedge \phi^2.$$

We find

(29) $$d\pi = A\phi^1 \wedge \phi^2,$$

where A is of the form $A(u^1, u^2)/\lambda^2$. The forms ϕ^1, ϕ^2, π define a connection in the associated principal bundle with the group NG.

The structure can be interpreted as a three-web of plane curves in the sense of Blaschke [**24**], the curves being defined respectively by the equations

(30) $$\theta^1 = 0, \qquad \theta^2 = 0, \qquad \theta^1 + \theta^2 = 0.$$

The condition $A = 0$ is a necessary and sufficient condition that the curves can be locally mapped into three families of parallel straight lines. It can also be interpreted geometrically as the condition for a certain hexagonal configuration to exist.

EXAMPLE 5. G is the group of all matrices of the form

$$\begin{pmatrix} \lambda & -\mu \\ \mu & \lambda \end{pmatrix},$$

$\lambda^2+\mu^2\neq 0$. Again put

$$\phi^1 = \lambda\theta^1 + \mu\theta^2, \quad (31)$$
$$\phi^2 = -\mu\theta^1 + \lambda\theta^2.$$

There are linear differential forms π_1, π_2 satisfying the equations

$$d\phi^1 = \pi_1 \wedge \phi^1 + \pi_2 \wedge \phi^2, \quad (32)$$
$$d\phi^2 = -\pi_2 \wedge \phi^1 + \pi_1 \wedge \phi^2.$$

But they are not completely determined by these conditions. The structure is given by $(\theta^1)^2+(\theta^2)^2$, determined up to a positive factor. It is therefore the two-dimensional conformal structure. No connection can be introduced, which will depend on this structure alone; for otherwise it would mean that the local homeomorphisms leaving the conformal structure invariant would depend on a finite number of constants. This is a first example of a pseudo-group structure.

The general problem of equivalence is a problem on exterior differential systems of a certain type. In the real analytic case the theorem of Cartan-Kuranishi [**86**] says that in a finite number of steps the system can be prolonged either to a system without solution or to a system in involution. They correspond respectively to the cases that the two G-structures are locally inequivalent or equivalent. In the latter case there may be a pseudo-group of local homeomorphisms, other than the identity, which leaves the structures invariant. To carry out the general program of Cartan-Kuranishi in a particular case is not always an easy problem and very frequently leads to complicated calculations. We mention two simple instances where the problem of equivalence is not solved: (1) Almost complex structure $G=\mathrm{GL}(m, C)$, $n=2m$. The problem of equivalence was solved by Libermann for $m=2$ [**93**]. (2) Symplectic structure, where G is the linear group in a space of dimension $2m$ consisting of all transformations leaving invariant an exterior quadratic form of maximum rank. This is in other words the local classification of antisymmetric covariant tensor fields $a_{ij}=-a_{ji}$ of order two and maximum rank.

Two classical cases where the problem of equivalence is solved and a connection attached to the structure are the following:

EXAMPLE 6. $G=O(n)$. This is the riemannian structure where the Levi-Civita connection can be attached.

EXAMPLE 7. G is the group of all matrices of order n ($\geqq 3$) of the form λA, $\lambda>0$, where A is orthogonal. This is the conformal structure, to which a normal conformal connection, uniquely determined, can be attached.

There are various advantages to be gained in attaching a connection to a G-structure, if this is possible. First the structure is put in a general setting, whose results will then be applicable. Secondly, it will be possible to introduce geometrical concepts. In fact, the geometry of a connection along a curve is the same, whether the curvature form is zero or not (Fermi's theorem). Thus the geometry appears more clearly as a generalization of classical geometry. Thirdly, G-structures equivalent to a connection are in a sense simpler. For instance, the group of automorphisms leaving the structure invariant will be a Lie group.

It appears from the examples that it will be desirable to consider the associated principal bundle with the group NG. We are actually considering a bundle with the group NG with a cross-section into the associated bundle of homogeneous spaces NG/G, making it a bundle with the structural group G. In such a bundle there is a torsion form associated to a connection with the structural group NG. The question arises as to whether it is possible to attach uniquely a connection by properties of this torsion form. It can be proved that such a connection exists whenever G is compact [**43**]. In the particular case of the riemannian structure the Levi-Civita connection is completely determined by the vanishing of the torsion form. For general compact G there may be different ways of attaching the connection which will be characterized by different properties of the torsion form. In this context it would be a problem of some interest to determine the properties of the noncompact G, for which the answer to the above question is affirmative. (Cf. §13 on the Weyl-Cartan theorem.)

An almost complex structure whose torsion form (this can be defined, although there is no connection) is zero is called integrable. Newlander and Nirenberg proved that such an almost complex structure is subordinate to a complex structure [**101**], [**103**]. It follows that an almost complex manifold whose torsion form vanishes identically is a complex manifold. This is an example of a pseudo-group structure. No connection can be attached to a complex manifold from its complex structure alone.

Among the G-structures those to which a connection can be attached and those which are subordinate to a pseudo-group structure seem to be the extreme cases. They are among the most important G-structures. The others are probably too complicated to be of much geometrical interest.

A further generalization is to structures which involve elements of contact of higher order. An example is the projective geometry of paths of Veblen and T. Y. Thomas. Much work on the foundations of this generalization has been done by Ehresmann and his school.

5. **Harmonic forms** [64], [68], [119], [134]. A general method which gives global implications on the existence of geometrical structures satisfying differential conditions is the theory of harmonic differential forms of Hodge. Let M be a compact oriented n-dimensional riemannian manifold. Let Δ be the Laplace-de Rham operator on M. A differential form η is called harmonic, if $\Delta\eta=0$. Since M is compact, every harmonic form is closed. All harmonic forms of degree q constitute a real vector space H^q. The fundamental theorem on harmonic forms says that the linear mapping $\rho: H^q \to H^q(M, R)$ ($=q$-dimensional cohomology space of M with real coefficients) defined by sending a harmonic form to its cohomology class in the sense of de Rham's theorem is a one-one isomorphism. We are going to consider the existence of certain structures on M, which will allow a finer analysis of the cohomology ring of M with real coefficients through the application of harmonic forms.

Let G be a subgroup of SO(n). Then a G-structure on M defines an orientation of M and a riemannian structure, to be called the associated riemannian structure. Such a G-structure is called holonomic, if it satisfies one of the following three conditions, which are equivalent [43], [84], [135]:

(1) The homogeneous holonomy group of the Levi-Civita connection (of the associated riemannian structure) is G or a subgroup of G.

(2) Under the Levi-Civita parallelism permissible frames remain permissible.

(3) There is a connection with the structural group NG, whose torsion form is zero.

It seems to us that these are the structures to which the harmonic forms can be applied most advantageously to derive global implications.

Particular cases of a holonomic G-structure include the following:

EXAMPLE 1. $G=U(m)$, $m=n/2$. This is essentially the Kähler structure on a complex manifold in its real formulation.

EXAMPLE 2. $G=\mathrm{SO}(p)\times\mathrm{SO}(n-p)$, $1\leq p\leq n-1$. This means that the manifold has a field of p-dimensional plane elements parallel with respect to the Levi-Civita connection.

A G-structure on M allows a finer classification of its exterior differential forms. (This applies also to other tensor fields.) In fact, G acts on T^* and has an induced representation on $\Lambda^q(T^*)$, $0\leq q\leq n$. Let W be an invariant subspace of $\Lambda^q(T^*)$ under the action of G. A q-form on M is said to be of type W if the element it associates to every point $x\in M$ belongs to the corresponding subspace W_x. If $G\subset O(n)$, then every invariant subspace W has an orthogonal space $W^\perp$ which is also invariant and we can define to a q-form η its orthogonal projection $P_W\eta$ into a q-form of type W. More generally, if $Q: W \to \Lambda^r(T^*)$ is a linear mapping which commutes with the action of G, we define an operator Q on q-forms of type W, the image being then an r-form. An example of such a mapping Q, besides P_W, is the exterior multiplication by an exterior $(r-q)$-form which is invariant under G. When there are such operators, it is of importance to find the conditions under which they commute with the Laplace-de Rham operator, and we have the theorem:

Let M have a holonomic G-structure. Then $P_W\Delta = \Delta P_W$. Moreover, if Q is the multiplication by an invariant exterior differential form, we have also $Q\Delta = \Delta Q$.

The proof of this theorem makes use of an explicit formula of R. Weitzenböck for Δ. (The formula allows a simple derivation of Bochner's theorem on the relation between curvature and Betti numbers of a compact riemannian manifold. Considering the importance of the operator Δ, it should have further applications.) It follows from the theorem that if η is harmonic, then $P_W\eta$ is also harmonic. Hence if $\Lambda^q(T^*)$ is a direct sum of the invariant subspaces $W_1, \cdots, W_k$, the space H^q of harmonic forms of degree q is a direct sum of the spaces of harmonic forms of types $W_1, \cdots, W_k$ respectively. The second part of the theorem leads to isomorphisms of subspaces of harmonic forms of different degrees. The analysis of the cohomology groups (over the real field) of M is reduced to a purely algebraic problem (namely, that of studying the induced representation of G on $\Lambda(T^*)$). From this one derives various global implications from the existence of a holonomic G-structure.

EXAMPLE 1 (CONTINUED). The group G leaves invariant an exterior form of degree 2 and maximum rank. Denote by Ω the corresponding differential form on M. Introduce on the exterior differential forms η of M the operators

$$L\eta = \Omega \wedge \eta, \qquad \Lambda = *^{-1}L*. \tag{33}$$

A form η is called primitive, if $\Lambda\eta = 0$. It follows from our commutativity theorem that every harmonic form η can be written uniquely as

$$\eta = \sum_{r \geq \max(0, m-p)} L^r\eta_r, \qquad p = \deg \eta \tag{34}$$

where η_r are harmonic and primitive (Hodge's decomposition theorem). For compact orientable manifolds with such a holonomic G-structure we can derive the following global properties:

(1) Every odd-dimensional Betti number is even.

(2) Let u be the cohomology class determined by Ω. The homomorphism $H^{r-2}(M, R)\to H^r(M, R)$ defined by $\alpha\to\alpha\cup u$ (=cup product of α, u), $\alpha\in H^{r-2}(M, R)$, is, for $r\leqq m$, an isomorphism.

(3) The homomorphism $H^r(M, R)\to H^{2m-r}(M, R)$, $r\leqq m$, defined by $\alpha\to\alpha\cup u^{m-r}$, $\alpha\in H^r(M, R)$, is a one-one isomorphism.

It is worth noting that these properties can be derived without reference to the complex structure.

EXAMPLE 2 (CONTINUED). In this case we can derive that the p-dimensional Betti number is $\geqq 1$. For $p=1$, a further analysis of the situation gives the inequality $b^2\geqq b^1-1$, b^1 and b^2 being respectively the one and two-dimensional Betti numbers. It would be natural to ask whether for $p=1$ the conditions are now sufficient.

The use of harmonic forms also gives information on the multiplicative structure of the cohomology ring $H^*(M, R)$, particularly the index $\tau(M)$. The latter is defined as follows: $\tau(M)=0$, if $\dim M\not\equiv 0$, mod 4. If $\dim M=4k$, we consider in the real vector space $H^{2k}(M, R)$ the function $f(u, v)=(u\cup v)M$, $u, v\in H^{2k}(M, R)$, which is the value of the cohomology class $u\cup v$ on the fundamental class M. This function is a symmetric nondegenerate bilinear form. $\tau(M)$ is defined to be the number of its positive eigenvalues minus the number of its negative eigenvalues.

Consider now the space $\Lambda^{2k}(T^*)=W_1\oplus W_2$, where W_1 (resp. W_2) is the subspace of all elements invariant under $*$ (resp. transformed to its negative by $*$). Let h_i be the dimension of the space of harmonic forms of degree $2k$ and type W_i, $i=1, 2$. Then a study of the multiplicative properties of harmonic forms gives the theorem: $\tau(M)=h_1-h_2$. From this theorem Hodge's index theorem on compact Kähler manifolds can be derived by purely algebraic considerations.

We will mention another application of the above index theorem: Suppose that a compact orientable manifold of dimension $4k$ has $2k$ linearly independent vector fields which are parallel with respect to a riemannian metric. Then its index is zero. Examples (due to Atiyah, private communication) exist of four-dimensional compact manifolds with nonzero index but with two linearly independent vector fields (nonparallel with respect to any riemannian metric).

6. **Leaved structure** [51], [60], [116], [117]. From the point of view of §4 a leaved structure is a reduction of the structural group $G(n, R)$ of the tangent bundle of a manifold of dimension n to the subgroup

$$\begin{pmatrix} G(p, R) & 0 \\ * & * \end{pmatrix}, \qquad 1 \leqq p \leqq n-1,$$

such that a local differential condition is satisfied. More precisely, if the structural group of the tangent bundle is so reduced, that of the cotangent bundle will be reduced accordingly, and a field of permissible coframes will be given locally by the linear differential forms $(\theta^1, \cdots, \theta^n)$, which are linearly independent and are such that $\theta^{p+1}, \cdots, \theta^n$ are determined up to a transformation of $GL(n-p, R)$. A leaved structure is such a reduction satisfying the further conditions

$$d\theta^i \wedge \theta^{p+1} \wedge \cdots \wedge \theta^n = 0, \qquad p+1 \leqq i \leqq n. \tag{35}$$

By Frobenius' theorem this means that the pfaffian system

$$\theta^{p+1} = \cdots = \theta^n = 0 \tag{36}$$

is completely integrable, i.e., there exists at each point a local coordinate system $(x^1, \cdots, x^n)$ relative to which the system becomes

$$dx^{p+1} = \cdots = dx^n = 0. \tag{37}$$

The integer p is called the dimension of the leaved structure, and $n-p$ its codimension. A leaf is a "maximal" integral submanifold.

The simplest example of a leaved structure is given by a differential equation of the first order in the plane. A consequence of the Poincaré-Bendixon theory says that the leaves go to infinity in both directions. The problem of classifying the leaved structures in the plane was solved by W. Kaplan [69]. In particular, Kaplan proved that to a leaved structure in the plane there exists a continuous real-valued function in the plane, which has neither a maximum nor a minimum and which is constant on the leaves. In this respect it may be of interest to mention that Wazewski gave examples of C^∞-leaved structures in the plane such that any differentiable function in the plane which is constant on the leaves is a constant.

Another example is the leaved structures (of dimension 1) on a torus [125]. Two extreme cases are: (1) All leaves are closed. (2) The leaves are ergodic. If the torus is taken to be a unit square with opposite sides identified, the second case occurs when the directions of the field makes a constant angle $\alpha\pi$ (α irrational) with the sides. The fundamental theorem of Denjoy says that (under sufficient smoothness hypotheses) if there is no closed leaf, then every leaf is ergodic. Denjoy's work has been extended by A. J. Schwartz to arbitrary closed surfaces [120]. If we interpret a leaved structure of dimension 1 as the action of a surface by the real line, we define a minimal set of this action to be a nonempty closed invariant set, which contains no proper subset with the same property. The Denjoy-Schwartz theorem says that if a compact connected two-dimensional surface is under the action of the real line, then a minimal set is: (1) either a single point; (2) or a closed curve; (3) or the whole surface, the last case happening only for the torus.

Another simple example of a leaved structure is given by a coset decomposition of a Lie group relative to a subgroup.

A notable example of a leaved structure of codimension one on the three-sphere S^3 was given by G. Reeb [**116**]. Geometrically this can be described as follows: Consider S^3 as the union of two solid tori with a two-dimensional torus as common boundary. In each solid torus take a leaved structure of codimension one whose integral submanifolds are like paraboloids with the boundary torus as a limiting surface. These leaved structures can be fitted together to give a leaved structure on S^3. The same construction cannot be extended to a sphere of higher odd dimension, and it is an open problem whether S^{2m+1}, $m>1$, has a leaved structure of codimension one.

Reeb's leaved structure is not analytic and he raised the question whether an analytic one (of codimension one) exists on S^3. The answer was given as negative by A. Haefliger, who proved the following theorem [**59**]: *Let M be a compact real analytic manifold which has a real analytic leaved structure of codimension one. Then the fundamental group $\pi_1(M)$ is not finite.*

If a manifold has a leaved structure of dimension p, it must have a G-structure, with $G=\mathrm{GL}(p, R)$. The problem, called fundamental by Reeb, is whether this condition is sufficient. In fact, it is not known whether any $\mathrm{GL}(p, R)$-structure is homotopic to a leaved structure of dimension p, i.e., whether they can be connected by a differentiable family of $\mathrm{GL}(p, R)$-structures. A compact orientable three-dimensional manifold is parallelisable, but it is an open question whether it has a leaved structure of codimension one.

When a manifold has a leaved structure, an important invariant is the holonomy group in the sense of Ehresmann. It gives an accurate description of the behavior of the leaves in the neighborhood of a leaf. An important problem is whether a leaved structure has a compact leaf. On the three-sphere S^3 the existence of a compact leaf for a leaved structure of dimension one (respectively two) was conjectured by H. Seifert (respectively by H. Kneser) [**121**]. A proof of Kneser's conjecture has been announced by S. P. Novikov [**106**].

Let $X_1, \cdots, X_p$ be vector fields, which span the tangent spaces of the leaves. Then condition (35) is equivalent to the condition that the brackets $[X_i, X_j]$ are linear combinations of $X_1, \cdots, X_p$. Milnor defined as the rank of a manifold the maximum number of vector fields $X_1, \cdots, X_p$, everywhere linearly independent, such that $[X_i, X_j]=0$, $1 \leqq i, j \leqq p$. Lima proved that S^3 has rank one [**95**]. This was generalized by Lima and H. Rosenberg to the result that the rank of a compact simply connected manifold of dimension n is at most $n-2$. The problem is closely related to the study of the action of noncompact transformation groups on a manifold. Except when the transformation group is one-dimensional (flows on a manifold), very little is known.

An important notion on the leaved structure is that of "structural stability," which was initiated by Andronov and Pontrjagin [**7**] in the case of a disk and whose study has so far been restricted to one-dimensional leaves. A structural stable vector field is one such that the topological properties of the trajectories remain invariant under a small perturbation of the vector field (relative to some topology). Peixoto proved that the structurally stable vector fields on a compact two-dimensional surface are open and dense in the set of all vector fields on the surface [**109**]. The study in the higher-dimensional case is wide open. In fact, Smale gave examples of manifolds in which the structurally stable vector fields are not dense in the set of all vector fields. What makes the theory in higher dimensions much more difficult is the presence of recurrence. An important recent result is the "closing lemma" of C. C. Pugh [**113**]. Among its consequences is the theorem that the closure of the set of all nonwandering points is the closure of the set of all closed trajectories.

7. **Complex structure** [**66**], [**134**]. The existence of a complex structure on a manifold is a nontrivial fact, so that an understanding of complex manifolds should begin with some examples. Two obvious necessary conditions are even dimensionality and orientability. An orientable manifold of two (real) dimensions always has a complex structure. The difficulty in proving this is the local theorem of Korn and Lichtenstein to the effect that a two-dimensional (positive definite) C^∞-riemannian manifold is locally conformal to the euclidean plane. For higher dimensions the simplest examples are the m-dimensional complex euclidean space $E_m(C)$ (or simply E_m) and the m-dimensional complex projective space $P_m(C)$ (or P_m). (E_1 is known as the gaussian plane and P_1 as the Riemann sphere in complex function theory.) The next examples are to be their quotient manifolds and submanifolds. Compact submanifolds of P_m are, by a theorem of Chow, the same as the nonsingular algebraic varieties. Of course there are no compact submanifolds in E_m, other than points, as there is no nonconstant analytic function on a compact complex manifold.

An example of a quotient manifold of E_m is E_m/Δ, where Δ is the discrete group generated by $2m$ translations linear independent over the reals. This is called a complex torus. Depending on the choice of Δ, there are some (those satisfying the Riemann conditions) which are isomorphic to algebraic varieties and there are also some on which there exists no nonconstant meromorphic function. The former are called abelian varieties.

It is equally possible to take an open submanifold of E_m and consider a quotient space of it. If $z_1, \cdots, z_m$ are the coordinates of E_m and $0=(0, \cdots, 0)$, then the discrete group Δ_1 generated by the transformations

$$z_k' = 2z_k, \qquad 1 \leqq k \leqq m, \tag{38}$$

has no fixed point in E_m-0 and $(E_m-0)/\Delta_1$ is a complex manifold. This is called a Hopf manifold. Topologically it is homeomorphic to $S^1 \times S^{2m-1}$. For $m>1$ a Hopf manifold cannot be given a Kähler structure and is therefore not algebraic, because the latter must have its second Betti number $\geqq 1$.

To find further complex manifolds an obvious process is to form cartesian products. More significant are the blowing-up process (σ-process of H. Hopf) and fiber space constructions.

An example of blowing-up is as follows: In the projective plane P_2 let a line element (p, L) be the pair consisting of a point p and a line L through p. Consider all line elements (p, L) such that L passes through a given point b of P_2. They form a complex manifold M_2 of two (complex) dimensions. There is a complex analytic mapping f: $M_2 \rightarrow P_2$, defined by $f(p, L)=p$. Then f is one-to-one for $M_2-f^{-1}(b)$ and $f^{-1}(b)$ is isomorphic to P_1. We describe this geometrically by saying that M_2 is obtained from P_2 by blowing up a point. Generally we can blow up a point b in any complex manifold M_m, leaving the points of M_m-b unchanged and replacing b by P_{m-1}. The same process generalizes to the blowing-up of any nonsingular submanifold. The process has its origin in the birational transformation of algebraic varieties, where it is known as a quadratic transformation. It exists only for $m>1$. Blanchard proved that if a compact Kähler manifold is blown up along a nonsingular submanifold, the resulting manifold is again Kählerian [23]. The process is of great importance in the theory of complex manifolds.

A simple application is to a construction due to Andreotti. Let T be a complex torus group of dimension $m>1$. (The difference between a complex torus and a complex torus group is that the latter has the additive group structure, i.e., a certain point singled out as the neutral element of the group.) Let α be the automorphism $x \rightarrow -x$, $x \in T$, and consider the quotient space T/α. The latter has singularities, 2^{2m} in number, arising from the fixed points of α, which are of a very simple type. They can be resolved by the blowing-up process, leading to a nonsingular complex manifold K, called the Kummer manifold. K is simply connected (Spanier [127]). If $m=2$ and T is an abelian variety, K is the classical Kummer surface. If T has no nonconstant meromorphic function, the same is true of K. Thus we get an example of a simply-connected compact Kähler manifold which has no meromorphic functions other than constants.

If we are interested in compact complex manifolds, the simplest class of fiber spaces consists of those of complex tori. In fact, a Hopf manifold is a fiber bundle of one-dimensional complex tori over a complex projective space, with transition functions which are holomorphic. In order to have a sufficiently wide class of complex manifolds, it is desirable to relax the condition of a fiber space, in the sense that the fibers are not required to be isomorphic to each other as complex manifolds. Following Blanchard, Calabi, Atiyah, Bott, a class of fiber spaces of complex tori can be constructed as follows [9]: In E_{2m} let G be the Grassmann manifold of all m-dimensional vector spaces through the origin. Let A be the real vector space of dimension $2m$ imbedded as a subset of E_{2m}. Consider the subset $J\subset G$ consisting of those m-dimensional vector spaces which have only the zero vector in common with A; J is an open subset of G. Over J there is the universal vector bundle of dimension m, attaching to each point of J the corresponding vector space. There is also the trivial vector bundle of dimension $2m$ over J. Let E' be their quotient bundle. The basis vectors of E_{2m} give by projection in each fiber of E' $2m$ vectors which are linearly independent over the reals. The quotient space of E' by these $2m$ vector fields is then a fiber space of complex tori over J. One can construct a compact complex manifold out of it by considering its restriction to a compact submanifold of J. Such a compact submanifold is, for instance, the manifold of all the m-dimensional vector spaces through the origin and lying on the quadric cone

$$z_1^2 + \cdots + z_{2m}^2 = 0,$$

where $z_1, \cdots, z_{2m}$ are the coordinates in E_{2m}. It can be seen that this is the homogeneous space $O(2m)/U(m)$. By a theorem of Blanchard [23] the compact complex manifold so obtained (the fiber space!) which, as a differentiable manifold, is the product $O(2m)/U(m) \times T^{2m}$ (T^{2m} is a real torus of real dimension $2m$), has no Kählerian structure. On the other hand, $O(2m)/U(m)\times T^{2m}$ has a Kählerian complex structure derived from those of the two factors. We get in this way a manifold which has both Kählerian and non-Kählerian complex structures. The problem is unsolved whether there exists a simply-connected compact manifold with the same property.

Among the nonsingular algebraic varieties a notable family consists of the ruled surfaces, and in fact ruled surfaces with a directrix line. In P_{n+3} with the homogeneous coordinates $(x_0, x_1, \cdots, x_{n+3})$ consider the normal curve

$$x_0 = x_1 = 0, \qquad x_2 = 1, \qquad x_3 = t, \cdots, x_{n+3} = t^{n+1}$$

in the linear subspace of dimension $n+1$ with the equation $x_0 = x_1 = 0$. The lines joining the point of the normal curve with the parameter t to the point $(1, t, 0, \cdots, 0)$ generates a ruled surface without singularities. We denote it by Σ_n. As a real manifold it is a bundle of S^2 over S^2. Topologically there are only two classes of bundles of S^2 over S^2: the cartesian product $S^2 \times S^2$ and another. It can be shown that Σ_n is differentiably the product $S^2 \times S^2$ if and only if n is even. As complex manifolds the Σ_n are distinct from each other (a theorem of Hirzebruch [**65**]). Thus $S^2 \times S^2$ is a simply connected manifold which admits an infinite number of complex structures.

Another class of complex manifolds arises from a different context, namely those which admit a transitive group of complex analytic automorphisms, the so-called homogeneous manifolds. If G is a connected compact Lie group and T a maximal toroid of G, Borel observed that G/T has a complex structure [**25**]. Goto, Borel, and Weil proved that it is an algebraic variety and Goto proved that it is rational [**56**], [**124**]. Wang determined all the compact homogeneous complex manifolds with a finite fundamental group, among which are the simply connected even-dimensional compact Lie groups [**133**]. Many of these manifolds are non-Kählerian, and are thus far from being algebraic. The problem requires a detailed analysis of Lie algebras.

The question of the complex structure on a manifold can be generally divided into four stages:

(a) Existence of an almost complex structure. This is mainly a question on fiber bundles.

(b) Existence of a complex structure on an almost complex manifold. If a manifold is known to have an almost complex structure, then the methods of (a) do not give further information. The integrability conditions will easily decide whether the given almost complex structure is subordinate to a complex structure. If it is not, no general method is known to find out whether there is a complex structure (except of course when one is given). In this respect the question whether S^6 has a complex structure remains one of the most urgent problems on complex manifolds. A remarkable recent result of van de Ven, not yet published, gives an example of a compact four-dimensional almost complex manifold which has no complex structure. The absence of a complex structure on this manifold follows from the Riemann-Roch-Hirzebruch formula for arbitrary compact complex manifolds, which in turn is a consequence of the Atiyah-Singer index theorem (see §10).

(c) Determination of all complex structures on a manifold. Riemann's mapping theorem says that S^2 has a uniquely determined complex structure. Hirzebruch's example of $S^2 \times S^2$ has an infinite number of complex structures. It is not known whether $S^2 \times S^2$ has other complex structures than those given by Hirzebruch. Andreotti proved that there are no other algebraic structures [6]. Andreotti also proved that there is only one Kähler structure on P_2 (the standard one); Hirzebruch and Kodaira proved that there is only one Kähler structure on P_m for odd m [67].

(d) If a manifold has a continuum of complex structures, it is important to give a structure to this continuum in a natural way. The classical moduli problem on compact Riemann surfaces consists in making an analytic space of dimension $3g-3$ out of the set of complex structures on a Riemann surface of genus $g>1$ [19]. In high dimensions Frölicher and Nijenhuis proved that the complex structure on a compact complex manifold M is rigid, if $H^1(M, \Theta)=0$, where $H^1(M, \Theta)$ is the one-dimensional cohomology group of M over the sheaf of germs of holomorphic vector fields (cf. §8) [53]. Subsequently Kodaira and Spencer made extensive studies of the deformation of complex structures [83]. The crowning achievement is the theorem of M. Kuranishi, which says that, for any compact complex manifold, there exists a universal family of deformations [87], [88].

8. **Sheaves** [31], [38], [54], [58], [123]. The theory of sheaves is one of the most basic and natural concepts on manifolds with a structure. Let us begin by its definition:

Let M be a topological space. A sheaf of abelian groups on M, or simply a sheaf, is given by: (1) A function $x \to S_x$, which associates to every point $x \in M$ an abelian group S_x (to be written additively); (2) A topology (not necessarily satisfying separation axioms) in $S = \bigcup_{x \in M} S_x$, such that the following conditions are satisfied (ψ denoting the mapping $S \to M$, defined by $\psi(S_x)=x$):

(S_1) The mapping $\sigma \to -\sigma$ which, to each $\sigma \in S$ associates the inverse $-\sigma$ of σ in the group $S_{\psi(\sigma)}$, is a continuous mapping of S into itself. The mapping $(\sigma, \tau) \to \sigma + \tau$ defined for the set $R = \{(\sigma, \tau) \mid \psi(\sigma) = \psi(\tau)\}$ is a continuous mapping of R into S.

(S_2) ψ is a local homeomorphism, i.e., every $\sigma \in S$ has a neighborhood V such that the restriction of ψ to V is a homeomorphism of V onto an open subset of M.

EXAMPLE 1. THE CONSTANT SHEAF. Let G be an abelian group, and let $S = M \times G$.

The following notions are defined in an obvious way: subsheaf, quotient sheaf, homomorphism of sheaves, exact sequence of sheaves. Also we can define in the same way sheaves of other algebraic structures, such as rings, ideals, modules and even nonabelian groups.

In practice a sheaf is given by a presheaf, which is a family of abelian groups S_U associated to the members of an open covering $\{U, V, \cdots\}$ of M, such that the following conditions are satisfied: (1) To every pair (U, V) with $V \subset U$ there is a homomorphism $f_{VU}: S_U \to S_V$; (2) To every triple (U, V, W) with $W \subset V \subset U$ we have $f_{WU} = f_{WV} \circ f_{VU}$. To $x \in M$ we define S_x to be the inductive limit of S_U with $U \ni x$, etc.

EXAMPLE 2. Let S_U be the additive group of continuous real-valued functions in U. To $V \subset U$ define f_{VU} to be the restriction. In this way we get the sheaf of germs of continuous real-valued functions. Similarly, there is the sheaf of germs of differentiable functions on a differentiable manifold, holomorphic functions on a complex manifold, etc.

Let M be a topological space and S a sheaf of abelian groups over M. For every $q \geq 0$ we can define by the standard Čech theory a cohomology group $H^q(M, S)$, S being called the coefficient sheaf. To an open set $U \subset M$ denote by $\Gamma(U, S)$ the group of all sections over U (a section over U is a continuous mapping $s: U \to S$ such that $\psi \circ s = \text{identity}$). Let $\{U_i\}$ be an open covering of M. Then a cochain of dimension q is an element $f_{i_0 \cdots i_q} \in \Gamma(U_{i_0} \cap \cdots \cap U_{i_q}, S)$, which is alternating in its indices and is zero if $U_{i_0} \cap \cdots \cap U_{i_q} = \emptyset$. The coboundary is defined in a standard way. The cohomology group of the covering is the quotient group of the group of all cocycles of dimension q over the subgroup of coboundaries. The inductive limit of the cohomology groups of the coverings is the cohomology group $H^q(M, S)$.

A homomorphism of sheaves $S \to S'$ induces a homomorphism $H^q(M, S) \to H^q(M, S')$ of the corresponding cohomology groups. From now on assume M to be paracompact. To a subsheaf R of S we can define the natural homomorphism

$$\delta^q: H^q(M, S/R) \to H^{q+1}(M, R).$$

Then the fundamental property of cohomology is the following: Let

$$0 \to R \to S \to T \to 0$$

be an exact sequence of sheaves on M. The sequence of cohomology groups

$$0 \to H^0(M, R) \to H^0(M, S) \to H^0(M, T) \to H^1(M, R) \to H^1(M, S) \to \cdots$$

is exact. (The homomorphisms are those defined above.)

EXAMPLE 3. Meaning of $H^0(M, S)$. Let $\{U_i\}$ be a covering of M. A 0-cochain is formed by the sections s_i: $U_i \to S$. It is a cocycle if and only if $s_i = s_j$ in $U_i \cap U_j \neq \emptyset$. Hence $H^0(M, S)$ is the group of all global sections of S.

EXAMPLE 4. Meaning of $H^1(M, S)$. A one-cochain is given by the sections s_{ij}: $U_i \cap U_j \to S$, whenever $U_i \cap U_j \neq \emptyset$. It is a cocycle if and only if $s_{ij} + s_{jk} + s_{ki} = 0$ for $U_i \cap U_j \cap U_k \neq \emptyset$. If S is the sheaf of germs of mappings into a topological group Γ (Γ not necessarily abelian), $H^1(M, S)$ is the set of all equivalence classes (in the sense of bundles) of bundles with the group Γ.

EXAMPLE 5. DE RHAM'S THEOREM. *Let M be a compact differentiable manifold. Let A^p (resp. C^p) be the sheaf of germs of exterior differential forms (resp. closed ext. diff. forms) of degree p. In particular, C^0 is the constant sheaf of real numbers. Then the sequence*

$$0 \to C^p \xrightarrow{i} A^p \xrightarrow{d} C^{p+1} \to 0,$$

where i is the injection and d is the exterior differentiation, is exact. (The essential point in the proof of this exactness is the so-called Poincaré lemma, which says that any closed differential form is locally an exterior derivative.) From this follows the exactness of the corresponding sequence of cohomology groups. But A^p is a fine sheaf and $H^q(M, A^p) = 0$, $q \geqq 1$. (A fine sheaf means essentially that its global sections can be localized.) It follows from the exact sequence that

$$H^q(M, C^p) \cong H^{q+1}(M, C^{p-1}), \qquad q > 0.$$

By applying this isomorphism successively, we get

$$H^{p+q}(M, R) \cong H^0(M, C^{p+q})/dH^0(M, A^{p+q-1})$$

which is precisely the statement of de Rham's theorem.

EXAMPLE 6. It is in complex manifolds (a "richer" structure) that the sheaf theory is most useful. The simplest example is Dolbeault's theorem. Let M be a compact complex manifold. Let $A^{p,q}$ (resp.

$B^{p,q}$) be the sheaf of germs of exterior differential forms (resp. d''-closed) of type (p, q). Then the sequence of sheaves

$$0 \rightarrow B^{p,q} \xrightarrow{i} A^{p,q} \xrightarrow{d''} B^{p,q+1} \rightarrow 0$$

is exact. (The proof that d'' is onto is here more difficult.) The sheaf $A^{p,q}$ is fine, while the sheaf $B^{p,0}$ is the sheaf of germs of holomorphic differential forms of degree p. As in Example 3, we derive from the exact sequence of cohomology groups the isomorphism

$$H^r(M, B^{p,0}) \cong H^0(M, B^{p,r})/d''H^0(M, A^{p,r-1})$$

which is the statement of Dolbeault's theorem.

EXAMPLE 7. COUSIN'S PROBLEM. In a complex manifold M let S be the sheaf of germs of meromorphic functions and Ω the subsheaf of germs of holomorphic functions. A section of the quotient sheaf S/Ω is a system of principal parts. The classical additive Cousin problem consists in deciding whether such a system of principal parts is that of a global meromorphic function. It follows from the exact sequence of cohomology groups that this Cousin problem always has a solution if $H^1(M, \Omega) = 0$.

EXAMPLE 8. Kodaira-Spencer's classification of complex analytic line bundles over a compact Kähler manifold [82]. The structural group is the multiplicative group C^* of nonzero complex numbers. If Ω^* denotes the sheaf of germs of nonzero holomorphic functions, the group of complex line bundles over a complex manifold M is $H^1(M, \Omega^*)$. To determine this group we consider the exact sequence of sheaves

$$0 \rightarrow Z \xrightarrow{j} \Omega \xrightarrow{e} \Omega^* \rightarrow 0,$$

where j is the injection and e is defined by $e(f(z)) = \exp(2\pi i f(z))$. If M is compact, we get the exact sequence

$$0 \rightarrow H^1(M, Z) \xrightarrow{j} H^1(M, \Omega) \xrightarrow{e} H^1(M, \Omega^*) \xrightarrow{\delta} H^2(M, Z) \rightarrow \cdots .$$

The homomorphism δ associates to every complex line bundle its characteristic class. The subgroup of all complex line bundles with characteristic class zero is therefore isomorphic to $H^1(M, \Omega)/jH^1(M, Z)$. If M is a compact Kähler manifold, this can be proved to be a complex torus, which is the Picard variety of M.

This example involves the second cohomology group.

In Examples 6, 7, 8 we have seen some applications of sheaf theory to complex manifolds. More significant applications arise from the notion of an analytic coherent sheaf. Let M be a complex analytic manifold. Let $\lambda: \Omega \to M$ be the sheaf of germs of holomorphic functions. A sheaf $\psi: S \to M$ is called analytic, if, for every $x \in M$, S_x has a module structure over the ring Ω_x, such that the mapping $(f, \alpha) \to f\alpha$ defined for the subset $R = \{(f, \alpha) | \lambda(f) = \psi(\alpha)\}$, is a continuous mapping of R into S.

EXAMPLE 9. Let Ω^q be q copies of the sheaf Ω; an element of Ω_x^q is an ordered set $(f_1, \cdots, f_q)$ of q germs of holomorphic functions at x. Let Ω_x act on Ω_x^q according to the formula

$$f(f_1, \cdots, f_q) = (ff_1, \cdots, ff_q).$$

Then Ω^q is an analytic sheaf.

Let S and S' be two analytic sheaves on M. A homomorphism $f: S \to S'$ of sheaves is called analytic if, for every $x \in M$, the homomorphism $f_x: S_x \to S'_x$ is compatible with the operations of Ω_x. The kernel, the image and the cokernel of f are then all analytic sheaves.

An analytic sheaf S over M is called coherent if every $x \in M$ has an open neighborhood U such that the induced analytic sheaf $S(U)$ is isomorphic to the cokernel of an analytic homomorphism $f: \Omega^q(U) \to \Omega^r(U)$ (q and r integers). In particular, an analytic sheaf is called locally free if every $x \in M$ has an open neighborhood U such that the induced sheaf $S(U)$ is isomorphic to $\Omega^q(U)$ for a certain integer q.

Stein manifolds. The Stein manifolds (of dimension >0) are noncompact complex manifolds which generalize the domains of holomorphy and which possess a sufficiently large number of holomorphic functions. Precisely speaking, a Stein manifold is a complex manifold with a countable base, satisfying the following conditions:

(1) To any two points $x, y \in M$, $x \neq y$, there exists a holomorphic function f in M, such that $f(x) \neq f(y)$.

(2) To every point $x \in M$ there exist holomorphic functions in M, which form a local coordinate system at x.

(3) M is holomorphically convex.

The following fundamental theorem (due to Oka, Cartan, and Serre) accounts for most of the properties of Stein manifolds:

A complex manifold M is a Stein manifold if and only if the following two properties hold:

(A) *For any coherent sheaf S over M the module of global sections of S over M generates at every point $x \in M$ the module of local sections.*

(B) *For any coherent sheaf S over M, $H^q(M, S) = 0$, $q \geq 1$.*

The proof of the direct part of the theorem, i.e., that a Stein manifold has the properties (A) and (B), is difficult. The theorem has many consequences of which we mention the following:

(1) The additive Cousin problem always has a solution on a Stein manifold.

(2) The second Cousin problem, the problem whether a given divisor is the divisor of a meromorphic function, has a solution on a Stein manifold if $H^2(M, Z)=0$.

(3) Every meromorphic function on a Stein manifold is the quotient of two holomorphic functions.

Compact complex manifolds. When M is compact, Cartan and Serre proved that $H^q(M, S)$ is of finite dimension for any analytic coherent sheaf S.

Let M be of complex dimension m. Let W be a holomorphic vector bundle over M and let W^* be its dual bundle. Denote by $\Lambda^r(W)$ the sheaf of germs of holomorphic differential forms of degree r with values in W. Serre's duality theorem [122] says that the vector spaces

$$H^q(M, \Lambda^r(W)) \quad \text{and} \quad H^{m-q}(M, \Lambda^{m-r}(W^*))$$

are in duality and hence have the same dimension. Actually Serre's duality theorem is valid for noncompact complex manifolds; we state it for compact manifolds for simplicity.

By introducing an hermitian scalar product in the vector bundle W, Kodaira extended Hodge's harmonic forms and proved that $H^q(M, \Lambda^p(W))$ is isomorphic to the space $H^{p,q}$ of harmonic forms of type (p, q) and with values in W [79]. He, and later Akizuki and Nakano, gave sufficient conditions of a differential-geometric nature for the vanishing of the cohomology groups $H^q(M, \Lambda^p(W))$ (for the particular case of a line bundle W) [2], [80]. The method is a generalization of Bochner's method on riemannian manifolds.

9. **Characteristic classes** [12], [25], [64], [96]. A systematic theory of characteristic classes begins with the universal bundle theorem. It says that with a given compact manifold M as base space the fiber bundles with a structural group G (Lie group) are in one-one correspondence with the homotopy classes of mappings of M into a classifying space B_G; the latter depends only on G and on the dimension of M. Over B_G there is a universal bundle with the group G and the correspondence is established by the condition that the given bundle over M is induced by a mapping $f\colon M\to B_G$, which is defined up to a homotopy. With a coefficient ring A the induced homomorphism

$$f^*\colon H^*(B_G, A) \to H^*(M, A)$$

is completely determined by the bundle and is called the characteristic homomorphism. The image elements of this homomorphism are called the characteristic classes. The product bundle is induced by the constant mapping, for which all the characteristic classes except

1 are zero. Thus the characteristic classes are the first invariants which describe the deviation of a fiber bundle from a product bundle. From the definition it follows that the characteristic classes have the naturality property, i.e., that they are contravariant functors under bundle maps.

The first problem is to find the classifying spaces B_G for given groups G and to determine their cohomology rings. For $G=U(q)$, $B_{U(q)}$ can be chosen to be the Grassmann manifold $G(q, N, C)$ of all linear subspaces of dimension q through the origin of the complex Euclidean space $E_{q+N}(C)$ of dimension $q+N$, with N sufficiently large. Similarly, $B_{O(q)}$ can be chosen to be the Grassmann manifold $G(q, N, R)$ (resp. $\tilde{G}(q, N, R)$) of all linear subspaces (resp. oriented linear subspaces) of dimension q through the origin of the real Euclidean space $E^{q+N}(R)$ of dimension $q+N$, with N large. The cohomology groups of these Grassmann manifolds have been determined by Ehresmann. In particular, there are elements of $H^{2i}(G(q, N, C), Z)$, $0 \leqq i \leqq q$, which can, for instance, be completely described by the Schubert varieties and which generate the cohomology ring $H^*(G(q, N, C), Z)$, such that their images under f^* are the characteristic classes $c_i \in H^{2i}(M, Z)$ of the $U(q)$-bundle. Similarly, we have, for an $O(q)$-bundle or an SO(q)-bundle, the characteristic classes $W^i \in H^i(M, Z_2)$, $0 \leqq i \leqq q$, and $p_k \in H^{4k}(M, Z)$, $0 \leqq k \leqq [q/4]$, called respectively the Stiefel-Whitney classes and the Pontrjagin classes. A finer analysis of the cohomology of $G(q, N, R)$ allows a definition of W^{2i+1} as an element of $H^{2i+1}(M, Z)$. Also, for an SO(q)-bundle, the highest dimensional Stiefel-Whitney class W^q can be defined to be an element of $H^q(M, Z)$, and we will call it the Euler class.

From this definition one can immediately derive necessary conditions on characteristic classes for the reduction of the structural group of a fiber bundle. Let G' be a subgroup of G. The universal bundle $E_{G'} \to B_{G'}$ can be considered to be a G-bundle and is hence induced by a mapping $h: B_{G'} \to B_G$ of the respective classifying spaces. A bundle over M induced by a mapping $f: M \to B_G$ is a G'-bundle, if and only if a mapping f' exists such that the following diagram is commutative:

$$\begin{array}{ccc} M & \xrightarrow{f} & B_G \\ & f' \searrow \quad \nearrow h & \\ & B_{G'} & \end{array}$$

i.e., $f=h \circ f'$. In terms of the cohomology rings of these spaces this implies the commutativity of the diagram:

$$\begin{array}{ccc} H^*(M, A) & \xleftarrow{f^*} & H^*(B_G, A) \\ & {}_{f'^*}\nwarrow \quad \swarrow_{h^*} & \\ & H(B_{G'}, A) & \end{array}$$

The latter generally implies relations between the characteristic classes of the G-bundle.

EXAMPLE 1. $G = \mathrm{SO}(q)$, $G' = \mathrm{SO}(q-1)$. The reduction of a G-bundle to a G'-bundle is in this case equivalent to the existence of a cross-section of the associated bundle of spheres $S^{q-1} = \mathrm{SO}(q)/\mathrm{SO}(q-1)$. To define the mapping

$$h: \tilde{G}(q-1, N, R) \to \tilde{G}(q, N, R)$$

geometrically we consider $\tilde{G}(q, N, R)$ to be the Grassmann manifold of oriented q-dimensional linear spaces through the origin of $E^{q+N}(R)$ as defined above and $\tilde{G}(q-1, N, R)$ to be that of oriented $(q-1)$-dimensional linear spaces through the origin of $E^{q+N-1}(R) \subset E^{q+N}(R)$. Let x_0 be a nonzero vector through the origin of $E^{q+N}(R)$ and perpendicular to $E^{q+N-1}(R)$. To an element of $\tilde{G}(q-1, N, R)$ its image under h is the linear space spanned by it and x_0. By studying this mapping h one derives easily that a necessary condition for the reduction of the G-bundle to a G'-bundle is that the Euler class should be zero. (In this case it can be proved that the condition is also sufficient.)

EXAMPLE 2. $G = \mathrm{SO}(2m)$, $G' = U(m)$. For the case of the tangent bundle of a manifold the reduction to a G'-bundle is equivalent to the existence of an almost complex structure. The mapping

$$h: G(m, N, C) \to G(2m, 2N, R)$$

is defined by taking, to an m-dimensional complex vector space of $E_{m+N}(C)$, its underlying real vector space. A study of the homomorphism on the cohomology rings induced by h gives the conditions [139]

$$W^i = 0, \qquad (i \text{ odd})$$

$$\sum_{0 \le i \le [m/2]} (-1)^i p_i = \sum_{0 \le i \le m} (-1)^i c_i \sum_{0 \le j \le m} c_j, \tag{39}$$

where p_i are the Pontrjagin classes of M and c_i are the characteristic classes of the reduced $U(m)$-bundle. If $M = S^{4k}$, all the Pontrjagin classes will be zero and the formula gives $c_{2k} = 0$. But c_{2k} can be proved to be the Euler class of the tangent bundle, so that we get a contradiction. This proves that S^{4k} has no almost complex structure.

These examples show the importance of finding relations between the characteristic classes of a G-bundle and those of its reduced G'-bundle, $G' \subset G$.

Further necessary conditions are found by considering the cohomology operations (Bockstein's operations, Steenrod's cohomology operations, Pontrjagin-Thomas operations, etc.). By using the for mulas expressing $p_p^t c_j$ as a polynomial of $c_1, \cdots, c_m$ (p_p^t is the ford reduced power operation), Borel and Serre proved that S^{2m} is not almost complex for $m \geqq 4$ [26].

Operations on vector bundles. Operations on fibers which commute with the action of the structural group can be extended to operations on bundles. For bundles with the group GL(q, R) or GL(q, C) (real or complex vector bundles), two such operations are particularly important: (1) Whitney sum; (2) tensor product. (There are other operations.) To define them consider GL(q) (over the real or complex field) to be the group of all $q \times q$ nonsingular matrices. Define

$$w\colon \mathrm{GL}(q_1) \times \mathrm{GL}(q_2) \to \mathrm{GL}(q_1 + q_2)$$

by

$$w(A_1, A_2) = \begin{pmatrix} A_1 & 0 \\ 0 & A_2 \end{pmatrix}, \quad A_i \in \mathrm{GL}(q_i), \quad i = 1, 2.$$

If over M there are GL(q_1) and GL(q_2) bundles which, relative to a covering $\{U, V, \cdots\}$, are given by the transition functions γ'_{UV}, γ''_{UV}, their Whitney sum has the transition functions $w(\gamma'_{UV}, \gamma''_{UV})$. Similarly, the tensor product of matrices defines a mapping

$$\mathrm{GL}(q_1) \times \mathrm{GL}(q_2) \to \mathrm{GL}(q_1 q_2),$$

from which one gets the tensor product of vector bundles.

It is possible to express the characteristic classes of a Whitney sum (resp. tensor product) in terms of those of the summands (resp. factors). The relations are particularly simple for complex vector bundles. For a GL(q, C)-bundle W let

$$c(W) = 1 + c_1(W) + \cdots + c_q(W). \tag{40}$$

Then

$$c(W_1 \oplus W_2) = c(W_1)c(W_2), \tag{41}$$

where $W_1 \oplus W_2$ is the Whitney sum. This relation (41) is usually called the Whitney duality theorem. Write

$$c(W) = \prod_{1 \leq i \leq q} (1 + \gamma_i(W)). \tag{42}$$

Then the characteristic classes of a tensor product are given by the formula

$$c(W_1 \otimes W_2) = \prod_{1 \leq i \leq q_1;\, 1 \leq j \leq q_2} (1 + \gamma_i(W_1) + \gamma_j(W_2)). \tag{43}$$

Characteristic classes and curvature. The characteristic classes with real coefficients are closely related to the curvature of a connection. Such relations include some of the classical results on global differential geometry. Let $E \rightarrow M$ be a principal fiber bundle with a structural group which is compact and connected. Consider a real valued multilinear function $F(X_1, \cdots, X_s)$, with arguments X_i in the Lie algebra $L(G)$, such that it satisfies the conditions: (1) It is symmetric in any two arguments; (2) It is invariant under the action of the adjoint group, i.e.,

$$F(\mathrm{ad}(\gamma)X_1, \cdots, \mathrm{ad}(\gamma)X_s) = F(X_1, \cdots, X_s), \qquad \gamma \in G. \tag{44}$$

Suppose there be a connection in the bundle, with the curvature form Φ, which is $L(G)$-valued. To the arguments of F we substitute Φ, putting

$$F(\Phi) = F(\Phi, \cdots, \Phi). \tag{45}$$

Then $F(\Phi)$ is a closed exterior differential form of degree $2s$ in M. A theorem of Weil states that the cohomology class determined by $F(\Phi)$ is independent of the choice of the connection and depends only on the bundle [**40**]. In this way one can identify such functions F (which depends only on the group G) with the characteristic classes of the bundle.

EXAMPLE 3. $G = U(q)$. $L(U(q))$ is the space of all complex-valued $q \times q$ skew-hermitian matrices. Let $(\Phi_{ij}) = -{}^t(\overline{\Phi}_{ij})$ be the curvature form, and let

$$\det\left(\delta_{ij} + \frac{1}{2\pi\sqrt{-1}}\,\Phi_{ij}\right) = 1 + \Phi_1 + \cdots + \Phi_q, \tag{46}$$

where Φ_k is an exterior differential form of degree $2k$ in M. The cohomology class determined by Φ_k is the characteristic class c_k.

EXAMPLE 4. $G = \mathrm{SO}(2q)$. The integrand in the Gauss-Bonnet formula (2) is the Euler class.

Divisibility. For a $\mathrm{GL}(q, C)$-bundle the characteristic class c_k is an element of $H^{2k}(M, Z)$. If M is oriented and of even dimension $n=2m$, a polynomial of the form

$$(47) \qquad \sum_{k_1+\cdots+k_s=m} \alpha_{k_1\cdots k_s} c_{k_1} \cdots c_{k_s}$$

with the integer coefficients $\alpha_{k_1\cdots k_s}$ is an element of $H^{2m}(M, Z)$ and has an integral value over the fundamental homology class of M. Such an integer is called a characteristic number. An important divisibility property is given by the following theorem of Bott [28]: *The characteristic number c_n of any complex vector bundle over the sphere S^{2n} is divisible by $(n-1)!$*. This theorem has many geometrical applications. For instance, if the vector bundle has the tangent bundle of S^{2n} as the underlying real vector bundle, we would have $c_n=2$ $=$Euler-Poincaré characteristic of S^{2n}. This is imossible for $n\geqq 4$, which gives a new proof of the Borel-Serre theorem that S^{2n} has no almost complex structure for $n\geqq 4$.

Generally speaking, the characteristic classes c_k, being cohomology classes with integer coefficients, seem to characterize a complex vector bundle quite strongly. A theorem of Peterson says that if M has in dimension $2k$ only torsion coefficients which are relatively prime to $(k-1)!$, then a complex vector bundle of sufficiently large dimension over M is a product bundle if all its characteristic classes are zero [110].

10. **Riemann-Roch, Hirzebruch, Grothendieck, and Atiyah-Singer Theorems** [11], [13], [27], [64], [108]. The classical Riemann-Roch theorem for compact Riemann surfaces has received important developments through the recent works of Hirzebruch, Grothendieck, and Atiyah-Singer. Because of their relations with different fields of mathematics we will follow the historical development.

Let M be a compact Riemann surface and f a meromorphic function over M. At a point p with the local coordinate $z(z(p)=0)$ we have

$$(48) \qquad f = a_n z^n + a_{n+1} z^{n+1} + \cdots, \quad a_n \neq 0.$$

The integer $n(p)$ is independent of the choice of the local coordinate and is called the order of f at p. The point p is a pole if $n(p)<0$. We call

$$(49) \qquad \operatorname{div}(f) = \sum_p n(p)p$$

the divisor of the meromorphic function f. The sum is a finite sum, because the zeros and poles are finite in number.

Generally a finite sum

$$D = \sum_p m(p)p, \qquad m(p) \in Z, \tag{50}$$

is called a divisor. The integer $d(D) = \sum m(p)$ is called its degree. All the divisors form an additive group. The divisor D is said to be positive ($\geqq 0$), if all $m(p) \geqq 0$.

The Riemann-Roch problem is to study the dimension of the complex vector space of all meromorphic functions f such that $\operatorname{div}(f) + D \geqq 0$, for a given divisor D. The latter condition means that f has a pole of order $\leqq +m(p)$ if $m(p) > 0$, and a zero of order $\geqq -m(p)$ if $m(p) < 0$, and is regular at all other points. We denote the dimension of the complex vector space in question by $l(D)$, and the classical Riemann-Roch theorem is given by the formula

$$l(D) = d(D) - g + i + 1, \tag{51}$$

where g is the genus of M and $i \geqq 0$ is the index of specialty of D.

The index of specialty i is defined as follows: First we say that two divisors D_1, D_2 are linearly equivalent, if there exists a meromorphic function f such that $D_1 - D_2 = \operatorname{div}(f)$. Since $\operatorname{div}(fg) = \operatorname{div}(f) + \operatorname{div}(g)$, $\operatorname{div}(f^{-1}) = -\operatorname{div}(f)$, linear equivalence is an equivalence relation. Let $\omega = b(z)dz$ be a meromorphic differential form. Locally we can write

$$b(z) = b_n z^n + b_{n+1} z^{n+1} + \cdots, \quad b_n \neq 0.$$

Then $n(p)$ is called the order of ω at p, and we define the divisor of ω to be $\operatorname{div}(\omega) = \sum n(p)p$. The linear equivalence class of $\operatorname{div}(\omega)$, to be denoted by K, is independent of ω and depends only on M, because the ratio of two meromorphic differential forms is a meromorphic function. K is called the canonical divisor class. The index of specialty i is by definition

$$i = \dim H^0(M, \Omega(K - D)). \tag{52}$$

(For a divisor D, $\Omega(D)$ is the sheaf of germs of mermorphic functions f such that $\operatorname{div}(f) + D \geqq 0$.) By Serre's duality theorem we have the isomorphism

$$H^0(M, \Omega(K - D)) \cong H^1(M, \Omega(D)). \tag{53}$$

Then the Riemann-Roch theorem can be written

$$\dim H^0(M, \Omega(D)) - \dim H^1(M, \Omega(D)) = d - g + 1. \tag{54}$$

This formula was generalized by Hirzebruch to higher dimensions in the following way: Let M be a compact complex manifold of complex dimension m, and W an analytic vector bundle with the structural group $GL(q, C)$. W is called a line bundle if $q=1$. (A line bundle is essentially the differential-geometric version of a divisor. For a divisor is given in a coordinate neighborhood U by an equation $\phi_U=0$, and it defines a line bundle with the transition functions $g_{UV}=\phi_U/\phi_V$. Conversely, it was proved by Kodaira and Spencer that every line bundle over a nonsingular algebraic variety is defined by a divisor.) Let $\Omega(W)$ be the sheaf of germs of holomorphic sections of W, and let

$$\chi(M, W) = \sum_{0 \leq i \leq m} (-1)^i \dim H^i(M, \Omega(W)). \tag{55}$$

Let c_i, $1 \leq i \leq m$, be the characteristic classes of the tangent bundle of M and d_j, $1 \leq j \leq q$, those of the bundle W. Write

$$\begin{aligned} 1 + \sum c_i &= \prod (1 + \gamma_i), \\ 1 + \sum d_j &= \prod (1 + \delta_j), \end{aligned} \tag{56}$$

and put

$$T(M, W) = \left\langle (e^{\delta_1} + \cdots + e^{\delta_q}) \prod_i \frac{\gamma_i}{1 - \exp(-\gamma_i)}, M \right\rangle, \tag{57}$$

where the right-hand side denotes the value of the cohomology class in question over the fundamental class of M, the value being zero for every summand of dimension $\neq 2m$. Hirzebruch's formula says that, for nonsingular algebraic varieties, we have

$$\chi(M, W) = T(M, W). \tag{58}$$

If W is not involved, then

$$\chi(M) = T(M). \tag{59}$$

$\chi(M)$ is called the arithmetic genus and $T(M)$ the Todd genus. If $m=1$ and W is a line bundle defined by a divisor D, then $d_1 \cdot M = d(D)$, $c_1 \cdot M = 2-2g$, and we get immediately the classical Riemann-Roch theorem from Hirzebruch's formula. Hirzebruch's proof of his formula makes use of Thom's theory of "cobordism."

Grothendeick formulates the Riemann-Roch problem as a problem on mappings Let $F(M)$ be the free abelian group generated by the set of all isomorphy classes of complex vector bundles over M. M need not be connected, in which case it is not assumed that the bundle has the same fiber dimension over different components of M. An element $x \in F(M)$ is thus a formal finite linear combination

$$x = \sum n_i W_i, \qquad n_i \in Z, \quad W_i = \text{complex vector bundle}. \tag{60}$$

Let

$$0 \to W' \to W \to W'' \to 0$$

be a short exact sequence of complex vector bundles. Let $K(M)$ be the quotient group of $F(M)$ modulo the subgroup generated by $W-W'-W''$ for all short exact sequences. (In other words, we consider all extensions of complex vector bundles to be trivial. This is a natural idea, because by the Whitney duality theorem all the characteristic classes of $W-W'-W''$ are zero.) With the Whitney sum as addition and the tensor product as multiplication, $K(M)$ becomes a ring, called the Grothendieck ring.

Without danger of confusion, if W is a complex vector bundle, we will denote also by W the element it determines in $K(M)$. We define the character

$$\text{ch}(W) = \sum_{1 \leq j \leq q} e^{\delta_j} \in H^*(M, Q), \tag{61}$$

where $H^*(M, Q)$ is the cohomology ring of M with coefficients in the rational number field Q. Then

$$\text{ch}: K(M) \to H^*(M, Q) \tag{62}$$

is a ring homomorphism.

Generalizing the Todd genus we define the Todd cohomology class

$$t(M) = \prod_i \frac{\gamma_i}{1 - \exp(-\gamma_i)} \in H^*(M, Q). \tag{63}$$

Let f: $M \to N$ be a continuous mapping, where M, N are compact oriented manifolds of real dimensions μ, ν respectively. The diagram

$$\begin{array}{ccc} H_p(M, Q) & \xrightarrow{f_*} & H_p(N, Q) \\ \uparrow & & \uparrow \\ H^{\mu-p}(M, Q) & \xrightarrow{f_*} & H^{\nu-p}(N, Q) \end{array}$$

where the first row is the induced linear mapping on the homology vector spaces and the vertical rows are isomorphisms established by Poincaré's duality, defines a linear mapping of the last row, which we also denote by f_*.

Now let M, N be nonsingular algebraic varieties, and f: $M \to N$ be a holomorphic mapping. An additive homomorphism $f_!: K(M) \to K(N)$ can be defined as follows: Let $W \in K(M)$. To every open set $V \subset N$ assign the group $H^q(f^{-1}(V), \Omega(W))$. This is the presheaf of a sheaf

$R^q\Omega$. Define $f_!(W) = \sum_i (-1)^i R^i\Omega$. For algebraic variety, the algebraic coherent sheaves and complex vector bundles can be identified. Hence $f_!(W) \in K(N)$.

Grothendieck's theorem is then the formula:

$$f_*(\mathrm{ch}(W)t(M)) = \mathrm{ch}(f_! W)t(N). \tag{64}$$

Frow this Hirzebruch's formula follows by taking N to be a point. Another important consequence of the theorem is that the Todd genus of a nonsingular algebraic variety is a birational invariant.

A remark could be made as to the reason of the form of the factors in the Todd class. This form automatically appears when one considers the simple case of a hypersurface in projective space. From this one extends it to algebraic varieties which are complete intersections of hypersurfaces and then to the most general algebraic varieties.

Atiyah and Hirzebruch generalized a weakened version of Grothendieck's theorem to real manifolds. It starts with the observation that the Todd class can be written

$$t(M) = \exp(c_1/2)\alpha(M), \tag{65}$$

where

$$\alpha(M) = \prod \frac{\dfrac{\gamma_i}{2}}{\sinh \dfrac{\gamma_i}{2}} \in H^*(M, Q), \tag{66}$$

to be called the A-class, depends on the Pontrjagin classes of M only. Since $c_1 \equiv W^2$, mod 2, for the Todd class to be defined the real manifold must have the property that its Stiefel-Whitney class W^2 is the reduction mod 2 of an integral class. A necessary and sufficient condition for this is that $W^3 = 0$. (Note that W^3 is a class with integer coefficients.)

Now let M be a real manifold with $W^3 = 0$. Let $d \in H^2(M, Z)$ be such that its reduction mod 2 is W^2. We call

$$R(M) = \mathrm{ch}(K(M)) \exp(d/2)\alpha(M) \in H^*(M, Q) \tag{67}$$

the Riemann-Roch group of M. It is independent of the choice of d and is isomorphic to $\mathrm{ch}(K(M))$ as groups. A weakened version of Grothendieck's theorem states that

$$f_*(R(M)) \subset R(N), \tag{68}$$

if M, N are nonsingular algebraic varieties and $f: M \to N$ is a holomorphic mapping. The generalization by Atiyah-Hirzebruch is the following theorem.

Let M and N be compact oriented differentiable manifolds with $W^3(M) = W^3(N) = 0$ and $\dim M - \dim N \equiv 0$, mod 2. *Let $f: M \to N$ be a continuous mapping. Then $f_*(R(M)) \subset R(N)$.*

The theorem has various applications. We mention the following:

(1) The Todd genus of a compact almost complex manifold is an integer (Milnor).

(2) If $W^2(M) = 0$, then the A-genus is an integer.

(3) From their theorem Atiyah and Hirzebruch derived sharp lower bounds of the dimension of the sphere into which the real and complex projective spaces can be differentiably imbedded.

Atiyah and Singer considered elliptic differential operators on a compact (orientable) real differentiable manifold and obtained an important theorem which contains the Riemann-Roch-Hirzebruch formula (58) as a special case. To an elliptic differential operator D (from the sections of one complex vector bundle to those of another) one associates, through the character defined in (61), a cohomological invariant $\mathrm{ch}(D) \in H^*(M, Q)$, the character of D. Moreover, the definition of the Todd class in (63) can be extended to a real manifold by simply taking the complexification of the tangent bundle of M. It will then be expressible as a polynomial of the Pontrjagin classes of M. The rational number

$$i_t(D) = \langle \mathrm{ch}(D) t(M), M \rangle, \tag{69}$$

where the right-hand side denotes the pairing of a cohomology class and the fundamental class M, is called the topological index of the elliptic operator D.

On the other hand, D being an elliptic operator on a compact manifold, the spaces $\mathrm{Ker}(D)$ (i.e., the null space) and Coker (D) are both finite dimensional, and we define the analytical index of D to be the integer

$$i_a(D) = \dim \ker D - \dim \operatorname{coker} D. \tag{70}$$

The Atiyah-Singer theorem says that $i_t(D) = i_a(D)$.

Among the consequences of the Atiyah-Singer theorem are:

(1) The Gauss-Bonnet formula (2), when applied to the operator $d + \delta$, where δ is the codifferential defined by $\delta = \pm * d *$.

(2) The Hirzebruch index theorem, expressing the index $\tau(M)$ in terms of the Pontrjagin classes of M (cf. §5).

(3) Extension of the Riemann-Roch-Hirzebruch formula (58) to an arbitrary compact complex manifold.

It will be of great importance to find a theorem on proper maps of compact complex manifolds, which will imply as consequences both the Grothendieck and the Atiyah-Singer theorems.

11. **Holomorphic mappings of complex analytic manifolds [1], [45], [137]**. A holomorphic mapping $f: V \to M$ of a complex manifold V into a complex manifold M is a continuous mapping which is locally defined by expressing the local coordinates of the image point as holomorphic functions of those of the original point.

I. *Compact manifolds.* (All manifolds in this part of the discussion are compact.)

The condition of a holomorphic mapping is so strong that it is generally not clear whether, for given manifolds V, M, such a mapping exists, which is not a constant. In fact, it is well known that if M is a complex torus and V satisfies the condition $H^1(V, R) = 0$, then every holomorphic mapping of V into M is constant. Similarly, let $h_{r0}(V)$ (resp. $h_{r0}(M)$) be the dimension of the vector space of holomorphic exterior differential forms of type $(r, 0)$ of V (resp. M). If $\dim V = \dim M$ and if the Jacobian of f is not identically zero, then $h_{r0}(V) \geqq h_{r0}(M)$. It follows that a nonconstant holomorphic mapping f exists from a Riemann surface (=one-dimensional complex manifold) V into a Riemann surface M only when $h_{10}(V) \geqq h_{10}(M)$, where the two quantities are now the genera of V and M respectively.

The celebrated theorems of Chow and Kodaira [**58**], [**81**] can be interpreted as assertions on holomorphic mappings. In fact, Chow's theorem implies that if V is compact, the image set $f(V)$ of a holomorphic mapping $f: V \to P_m$ (=projective space of dimension m) is an algebraic variety. Kodaira's theorem says that if V is a Kähler manifold whose fundamental two-form has rational periods, then there exists a holomorphic mapping $f: V \to P_m$ which is one-to-one.

Suppose that a holomorphic mapping $f: V \to M$ exists. An important problem is to find relations between the invariants (such as the characteristic numbers) of V, M and quantities which depend on the mapping (for instance, the degree when $\dim V = \dim M$). A classical relation of this nature is the Riemann-Hurwitz formula: Let V, M be Riemann surfaces with Euler characteristics $\chi(V)$, $\chi(M)$ respectively. Let $f: V \to M$ be a holomorphic mapping of degree d. Then we have

$$\chi(V) + w = d\chi(M), \tag{69}$$

where w is the index of ramification, i.e., the sum of the orders of the points of ramification.

Another set of relations of this nature consists of the Plücker formulas for an algebraic curve. Let an algebraic curve be defined by a holomorphic mapping $f: V \to P_m$, where V is a Riemann surface. Suppose that the algebraic curve is nondegenerate, i.e., that the image $f(V)$ does not belong to a linear space of dimension $\leqq m-1$. To this curve is defined the pth associated curve $f^p: V \to \mathrm{Gr}(m, p)$, $0 \leqq p \leqq m-1$, formed by the osculating projective spaces of dimension p, where $\mathrm{Gr}(m, p)$ is the Grassmann manifold of all p-dimensional projective spaces in P_m (thus $\mathrm{Gr}(m, 0) = P_m$; notice the relation with the Grassman manifolds in §9 and the difference of notations). $f^p(V)$ defines a cycle in $\mathrm{Gr}(m, p)$, which is homologous to a positive integral multiple ν_p of the fundamental two-cycle of $\mathrm{Gr}(m, p)$. The integer $\nu_p > 0$ is called the order of rank p of our algebraic curve. Geometrically it is the number of points of the curve at which the osculating spaces of dimension p meet a generic linear space of dimension $m-p-1$ of P_m. A stationary point of order p is one at which the pth associated curve has a tangent with a contact of higher order. The stationary points are isolated and a positive index can be associated to each of them. Let w_p be the sum of indices at the stationary points of order p. Then Plücker's formulas are

$$-w_p - \nu_{p-1} + 2\nu_p - \nu_{p+1} = \chi(V), \qquad 0 \leqq p \leqq m-1. \tag{70}$$

Here the right-hand side is an invariant of V, while the left-hand side involves quantities which depend on the mapping. The special cases of the Plücker formula for $m=1$ and the Riemann-Hurwitz formula for $M = P_1$ give the same relation, as to be expected.

For nonsingular algebraic varieties a much more profound relation between invariants of manifolds and quantities depending on a holomorphic mapping is given by Grothendieck's Riemann-Roch theorem (cf. §10). Here one utilizes the classification of singularities by Thom. Applying the results of Grothendieck and Thom, I. R. Porteous [111] derived relations between the characteristic classes for the following cases: (a) dilatations; (b) ramified coverings with singularities of a relatively simple type.

If V and M are of the same dimension, an important invariant of the mapping f is its degree $d(f) \geqq 0$. It is equal to the number $n(\alpha)$ of times that any point $\alpha \in M$ is covered by the image of the mapping. Define an hermitian metric on M such that the total volume of M is 1. Then we have

$$d(f) = n(\alpha) = \operatorname{vol} f(V). \tag{71}$$

II. *Equidistribution*. The most important case of a holomorphic mapping $f: V \to M$ is when V is the euclidean line E_1 and M is the projective line P_1, in which case the mapping is nothing else but a meromorphic function. The famous theorem of Picard says that if f is not a constant, the set $P_1 - f(E_1)$ cannot contain more than two points. Much sharper results generalizing Picard's theorem were obtained by R. Nevanlinna in the form of his defect relations. The general aim is to ascertain, whenever it is true, that the image set is large and that, in a suitable sense it is "equidistributed."

As with meromorphic functions, the case which will yield interesting results is when V is noncompact and M is compact. This assumption we will make throughout Part II except in the last section. The basic idea is the realization that results which are true for compact V can be extended to those V, which are "complex analytically large" (such as a parabolic open Riemann surface or a high-dimensional domain with a pseudo-concave exhaustion). The manifold V is to be exhausted by a sequence of compact polyhedra with boundaries, so that an intermediate step is to study the mapping $f: D \to M$ where D is a compact polyhedron with boundary. Generalizations of the results of Part I to this case are the necessary prerequisites. What remains consists in the estimate and the study of the asymptotic behavior of the boundary integrals, which are usually delicate considerations.

So far the deepest result in this direction is a theorem of Ahlfors on holomorphic curves in projective space, whose study was initiated by H. and J. Weyl. By a holomorphic curve is meant a holomorphic mapping $f: V \to P_m$ (our holomorphic curve is the same as a meromorphic curve in Weyl's terminology). It is said to be nondegenerate, if $f(V)$ does not belong to a linear subspace of P_m. Ahlfors' theorem, in a slightly generalized form, is the following:

Let $f: V \to P_m$ be a nondegenerate holomorphic curve such that V can be compactified as a Riemann surface by the addition of a finite number of points. Given

$$\binom{m+1}{p+1} + 1$$

linear spaces of dimension $m-p-1$ in general position, one of them must meet an osculating space of dimension p of the curve.

Considering the beauty of the theorem, it is natural to ask whether the image space P_m can be replaced by a more general space, such as the Grassmann manifold $\mathrm{Gr}(m, p)$ of all p-dimensional (projective) linear subspaces in P_m $(0 < p < m)$. By imbedding $\mathrm{Gr}(m, p)$ in P_N, with

$$N = \binom{m+1}{p+1} - 1,$$

one derives from the Ahlfors' theorem results of the Picard type on the relative position of a holomorphic curve in $\mathrm{Gr}(m, p)$ with respect to a finite number of prime divisors in general position on $\mathrm{Gr}(m, p)$. However the bound so obtained is generally not sharp. To put ourselves in a specific position we would like to ask the following question: Suppose $f: C \to \mathrm{Gr}(2m-1, m-1)$ be a holomorphic curve. To each $\zeta \in C$ let $f(\zeta)$ be spanned by m points whose homogeneous coordinate vectors are $Z_1, \cdots, Z_m$. The curve is called nondegenerate, if the determinant

$$\left(Z_1, \cdots, Z_m, \frac{dZ_1}{d\zeta}, \cdots, \frac{dZ_m}{d\zeta}\right)$$

does not vanish identically. Is it true that, given $2m+1$ linear subspaces of dimension $m-1$ of P_{2m-1} in general position, one of them will intersect $f(\zeta)$ for a certain $\zeta \in C$? The affirmative answer to this for $m=1$ is the classical Picard theorem.

When $\dim V > 1$, value distribution in the strict sense fails completely. In fact, the classical example of Fatou and Bieberbach is a holomorphic mapping $f: E_2 \to E_2$, such that the Jacobian determinant is identically equal to one, while $E_2 - f(E_2)$ contains an open subset. The problem should thus be looked at from a more general viewpoint.

When $\dim V = n$, and $M = P_m$, $n \leqq m$, the first main theorem can be generalized, as follows: Let A be a linear subspace of dimension $m-n$ in P_m. Let $A^\perp$ be the linear subspace of dimension $n-1$ which is orthogonal to A (relative to the standard hermitian metric in P_m). To $Z \in P_m - A$, the space spanned by Z and A meets $A^\perp$ in a unique point. This defines a mapping $\psi: P_m - A \to A^\perp$. Denote by $s(Z, A)$ the distance between Z and A. Let Ω be the fundamental two-form of P_m, so normalized that $\int_{P_m} \Omega^m = 1$. Let $j: A^\perp \to P_m$ be the identity mapping, and let $\Phi = (j \circ \psi)^* \Omega$. Then

$$\Lambda = \frac{1}{2\pi i}(d' - d'') \log \sin s(Z, A) \wedge \left(\sum_{0 \leq k \leq n-1} \Phi^k \wedge \Omega^{n-k-1} \right) \tag{72}$$

is a real differential form of degree $2n-1$ in $P_m - A$. If A is generic, $f(D)$ cuts A in a finite number $n(A, D)$ of points, and we have the formula

$$n(A, D) - v(D) = \int_{f(\partial D)} \Lambda, \tag{73}$$

provided that $f^{-1}(A)\cap\partial D=\emptyset$ ($v(D)$ is the volume of $f(D)$ and is equal to $\int_{f(D)}\Omega^n$). This formula, to be called the first main theorem, is due to H. Levine [92].

It turns out that this first main theorem leads to a geometrical conclusion, in the special case that V is the coordinate space of n dimensions, with the coordinates $z_1, \cdots, z_n$. We exhaust V by the balls D_r defined by $z_1\bar{z}_1+\cdots+z_n\bar{z}_n\leqq r^2$. Relative to the parameter r the first main theorem can be integrated. Let

$$(74)\quad v_k(r)=\int_{D_r} f^*\Omega^{n-k}\wedge\Omega_0^k,\quad 0\leqq k\leqq n,\qquad \Omega_0=\frac{i}{2}\sum_{1\leq l\leq n} dz_l\wedge d\bar{z}_l.$$

By applying integral geometry to the integrated form of the first main theorem, we obtain the following theorem [46]:

Let $f\colon V\to P_m$ be a holomorphic mapping of the euclidean space V of dimension n into P_m. Assume that the following conditions are fulfilled, as $r\to\infty$:

(1) *The order function*

$$T(r)=\int_{r_0}^{r}\frac{v_0(t)\,dt}{t^{2n-1}}\longrightarrow\infty,$$

$$(2)\qquad \int_{r_0}^{r}(v_1'(t)\,dt)/t^{2n}=o(T(r)).$$

Then the set of linear spaces of dimension $m-n$ which do not meet $f(V)$ is of measure zero in the Grassmann manifold Gr$(m, m-n)$ *of linear subspaces of dimension $m-n$ in P_m, $m\geqq n$.*

A special but important case of holomorphic mappings is that of the holomorphic sections of a holomorphic vector bundle. The study of the zeroes of a section or of points at which several sections are linearly dependent is closely related to the value distribution problem and is in many cases equivalent to it. The problem is to describe those properties which depend on the bundle only and are independent of the sections, for the case that the base manifold is not necessarily compact. The algebraic aspect was recently studied by Bott and Chern [30], namely the properties of the differential forms in the vector bundle relative to the operator $D=id'd''$. This could be regarded as a refinement of the theory of characteristic classes for the complex analytic case. From it an equidistribution theorem was derived, but much remains to be done.

12. **Isometric mappings of Riemannian manifolds.** A mapping of a riemannian manifold into another is called an isometry, if it preserves

the lengths of the tangent vectors. If an isometry is present, the basic local invariant is the second fundamental form, which is a quadratic differential form, with value in the normal bundle. The second fundamental form is related to the curvature forms of both riemannian manifolds by the (generalized) Gauss equations, but until now the exact relationship (a purely algebraic problem) has not been sufficiently clarified.

The problem of isometric mappings is almost as old as differential geometry itself, beginning with the theory of curves and surfaces. If V and M are real analytic riemannian manifolds of dimensions n and $m=n(n+1)/2$ respectively, then the theorem of Janet-Cartan says that there is locally an isometric mapping of V into M [**35**]. Without analyticity the same result has only been proved for $n=2$, and in fact only under the additional assumption that the gaussian curvature of V keeps a constant sign in a neighborhood [**61**]. For $n>2$ the generalization of the Janet-Cartan theorem to C^∞-data seems to be difficult, even allowing additional conditions in the form of inequalities between the curvatures.

Perhaps the first global isometric imbedding theorem is the solution of the Weyl problem, which is to find an isometric imbedding of a two-dimensional compact riemannian manifold with positive gaussian curvature into the three-dimensional euclidean space [**138**]. Weyl's work was completed by H. Lewy, L. Nirenberg, A. D. Alexandrow, and A. V. Pogorelov [**3**], [**49**], [**91**], [**104**]. A bold isometric imbedding theorem of an arbitrary riemannian manifold into an euclidean space was proved by J. Nash. For a C^1-riemannian manifold V of dimension n Nash's theorem, improved by N. H. Kuiper, says that V can be C^1-isometrically imbedded in an euclidean space of dimension $2n+1$ [**85**], [**99**]. Among other results Kuiper also found an isometric imbedding of the hyberbolic plane as a closed subset in euclidean three-space. However, such imbeddings could be pathological. Nash's imbedding theorem is still true, if more smoothness is imposed, but the dimension of the receiving euclidean space will have to be higher and the proof is more difficult [**89**], [**97**], [**100**]. For example, Nash proved that a compact two-dimensional riemannian manifold can be isometrically imbedded in an euclidean space of dimension 17. Whether this bound can be improved should be a very interesting question in differential geometry.

If a riemannian manifold can be isometrically imbedded in an euclidean space E^m of dimension m there arises the question as to how far it is determined, modulo the isometries in E^m. In the general case practically nothing is known. A classical theorem of Cohn-

Vossen, complemented by a remark of K. Voss, says that a closed convex surface (i.e., Gaussian curvature $\geqq 0$) in an euclidean three-space E^3 is completely determined by its metric, modulo the isometries of E^3 [**131**].

To restrict the submanifolds under consideration and thus to make a Cohn-Vossen type uniqueness theorem more plausible, the notion of total absolute curvature seems to deserve attention. This is the volume of the image of the unit normal bundle of the submanifold under the Gauss mapping, so normalized that the volume of the unit sphere in E^m is 1. Chern and Lashof proved that a compact immersed submanifold in E^m has total curvature $\geqq 2$, and if it is equal to 2, it is a convex hypersurface imbedded in an euclidean space of one dimension higher [**44**]. For a given compact manifold the minimum of the total curvatures for all possible immersions is a differential topological invariant and is an integer. The so-called tight immersions, i.e., those for which the minimum total curvature is attained, can thus be considered as generalizations of convex hypersurfaces. It is not known whether there exist two tightly immersed compact submanifolds, which are isometric but which do not differ by an isometry of E^m.

A. D. Alexandrow proved that two tightly imbedded analytic compact surfaces of genus one in euclidean three-space E^3 differ by an isometry of E^3, if they are isometric. To remove the assumption of analyticity Nirenberg has to put in some additional hypotheses [**105**], so that the question whether the analyticity assumption in Alexandrow's theorem could be removed remains unsettled. On the other hand, the notion of a tight imbedding can be easily extended to polyhedral submanifolds, and T. Banchoff gave examples of two polyhedral surfaces of genus one in euclidean three-space E^3 such that their corresponding faces are congruent, but they do not differ by a motion or a reflection of E^3 [**14**].

An important class of submanifolds in E^m consists of the minimal submanifolds. This is an isometric immersion $V \rightarrow E^m$ such that the coordinate functions are harmonic functions on V. Equivalently, we can also say that the trace vector of the second fundamental form is identically zero or that it solves locally the Plateau problem, being the submanifold of the least area with a given boundary. A complete minimal submanifold in E^m is never compact. It is, however, not known whether it is unbounded. So far the study has been mostly confined to minimal surfaces (i.e., dimension two); the reason lies in the intimate relationship of this case to complex function theory. In fact, the surface, with a fixed orientation, has an underlying complex structure, and the Grassmann manifold of all oriented planes through a fixed point 0 of E^m is a homogeneous complex manifold. A surface in E^m is minimal, if and only if the Gauss mapping, which sends a point p of the surface to the plane through 0 parallel to the tangent plane at p, is anti-holomorphic. An important problem is to determine the type of the surface V (in the sense of complex function theory) from the properties of the Gauss mapping. A classical theorem of S. Bernstein, as generalized by R. Osserman, can be interpreted as a uniqueness theorem. It says that a complete minimal surface in E^3 is a plane, if its image under the Gauss mapping omits an open subset of the Grassmann manifold (which in this case is the unit sphere). This theorem has a generalization to minimal surfaces in E^m [47], [107].

13. **General theory of G-structures** [32], [34], [37], [77], [126]. The general theory of G-structures is concerned primarily with a local problem, i e., the equivalence problem formulated in §4. The first local invariant is the first-order structure function, whose definition is a little complicated and we will not give it here. A G-structure is called locally flat, if it is equivalent to one given by the differentials of the local coordinates ($\theta_U^i = du^i$ in the notation of §4). The structure function is zero for a locally flat structure. In some cases the converse of this statement is true. For an almost complex structure the vanishing of the structure function is equivalent to the fulfillment of the integrability conditions. That this implies local flatness is an easy theorem in the real analytic case. Without analyticity this is a theorem of Newlander and Nirenberg, whose proof involves delicate considerations in partial differential equations [101], [103].

Given a G-structure, its local automorphisms may depend on a finite number of constants or on arbitrary functions. The G-structure is then said respectively to be of finite or infinite type. For an irreducible linear group G, Kobayashi and Nagano found a necessary and sufficient condition for a G-structure to be of finite type in terms of their theory of filtered Lie algebras [77]. The case of infinite type leads to structures defined by infinite pseudo-groups. In this respect Cartan's determination of the simple infinite pseudo-groups (over the complex field) has not been completely verified, and its clarification should be one of the outstanding problems in the theory of infinite pseudo-groups.

Within this framework is a problem studied by Weyl and Cartan on the so-called Pythagorasian nature of metric. This is to find those G such that a unique affine connection exists, which preserves the admissible frames and which has a given torsion form. For dimensions $\geqq 3$ this implies that G is the orthogonal group (of arbitrary signature). The problem has recently been studied and extended by Klingenberg and Kobayashi-Nagano [70], [76].

Bibliography

Note. With the exception of references of a general nature, such as books or review articles, this Bibliography contains only papers referred to in the text.

1. L. Ahlfors, *The theory of meromorphic curves*, Acta Soc. Sci. Fenn. Nova Ser. A **3** (1941), no. 4, 1–31.

2. Y. Akizuki and S. Nakano, *Note on Kodaira-Spencer's proof of Lefschetz theorems*, Proc. Japan Acad. **30** (1954), 266–272.

3. A. D. Alexandrow, *Die innere Geometrie der konvexen Flächen*, Akademie Verlag, Berlin, 1955.

4. C. B. Allendoerfer and A. Weil, *The Gauss-Bonnet theorem for Riemannian polyhedra*, Trans. Amer. Math. Soc. **53** (1943), 101–129.

5. W. Ambrose and I. M. Singer, *A theorem on holonomy*, Trans. Amer. Math. Soc. **75** (1953), 428–443.

6. A. Andreotti, *On the complex structures of a class of simply-connected manifolds*, Algebraic geometry and topology (Lefschetz Jubilee Volume), Princeton Univ. Press, Princeton, N. J., 1957, pp. 53–77.

7. A. Andronov and L. Pontrjagin, *Systèmes grossiers*, C.R. Acad. Sci. Paris **14** (1937), 247–250.

8. M. F. Atiyah, *Complex analytic connections in fiber bundles*, Trans. Amer. Math. Soc. **85** (1957), 181–207.

9. ———, *Some examples of complex manifolds*, Bonn. Math. Schr. **6** (1958), 1–28.

10. M. F. Atiyah and R. Bott, *On the periodicity theorem for complex vector bundles*, Acta Math. **112** (1964), 229–247.

11. M. F. Atiyah and F. Hirzebruch, *Riemann-Roch theorem for differentiable manifolds*, Bull. Amer. Math. Soc. **65** (1959), 276–281.

12. ———, *Charakteristiche Klassen und Anwendungen*, Enseignement Math. **7** (1961), 188–213.

13. M. F. Atiyah and I. M. Singer, *The index of elliptic operators on compact manifolds*, Bull. Amer. Math. Soc. **69** (1963), 421–433.

14. T. F. Banchoff, *Tightly embedded two-dimensional polyhedral manifolds*, Amer. J. Math. **87** (1965), 462–472.

15. M. Berger, *Sur quelques variétés riemanniennes suffisamment pincées*, Bull. Soc. Math. France **88** (1960), 57–71.

16. ———, *Les variétés riemanniennes normales simplement connexes à courbure strictement positive*, Ann. Scuola Norm. Sup. Pisa **15** (1961), 179–246.

17. ———, *Les variétés riemanniennes dont la courbure satisfait certaines conditions*, pp. 447–456, Proc. Internat. Congress Math., 1962.

18. D. Bernard, *Sur la géométrie différentielle de G-structures*, Ann. Inst. Fourier (Grenoble), **10** (1960), 151–270.

19. L. Bers, *Spaces of Riemann surfaces*, pp. 349–361, Proc. Internat. Congress Math., 1958.

20. R. L. Bishop and R. J. Crittenden, *Geometry of manifolds*, Academic Press, New York, 1964.

21. R. L. Bishop and S. I. Goldberg, *On the second cohomology group of a Kähler manifold of positive curvature*, Proc. Amer. Math. Soc. **16** (1965), 119–122.

22. ———, *Rigidity of positively curved Kähler manifolds*, (to appear).

23. A. Blanchard, *Sur les variétés analytiques complexes*, Ann. Sci. École Norm. Sup. Paris **73** (1956), 157–202.

24. W. Blaschke and G. Bol, *Geometrie der Gewebe*, Springer, Berlin, 1938.

25. A. Borel, *Sur la cohomologie des espaces fibrés principaux de groupes de Lie compacts et des espaces homogènes*, Ann. of Math. **57** (1953), 115–207.

26. A. Borel and J.-P. Serre, *Groupes de Lie et puissances réduits de Steenrod*, Amer. J. Math. **75** (1953), 409–448.

27. ———, *Le théorème de Riemann-Roch d'après Grothendieck*, Bull. Soc. Math. France **86** (1958), 97–136.

28. R. Bott, *The space of loops on a Lie group*, Michigan Math. J. **5** (1958), 35–61.

29. ———, *The stable homotopy of the classical groups*, Ann. of Math. **70** (1959), 313–337.

30. R. Bott and S. S. Chern, *Hermitian vector bundles and the equidistribution of the zeroes of their holomorphic sections*, Acta Math. **114** (1965), 71–112.

31. E. Calabi, *On compact riemannian manifolds with constant curvature*. I, pp. 155–180, Differential geometry, Proc. Sympos. Pure Math., Vol. 3, Amer. Math. Soc., Providence, R. I., 1961.

32. E. Cartan, *Sur la structure des groupes infinis de transformations*, Ann. Sci. École Norm Sup. Paris **21** (1904), 153–206; ibid., **22** (1905), 219–308(=Oeuvres complètes, Partie II, pp. 571–714).

33. ———, *Les sousgroupes des groupes continus de transformations*, Ann. Sci. École Norm Sup. Paris **25** (1908), 57–194 (=Oeuvres complètes, Partie II, pp. 719–856).

34. ———, *Les groupes de transformations continus, infinis, simples*, Ann. Sci. École Norm Sup. Paris **26** (1909), 93–161 (=Oeuvres complètes, Partie II, pp. 857–925).

35. ———, *Sur la possibilité de plonger un espace riemannien donné dans un espace euclidien*, Ann. Soc. Polon. Math. **6** (1927), 1–7 (=Oeuvres complètes, Partie III, pp. 1091–1097).

36. ———, *Les problèmes d'équivalence*, Oeuvres complètes, Partie II, 1311–1334.

37. ———, *La structure des groupes infinis*, Oeuvres complètes, Partie II, 1335–1384.

38. H. Cartan, *Variétés analytiques complexes et cohomologie*, Colloque tenu à Bruxelles, 1953, pp. 41–55; Thone, Liège, 1953.

39. S. S. Chern, *On the curvatura integra in a riemannian manifold*, Ann. of Math. **46** (1945), 674–684.

40. ———, *Differential geometry of fiber bundles*, Vol. II, pp. 397–411. Proc. Internat. Congress Math., 1950.

41. ———, *Pseudo-groupes continus infinis. Geometrie differentielle*, Colloquium, Strasbourg, pp. 119–136, 1953, Centre National de la Recherche Scientifique, Paris, 1953.

42. ———, *On curvature and characteristic classes of a Riemann manifold*, Abh. Math. Sem. Univ. Hamburg **20** (1955), 117–126.

43. ———, *On a generalization of Kähler geometry*, Algebraic geometry and topology (Lefschetz Jubilee Volume), Princeton Univ. Press, Princeton, N.J., 1957, pp. 103–121.

44. S. S. Chern and R. Lashof, *On the total curvature of immersed manifolds*, Amer. J. Math. **79** (1957), 302–318; II, Michigan Math. J **5** (1958), 5–12.

45. S. S. Chern, *Complex analytic mappings of Riemann surfaces*. I, Amer. J. Math. **82** (1960), 323–337.

46. ———, *The integrated form of the first main theorem for complex analytic mappings in several complex variables*, Ann. of Math. **71** (1960), 536–551.

47. ———, *Minimal surfaces in an euclidean space of n dimensions*, Differential and combinatorial topology (More Jubilee Volume), Princeton University Press, Princeton, N.J., 1965, pp. 187–198.

48. M. doCarmo, *The cohomology ring of certain Kählerian manifolds*, Ann. of Math. **81** (1965), 1–14.

49. N. W. Efimov, *Flächenverbiegung im Grossen*, Academie Verlag, Berlin, 1957.

50. C. Ehresmann, *Les connexions infinitésimales dans un espace fibré différentiable*, Colloque de topologie, Bruxelles, pp. 29–55, 1950; Thone, Liège, 1951.

51. ———, *Sur la théorie des variétés feuilletées*, Univ. Roma Ist. Naz. Alta Rend Mat. e Appl. (5) **10** (1951), 64–82.

52. T. Frankel, *Manifolds with positive curvature*, Pacific J. Math. **11** (1961), 165–174.

53. A. Fröhlicher and A. Nijenhuis, *A theorem on stability of complex structures*, Proc. Nat. Acad. Sci. U.S.A. **43** (1957), 239–241.

54. R. Godement, *Théorie des faisceaux*, Actualités Sci Indust. No. 1252, Hermann, Paris, 1958.

55. S. I. Goldberg, *Curvature and homology*, Academic Press, New York, 1962.

56. M. Goto, *On algebraic homogeneous spaces*, Amer. J. Math. **76** (1954), 811–818.

57. D. Gromoll, *Differenzierbare Strukturen und Metriken positiver Krümmung auf Sphären*, Dissertation, Univ. of Bonn, Bonn, 1964.

58. R. Gunning and H. Rossi, *Analytic functions of several complex variables*, Prentice-Hall, Englewood Cliffs, N. J., 1965.

59. A. Haefliger, *Structures feuilletées et cohomologie à valeur dans un faisceau de groupoides*, Comment. Math. Helv. **32** (1958), 248–329.

60. ———, *Variétés feuilletées*, Ann. Scuola Norm. Sup. Pisa. **16** (1962), 367–397.

61. P. Hartman and A. Wintner, *On the embedding problem in differential geometry*, Amer. J. Math. **72** (1950), 553–564.

62. S. Helgason, *Differential geometry and symmetric spaces*, Academic Press, New York, 1962.

63. N. J. Hicks, *Notes on differential geometry*, Math. Studies No. 3, Van Nostrand, New York, 1965.

64. F. Hirzebruch, *Neue topologische Methoden in der algebraischen Geometrie*, Ergebnisse der Math und ihrer Grenzgebiete, 1962.

65. ———, *Über eine Klasse von einfach-zusammenhängenden komplexen Mannigfaltigkeiten*, Math Ann. **124** (1957), 77–86.

66. ———, *Komplexe Mannigfaltigkeiten*, pp. 119–136, Proc. Internat. Congress Math., 1958.

67. F. Hirzebruch and K. Kodaira, *On the complex projective spaces*, J. Math. Pures Appl. **36** (1957), 201–216.

68. W. V. D. Hodge, *The theory and applications of harmonic integrals*, Cambridge Univ. Press, New York, 1952.

69. W. Kaplan, *Regular curve-families filling the plane*, I, Duke Math. J. **7** (1940), 154–185; II, Duke Math. J. **8** (1941), 11–46.

70. W. Klingenberg, *Eine Kennzeichnung der Riemannschen sowie der Hermitschen Mannigfaltigkeiten*, Math. Z. **70** (1959), 300–309.

71. ———, *Contributions to riemannian geometry in the large*, Ann. of Math. **69** (1959), 654–666.

72. ———, *Über Riemannsche Mannigfaltigkeiten mit positiver Krümmung*, Comment. Math. Helv. **35** (1961), 47–54.

73. ———, *On the topology of riemannian manifolds where the conjugate points have a similar distribution as in symmetric spaces of rank* 1, Bull. Amer. Math. Soc. **69** (1963), 95–100.

74. ———, *Neue Methoden und Ergebnisse in der Riemannschen Geometrie*, Jber. Deutsch. Math.-Verein. **66** (1964), 85–94.

75. S. Kobayashi, *Topology of positively pinched Kähler manifolds*, Tôhoku Math. J. **15** (1963), 121–139.

76. S. Kobayashi and T. Nagano, *On a fundamental theorem of Weyl-Cartan on G-structures*, J. Math. Soc. Japan **17** (1965), 84–101.

77. ———, *On filtered Lie algebras and geometric structures*. I, J. Math. Mech. **13** (1964), 875–908; II, J. Math. Mech. **14** (1965), 513–521; III, J. Math. Mech. **14** (1965), 679–706; IV, J. Math. Mech. **15** (1966), 163–175; V, (to appear).

78. S. Kobayashi and K. Nomizu, *Foundations of differential geometry*, Vol. 1, Interscience, New York, 1963.

79. K. Kodaira, *On cohomology groups of compact analytic varieties with coefficients in some analytic faisceaux*, Proc. Nat. Acad. Sci. U.S.A. **39** (1953), 865–868.

80. ———, *On a differential geometric method in the theory of analytic stacks*, Proc. Nat. Acad. Sci. U.S.A. **39** (1953), 1268–1273.

81. ———, *On Kähler varieties of restricted type*, Ann. of Math. **60** (1954), 28–48.

82. K. Kodaira and D. C. Spencer, *Groups of complex line bundles over compact Kähler varieties*, Proc. Nat. Acad. Sci. U.S.A. **39** (1953), 868–872.

83. ———, *On deformations of complex analytic structures*. I, II, Ann. of Math. **67** (1958), 328–466; III, Ann. of Math. **71** (1960), 43–76.

84. V. Yoh Kraines, *Topology of quaternionic manifolds*, Bull. Amer. Math. Soc. **71** (1965), 526–527.

85. N. H. Kuiper, *On C'-isometric imbeddings*. I, Indag. Math. **17** (1955), 545–556; II, Indag. Math. **17** (1955), 683–689.

86. M. Kuranishi, *On E. Cartan's prolongation theorem of exterior differential systems*, Amer. J. Math. **79** (1957), 1–47.

87. ———, *On the locally complete families of complex analytic structures*, Ann. of Math. **75** (1962), 536–577.

88. ———, *New proof for the existence of locally complete families of complex structures*, pp. 142–154, Proc. Conf. Comp. Anal., Minneapolis, 1964, Springer, Berlin, 1965.

89. S. Lang, *Fonctions implicites et plongements riemanniens*, Séminaire Bourbaki, No. 237, 1961–1962.

90. ———, *Introduction to differentiable manifolds*, Interscience, New York, 1962

91. H. Lewy, *On the existence of a closed convex surface realizing a given riemannian metric*, Proc. Nat. Acad. Sci. U.S.A. **24** (1938), 104–106.

92. H. Levine, *A theorem on holomorphic mappings into complex projective space*, Ann. of Math. **71** (1960), 529–535.

93. P. Liebermann, *Problèmes d'équivalence relatifs à une structure presque complexe sur une variété à quatre dimensions*, Acad. Roy. Belg. Bull. Cl. Sci. (5) **36** (1950), 742–755.

94. A. Lichnerowicz, *Théorie globale des connexions et des groupes d'holonomie*, Edizioni Cremonese, Rome, 1957.

95. E. L. Lima, *Commuting vector fields on* S^3, Ann. of Math. **81** (1965), 70–81.

96. J. Milnor, *Lectures on characteristic classes*, mimeographed notes, Univ. of Princeton, Princeton, N.J., 1958 (to appear).

97. J. Moser, *A new technique for the construction of solutions of nonlinear differential equations*, Proc. Nat. Acad. Sci. U.S.A. **47** (1961), 1824–1831.

98. S. Myers, *Riemannian manifolds with positive mean curvature*, Duke Math. J. 8 (1941), 401–404.

99. J. Nash, *C^1-isometric imbeddings*, Ann. of Math. 60 (1954), 383–396.

100. ———, *The embedding problem for riemannian manifolds*, Ann. of Math. 63 (1956), 20–63.

101. A. Newlander and L. Nirenberg, *Complex analytic coordinates in almost complex manifolds*, Ann. of Math. 65 (1957), 391–404.

102. J. C. C. Nitsche, *On new results in the theory of minimal surfaces*, Bull. Amer. Math. Soc. 71 (1965), 195–270.

103. A. Nijenhuis and W. B. Woolf, *Some integration problems in almost complex and complex manifolds*, Ann. of Math. 77 (1963), 424–489.

104. L. Nirenberg, *The Weyl and Minkowski problems in differential geometry in the large*, Comm. Pure Appl. Math. 6 (1953), 337–394.

105. ———, *Rigidity of a class of closed surfaces*, Nonlinear problems, R. E. Langer (Editor) 1963, pp. 177–193.

106. S. P. Novikov, *Foliations of codimension* 1 *on manifolds*, Soviet Math. Dokl. 5 (1964), 540–544.

107. R. Osserman, *Global properties of minimal surfaces in E^3 and E^n*, Ann. of Math. 80 (1964), 340–364.

108. R. Palais, *Seminar on the Atiyah-Singer index theorem*, Annals of Mathematics Studies, Princeton Univ. Press, Princeton, N. J., 1965.

109. M. M. Peixoto, *Structural stability on two-dimensional manifolds*, Topology 1 (1962), 101–120.

110. F. Peterson, *Some remarks on Chern classes*, Ann. of Math. 69 (1959), 414–420.

111. I. R. Porteous, *Blowing up Chern classes*, Proc. Cambridge Philos. Soc. 56 (1960), 118–124.

112. A. Preissmann, *Quelques propriétés globales des espaces de Riemann*, Comment. Math. Helv. 15 (1942–1943), 175–216.

113. C. C. Pugh, *The closing lemma*, Amer. J. Math. (to appear).

114. H. E. Rauch, *A contribution to differential geometry in the large*, Ann. of Math. 54 (1951), 38–55.

115. ———, *Geodesics and curvature in differential geometry in the large*, Yeshiva Univ., New York, 1959.

116. G. Reeb, *Sur certaines propriétés topologiques des variétés feuilletées*, Actualités Sci. Indust. No. 1183, Hermann, Paris, 1952.

117. ———, *Sur la théorie générale des systèmes dynamiques*, Ann. Inst. Fourier (Grenoble) 6 (1952), 89–115.

118. ———, *Sur une généralisation d'une théorème de M. Denjoy*, Ann. Inst. Fourier (Grenoble) 11 (1961), 185–200.

119. G. de Rham, *Variétés différentiables*, Hermann, Paris, 1955.

120. A. J. Schwartz, *A generalization of a Poincaré-Bendixson theorem to closed two-dimensional manifolds*, Amer. J. Math. 85 (1963), 453–458.

121. H. Seifert, *Closed integral curves in three-space and isotopic two-dimensional deformations*, Proc. Amer. Math. Soc. 1 (1950), 287–302.

122. J.-P. Serre, *Un théorème de dualité*, Comment. Math. Helv. 29 (1955), 9–26.

123. ———, *Faisceaux algébriques cohérents*, Ann. of Math. 61 (1955), 197–278.

124. ———, *Représentations linéaires et espaces homogènes kähleriens des groupes de Lie compacts*, Séminaire Bourbaki, No. 100, 1954.

125. C. L. Siegel, *Note on differential equations on the torus*, Ann. of Math. 46 (1945), 423–428.

126. I. M. Singer and S. Sternberg, *The infinite group of Lie and Cartan*, J. Analyse Math. **15** (1965), 1–114.

127. E. H. Spanier, *The homology of Kummer manifolds*, Proc. Amer. Math. Soc. **7** (1956), 155–160.

128. N. Steenrod, *The topology of fibre bundles*, Princeton Univ. Press, Princeton, N. J., 1951.

129. S. Sternberg, *Lectures on differential geometry*, Prentice-Hall, Englewood Cliffs, N. J., 1964.

130. O. Veblen and J. H. C. Whitehead, *The foundations of differential geometry*, Cambridge Univ. Press, New York, 1932.

131. K. Voss, *Differentialgeometrie geschlossener Flächen im euklidischen Raum.* I, Jber. Deutscher. Math.-Verein **63** (1960–1961), 117–136.

132. H. C. Wang, *Two-point homogeneous spaces*, Ann. of Math. **55** (1952), 177–191.

133. ———, *Closed manifolds with homogeneous complex structure*, Amer. J. Math. **76** (1954), 1–32.

134. A. Weil, *Introduction a l'étude des variétés kählériennes*, Hermann, Paris, 1958.

135. ———, *Un théorème fondamental de Chern en géométrie riemannienne*, Seminaire Bourbaki, No. 239, 1961–1962.

136. H. Weyl, *Raum, Zeit, Materie*, 5th ed., Berlin, 1923; English transl., Dover, New York, 1950.

137. ———, *Meromorphic functions and analytic curves*, Annals of Mathematics Studies No. 12, Princeton Univ. Press, Princeton, N. J., 1943.

138. ———, *Über die Bestimmung einer geschlossenen konvexen Fläche durch ihr Linienelement*, Vierteljarsch. Naturforsch. Ges. Zürich **61** (1916), 40–72; *Selecta Hermann Weyl*, pp. 148–178, Birkhäuser Verlag, Basel, 1956.

139. W. T. Wu, *Sur les classes caractéristiques des structures fibrées sphériques*, Actualités Sci. Indust., Hermann, Paris, 1952.

140. H. Yamabe, *On deformation of riemannian structures on compact manifolds*, Osaka Math. J. **12** (1960), 21–37.

141. K. Yano and S. Bochner, *Curvature and Betti numbers*, Annals of Mathematics Studies No. 32, Princeton Univ. Press, Princeton, N. J , 1953.

Reprinted from
Bull. Amer. Math. Soc. **72** (1966) 167–219

On the Kinematic Formula in Integral Geometry

SHIING-SHEN CHERN

1. Introduction. The kinematic density in euclidean space was first introduced by Poincaré. In modern terminology it is the Haar measure of the group of motions which acts on the space. One of the basic problems in integral geometry is to find explicit formulas for the integrals of geometric quantities over the kinematic density in terms of known integral invariants. An important example is the kinematic formula of Blaschke, as follows [1]: Let E^3 be the euclidean three-space, and let dg be the kinematic density, so normalized that the measure of all positions about a point is $8\pi^2$. (In other words, the measure of all positions of a domain D with volume V such that D contains a fixed point is $8\pi^2 V$.) Let D_i , $i = 1, 2$, be a domain with smooth boundary, of which V_i , $M_1^{(i)}$, $M_2^{(i)}$, χ_i are respectively the volume, the area of the boundary, the integral of mean curvature of the boundary, and the Euler characteristic. Then, if D_1 is fixed and D_2 moves with kinematic density dg, we have

$$\int \chi(D_1 \cap g\, D_2)\, dg = 8\pi^2(4\pi\chi_1 V_2 + 4\pi\chi_2 V_1 + M_1^{(1)} M_2^{(2)} + M_2^{(1)} M_1^{(2)}), \tag{1}$$

where $\chi(D_1 \cap gD_2)$ is the Euler characteristic of the intersection $D_1 \cap gD_2$. Formula (1) contains as special or limiting cases many of the formulas in integral geometry in euclidean three-space. It was generalized to euclidean n-space by C. T. Yen and the present author [3] and to the non-euclidean spaces by Santalo [4].

This paper will be concerned with a pair of compact submanifolds (without boundary) M^p, M^q of arbitrary dimensions p, q in euclidean n-space E^n and with the integration with respect to dg of certain geometrical quantities of the submanifold $M^p \cap g(M^q)$. The latter depend only on the induced riemannian metric of $M^p \cap g(M^q)$ and are defined as follows:

Let X be a riemannian manifold of dimension k. In the bundle B of orthonormal frames over X we have the coframes, which consist of k linearly independent linear differential forms φ_α , such that the riemannian metric in X is

$$ds^2 = \sum_\alpha \varphi_\alpha^2 . \tag{2}$$

Journal of Mathematics and Mechanics, Vol. 16, No. 1 (1966).

(In this and other formulas in this section we have $1 \leqq \alpha, \beta, \gamma, \delta \leqq k$.) Let

$$\varphi_{\alpha\beta} = -\varphi_{\beta\alpha} \tag{3}$$

be the connection forms of the Levi-Civita connection. They are linear differential forms in B and satisfy the "structural equations"

$$\begin{aligned} d\varphi_\alpha &= \sum_\beta \varphi_\beta \wedge \varphi_{\beta\alpha}, \\ d\varphi_{\alpha\beta} &= \sum_\gamma \varphi_{\alpha\gamma} \wedge \varphi_{\gamma\beta} + \Phi_{\alpha\beta}, \end{aligned} \tag{4}$$

where

$$\Phi_{\alpha\beta} = \tfrac{1}{2} \sum_{\gamma,\delta} S_{\alpha\beta\gamma\delta}\varphi_\gamma \wedge \varphi_\delta . \tag{5}$$

The coefficients $S_{\alpha\beta\gamma\delta}$ are functions in B and have the symmetry properties

$$\begin{aligned} S_{\alpha\beta\gamma\delta} &= -S_{\alpha\beta\delta\gamma} = -S_{\beta\alpha\gamma\delta}, \\ S_{\alpha\beta\gamma\delta} &= S_{\gamma\delta\alpha\beta}, \\ S_{\alpha\beta\gamma\delta} &+ S_{\alpha\gamma\delta\beta} + S_{\alpha\delta\beta\gamma} = 0. \end{aligned} \tag{6}$$

From these functions we construct the following scalar invariants in X:

$$I_e = \frac{(-1)^{e/2}(k-e)!}{2^{k/2}k!} \sum \delta\begin{pmatrix}\alpha_1 & \cdots & \alpha_e \\ \beta_1 & \cdots & \beta_e\end{pmatrix} S_{\alpha_1\alpha_2\beta_1\beta_2} \cdots S_{\alpha_{e-1}\alpha_e\beta_{e-1}\beta_e}, \tag{7}$$

where e is an even integer satisfying $0 \leqq e \leqq k$,

$$\delta\begin{pmatrix}\alpha_1 & \cdots & \alpha_e \\ \beta_1 & \cdots & \beta_e\end{pmatrix}$$

is equal to $+1$ or -1 according as $\alpha_1, \cdots, \alpha_e$ is an even or odd permutation of $\beta_1, \cdots, \beta_e$, and is otherwise zero, and the summation is over all $\alpha_1, \cdots, \alpha_e$ and $\beta_1, \cdots, \beta_e$, independently from 1 to k. When X is oriented and compact, we let

$$\mu_e(X) = \int_X I_e \, dv, \tag{8}$$

where dv is the volume element. Thus $\mu_e(X)$ are integral invariants of X, with $\mu_0(X)$ equal to the total volume. If k is even, then the Gauss–Bonnet formula says that [2]

$$\mu_k(X) = \frac{(2\pi)^{k/2}}{(k-1)(k-3)\cdots 1}\chi(X), \tag{9}$$

where $\chi(X)$ is the Euler–Poincaré characteristic of X. The numerical coefficient before the summation in (7) is so chosen that $I_e(X) = 1$, when X is the unit k-sphere in E^{k+1} with the induced metric.

These integral invariants appear in a natural way in Weyl's formula for the volume of a tube [8]. In fact, suppose X be imbedded in E^n and suppose T_ρ be a tube of radius ρ about X, *i.e.*, the set of all points at a distance $\leqq \rho$ from X. Then, for ρ sufficiently small, Weyl proved that the volume of T_ρ is given by the formula

$$V(T_\rho) = O_m \sum_e \frac{(e-1)(e-3)\cdots 1}{(m+e)(m+e-2)\cdots m} \mu_e(X)\rho^{m+e}, \tag{10}$$

$$e \text{ even}, \quad 0 \leqq e \leqq k, \quad m = n - k,$$

where O_m is the volume of the unit sphere (of dimension $m - 1$) in E^m, and its value is given by

$$O_m = 2\pi^{m/2} \Big/ \Gamma\left(\frac{m}{2}\right). \tag{11}$$

This formula is also valid for large values of ρ, provided that the volume is counted with multiplicities for domains where T_ρ intersects itself.

The main result of this paper is the kinematic formula

$$\int \mu_e(M^p \cap gM^q)\, dg = \sum_{\substack{0 \leqq i \leqq e \\ i \text{ even}}} c_i \mu_i(M^p)\mu_{e-i}(M^q), \tag{12}$$

$$e \text{ even}, \quad 0 \leqq e \leqq p + q - n,$$

where c_i, to be given below (formula (79)), are numerical constants depending on n, p, q, e.

2. Preliminaries. Let E^n be the euclidean space of dimension n, and $xe_1 \cdots e_n$ an orthonormal frame, or simply a frame, in E^n, so that $x \in E^n$ and $e_1, \cdots, e_n$ are vectors through x, whose scalar products satisfy the relations

$$(e_A, e_B) = \delta_{AB}, \quad 1 \leqq A, B, C \leqq n. \tag{13}$$

The kinematic density is the volume element in the space of all frames, and is given by

$$dg = \bigwedge_{\substack{A,B,C \\ B<C}} (dx, e_A)(de_B, e_C), \tag{14}$$

where the right-hand side is the exterior product of the differential forms in question. By (14), dg is normalized. The factor

$$dg_0 = \bigwedge_{B<C} (de_B, e_C) \tag{15}$$

in this product is the kinematic density about a point, and we have

$$dg = dx\, dg_0, \tag{16}$$

where dx is the volume element in E^n.

To define the density dE^p in the space of all p-dimensional linear subspaces

E^p in E^n consider the frames $xe_1 \cdots e_n$, such that $x \varepsilon E^p$ and $e_1, \cdots, e_p$ are parallel to E^p (so that $e_{p+1}, \cdots, e_n$ are normal vectors of E^p). Then

$$(17) \qquad dE^p = \bigwedge_{h,\lambda,\mu} (dx, e_\lambda)(de_h, e_\mu), \qquad 1 \leqq h \leqq p, \qquad p+1 \leqq \lambda, \mu \leqq n.$$

Similarly, the density in the space of all E^p through a fixed point O of E^n is

$$(18) \qquad dE^p(O) = \bigwedge_{h,\mu} (de_h, e_\mu), \qquad 1 \leqq h \leqq p, \qquad p+1 \leqq \mu \leqq n,$$

where $e_1, \cdots, e_p$ are vectors of the frame, which span E^p.

We will also consider the space of all (x, E^p), such that E^p is a p-dimensional linear subspace of E^n and $x \varepsilon E^p$. Its density will be

$$(19) \qquad d(x, E^p) = \bigwedge_{A,h,\mu} (dx, e_A)(de_h, e_\mu),$$

and we have

$$(20) \qquad d(x, E^p) = dx\,(E^p)\,dE^p,$$

where $dx(E^p)$ is the volume element in E^p.

The total volumes

$$\int dg_0, \qquad \int dE^p(O)$$

are finite, and we will give their values. Let $e_1, \cdots, e_n$ be a frame through O. By mapping it to the last vector e_n, we get a fibering of the space $SO(n)$ of all frames through O, whose base space is the unit sphere in E^n and whose fiber is $SO(n-1)$. It follows by induction that

$$(21) \qquad \int_{SO(n)} dg_0 = O_n O_{n-1} \cdots O_2 .$$

The Grassmann manifold of all oriented E^p through O is the quotient space $SO(n)/SO(p) \times SO(n-p)$, whose total volume is $O_n \cdots O_2/O_p \cdots O_2\, O_{n-p} \cdots O_2$. It follows that the total volume of the Grassmann manifold $Gr(p)$ of all unoriented E^p through O is

$$(22) \qquad \int_{Gr(p)} dE^p(O) = O_n \cdots O_1/O_p \cdots O_1 O_{n-p} \cdots O_1 .$$

Let $E_0^p \varepsilon Gr(p)$ be a fixed linear subspace of dimension p through O. Let E_0^p and E^p be spanned by the vectors e_h^0 and e_k respectively $(1 \leqq h, k \leqq p)$, such that

$$(23) \qquad (e_h^0, e_k^0) = (e_h, e_k) = \delta_{hk} .$$

Then $\Delta(E_0^p, E^p) = |\det(e_h^0, e_k)|$ depends only on E_0^p and E^p. Since $dE^p(O)$ is invariant under the rotations about O, the integral $\int_{Gr(p)} \Delta^N(E_0^p, E^p)\, dE^p(O)$ is independent of E_0^p and is a numerical constant, to be denoted by $c(n, p, N)$.

We wish to show that its value is given by

$$(24) \qquad c(n, p, N) = \int_{Gr(p)} \Delta^N \, dE^p(O) = \frac{O_{n+N} \cdots O_1}{O_{p+N} \cdots O_1} \frac{O_N \cdots O_1}{O_{n-p+N} \cdots O_1}.$$

To prove this we need a formula for $dE^p(O)$, which is also useful for other purposes. In fact, let E_0^q be a fixed q-dimensional linear subspace through O, with $q \leqq p$. Let $Gr^*(p)$ be the subset of $Gr(p)$, whose E^p satisfy the condition $\dim (E^p \cap (E_0^q)^\perp) = p - q$, where $(E_0^q)^\perp$ is the $(n - q)$-dimensional linear subspace through O orthogonal to E_0^q. $Gr^*(p)$ differs from $Gr(p)$ by a set of measure zero. We choose the frames $Oe_1 \cdots e_n$ associated to $E^p \in Gr^*(p)$, such that $e_1, \cdots, e_{p-q}$ span $E^p \cap (E_0^q)^\perp$ and $e_1, \cdots, e_p$ span E^p. Let $Oe_1 \cdots e_{p-q} f_{p-q+1} \cdots f_n$ be frames satisfying the condition that $f_{n-q+1}, \cdots, f_n$ are constant vectors spanning E_0^q. Then e_i differ from f_j by a proper orthogonal transformation

$$(25) \qquad e_i = \sum_j u_{ij} f_j, \qquad p - q + 1 \leqq i, j \leqq n,$$

where

$$(26) \qquad u_{ij} = (e_i, f_j).$$

Since $f_{n-q+1}, \cdots, f_n$ are constant vectors, we have

$$(de_\alpha, f_\rho) = -(df_\rho, e_\alpha) = 0, \qquad 1 \leqq \alpha \leqq p - q, \qquad n - q + 1 \leqq \rho \leqq n.$$

It follows that

$$dE^p(O) = \bigwedge_{\alpha, r, \lambda, \mu} (de_\alpha, e_\lambda)(de_r, e_\mu) = D^{p-q} \bigwedge_{\alpha, r, s, \mu} (de_\alpha, f_s)(de_r, e_\mu),$$

$$p + 1 \leqq \lambda, \mu \leqq n; \qquad p - q + 1 \leqq r \leqq p; \qquad p - q + 1 \leqq s \leqq n - q,$$

where

$$(27) \qquad D = \det (u_{\lambda\mu}).$$

We write the last formula as

$$(28) \qquad dE^p(O) = D^{p-q} \, dE^{p-q}((E_0^q)^\perp) \, dE^p(E^{p-q}),$$

where

$$(29) \qquad dE^p(E^{p-q}) = \bigwedge_{r, \mu} (de_r, e_\mu)$$

is the density in the space of all E^p about $E^{p-q} = E^p \cap (E_0^q)^\perp$, and

$$(30) \qquad dE^{p-q}((E_0^q)^\perp) = \bigwedge_{\alpha, s} (de_\alpha, f_s)$$

is the density in the space of all E^{p-q} through O and in $(E_0^q)^\perp$.

We now integrate both sides of the equation (28) over $Gr^*(p)$. In integrating its right-hand side, we first hold E^{p-q} fixed and then integrate over all E^{p-q}

through O and in $(E_0^q)^\perp$. The result is

$$c(n - p + q, q, p - q) = \frac{O_n \cdots O_1 O_{p-q} \cdots O_1}{O_p \cdots O_1 O_{n-q} \cdots O_1}.$$

This reduces to (24) by a change of notation.

In the same way we can derive for dE^p a formula analogous to (28). In fact, let E_0^{n-q} be a fixed space of dimension $n - q$ in E^n, not necessarily through O, with $q \leqq p$. Consider those E^p such that dim $(E^p \cap E_0^{n-q}) = p - q$, and let $E^{p-q} = E^p \cap E_0^{n-q}$. Let $(E_0^{n-q})^\perp$ (respectively $E^{p-q}(E^p)^\perp$) be the space through O orthogonal to E_0^{n-q} (respectively the space through O orthogonal to E^{p-q} and parallel to E^p). Both are of dimension q and the invariant Δ is defined. Then we have the following density formula

$$dE^p = \Delta^{p-q+1}\, dE^{p-q}(E_0^{n-q})\, dE^p(E^{p-q}), \tag{31}$$

where $dE^{p-q}(E_0^{n-q})$ (respectively $dE^p(E^{p-q})$) is the density in the space of all E^{p-q} in E_0^{n-q} (respectively all E^p through E^{p-q}). The difference between (28) and (31) lies in the fact that in (28) all the linear subspaces under consideration are through O, while this condition is not imposed in (31).

Let $f(E^p)$ be an integrable function which depends only on $E^{p-q} = E^p \cap E^{n-q}$. Then the integration with respect to $dE^p(E^{p-q})$ can be carried out, and we have

$$\begin{aligned} \int f(E^p)\, dE^p &= c(n - p + q, q, p - q + 1) \int f(E^{p-q})\, dE^{p-q}(E_0^{n-q}) \\ &= \frac{O_{n+1} \cdots O_1 O_{p-q+1} \cdots O_1}{O_{p+1} \cdots O_1 O_{n-q+1} \cdots O_1} \int f(E^{p-q})\, dE^{p-q}(E_0^{n-q}). \end{aligned} \tag{32}$$

3. A differential formula. Although we will need the formula only in the special case of the euclidean space, it is desirable to formulate the problem for a general homogeneous space, as follows: Let G be a Lie group, acting on a left coset space G/H by left multiplication. Let M^r, M^s be submanifolds in G/H, of dimensions r, s respectively. Consider the space

$$Y = \bigcup_{g \in G} (M^r \cap gM^s) \times g, \tag{33}$$

with an obvious topology. A point in this space can be written

$$(x = gy, g \mid x \in M^r, y \in M^s).$$

By sending this point to $(x, y) \in M^r \times M^s$, we get a mapping

$$\psi : Y \to M^r \times M^s. \tag{34}$$

By definition we have $\psi^{-1}(x, y) = (x = gH_y y, gH_y)$, where H_y is the group of isotropy at y, *i.e.*, the subgroup of all transformations of G, which leave y fixed. Suppose that G/H has a riemannian structure invariant under G. Then $M^r \cap gM^s$ has a volume element Φ_g, and Y has the volume element $\Phi_g dg$. The problem is to express $\Phi_v dg$ as a multiple of $dh_v \Omega_1 \Omega_2$, where Ω_1, Ω_2 are respectively the volume elements of M^r, M^s, and dh_v that of H_v.

We wish to solve this problem for a special case: By a k-field in E^n is meant a point $x \in E^n$, together with an ordered set of vectors $e_1, \cdots, e_k$, satisfying

$$(e_\gamma, e_\tau) = \delta_{\gamma\tau}, \qquad 1 \leqq \gamma, \tau \leqq k.$$

Let $F_{n,k}$ be the space of all k-fields of E^n, so that $F_{n,0} = E^n$. We can write $F_{n,k}$ as a left coset space G/H, where G is the group of all proper motions in E^n and H is isomorphic to $SO(n-k)$.

Let M^p be a submanifold of dimension p of E^n, and M^r be the submanifold of $F_{n,k}$, which consists of the k-fields $xe_1 \cdots e_k$, such that $x \in M^p$ and $e_1, \cdots, e_k$ are tangent to M^p at $x (k \leqq p)$. Then

$$r = p + (p-1) + \cdots + (p-k) = \tfrac{1}{2}(k+1)(2p-k). \tag{35}$$

To define a density Θ on M^r choose the vectors $e_{k+1}, \cdots, e_p$, so that $e_1, \cdots, e_p$ constitute an orthonormal frame of tangent vectors to M^p at x. Then we define

$$\Theta = \bigwedge_{L,\gamma,\delta} (dx, e_L)(de_\gamma, e_\delta), \quad 1 \leqq L \leqq p, \quad 1 \leqq \gamma \leqq k, \quad \gamma + 1 \leqq \delta \leqq p, \tag{36}$$

which is clearly independent of the choice of $e_{k+1}, \cdots, e_p$.

Similarly let M^s be the manifold of all tangent k-fields of a submanifold M^q of dimension q in E^n. Since M^p amd gM^q generally intersect in a submanifold of dimension $p + q - n$, our problem is meaningful, only when $k \leqq p + q - n$, which we suppose from now on. We will solve our problem for the submanifolds M^r, M^s in $F_{n,k}$.

As we will be using indices over different ranges, let us list them in the following table, which will be strictly followed through the rest of this section:

$$\begin{gathered} 1 \leqq A, B, C \leqq n; \qquad 1 \leqq \alpha, \beta \leqq p + q - n; \\ p + q - n + 1 \leqq i, j \leqq n; \qquad p + 1 \leqq \lambda, \mu \leqq n; \\ q + 1 \leqq \rho, \sigma \leqq n; \qquad p + q - n + 1 \leqq a, b \leqq p; \\ p + q - n + 1 \leqq h, l \leqq q; \qquad k + 1 \leqq \zeta \leqq n; \\ 1 \leqq \gamma, \tau, \xi \leqq k; \qquad \gamma + 1 \leqq \delta \leqq p + q - n. \end{gathered} \tag{37}$$

Let Oa_A be a fixed frame and $O'a'_A$ the moving frame in E^n, *i.e.*, the frame obtained from the former by a proper motion g. By (14) the kinematic density is

$$dg = \bigwedge_{A;B<C} (dO', a'_A)(da'_B, a'_C). \tag{38}$$

By gM^q we mean the submanifold obtained from M^q by the motion g, so that gM^q has the same relative position to $O'a'_A$ as M^q is to Oa_A.

Let xe_A be orthonormal frames, so that $x \in M^p$ and $e_1, \cdots, e_p$ are tangent vectors to M^p at x. Then

$$(dx, e_\lambda) = 0 \tag{39}$$

and it is well-known that

(40) $$de_\lambda \equiv 0, \text{ mod } dx, e_\mu .$$

Similarly, let $x'e'_A$ be frames, such that $x' \varepsilon gM^q$ and $e'_1, \cdots, e'_q$ are tangent vectors to gM^q at x'. Let d' be the differential relative to the moving frame $O'a'_A$, so that

(41) $$d'O' = d'a'_A = 0.$$

Then, as above,

(39a) $$(d'x', e'_\rho) = 0$$

and

(40a) $$d'e'_\rho \equiv 0, \text{ mod } d'x', e'_\sigma .$$

Suppose g be generic, so that $M^p \cap gM^q$ is of dimension $p + q - n$. We restrict the above families of frames by the condition

(42) $$x = x', \qquad e_\alpha = e'_\alpha .$$

Geometrically the latter means that $x \varepsilon M^p \cap gM^q$ and e_α are the tangent vectors to $M^p \cap gM^q$ at x. The two submanifolds M^p and gM^q have at x a scalar invariant, which is

(43) $$\Delta = |\det (e_a, e'_\rho)| = |\det (e_\lambda, e'_h)|.$$

This is the invariant introduced in §2 for the spaces spanned by e_a, e'_ρ respectively in the orthogonal complement of the space spanned by e_α. For a pair of hypersurfaces ($p = q = n - 1$) it is the absolute value of the sine of the angle between their normal vectors. Using these frames we have

(44) $$\Phi_g = \bigwedge (dx, e_\alpha)(de_\gamma, e_\delta).$$

Let us write

$$x = x' = O' + \sum x'_A a'_A .$$

By differentiation it follows that

$$dO' \equiv dx - d'x', \qquad \text{mod } da'_A .$$

Mod da'_A we have therefore

$$\begin{aligned}
\bigwedge (dx, e_\alpha)(dO', a'_A) &= \bigwedge (dx, e_\alpha)(dO', e'_A) \\
&\equiv \bigwedge (dx, e_\alpha)(dx - d'x', e'_A) \\
&= \bigwedge (dx, e_\alpha)(dx - d'x', e_\alpha)(dx - d'x', e'_i) \\
&= \pm \bigwedge (dx, e_\alpha)(d'x', e_\alpha)(dx - d'x', e'_h)(dx, e'_\rho) \\
&= \pm \bigwedge (dx, e_a)(dx, e'_\rho)(d'x', e_\alpha)(d'x', e'_h) \\
&= \pm\Delta \bigwedge (dx, e_\alpha)(dx, e_a)(d'x', e_\alpha)(d'x', e'_h).
\end{aligned}$$

In these exterior products the indices run over the ranges agreed on in (37). In the reduction the first step follows from the fact that a'_A differ from e'_A by a proper orthogonal transformation, the fourth step makes use of (39a), and the fifth step uses the fact that an exterior product is zero, if it contains more than p factors involving dx. Observe that the product in the last expression is the product of the volume elements of M^p and M^q.

Let

$$e'_A = \sum_B u'_{AB} a'_B ,$$

so that (u'_{AB}) is a proper orthogonal matrix. Then

$$(d - d')e'_A = \sum_B u'_{AB}\, da'_B ,$$

and we have

$$\bigwedge_{A<B} (da'_A , a'_B) = \bigwedge_{A<B} (de'_A - d'e'_A , e'_B).$$

It follows that

$$\begin{aligned}
&\bigwedge_{A<B} (de_\gamma , e_\delta)(da'_A , a'_B) \\
&\qquad = \bigwedge_{A<B} (de_\gamma , e_\delta)(de'_A - d'e'_A , e'_B) \\
&\qquad = \bigwedge_{\substack{\tau<A \\ \zeta<B}} (de_\gamma , e_\delta)(de_\tau - d'e_\tau , e'_A)(de'_\zeta - d'e'_\zeta , e'_B) \\
&\qquad = \pm \bigwedge_{\zeta<B} (de_\gamma , e_\delta)(d'e_\gamma , e_\delta)(de_\tau - d'e_\tau , e'_i)(de'_\zeta - d'e'_\zeta , e'_B) \\
&\qquad \equiv \pm \bigwedge_{\zeta<B} (de_\gamma , e_\delta)(d'e_\gamma , e_\delta)(de_\tau - d'e_\tau , e'_h)(de_\tau , e'_\rho)(de'_\zeta - d'e'_\zeta , e'_B) \\
&\qquad = \pm\Delta^k \bigwedge_{\zeta<B} (de_\gamma , e_\delta)(de_\gamma , e_a)(d'e_\gamma , e_\delta)(d'e_\tau , e'_h)(de'_\zeta - d'e'_\zeta , e'_B).
\end{aligned}$$

Here the congruence is mod $d'x'$ and follows from (40a) in view of the relation $(d'e_\tau , e'_\rho) = -(d'e'_\rho , e_\tau)$.

The densities on M^r, M^s are by definition respectively

$$\text{(45)} \qquad \begin{aligned}
\Theta_1 &= \bigwedge (dx, e_\alpha)(dx, e_a)(de_\gamma , e_\delta)(de_\tau , e_a), \\
\Theta_2 &= \bigwedge (d'x', e'_\alpha)(d'x', e'_h)(d'e_\gamma , e_\delta)(d'e_\tau , e'_h).
\end{aligned}$$

Remembering the meaning of the operator $d - d'$, we also see that

$$\text{(46)} \qquad dh_y = \bigwedge_{\zeta<B} (de'_\zeta - d'e'_\zeta , e'_B).$$

Multiplying the expressions in (38), (44) and making use of the two long equations above, we get

$$\text{(47)} \qquad \Phi_g\, dg = \pm\Delta^{k+1}\, dh_y \Theta_1 \Theta_2 .$$

This is the density formula desired, the factor being Δ^{k+1}.

4. Higher sectional curvatures and mixed curvatures of a pair of riemannian manifolds. The invariants I_e discussed in §1 can be generalized to the notion of a mixed curvature of a pair of riemannian manifolds. Let M^p, M^q be two riemannian manifolds, of dimensions p, q respectively. Let B_1, B_2 be respectively the bundles of k-fields over M^p, M^q. Thus B_1 (resp. B_2) is the space of all $xe_1 \cdots e_k$ (resp. $yf_1 \cdots f_k$) such that $x \in M^p$ (resp. $y \in M^q$) and $e_1, \cdots, e_k$ (resp. $f_1, \cdots, f_k$) are k mutually perpendicular unit tangent vectors at x (resp. at y). Let $\tilde{B}_1$ and $\tilde{B}_2$ be the principal bundles of orthonormal frames of M^p and M^q respectively. Then we have

$$\tilde{B}_1 \times \tilde{B}_2 \xrightarrow{\pi} B_1 \times B_2 \xrightarrow{\psi} M^p \times M^q, \tag{48}$$

where the mapping π is defined by taking the first k vectors of the frames. As in §1 we regard the curvature functions

$$P_{lmrs}, Q_{\lambda\mu\rho\sigma}, 1 \leqq l, m, r, s \leqq p; \qquad 1 \leqq \lambda, \mu, \rho, \sigma \leqq q$$

of M^p and M^q as scalar functions in $\tilde{B}_1 \times \tilde{B}_2$. Following the definition of I_e we set

$$I(f, g) = \frac{(k-f-g)!}{(-2)^{f+g/2}k!} \sum \delta\begin{pmatrix}\alpha_1 & \cdots & \alpha_{f+g}\\ \beta_1 & \cdots & \beta_{f+g}\end{pmatrix} \tag{49}$$

$$\cdot P_{\alpha_1\alpha_2\beta_1\beta_2} \cdots P_{\alpha_{f-1}\alpha_f\beta_{f-1}\beta_f} Q_{\alpha_{f+1}\alpha_{f+2}\beta_{f+1}\beta_{f+2}} \cdots Q_{\alpha_{f+g-1}\alpha_{f+g}\beta_{f+g-1}\beta_{f+g}},$$

where f, g are even integers, and the α's and β's run from 1 to k. Clearly $I(f, g)$ is a scalar function in $B_1 \times B_2$ and is thus defined for a pair of k-fields of M^p and M^q respectively. When $M^p = M^q$ and the two k-fields are chosen to be identical, $I(f, g) = I_{f+g}$, $k = f + g$, will depend only on the k-plane spanned by the k-field. It is then the "sectional curvature for a k-plane" of a riemannian manifold, generalizing the riemannian sectional curvature for a two-plane. This high-dimensional sectional curvature has recently been studied by various authors [6], in connection with some applications of the Gauss–Bonnet formula.

The density Θ in (36), which we will now denote by Θ_1, defines a volume element in B_1. The product $\Theta_1\Theta_2$, where Θ_2 is the volume element in B_2, gives a volume element in $B_1 \times B_2$. We wish to prove the formula

$$\int_{B_1 \times B_2} I(f, g)\Theta_1\Theta_2 = O_p \cdots O_{p-k+1} O_q \cdots O_{q-k+1} \mu_f(M^p)\mu_g(M^q), \tag{50}$$

provided that M^p and M^q are both compact.

To prove this formula we fix a point $(x, y) \in M^p \times M^q$ and integrate over the fiber $\psi^{-1}(x, y)$, *i.e.*, over all k-fields with origins at x and y. Let e_l^0 amd f_λ^0 be fixed orthonormal frames at x, y respectively and let P_{lmrs}^0, $Q_{\lambda\mu\rho\sigma}^0$ be the values of the curvature functions for these frames. For the k-fields given by

$$e_\alpha = \sum_l u_{\alpha l} e_l^0, \qquad f_\alpha = \sum_\lambda v_{\alpha\lambda} f_\lambda^0,$$

we have

$$P_{\alpha\beta\gamma\delta} = \sum u_{\alpha l} u_{\beta m} u_{\gamma r} u_{\delta s} P^0_{lmrs}$$

$$Q_{\alpha\beta\gamma\delta} = \sum v_{\alpha\lambda} v_{\beta\mu} v_{\gamma\rho} v_{\delta\sigma} Q^0_{\lambda\mu\rho\sigma} .$$

Thus the integrand is a polynomial F in $u_{\alpha l}$, $v_{\beta\lambda}$, of even degrees in each of these sets of variables, and no $u_{\alpha l}$ or $v_{\beta\lambda}$ appears in a power higher than the second. If $e_1, \cdots, e_k$ is a k-field at x, replacement of $u_{\alpha l}$ by $-u_{\alpha l}$, for a fixed l and all α, will give a new k-field at x. This means that the terms in F, which give a non-zero value to the integral, must contain each $u_{\alpha l}$ (and similarly each $v_{\beta\lambda}$) exactly in its square. It follows that the only terms of F to be concerned are those for which $\alpha_1, \cdots, \alpha_f$ and $\alpha_{f+1}, \cdots, \alpha_{f+g}$ are permutations of $\beta_1, \cdots, \beta_f$ and $\beta_{f+1}, \cdots, \beta_{f+g}$ respectively and hence that the integral over the fiber $\psi^{-1}(x, y)$ is a constant multiple of $I_f(x)I_g(y)$, where each factor stands for the invariant defined in (7) for the manifolds M^p and M^q respectively. This proves (50), except for the value of the constant factor. The latter is determined by taking M^p and M^q to be the unit spheres in E^{p+1} and E^{q+1} respectively. Then the integrand at the left-hand side of (50) is equal to 1 and the integral itself equal to

$$O_{p+1} \cdots O_{p-k+1} O_{q+1} \cdots O_{q-k+1} .$$

5. A Meusnier-Euler type theorem. We will pursue the situation studied in §3, where M^p and gM^q intersect in a submanifold M^{p+q-n} of dimension $p + q - n$. We will use the same notations, and in particular the ranges of indices in the table (37). The calculation of $\mu_e(M^{p+q-n})$ in the main formula (12) needs a knowledge of I_e, which in turn depends on the induced riemannian metric of M^{p+q-n}. Our first problem is therefore to express the second fundamental forms of M^{p+q-n} in terms of those of M^p and gM^q and the relative position of their tangent spaces.

For the families of frames xe_A and xe'_A introduced in §3, let

$$\omega_A = (dx, e_A), \qquad \omega'_A = (d'x', e'_A),$$
$$\omega_{AB} = (de_A, e_B), \qquad \omega'_{AB} = (d'e'_A, e'_B), \tag{51}$$

so that

$$\omega_{AB} + \omega_{BA} = 0, \qquad \omega'_{AB} + \omega'_{BA} = 0. \tag{52}$$

Equations (39) and (39a) can be written

$$\omega_\lambda = 0, \qquad \omega'_\rho = 0. \tag{53}$$

while (40) and (40a) allow us to put, *when restricted to* M^{p+q-n},

$$\omega_{\alpha\lambda} = \sum h_{\alpha\lambda\beta}\omega_\beta , \qquad \omega'_{\alpha\rho} = \sum h'_{\alpha\rho\beta}\omega_\beta , \tag{54}$$

where

$$h_{\alpha\lambda\beta} = h_{\beta\lambda\alpha}\,, \qquad h'_{\alpha\rho\beta} = h'_{\beta\rho\alpha}\,. \tag{55}$$

The second fundamental forms of M^p and gM^q restricted to their intersection M^{p+q-n} are respectively

$$\begin{aligned} II_\lambda &= (e_\lambda\,, d^2x) = \sum_{\alpha,\beta} h_{\alpha\lambda\beta}\omega_\alpha\omega_\beta \\ II'_\rho &= (e'_\rho\,, d'^2x) = \sum_{\alpha,\beta} h'_{\alpha\rho\beta}\omega_\alpha\omega_\beta\,. \end{aligned} \tag{56}$$

We wish to express $(e_a\,, d^2x)$ as a linear combination of II_λ and II'_ρ.

For this purpose set

$$e'_\rho = \sum_a u_{\rho a}e_a + \sum_\lambda u_{\rho\lambda}e_\lambda\,, \tag{57}$$

so that

$$u_{\rho a} = (e'_\rho\,, e_a). \tag{58}$$

Under our hypothesis $\Delta = |\det\,(u_{\rho a})| \neq 0$, let $(v_{b\sigma})$ be the inverse matrix of $(u_{\rho a})$, so that

$$\sum_\sigma v_{b\sigma}u_{\sigma a} = \delta_{ba}\,, \qquad \sum_a u_{\rho a}v_{a\sigma} = \delta_{\rho\sigma}\,. \tag{59}$$

Then we have

$$e_a = \sum_\rho v_{a\rho}e'_\rho + \sum_\lambda v_{a\lambda}e_\lambda\,, \tag{60}$$

where

$$v_{a\lambda} = -\sum_\rho v_{a\rho}u_{\rho\lambda}\,. \tag{61}$$

The condition $(e'_\rho\,, e'_\sigma) = \delta_{\rho\sigma}$ is expressed by

$$\delta_{\rho\sigma} = \sum_a u_{\rho a}u_{\sigma a} + \sum_\lambda u_{\rho\lambda}u_{\sigma\lambda}\,. \tag{62}$$

From (60) we have

$$II_a = (e_a\,, d^2x) = \sum_{\alpha,\beta} h_{\alpha a\beta}\omega_\alpha\omega_\beta = \sum_\rho v_{a\rho}II'_\rho + \sum_\lambda v_{a\lambda}II_\lambda\,. \tag{63}$$

For $n = 3$, $p = 2$, and gM^q a plane, this is essentially the classical Meusnier's theorem.

It follows from the structure equations of the euclidean space that the curvature forms of the induced riemannian metric on M^{p+q-n} and its second fundamental forms are related by

$$\Phi_{\alpha\beta} = \tfrac{1}{2}\sum_{\gamma,\delta} S_{\alpha\beta\gamma\delta}\omega_\gamma \wedge \omega_\delta = -\sum_i \omega_{\alpha i} \wedge \omega_{\beta i}\,, \tag{64}$$

whence

$$S_{\alpha\beta\gamma\delta} = -\sum_i (h_{\alpha i\gamma} h_{\beta i\delta} - h_{\alpha i\delta} h_{\beta i\gamma}). \tag{65}$$

To express $S_{\alpha\beta\gamma\delta}$ in terms of $h_{\alpha\lambda\beta}$, $h'_{\alpha\rho\beta}$, we introduce the quantities

$$\begin{aligned} P_{\alpha\beta\gamma\delta} &= -\sum_\lambda (h_{\alpha\lambda\gamma} h_{\beta\lambda\delta} - h_{\alpha\lambda\delta} h_{\beta\lambda\gamma}), \\ Q_{\alpha\beta\gamma\delta} &= -\sum_\rho (h'_{\alpha\rho\gamma} h'_{\beta\rho\delta} - h'_{\alpha\rho\delta} h'_{\beta\rho\gamma}), \end{aligned} \tag{66}$$

which are the curvature functions of M^p, gM^q respectively. Put also

$$\begin{aligned} H_{\alpha\beta\gamma\delta\lambda\mu} &= h_{\alpha\lambda\gamma} h_{\beta\mu\delta} - h_{\alpha\lambda\delta} h_{\beta\mu\gamma}, \\ H'_{\alpha\beta\gamma\delta\rho\sigma} &= h'_{\alpha\sigma\gamma} h'_{\beta\rho\delta} - h'_{\alpha\sigma\delta} h'_{\beta\rho\gamma}, \\ K_{\alpha\beta\gamma\delta\rho\lambda} &= h_{\alpha\lambda\gamma} h'_{\beta\rho\delta} + h_{\beta\lambda\delta} h'_{\alpha\rho\gamma} - h_{\alpha\lambda\delta} h'_{\beta\rho\gamma} - h_{\beta\lambda\gamma} h'_{\alpha\rho\delta}, \end{aligned} \tag{67}$$

and

$$L_{\alpha\beta\gamma\delta} = -\sum_{a,\lambda,\mu} v_{a\lambda} v_{a\mu} H_{\alpha\beta\gamma\delta\lambda\mu} + \sum_{a,\rho,\sigma,\lambda} v_{a\rho} v_{a\lambda} u_{\sigma\lambda} H'_{\alpha\beta\gamma\delta\rho\sigma} - \sum_{a,\rho,\lambda} v_{a\rho} v_{a\lambda} K_{\alpha\beta\gamma\delta\rho\lambda}. \tag{68}$$

Then we find, after a little calculation,

$$S_{\alpha\beta\gamma\delta} = P_{\alpha\beta\gamma\delta} + Q_{\alpha\beta\gamma\delta} + L_{\alpha\beta\gamma\delta}. \tag{69}$$

It is to be observed that each set of the quantities $P_{\alpha\beta\gamma\delta}$, $Q_{\alpha\beta\gamma\delta}$, $L_{\alpha\beta\gamma\delta}$ satisfies the symmetry relations (6). Formula (69) expresses the curvature functions of M^{p+q-n} in terms of those of M^p and gM^q and the "mixed curvatures" $L_{\alpha\beta\gamma\delta}$.

6. Proof of the main formula. In order to prove our formula (12) we take the differential formula (47) with $k = p + q - n$ and carry out the integrations after multiplying both sides by $I_e(M^{p+q-n})$. Integration of the left-hand side obviously gives a constant multiple of $\int \mu_e(M^p \cap gM^q)\, dg$. However, (47) is established under the assumption $\Delta \neq 0$, and a crucial question is that of the convergence of the integral at the right-hand side. By definition $v_{b\sigma}$ is equal to a cofactor of the matrix $(u_{\rho a})$ divided by Δ. From (68) and (61) it is seen that $L_{\alpha\beta\gamma\delta}$ is equal to $1/\Delta^2$ times a quantity which is bounded near $\Delta = 0$. The integrand is therefore equal to $\Delta^{p+q-n+1-e}$ times a bounded quantity, and is itself bounded near $\Delta = 0$, since $e \leqq p + q - n$. Since the set of g for which $\Delta = 0$ is of measure zero in the group of proper motions in E^n, it suffices to evaluate the integral at the right-hand side under the assumption $\Delta \neq 0$.

To integrate the right-hand side in question (*i.e.*, the right-hand side of (47) multiplied by $I_e(M^{p+q-n})$), we first keep fixed a frame in the tangent space T_x to M^{p+q-n} at x. H_y is isomorphic to $SO(2n - p - q)$. But the integrand depends only on the linear spaces spanned by e_a and e'_ρ respectively. Thus it suffices to integrate over the open subset defined by $\Delta \neq 0$ in the Grassmann manifold of all $(n - q)$-dimensional subspaces spanned by e'_ρ in $(T_x)^\perp$, the space of dimension $2n - p - q$ through x and orthogonal to T_x.

Actual evaluation of the integral seems to be difficult. In getting its expression we could be aided by the first main theorem on vector invariants ([7], p. 53). In fact, we observe the following:

1) The integral over H_ν, to be denoted by J, is a polynomial in $h_{\alpha\lambda\beta}$, $h'_{\alpha\rho\beta}$, and is a linear combination of summands of the form

$$\sum \delta\begin{pmatrix}\alpha_1 & \cdots & \alpha_e\\ \beta_1 & \cdots & \beta_e\end{pmatrix} h_{\alpha_1\lambda_1\beta_1} \cdots h_{\alpha_t\lambda_t\beta_t} h'_{\alpha_{t+1}\rho_{t+1}\beta_{t+1}} \cdots h'_{\alpha_e\rho_e\beta_e}.$$

2) J is invariant under the transformations

$$h_{\alpha\lambda\beta} \to \sum_\mu s_{\lambda\mu} h_{\alpha\mu\beta},$$

$$h'_{\alpha\rho\beta} \to \sum_\sigma s_{\rho\sigma} h'_{\alpha\sigma\beta},$$

where $(s_{\lambda\mu})$, $(s_{\rho\sigma})$ are arbitrary orthogonal matrices of orders $n - p$, $n - q$ respectively.

By the first main theorem on vector invariants it follows from 2) that J is equal to a polynomial in

$$\sum_\lambda h_{\alpha\lambda\beta} h_{\gamma\lambda\delta}, \qquad \sum_\rho h'_{\alpha\rho\beta} h'_{\gamma\rho\delta}.$$

By 1) we see further that J must be a linear combination of the mixed curvatures $I(f, g)$, $0 \leqq f, g \leqq e$, $f + g = e$, introduced in (49), with numerical coefficients. Application of (50) then gives (12).

It remains to determine the constants c_i in (12). This will be achieved by carrying out the integration in a special case, *i.e.*, the case when both M^p and M^q are spheres.

7. Determination of the constants. We wish to evaluate directly the integral

$$A_e = \int \mu_e(S^p \cap gS^q)\, dg, \tag{70}$$

where S^p, S^q are spheres of dimensions p, q and radii 1, a respectively. We begin by a special case:

Lemma. *Let S_0^{m-1} be a fixed unit sphere in E^m, and let S^{m-1} be the sphere of center x and radius R. Let*

$$B_e = \int \mu_e(S_0^{m-1} \cap S^{m-1})\, dx, \tag{71}$$

where dx is the volume element in E^m. Then

$$B_e = \frac{2O_m O_{m-1}}{O_{m-e}} R^{m-e-1} \sum_{\substack{0 \leqq 2\lambda+\mu \leqq e/2\\ 0 \leqq \lambda, \mu}} \frac{\left(\frac{e}{2}\right)!}{(2\lambda)!\,\mu!\left(\frac{e}{2} - 2\lambda - \mu\right)!} 2^{2\lambda} \frac{O_{2\lambda+m-e+1}}{O_{2\lambda+1}} R^{2\mu+2\lambda}. \tag{72}$$

The right-hand side is a polynomial of degree $m - 1$ in R, and we will introduce the constants b_{ij} according to the equation

$$B_e = b_{e,m-e-1}R^{m-e-1} + \cdots + b_{e,m-1}R^{m-1}. \tag{73}$$

To prove the lemma we suppose, without loss of generality, that $R < 1$. Let y be the distance of x from the center of S_0^{m-1}. Then the radius of the sphere $S^{m-1} \cap S_0^{m-1}$ is

$$\{4y^2 - (1 - R^2 + y^2)^2\}^{1/2}/2y,$$

and

$$\mu_e(S_0^{m-1} \cap S^{m-1}) = O_{m-1}\left\{\frac{4y^2 - (1 - R^2 + y^2)^2}{4y^2}\right\}^{(m-2-e)/2}$$

Since dx is equal to $y^{m-1}\, dy$ times the volume element of the unit sphere in E^m, we get

$$B_e = \frac{O_{m-1}O_m}{2^{m-e-2}} \int_{1-R}^{1+R} y^{e+1}\{4y^2 - (1 - R^2 + y^2)^2\}^{(m-e-2)/2}\, dy.$$

This integral is elementary and can be evaluated. In fact, put

$$t = y^2 - 1 - R^2.$$

We get

$$B_e = \frac{O_{m-1}O_m}{2^{m-e-1}} \int_{-2R}^{+2R} (t + 1 + R^2)^{e/2}(4R^2 - t^2)^{(m-e-2)/2}\, dt.$$

Expanding the first factor under the integral sign by the binomial theorem, we see that we will be concerned with integrals of the form

$$J(k, l) = \int_{-\alpha}^{\alpha} u^k(\alpha^2 - u^2)^{l/2}\, du. \tag{74}$$

Elementary calculus gives

$$\int_0^{\alpha} u^k(\alpha^2 - u^2)^{l/2}\, du = \frac{O_{k+l+3}}{O_{k+1}O_{l+2}}\alpha^{k+l+1}, \tag{75}$$

and hence

$$\begin{aligned} &J(k, l) = 0, \quad \text{if } k \text{ is odd},\\ &J(k, l) = \frac{2O_{k+l+3}}{O_{k+1}O_{l+2}}\alpha^{k+l+1}, \quad \text{if } k \text{ is even}. \end{aligned} \tag{76}$$

Formula (72) then follows by elementary reductions.

We now proceed to the evaluation of A_e. Let E^{p+1}, E^{q+1} be respectively the linear spaces which contain S^p, gS^q, and let x be the center of gS^q. Then gS^q

can be identified with the pair $\{x, E^{q+1} \mid x \varepsilon E^{q+1}\}$, and we have the fiberings

$$(77) \qquad G \xrightarrow{\pi} \{x, E^{q+1} \mid x \varepsilon E^{q+1}\} \xrightarrow{\psi} Gr,$$

where G is the space of all frames $xe_1 \cdots e_n$ in E^n and Gr is the Grassmann manifold of all $(q + 1)$-dimensional linear spaces in E^n; π is defined by taking E^{q+1} to be spanned by $e_{n-q}, \cdots, e_n$. The fibers of the first fibering are isomorphic to the group of isotropy of S^q and have the dimension

$$\{q(q + 1) + (n - q - 2)(n - q - 1)\}/2$$

and total volume

$$O_{q+1} \cdots O_2 O_{n-q-1} \cdots O_2 .$$

It follows that

$$A_e = 2O_{q+1} \cdots O_2 O_{n-q-1} \cdots O_2 \int \mu_e(S^p \cap gS^q)\, d(x, E^{q+1}),$$

where $d(x, E^{q+1})$ is the density in the homogeneous space of all $\{x, E^{q+1} \mid x \varepsilon E^{q+1}\}$. The factor 2 before the integral arises from the fact that our E^{q+1} is unoriented.

Let Gr^* be the subset of all E^{q+1}, such that

$$\dim (E^{p+1} \cap E^{q+1}) = p + q - n + 2.$$

Gr^* differs from Gr by a subset of measure zero, and it suffices to calculate the integral over $\psi^{-1}(Gr^*)$. Let $E^{p+1} \cap E^{q+1} = E^{p+q-n+2}$, and let H be the space of all $E^{n+q-n+2}$ in E^{p+1}. Then we have the natural mapping

$$\psi^{-1}(Gr^*) \rightarrow H,$$

and we calculate the integral by iteration with $E^{p+q-n+2}$ fixed. It suffices to restrict to those $E^{p+q-n+2}$, whose intersection with S^p is non-empty. The intersection $E^{p+q-n+2} \cap S^p = S^{p+q-n+1}$ is then a sphere of dimension $p + q - n + 1$ and radius r, $0 \leqq r \leqq 1$.

The density $d(x, E^{q+1})$ can be factored as in (20), with E^p replaced by E^{q+1}. We fix E^{q+1} and integrate over the volume element $dx(E^{q+1})$ of $x \varepsilon E^{q+1}$. Let $E^{n-p-1} = (E^{p+q-n+2})^\perp$ be the orthogonal complement of $E^{p+q-n+2}$ through the center of $S^{p+q-n+1}$. A point $x \varepsilon E^{q+1}$ can be coordinatized by its orthogonal projections $x_1 \varepsilon E^{p+q-n+2}$ and $x_2 \varepsilon E^{n-p-1}$, and

$$dx\,(E^{q+1}) = dx_1\,(E^{p+q-n+2})\, dx_2\,(E^{n-p-1}),$$

where the right-hand members are the volume elements of x_1, x_2 in their respective spaces. Let s be the distance from x to $E^{p+q-n+2}$. The intersection $S_1^{p+q-n+1} = gS^q \cap E^{p+q-n+2}$ is then a sphere of radius $\rho = (a^2 - s^2)^{1/2}$, and $dx_2(E^{n-p-1})$ is equal to $s^{n-p-2}\, ds$ times the volume element of the unit sphere in E^{n-p-1}. It follows that

$$A_e = O_{q+1} \cdots O_1 O_{n-q-1} \cdots O_2 O_{n-p-1}$$

$$\cdot \int dE^{q+1} \int_0^a s^{n-p-2}\, ds \int \mu_e(S^{p+q-n+1} \cap S_1^{p+q-n+1})\, dx_1\, (E^{p+q-n+2}).$$

The last integral has been evaluated in the lemma, which gives

$$\int \mu_e(S^{p+q-n+1} \cap S_1^{p+q-n+1})\, dx_1\, (E^{p+q-n+2}) = \sum_{\substack{0 \le i \le e \\ i \text{ even}}} b_{e,p+q-n+1-i} r^{p+q-n+1+i-e} \rho^{p+q-n+1-i}.$$

By (75) we get

$$\int_0^a s^{n-p-2}\, ds \int \mu_e(S^{p+q-n+1} \cap S_1^{p+q-n+1})\, dx_1\, (E^{p+q-n+2})$$

$$= \frac{1}{O_{n-p-1}} \sum_{\substack{0 \le i \le e \\ i \text{ even}}} b_{e,p+q-n+1-i} \frac{O_{q-i+2}}{O_{p+q-n+3-i}} r^{p+q-n+1-e+i} a^{q-i}.$$

It remains to integrate with respect to dE^{q+1}. This is facilitated by the observation that the integrand depends only on $E^{p+q-n+2}$. Applying (32), we get

$$A_e = \frac{O_{n+1} \cdots O_1 O_{p+q-n+3} \cdots O_1 O_{n-q-1} \cdots O_1}{O_1 O_{q+2} O_{p+2} \cdots O_1} \sum_{0 \le i \le e} b_{e,p+q-n+1-i} \frac{O_{q-i+2}}{O_{p+q-n+3-i}} a^{q-i}$$

$$\cdot \int r^{p+q-n+1-e+i}\, dE^{p+q-n+2}(E^{p+1}).$$

To integrate with respect to $dE^{p+q-n+2}(E^{p+1})$ let u be the distance of $E^{p+q-n+2}$ from the center of S^p, so that $r = (1 - u^2)^{1/2}$. Then, by (17) and (18),

$$dE^{p+q-n+2}(E^{p+1}) = dE^{p+q-n+2}(O) u^{n-q-2}\, du\, d\omega,$$

where $dE^{p+q-n+2}(O)$ is the volume element in the space of all $E^{p+q-n+2}$ through the center of S^p and $d\omega$ is the volume element on the unit sphere in the orthogonal complement $(E^{p+q-n+2})^\perp$ of $E^{p+q-n+2}$ through the center of S^p and in E^{p+1}. Performance of the integration gives the final results

$$(78) \qquad A_e = \frac{O_{n+1} \cdots O_1 O_{p+q-n+3}}{O_1 O_{p+2} O_{q+2}} \sum_{0 \le i \le e} \frac{O_{q+2-i} O_{p+2-e+i}}{O_{p+q-n+3-i} O_{p+q-n+3-e+i}} b_{e,p+q-n+1-i} a^{q-i}.$$

Comparing (12) and (78) and recalling that

$$\frac{1}{O_{p+1}} \mu_i(S^p) = 1, \qquad \frac{1}{O_{q+1}} \mu_i(S^q) = u^{q-i},$$

we get

$$(79) \qquad c_{e-i} = \frac{O_{n+1} \cdots O_1 O_{p+q-n+3} O_{q+2-i} O_{p+2-e+i}}{O_1 O_{p+2} O_{p+1} O_{q+2} O_{q+1} O_{p+q-n+3-i} O_{p+q-n+3-e+i}} b_{e,p+q-n+1-i},$$

where the b's are given by (72) and (73), with $m = p + q - n + 2$. Thus the exact values of the constants c_i in (12) are determined.

A particular case of (12) is the formula

$$\int \mu_0(M^p \cap gM^q)\, dg = \frac{O_{n+1} \cdots O_2 O_{p+q-n+1}}{O_{p+1} O_{q+1}} \mu_0(M^p)\mu_0(M^q), \tag{80}$$

which was first given by Santalo [4].

8. Integration over linear space; reproductivity of the invariants. It will be natural to consider the intersections $M^p \cap E^q$ and integrate over dE^q. By considerations much less elaborate than the above, we can derive the following integral formula

$$\int \mu_e(M^p \cap E^q)\, dE^q = \frac{O_{n+1} \cdots O_{n-q+1} O_{p+q-n+2} O_{p+q-n+1} O_{p+2-e}}{O_{q+1} \cdots O_1 O_{p+2} O_{p+1} O_{p+q-n+2-e}} \mu_e(M^p), \tag{81}$$

where M^p is a compact submanifold of E^n and $e \leqq p + q - n$. We will not go into the details here.

In conclusion we wish to remark that it would be of interest to characterize the invariants $\mu_e(M^p)$ by geometrical properties. One of them will perhaps be the "reproductive property" expressed by (81).

References

[1] Blaschke, W., *Integralgeometrie*, 3te Auflage, Berlin, 1955.

[2] Chern, S., On the curvatura integra in a riemannian manifold, *Ann. of Math.*, **46** (1945) 674–684.

[3] Chern, S., On the kinematic formula in the euclidean space of n dimensions, *Amer. J. Math.*, **74** (1952) 227–236.

[4] Santalo, L. A., *Geometria integral en espacios de curvatura constante*, Publicaciones de la Comision Nacional de la Energia Atomica, Buenos Aires, 1952.

[5] Santalo, L. A., *Introduction to integral geometry*, Paris, Hermann, 1953.

[6] Thorpe, J. A., On the curvatures of riemannian manifolds, to appear in *Illinois J. Math.*

[7] Weyl, H., *Classical groups*, Princeton, 1939.

[8] Weyl, H., On the volume of tubes, *Amer. J. Math.*, **61** (1939) 461–472.

University of California, Berkeley

Date Communicated: March 2, 1966

COMPLETE MINIMAL SURFACES IN EUCLIDEAN n-SPACE(*)

By

SHIING-SHEN CHERN AND ROBERT OSSERMAN

in Berkeley, California, U. S. A. *in Stanford, California, U. S. A.*

1. Introduction and statement of results.

We begin by recalling some basic facts concerning minimal surfaces in E^n the euclidean n-space. For further details, we refer to Chern [3] and Osserman [7].

Let S_0 be a Riemann surface and let $\alpha_1, \cdots, \alpha_n$ be analytic differentials on S_0, which we assume to be not all identically zero. Suppose that in terms of a local parameter ζ, we have $\alpha_k = \phi_k d\zeta$, $1 \leqq k \leqq n$. Then under the condition

$$(1) \qquad \sum_{1 \leqq k \leqq n} \phi_k^2 \equiv 0,$$

the surface $x(p): S_0 \to E^n$, defined by

$$(2) \qquad x_k = \operatorname{Re} \int \alpha_k,$$

assuming these integrals to have single-valued real parts, is called a *generalized minimal surface.* If furthermore

$$(3) \qquad \sum_{1 \leqq k \leqq n} |\phi_k|^2 \neq 0,$$

* Partial support to this work has been given to the first author by NSF grant GP 3990 and ONR contract 3656(14) and to the second author by Army Research Office grant 31–124–ARO(D)–170.

then the surface is called a *regular minimal surface*. We may note that in any case, equation (3) must hold everywhere except for at most isolated points. We shall use the expression *minimal surface* in the generalized sense, and add the qualification "regular" in those cases where condition (3) is imposed.

Let us remark that we are considering here only orientable minimal surfaces. For non-orientable surfaces one may always take the two-sheeted orientable covering surface, and deduce results for the original surface from those of the covering surface.

If we set $\zeta = \xi_1 + i\xi_2$, then we have

$$\phi_k = \frac{\partial x_k}{\partial \xi_1} - i\frac{\partial x_k}{\partial \xi_2}, \tag{4}$$

and if we denote by

$$g_{ij} = \frac{\partial x}{\partial \xi_i} \cdot \frac{\partial x}{\partial \xi_j}, \qquad 1 \leqq i,j \leqq 2, \tag{5}$$

the coefficients of the first fundamental form of the surface (2), then condition (1) becomes

$$g_{11} - g_{22} - 2ig_{12} = 0, \tag{6}$$

which means that ξ_1, ξ_2 are isothermal parameters. This condition may also be written in the form

$$g_{ij} = \lambda^2 \delta_{ij}, \qquad \lambda = \lambda(\zeta), \tag{7}$$

where

$$\lambda^2 = \left|\frac{\partial x}{\partial \xi_1}\right|^2 = \left|\frac{\partial x}{\partial \xi_2}\right|^2 = \frac{1}{2} \sum_{1 \leqq k \leqq n} |\phi_k|^2. \tag{8}$$

Thus, if a curve is defined in local coordinates on S_0 by $\zeta(t)$, $a \leqq t \leqq b$, then its image in E^n will have length

$$\int \lambda |d\zeta| = \int_a^b \lambda \left|\frac{d\zeta}{dt}\right| dt.$$

A path $p(t), 0 \leqq t < 1$, on S_0 is called *divergent* if for every compact set Q on S_0, there exists $t_0 < 1$ such that $p(t) \notin Q$ for $t > t_0$. The surface defined by (2) is *complete* if the image in E^n of every divergent path on S_0 has infinite length. Our main object of study will be complete minimal surfaces, and we shall show in particular that essentially all the results proved in [7] for the case $n = 3$ may be extended to arbitrary n.

Our principal tool in this investigation will be the *generalized Gauss map.* This is the map

$$G: S_0 \to Q_{n-2} \subset P^{n-1}(C), \tag{9}$$

where $P^{n-1}(C)$ is the complex projective space with the homogeneous coordinates $z_1, \cdots, z_n$, and Q_{n-2} is the quadric

$$z_1^2 + \cdots + z_n^2 = 0, \tag{10}$$

and the map (9) is defined by setting

$$z_k = \overline{\phi_k(\zeta)}, \qquad 1 \leqq k \leqq n. \tag{11}$$

If, in terms of another local parameter η, we have $\alpha_k = \psi_k d\eta$, $1 \leqq k \leqq n$, then $\phi_k = \psi_k d\eta / d\zeta$, and $(\bar{\phi}_1, \cdots, \bar{\phi}_n)$ and $(\bar{\psi}_1, \cdots, \bar{\psi}_n)$ define the same point in the projective space. By (1) this point lies on Q_{n-2}. The only difficulty may arise at points where (3) fails to hold, but if such a point corresponds to $\zeta = \zeta_0$, then one of the ϕ_k has a zero of the smallest order, say v, and factoring out $(\zeta - \zeta_0)^v$ we see that the map (9) extends continuously (and analytically) to these points also.

The quadric Q_{n-2} may be identified with the Grassmannian $G_{n,2}$ of oriented planes through the origin 0 of E^n. If $z_k = v_k + i w_k$, then to the point $z = (z_1, \cdots, z_n)$ corresponds the plane spanned by the vectors v, w, where by (10), v and w are orthogonal vectors of equal lengths. Thus the map G assigns to each point p of S_0 the plane through 0 parallel to the tangent plane at $x(p)$, spanned by $\partial x / \partial \xi_1$ and $\partial x / \partial \xi_2$.

If we denote by $\bar{G}$ the map of S_0 into $P^{n-1}(C)$ obtained by taking the complex conjugate $\bar{z}_k$, of each coordinate, then we obtain precisely an "analytic curve" in the terminology of Weyl [9]. The book of Weyl is devoted to a study of the subject from the point of view of "value distributions", i.e., the study of the image $\bar{G}(S_0)$ relative to the hyperplanes of $P^{n-1}(C)$. It develops the ideas and methods initiated by Ahlfors [1] in the case where S_0 is the complex plane; the work of Ahlfors was in turn inspired by an earlier paper of J. and H. Weyl. In his fundamental paper Ahlfors expressed the regret that applications were lacking. We hope to present here an application of this theory.

We have seen that each minimal surface gives rise to an analytic curve whose image lies in Q_{n-2}. Conversely, to each such curve corresponds by (2) a minimal surface in E^n. Note, however, that the surface is not determined uniquely, since if f is an arbitrary analytic function on S_0 (not identically zero), and if we set

$$\tilde{x}_k = \operatorname{Re} \int \beta_k , \qquad \beta_k = f\alpha_k ,$$

then we get a different minimal surface with the same Gauss map. All the surfaces so obtained will have the same tangent plane corresponding to a given point of S_0. They will also have other geometrical properties in common, as we shall see in the following.

The gaussian curvature K of the surface (2) at points where (3) holds is given in terms of the local parameter ζ by

$$K = -\frac{\Delta \log \lambda}{\lambda^2}. \tag{12}$$

Using (8), we find the expression

$$K = -\frac{4|\phi \wedge \phi'|^2}{|\phi|^6}, \tag{13}$$

where

$$\phi = (\phi_1, \cdots, \phi_n), \qquad |\phi|^2 = \sum_{1 \leq k \leq n} |\phi_k|^2 , \tag{14}$$

and

(15) $$|\phi \wedge \phi'|^2 = \sum_{1 \leqq j < k \leqq n} |\phi_j \phi'_k - \phi'_j \phi_k|^2 \quad .$$

Let D be a domain in the ζ-plane, and denote by S the corresponding part of the surface (2). Then the *area* of S is

(16) $$A(S) = \iint_D \lambda^2 d\xi_1 d\xi_2 = \frac{1}{2} \iint_D |\phi|^2 d\xi_1 d\xi_2,$$

and the *total curvature* of S is

(17) $$C(S) = \iint_D K dA = \iint_D K\lambda^2 d\xi_1 d\xi_2 = -2 \iint_D \frac{|\phi \wedge \phi'|^2}{|\phi|4} d\xi_1 d\xi_2.$$

If, instead of (2) we consider the surface $\tilde{S}$ defined by $\tilde{x}_k = \operatorname{Re} \int \beta_k$, where $\alpha_k = f\beta_k$, or $\phi_k(\zeta) = f(\zeta)\tilde{\phi}_k(\zeta)$, for some function $f(\zeta)$ analytic in D, then $\phi'_k = f\tilde{\phi}'_k + f'\tilde{\phi}_k$, and $\phi \wedge \phi' = f^2(\tilde{\phi} \wedge \tilde{\phi}')$, since $\tilde{\phi} \wedge \tilde{\phi} = 0$. Thus

$$|\phi \wedge \phi'|^2 / |\phi|^4 = |\tilde{\phi} \wedge \tilde{\phi}'|^2 / |\tilde{\phi}|^4,$$

and $C(S) = C(\tilde{S})$. In other words, the total curvature depends only on the Gauss map. In particular, if the surface is not regular, so that (3) fails to hold at some point, we may suppose that this point corresponds to $\zeta = 0$, and we may set $\phi_k = \zeta^\nu \tilde{\phi}_k$, $1 \leqq k \leqq n$, where $\tilde{\phi}_k(0) \neq 0$ for some k. Then the above calculation shows that even though the gaussian curvature may become infinite at a singular point, the total curvature over a compact domain remains finite, and in fact equals that of the regular surface defined by $\tilde{\phi}$.

The relation between the Gauss map and the total curvature can be made explicit by introducing the canonical riemannian metric on $G_{n,2}$; this may be obtained as the metric induced on Q_{n-2} by the standard hermitian metric on $P^{n-1}(C)$. The latter is defined up to a multiplicative factor by

$$|z \wedge dz|^2 / |z|^4.$$

This factor is most frequently chosen to be 1 or 4, but we shall set

(18) $$d\hat{s}^2 = 2\frac{|z \wedge dz|^2}{|z|^4},$$

so that in the case $n = 3$ the induced metric on Q_1 becomes that of the unit sphere, and the generalized Gauss map reduces to the classical one.

If D is a domain in the ζ-plane, and $\hat{S}$ is the image surface under the Gauss map (9), then the metric (18) takes the form

(19) $$d\hat{s}^2 = \hat{\lambda}^2 |d\zeta|^2, \qquad \hat{\lambda}^2 = 2\frac{|\phi \wedge \phi'|^2}{|\phi|^4},$$

and the area of $\hat{S}$ with respect to this metric is

(20) $$A(\hat{S}) = \iint_D \hat{\lambda}^2 d\xi_1 d\xi_2 = -C(S),$$

by (17). Equation (20) expresses the important property of the Gauss map: *the total curvature of any portion of the surface equals the negative of the area of its image under the Gauss map,* this area being computed in the metric (18).

We next use integral geometry to relate the area of the image to the "value distribution" of the analytic curve (9). Let us consider an arbitrary hyperplane

(21) $$H: \sum_{1 \leq k \leq n} a_k z_k = 0$$

in $P^{n-1}(C)$. To this hyperplane corresponds the analytic differential

(22) $$\omega = \sum_{1 \leq k \leq n} \bar{a}_k \alpha_k$$

on S_0. There are two possibilities. Either $\omega \equiv 0$, in which case the image of S_0 under the Gauss map lies entirely in the hyperplane H, or else the zeroes of ω are isolated and each of them has a given order. This order is called the *order of intersection* of $G(S_0)$ with H at the corresponding point. At a point where all the α_k vanish simultaneously one must first factor out the common zero, and then count the order of the zero of ω divided by this factor. For any compact subdomain Q of S_0, we may therefore define an integer $n(Q,a)$, *the total order of intersection of* $G(S_0)$ *with* H *at points of* Q, equal to the total order of zeroes of ω in Q (provided $G(S_0) \not\subset H$).

The space of all hyperplanes in $P^{n-1}(C)$ has a measure, to be denoted by da, which is invariant under the isometries of $P^{n-1}(C)$ with its standard hermitian metric. We normalize da so that the total measure

$$\int da = 1. \tag{23}$$

Then the area of the image $\hat{Q} = G(Q)$ can be expressed by the formula [8]:

$$A(\hat{Q}) = 2\pi \int n(Q,a)da, \tag{24}$$

where the integration is extended over all the hyperplanes not containing $G(Q)$ of $P^{n-1}(C)$.

Definition. The Gauss map (9) is called *algebraic* if the surface S_0 is conformally equivalent to a region D on a compact Riemann surface W, and if, when the differentials α_k are considered as analytic differentials on D, the ratios α_k/α_m extend to meromorphic functions on W, whenever $\alpha_m \not\equiv 0$. The Gauss map is called *degenerate* if the image $G(S_0)$ lies in a hyperplane of $P^{n-1}(C)$; otherwise it is called *non-degenerate.*

Theorem 1. *Let S be a minimal surface defined by a map* (2) *on a Riemann surface* S_0 *.Consider the following conditions*:

a) *S has finite total curvature*;

b) *there exists an integer N such that the image of* S_0 *under the Gauss map intersects at most N times all hyperplanes which do not contain it*;

c) *the Gauss map of* S_0 *is algebraic*;

d) *the surface* S_0 *is conformally equivalent to a compact surface W punctured at a finite number of points* $p_1, \cdots, p_r$, *and the differentials* α_k *are either regular or have a pole at each* p_j.

For an arbitrary surface S, d) $\Rightarrow$ c) $\Rightarrow$ b) $\Rightarrow$ a). *If S is a complete regular minimal surface, then all four conditions are equivalent.*

Theorem 2. *Let S be a complete regular minimal surface with Euler characteristic χ and r boundary components. Then*

$$C(S) \leqq 2\pi(\chi - r). \tag{25}$$

Theorem 3. *Let S be a simply-connected complete regular minimal surface in E^n whose Gauss map is non-degenerate. Then*

$$C(S) \leqq -2\pi(n-1). \tag{26}$$

Equality is attained for the surfaces obtained by setting

$$\phi_k = \begin{cases} a_m(\zeta^{m-1} + \zeta^{n-m}), & k = 2m-1, \\ ia_m(\zeta^{m-1} - \zeta^{n-m}), & k = 2m, \end{cases} \tag{27}$$

where, for $n = 2\nu$, $m = 1, \cdots, \nu$, and $\sum_{1 \leqq m \leqq \nu} a_m^2 = 0$; and for $n = 2\nu + 1$, $m = 1, \cdots, \nu$, $\phi_\nu = 2a_{\nu+1}\zeta^\nu$, and $\sum_{1 \leqq m \leqq \nu+1} a_m^2 = 0$. In both cases we assume $a_m \neq 0$ for all m.

The surfaces defined by (27) are complete regular minimal surfaces which are at the same time rational algebraic surfaces of order n^2. In the case $n = 3$, it reduces to Enneper's surface, which is the unique simply-connected complete minimal surface in E^3 for which equality holds in (26) [7, p. 360].

The classical Bernstein theorem is essentially concerned with the Gauss map: the size of its image and its equidistribution. When the minimal surface is of finite total curvature, a rather precise statement can be made. For this purpose we need following definition.

Definition. A set of hyperplanes in an m-dimensional projective space is *in general position* if each subset of k hyperplanes, with $k \leqq m$, has an $(m-k)$-dimensional intersection.

Theorem 4. *Let S be a complete regular minimal surface of finite total curvature in E^n, with the Gauss map defined by* (9). *Suppose that S is not a plane. Let L be the linear space of least dimension $m-1$ $(m \leqq n)$*

of $P^{n-1}(C)$, which contains $G(S_0)$. Then $G(S_0)$ can fail to intersect at most $(m-1)(m+2)/2$ hyperplanes in general position in L.

Even if S is not of finite total curvature, assertions can be made on the equidistribution of the image $G(S_0)$ under the Gauss map. We make the following definitions:

Definition. A point set D in $P^{n-1}(C)$ will be called *bounded* if the complement of D contains a hyperplane $\sum_{1 \leq k \leq n} a_k z_k = 0$, and if there exists $\varepsilon > 0$ such that for all points of D,

$$|\Sigma a_k z_k|^2 / \Sigma |z_k|^2 \geqq \varepsilon. \tag{28}$$

A subset D of $Q_{n-2} \subset P^{n-1}(C)$ is *of bounded type* if it does not intersect some hyperplane $\Sigma a_k z_k = 0$, and if there exists a pluriharmonic function h defined in D such that in D,

$$\log \Sigma |z_k|^2 - \log |\Sigma a_k z_k|^2 \leqq h. \tag{29}$$

Obviously a bounded set in Q_{n-2} is of bounded type, since we may choose for the function h simply the constant $-\log \varepsilon$. It would be interesting to have a geometrical description of sets of bounded type. For the case $n=3$, it follows from Lemma 2.1 of [7] that a set D lying on the quadric Q_1 is of bounded type if and only if $Q_1 - D$ has positive logarithmic capacity (where Q_1 is identified with the Riemann sphere).

Theorem 5. *Let S be a complete regular minimal surface defined by a map $x(p): S_0 \to E^n$. Suppose that S_0 is not conformally equivalent to a compact surface punctured at a finite number of points. Then for any compact subset R of S_0, the image of $S_0 - R$ under the Gauss map cannot lie in a set of bounded type.*

In particular, under the hypotheses, the image of $S_0 - R$ cannot lie in a bounded subset, and hence will intersect an everywhere dense set of hyperplanes.

Comparing the results of Theorem 4 and 5 and Lemma 3 (See below) we see that the hypotheses cover all possible cases of complete regular minima surfaces. We then have as special cases the two following results proved earlier.

Corollary 1. [6,7]. *If S is a complete regular minimal surface in E^3, then either S is a plane, or else the normals assume all directions with the possible exception of a set of capacity zero.*

Corollary 2. [3]. *If S is a complete regular minimal surface in E^n, then either S is a plane, or else the image of S under the Gauss map intersects a dense set of hyperplanes.*

2. Proofs of Theorems 1 and 2.

We begin by the lemma:

Lemma 1. *Let S be a minimal surface whose Gauss map G is algebraic. Then G extends to an antiholomorphic map of a compact surface W, and the image has a fixed finite order of intersection with every hyperplane which does not contain it.*

Proof. Let the surface S be defined by (2), where $\alpha_m \not\equiv 0$. Then

$$(\alpha_1, \cdots, \alpha_n) = \alpha_m(g_1, \cdots, g_n),$$

where $g_k = \alpha_k/\alpha_m$ extends to a meromorphic function on a compact surface W. The map $W \to P^{n-1}(C)$ defined by

$$p \to (\bar{g}_1(p), \cdots, \bar{g}_n(p))$$

extends the Gauss map to the whole surface W. It is understood that at a point where one or more of the g_k have a pole, one must factor out the highest order pole at the point, thus defining the map smoothly in a neighborhood of the point. Similarly, the order of intersection of the image with a given hyperplane $\Sigma a_k z_k = 0$ at a given point is determined by first factoring out the highest order pole or the highest order common zero of the g_k at the point, and then counting the order of the zero of $\Sigma \bar{a}_k g_k$ divided by that factor. Since $\Sigma \bar{a}_k g_k$, if not identically zero, is meromorphic on W, it follows that it has the same number of zeroes and poles, and hence the order of intersection with an arbitrary hyperplane $\Sigma a_k z_k = 0$ is the same as with the fixed hyperplane $z_m = 0$. But the latter is a finite number, since it can only occur at the poles of one of the g_k.

Proof of Theorem 1. That d) $\Rightarrow$ c) is obvious, and that c) $\Rightarrow$ b) follows immediately from Lemma 1. If b) holds, then by (20), (23), and (24), we have $-2\pi N \leqq C(S) \leqq 0$, so that a) holds. We shall prove that a) $\Rightarrow$ d) when S is complete and regular.

In this case it follows from a theorem of A. Huber [5, p. 61] that S_0 is finitely connected and is hence conformally equivalent to a compact Riemann surface from which a finite number of continua have been removed. Again using the property of finite total curvature, it follows from another theorem of Huber ([5, Theorem 15]; see also Finn [4, p. 27]) that each of these continua reduces to a point. Let us choose a doubly-connected neighborhood of one of these points so that it corresponds to the region $D: |\zeta| < \frac{1}{2}$ of the complex ζ -plane. We may set $\alpha_k = \phi_k(\zeta)d\zeta$ in this region, and the metric takes the form (8). Using the fact that the gaussian curvature is nowhere positive and the total curvature is finite, we may set

$$u(\zeta) = \frac{1}{2\pi} \iint_D \log \frac{1}{|w - \zeta|} \Delta \log \lambda(w) du_1 du_2, \qquad w = u_1 + iu_2, \tag{30}$$

and we have

$$\Delta \log \lambda \geqq 0, \qquad \log|w - \zeta| \leqq 0,$$

hence $u(\zeta) \geqq 0$ in D. By the Poisson formula, $\Delta u = -\Delta \log \lambda$, so that the function $v = u + \log \lambda$ is harmonic in D. By the completeness of S and the fact that $v \geqq \log \lambda$, we have $\int_C e^v |d\zeta| = \infty$ for every path C in D which tends to the origin. But this implies, by a result of Finn [4, p. 28] that $v(\zeta) = \beta \log|\zeta| + h(\zeta)$, where β is a constant and $h(\zeta)$ is harmonic at the origin. We therefore have

$$\Sigma |\phi_k(\zeta)|^2 = 2\lambda^2 \leqq 2e^{2v} \leqq M|\zeta|^{2\beta},$$

so that the functions ϕ_k can have at most a pole at the origin. Thus condition d) holds, and the theorem is proved.

Corollary. *The total curvature of a complete regular minimal surface is either* $-\infty$ *or* $-2\pi N$ *for some integer* $N \geqq 0$.

Proof. Since $K \leqq 0$, the total curvature is either $-\infty$ or else finite. In the latter case, by combining condition d) with Lemma 1 and formula (24), one sees that the area of the image of the compact surface W under the extended Gauss map is equal to $2\pi N$. Since the removal of a finite number of points does not change the area, the result follows immediately from (20).

Remark. A similar result was known to hold in E^3, but in that case the total curvature was always of the form $-4\pi N$. The apparently weaker result in higher dimensions turns out to be the correct one, and the reason for the difference is easy to explain. In the case $n = 3$, the "hyperplanes" of $P^2(C)$ are projective lines, and each of these intersects Q_1 with order 2. Q_1 is itself one-(complex-)dimensional, the image of a compact surface W under an algebraic Gauss map will be an algebraic covering surface with a fixed number of points, say N, lying over each point of Q_1. Thus the image will intersect each projective line with order $2N$, and the total curvature will be $-4\pi N$. For $n > 3$, Q_{n-2} contains projective lines. Suppose, in fact, that the line joining $(b_1, \cdots, b_n)$ and $(c_1, \cdots, c_n)$ be on Q_{n-2}. By setting

$$\phi_k(\zeta) = b_k + \zeta^N c_k, \qquad 1 \leqq k \leqq n,$$

condition (1) will be fulfilled and the surface defined by (2) will be a minimal surface whose Gauss map is algebraic of order N, and whose total curvature is $-2\pi N$.

Lemma 2. *Let S be a minimal surface which satisfies condition d) of Theorem* 1. *Let ζ be a local parameter on W in a neighborhood of one of the p_j, such that p_j corresponds to $\zeta = 0$. Let $\alpha_k = \phi_k(\zeta) d\zeta$ for $0 < |\zeta| < \varepsilon$. If S is complete, then*

(31) $$\sum_{1 \leqq k \leqq n} |\phi_k(\zeta)|^2 \sim \frac{a}{|\zeta|^{2m}}, \qquad a > 0, \ m \geqq 2.$$

Proof. At least one of the ϕ_k must have a pole, since otherwise the left-hand side of (31) would be bounded at the origin and by (8), the surface would not be complete. We therefore need only show $m \geqq 2$ in (31). If not, then each ϕ_k would have at most a simple pole at $\zeta = 0$, so that for suitable c_k, $\phi_k(\zeta) - c_k/\zeta$ is analytic. Then by (2), $x_k - \mathrm{Re}\{c_k \log \zeta\}$ is harmonic, and since x_k is single-valued, c_k must be real. But equation (1) implies that $\Sigma c_k^2 = 0$, hence each $c_k = 0$. This contradicts the fact that at least one ϕ_k has a pole, and the lemma is therefore proved.

Proof of Theorem 2. If $C(S) = -\infty$, there is nothing to prove. Otherwise we apply Theorem 1, and condition d) holds. If W is a compact surface of genus g, then $\chi = 2 - 2g - r$. For each $j = 1, \cdots, r$ let m_j be the maximum order of the poles of α_k at p_j. It is easy to see that one can choose $a_1, \cdots, a_n$ so that $\omega = \sum_k a_k \alpha_k$ has a pole of order m_j at p_j. By the Riemann relation, if ω has N zeros on W, then $N - \Sigma m_j = 2g - 2$. By Lemma 2, $m_j \geqq 2$ for $j = 1, \cdots, r$. Hence $N \geqq 2r + 2g - 2 = r - \chi$, and

$$C(S) = -2\pi N \leqq 2\pi(\chi - r).$$

Remark. This result was known earlier for the case $n = 3$ [7].

3, Proof of Theorem 3; decomposability of minimal surfaces.

When the minimal surface (2) undergoes an orthogonal transformation

(32) $$x_j = \sum_k b_{jk} \tilde{x}_k, \qquad 1 \leqq j, k \leqq n,$$

where (b_{jk}) is a real orthogonal matrix, its Gauss image $G(S_0)$ undergoes the same transformation. To study a minimal surface whose Gauss map is degenerate, i.e., belongs to a hyperplane

(33) $$a_1 z_1 + \cdots + a_n z_n = 0,$$

a preliminary problem is to classify the hyperplanes (33) under real orthogonal transformations of the z_k. This is easy, but we will not discuss it here. There are, however, two cases of geometrical interest:

Case 1. The a_k, up to a non-zero multiple, are real. In this case the surface S lies in a hyperplane of E^n.

Case 2. The hyperplane (33) is tangent to the quadric Q_{n-2}, or analytically $\sum_k a_k^2 = 0$. In this case the orthogonal transformation (32) can be so chosen that $\tilde{x}_1, \tilde{x}_2$ are conjugate harmonic functions on S_0.

Related to the notion of the degeneracy of the Gauss map is that of decomposability, to be defined as follows:

Definition. The minimal surface defined by equations (1) and (2) is *decomposable* if there exists a real orthogonal transformation

$$\phi_j = \sum_k b_{jk}\tilde{\phi}_k, \qquad 1 \leqq j, k \leqq n, \tag{34}$$

such that

$$\sum_{1 \leqq k \leqq m} \tilde{\phi}_k^2 = 0 \text{ for some } m < n. \tag{35}$$

By (1) it follows that if (35) holds, then also

$$\sum_{m+1 \leqq k \leqq n} \tilde{\phi}_k^2 = 0. \tag{36}$$

Decomposability of a minimal surface and degeneracy of its Gauss map are in general different notions. They coincide in the two cases discussed above and in the case $n = 3$. For $n = 3$ both mean that S lies on a plane.

If the Gauss map is degenerate, there is a projective subspace L of lowest dimension in $P^{n-1}(C)$, which contains the image $G(S_0)$. The map $G: S_0 \to L$ will then be non-degenerate and will be called the *reduced Gauss map*.

We now give a proof of Theorem 3. If $C(S) = -\infty$, then (26) holds trivially. If $C(S) > -\infty$, we may apply Theorem 1 and deduce that S_0 is the sphere punctured at one point. We may take S_0 to be the entire ζ-plane, so that the functions $\phi_k(\zeta)$ are polynomials. If the maximum degree of the ϕ_k is equal to N, then $C(S) = -2\pi N$, so that (26) holds unless $N < n-1$. But in that case a linear combination of the ϕ_k will be identically zero, and the Gauss map will be degenerate. As for the functions defined by (27), it is a direct verification that $\sum_k \phi_k^2 \equiv 0$, so that these equations may be used to define a minimal surface. The condition that all a_m are non-zero guarantees that this surface will be regular and, as is easily seen, that the Gauss map is nondegenerate. Finally the surface so defined is obviously simply connected and complete.

4. Singularities of analytic curves; proof of Theorem 4.

The proof of Theorem 4 will be based on the Plücker formulas for an algebraic curve in a complex projective space; *cf* [9], pp. 41–65. Let W be a compact Riemann surface of genus g and let $f: W \to P^{m-1}(C)$ be a nondegenerate algebraic curve. For a suitable choice of homogeneous coordinates $\zeta_0, \cdots, \zeta_{m-1}$ in $P^{m-1}(C)$, the equations of the curve can be put locally into the normal form

$$(37) \qquad \begin{aligned} \zeta_0 &= t^{\delta_0} + \cdots \\ &\cdots\cdots\cdots \\ \zeta_{m-1} &= t^{\delta_{m-1}} + \cdots, \end{aligned}$$

where

$$(38) \qquad 0 = \delta_0 < \delta_1 < \cdots < \delta_{m-1},$$

and where t is a local parameter on W. The integers

$$(39) \qquad v_q = \delta_{q+1} - \delta_q - 1, \qquad 0 \leqq q \leqq m-2,$$

are called the *stationary indices of order* q at the point $t = 0$. The stationary points, i.e., points with a non-zero stationary index, are isolated and hence are finite in number. We will denote by σ_q the sum of all stationary indices of order q. Let v_q, $0 \leqq q \leqq m-2$, be the *order of rank* q of the algebraic curve; geometrically this is the order of the associated curve of rank q, i.e., the curve formed by the osculating spaces of dimension q. Then Plücker's formulas are

(40) $$\sigma_q = 2\nu_q - \nu_{q+1} - \nu_{q-1} + 2(g-1), \quad 0 \leqq q \leqq m-2,$$

with the convention $\nu_{-1} = \nu_{m-1} = 0$. From (40) it follows that

(41) $$\sum_{1 \leqq h \leqq m-1} (m-h)\sigma_{h-1} = m\nu_0' + m(m-1)(g-1),$$

which we will use in the proof of Theorem 4.

Proof of Theorem 4. Under the hypotheses it follows from Theorem 1 that S_0 is conformally equivalent to a compact Riemann surface W punctured at the points p_j, $1 \leqq j \leqq r$, and that the Gauss map G extends to an anti-holomorphic map of W into $P^{n-1}(C)$. Let L be the linear subspace of lowest dimension of $P^{n-1}(C)$ which contains the image $G(W)$ and let $m-1$ be the dimension of L. Then $1 < m \leqq n$, and $G(W)$ is a non-degenerate algebraic curve in L to which the above considerations apply. Let Π_i, $1 \leqq i \leqq k_0$, be hyperplanes of L in general position which do not intersect $G(S_0)$. Then Π_i intersects $G(W)$ at certain of the points p_j, with a multiplicity which we denote by μ_{ij}. We have

(42) $$\sum_{1 \leqq j \leqq r} \mu_{ij} = \nu_0,$$

where ν_0 is the order of the algebraic curve $G(W)$.

Let (37) be the equations of $G(W)$ at p_j, whose parameter value is $t = 0$. At p_j the maximum possible value of μ_{ij} is $\delta_{m-1}(p_j)$, and that for the unique hyperplane $\zeta_{m-1} = 0$. A second hyperplane can intersect $G(W)$ at p_j with multiplicity at most $\delta_{m-2}(p_j)$, and a third, if in general position with respect to the first two, at most $\delta_{m-3}(p_j)$, etc. It follows that at most $m-1$ of the hyperplanes Π_i can intersect $G(W)$ at p_j and we have

$$\sum_{1 \leqq i \leqq k_0} \mu_{ij} \leqq \delta_1(p_j) + \cdots + \delta_{m-1}(p_j).$$

By (39) the right-hand side is equal to

$$\sum_{1 \leqq k \leqq m-1} (m-k)\nu_{k-1}(p_j) + \frac{1}{2} m(m-1).$$

Combining this with (41), (42), we get

$$(43)\quad k_0\nu_0 = \sum_{1\leqq i\leqq k_0,}\ \sum_{1\leqq j\leqq r} \mu_{ij} \leqq m\nu_0 + \frac{1}{2}\, m(m-1)\, \{2(g-1)+r\}.$$

On the other hand, by Theorem 2,

$$(44)\quad \nu_0 \geqq 2(r+g-1).$$

Eliminating g in the inequalities (43), (44), we get

$$\frac{1}{2}\, m(m-1)r \leqq \left\{\frac{1}{2}\, m(m+1) - k_0\right\} \nu_0.$$

Since the left-hand side is strictly positive, this gives

$$k_0 < \frac{1}{2}\, m(m+1),$$

which proves the theorem.

Remark 1. Theorem 4 expresses properties of the "defect values" of the reduced Gauss map. A corresponding statement in terms of the Gauss map itself follows, but its exact formulation does not seem to be simple. The problem is related to the following question of projective geometry: Let Π_i, $1 \leqq i \leqq N$, be hyperplanes in general position in $P^{n-1}(C)$ and let L be any other hyperplane. What is the largest number $\phi(n, N)$, independent of Π_i, L, such that at least $\phi(n, N)$ among the $\Pi_i \cap L$ are hyperplanes in general position in L? Very little is known about this number $\phi(n, N)$, which is $\leqq N$. In fact, the difference $N - \phi(n, N)$ could be relatively large. For instance, consider in $P^2(C)$ the configuration formed by the 9 points of inflection of a non-singular cubic curve. These points lie in threes on 12 lines such that every four of the lines pass through a point. Consider its dual configuration of 9 lines. It can be seen that at most four lines among them can be picked to be in general position, i.e., such that no three of them pass through the same point. It is possible to construct 9 planes in general position in $P^3(C$

whose intersections with a plane are the 9 lines of such a configuration. This proves that $\phi(4,9) \leqq 4$.

On the other hand, it is clear that, for a fixed n, $\phi(n, N)$ tends to infinity as N. It follows that there exists an integer $\psi(n)$, depending only on n, such that, under the hypotheses of Theorem 4, the image $G(S_0)$ under the Gauss map can fail to intersect at most $\psi(n)$ hyperplanes in general position in $P^{n-1}(C)$.

Remark 2. As remarked in §2 following the proof of Theorem 1, the case $n = 3$ is exceptional, and a stronger result can be deduced for this case using an additional argument. Suppose $n = 3$ and the hypotheses of Theorem 4 be satisfied. It follows from Theorem 4 that $G(S_0)$ can fail to intersect at most 5 lines in general position in $P^2(C)$. We wish to deduce the following stronger result (Theorem 3.3 in [7], p. 359): $G(S_0)$ *omits at most three points of* Q_1.

To prove this, we observe that ν_0 is even and that $G\colon W \to Q_1$ is a ramified covering of order $\nu_0/2$. Let $q \in Q_1$ and let $\pi\colon Q_1 \to L$ be the central projection of Q_1 from q onto a line L not through q. Then $\pi \circ G\colon W \to L$ is a ramified covering of order $\nu_0' = \nu_0/2$. Formula (43) gives, for $m = 2$,

$$k_0'\nu_0' \leqq 2\nu_0' + 2(g-1) + r$$

where k_0' is the number of points in $L - (\pi \circ G)S_0$. Combining this with (44), we get $k_0' \leqq 3$.

5. Proof of Theorem 5.

Lemma 3. *Let* W *be a compact Riemann surface, and let*

$$S_0 = W - \{p_1, \cdots, p_r\}.$$

Let S *be a minimal surface defined by* $x(p)\colon S_0 \to E^n$, *and let* $G\colon S_0 \to L$ *be the reduced Gauss map. Then either the Gauss map is algebraic or else, given any* $n+1$ *hyperplanes of* L *in general position,* $G(S_0)$ *must intersect at least one infinitely often.*

Proof. If the Gauss map is not algebraic, then it must fail to be algebraic at one of the p_j. If the dimension of L is $m-1$, then given any $m+1$ hyperplanes of L in general position, and given any neighborhood of this p_j, the image of this neighborhood must intersect at least one of the hyperplanes by the generalization of Borel's Theorem given in Cartan [2, p. 77]. Since $m \leqq n$, the result follows immediately by choosing a suitable sequence of neighborhoods.

Combining the results of Theorem 4 and Lemma 3, we find that we have very precise information concerning the Gauss map of a complete regular minimal surface whose base surface S_0 consists of a compact surface punctured at a finite number of points. Theorem 5 deals with a more general situation. We will first prove an elementary lemma:

Lemma 4. *A point set E in $P^{n-1}(C)$ intersects a dense set of hyperplanes if and only if E is not a bounded subset of $P^{n-1}(C)$.*

Proof. The set E will fail to intersect a dense set of hyperplanes if and only if there exists some hyperplane, which we may take to be the hyperplane $z_1 = 0$, such that E does not intersect any hyperplane in some neighborhood of $z_1 = 0$. This means that there exists $\delta > 0$ such that $\Sigma_k a_k z_k \neq 0$ for all $z \in E$ and for $a_1 = 1$, $\sum_{2 \leqq k \leqq n} |a_k|^2 < \delta$. But by elementary inequalities this is equivalent to the condition that

$$|z_1|^2 / \sum_{1 \leqq k \leqq n} |z_k|^2 \geqq \epsilon > 0,$$

so that E is bounded.

Proof of Theorem 5. Suppose that for some compact subset R of S_0, the image of $S_0 - R$ lies in a set of bounded type on Q_{n-2}. Then if we choose a local parameter ζ near a point of $S_0 - R$, the metric on S is given by $ds = \lambda |d\zeta|$, where

$$\lambda^2 = \frac{1}{2} \sum_k |\phi_k(\zeta)|^2 \leqq \frac{1}{2} \left| \sum_k \bar{a}_k \phi_k(\zeta) \right|^2 e^{h(G(\zeta))} = \tilde{\lambda}^2, \qquad 1 \leqq k \leqq n,$$

and where $\log \tilde{\lambda}$ is harmonic in ζ since h is pluriharmonic and $G(\zeta)$ is anti-holomorphic. The metric $d\tilde{s} = \tilde{\lambda} |d\zeta|$ is therefore a flat metric in $S_0 - R$, and if we extend it in an arbitrary smooth fashion over all of S_0, it will give a complete metric of finite total curvature. But we may then apply the results of Huber referred to earlier [5: p. 61, p. 71] (see also [7], p. 354) and conclude that S_0 must be conformally equivalent to a compact surface punctured at a finite number of points. This proves the theorem.

BIBLIOGRAPHY

1. L. V. Ahlfors, The theory of meromorphic curves, *Acta Soc. Sci. Fenn.* N.S.A **3**, No. 4(1941), 1–31.

2. H. Cartan, Sur les systèmes de fonctions holomorphes à variétés linéaires lacunaires et leurs applications, (Thèse) Gauthier-Villars, Paris 1928.

3. S. S. Chern, Minimal surfaces in an euclidean space of N dimensions, pp. 187–198 of Differential and Combinatorial Topology (A Symposium in Honor of Marston Morse), Princeton University Press, Princeton 1965.

4. R. Finn, On a class of conformal metrics, with application to differential geometry in the large, *Comment. Math. Helv.* **40** (1965), 1–30.

5. A. Huber, On subharmonic functions and differential geometry in the large, *Comment. Math. Helv.* **32** (1957), 13–72.

6. R. Osserman, Minimal surfaces in the large, *Comment. Math. Helv.* **35** (1961), 65–76.

7. R. Osserman, Global properties of minimal surfaces in E^3 and E^n, *Annals of Math.* (2) **80** (1964), 340–364.

8. L. A. Santaló, Integral geometry in Hermitian spaces, *Amer. J. Math.* **74** (1952), 423–434.

9. H. Weyl, Meromorphic functions and analytic curves, Annals Math. Studies, No. 12, Princeton 1943.

10. W. Wirtinger, Eine Determinantenidentität und ihre Anwendung auf analytische Gebilde in euklidischer und Hermitischer Massbestimmung, *Monatshefte für Math. und Physik* **44** (1936), 343–365.

UNIVERSITY OF CALIFORNIA AT BERKELEY
AND
STANFORD UNIVERSITY
STANFORD, CALIFORNIA, U. S. A.

(Rceived July 13, 1966)

Reprinted from
J. de l'Analyse Math. **19** (1967) 15–34

J. DIFFERENTIAL GEOMETRY
1 (1967) 21-31

EINSTEIN HYPERSURFACES IN A KÄHLERIAN MANIFOLD OF CONSTANT HOLOMORPHIC CURVATURE

SHIING-SHEN CHERN

Introduction

In his dissertation Brian Smyth studied the complete hypersurfaces in a complex space-form whose induced metric is einsteinian and proved that these are either totally geodesic or certain hyperquadrics of the complex projective space. We wish to show in this note that the corresponding *local* theorem is true:

Theorem. *Let* V *be a kählerian manifold of dimension* ≥ 3 *with constant holomorphic sectional curvature* K. *Let* f : M → V *be a holomorphically immersed hypersurface such that the induced metric is einsteinian. Then, if* K ≤ 0, M *is totally geodesic. If* K > 0 *and* V *is identified with the complex projective space,* M *is either totally geodesic or a hypersphere (cf. §3 for definition).*

1. Preliminaries on kählerian geometry

We will summarize the basic formulas of kählerian geometry. For details cf. [1].

In order to avoid repetitions it will be agreed that our indices have the following ranges throughout this paper:

$$(1) \qquad \begin{aligned} &1 \leq i, j, k, l \leq n, \\ &1 \leq \alpha, \beta, \gamma, \delta \leq n+1, \\ &0 \leq A, B, C, D \leq n+1. \end{aligned}$$

Let V be a kählerian manifold of complex dimension $n+1$. The metric defines an hermitian scalar product in the tangent spaces of V and a connection of type (1, 0) under whose parallelism the scalar product is preserved. More precisely, let $e_\alpha(x)$ be a field of unitary frames, defined for x in a neighborhood of V. Its dual coframe field consists of $n+1$ complex-valued linear differential forms θ_α of type (1, 0) such that the hermitian metric can be written

Communicated January 25, 1967. Work done under partial support by Guggenheim Foundation and NSF.

$$ds^2 = \sum_\alpha \theta_\alpha \bar{\theta}_\alpha . \tag{2}$$

The connection forms $\theta_{\alpha\beta}$ are characterized by the conditions

$$\theta_{\alpha\beta} + \bar{\theta}_{\beta\alpha} = 0 , \tag{3}$$

$$d\theta_\alpha = \sum_\beta \theta_\beta \wedge \theta_{\beta\alpha} , \tag{4}$$

and they can be interpreted geometrically as defining the covariant differential

$$De_\alpha = \sum_\beta \theta_{\alpha\beta} \otimes e_\beta . \tag{5}$$

The curvature forms $\Theta_{\alpha\beta}$ are then defined by

$$d\theta_{\alpha\beta} = \sum_\gamma \theta_{\alpha\gamma} \wedge \theta_{\gamma\beta} + \Theta_{\alpha\beta} , \tag{6}$$

and we have

$$\Theta_{\alpha\beta} = -\bar{\Theta}_{\beta\alpha} = \sum_{\gamma,\delta} R_{\alpha\beta\gamma\delta} \theta_\gamma \wedge \bar{\theta}_\delta . \tag{7}$$

The skew-hermitian symmetry of $\Theta_{\alpha\beta}$ expressed by the first equation of (7) is equivalent to the symmetry conditions

$$R_{\alpha\beta\gamma\delta} = \bar{R}_{\beta\alpha\delta\gamma} . \tag{8}$$

The Bianchi identities, which are relations obtained by exterior differentiation of (4) and (6), give the further symmetry relations

$$R_{\alpha\beta\gamma\delta} = R_{\gamma\beta\alpha\delta} = R_{\alpha\delta\gamma\beta} , \tag{9}$$

and the equation

$$d\Theta_{\alpha\beta} + \sum_\gamma \Theta_{\alpha\gamma} \wedge \theta_{\gamma\beta} - \sum_\gamma \theta_{\alpha\gamma} \wedge \Theta_{\gamma\beta} = 0 . \tag{10}$$

The metric on V is called *einsteinian*, if

$$d(\sum_\alpha \theta_{\alpha\alpha}) = \sum_\alpha \Theta_{\alpha\alpha} = \frac{R}{n+1} \sum_\alpha \theta_\alpha \wedge \bar{\theta}_\alpha , \tag{11}$$

where

$$R = \sum_{\alpha,\beta} R_{\alpha\alpha\beta\beta} = \bar{R} \tag{12}$$

is the scalar curvature.

The quantities $R_{\alpha\beta\gamma\delta}$ define the *holomorphic sectional curvature* to every tangent vector of V. In fact, let

$$\xi = \sum_\alpha \xi_\alpha e_\alpha \neq 0 \tag{13}$$

be a tangent vector at x. Then the holomorphic sectional curvature is defined to be

(14) $$R(x, \xi) = 2 \sum_{\alpha,\cdots,\delta} R_{\alpha\beta\gamma\delta}\xi_\alpha\xi_\gamma\bar{\xi}_\beta\bar{\xi}_\delta / (\sum_\alpha \xi_\alpha\bar{\xi}_\alpha)^2 .$$

Because of the symmetry relation (8), $R(x, \xi)$ is real.

V is said to be of *constant holomorphic sectional curvature* K if $R(x, \xi) = K$ for all (x, ξ). This is expressed by the condition

(15) $$R_{\alpha\beta\gamma\delta} = \frac{1}{4}(\delta_{\alpha\beta}\delta_{\gamma\delta} + \delta_{\alpha\delta}\delta_{\beta\gamma})K$$

or

(16) $$\Theta_{\alpha\beta} = \frac{1}{4}K(\theta_\beta \wedge \bar{\theta}_\alpha + \delta_{\alpha\beta} \sum_\gamma \theta_\gamma \wedge \bar{\theta}_\gamma) .$$

The above treatment depends on the choice of a frame field. As is well-known, the geometrical results which follow are independent of this choice. However, it is useful to know explicitly the effect of a change of the frame field on the various quantities. Let

(17) $$e^*_\alpha = \sum_\beta u_{\alpha\beta} e_\beta$$

be a new frame field defined in the neighborhood in question, where $u_{\alpha\beta}$ are complex-valued C^∞-functions such that $(u_{\alpha\beta})$ is a unitary matrix. Let θ^*_α, $\theta^*_{\alpha\beta}$ be the forms relative to the frame field e^*_α. Then, by definition and by (5), we have

(18) $$\theta^*_\alpha = \sum_\beta \bar{u}_{\alpha\beta}\theta_\beta$$

and

(19) $$\theta^*_{\alpha\beta} = \sum_\gamma du_{\alpha\gamma}\bar{u}_{\beta\gamma} + \sum_{\gamma,\delta} u_{\alpha\gamma}\theta_{\gamma\delta}\bar{u}_{\beta\delta} .$$

2. Hypersurfaces in a kählerian manifold

Let $f: M \to V$ be a holomorphic immersion, with $\dim M = n$, $\dim V = n + 1$. In a neighborhood of M we can choose a frame field in V such that $e_{n+1}(x)$, $x \in M$, is orthogonal to the tangent hyperplane to M at x. This is expressed analytically by the condition

(20) $$\theta_{n+1} = 0 .$$

Since M is an immersed hypersurface, the θ_i are linearly independent. Using (4), we get

$$0 = d\theta_{n+1} = \sum_i \theta_i \wedge \theta_{i,n+1} \,.$$

It follows by Cartan's lemma that

(21)
$$\theta_{i,n+1} = \sum_k a_{ik}\theta_k \,,$$

where

(22)
$$a_{ik} = a_{ki} \,.$$

M is *totally geodesic* if $a_{ik} = 0$.

The e_i define a unitary frame field in the tangent bundle of M, with θ_{ij} as the connection forms. Equation (6) gives

$$d\theta_{ij} = \sum_k \theta_{ik} \wedge \theta_{kj} - \theta_{i,n+1}\bar{\theta}_{j,n+1} + \Theta_{ij} \,,$$

so that

(23)
$$\tilde{\Theta}_{ij} = \Theta_{ij} - \theta_{i,n+1} \wedge \bar{\theta}_{j,n+1}$$

are the curvature forms of the induced metric on M.

Suppose now that V is of constant holomorphic sectional curvature, K, so that the equation (16) holds. Then

$$\sum_i \Theta_{ii} = \frac{1}{4}(n+1)K \sum_i \theta_i \wedge \bar{\theta}_i \,.$$

The condition that the induced metric on M is einsteinian can be expressed as

(24)
$$\sum_i \theta_{i,n+1} \wedge \bar{\theta}_{i,n+1} = \rho \sum_i \theta_i \wedge \bar{\theta}_i \,.$$

Using (21) this condition is equivalent to

(25)
$$\sum_i a_{ik}\bar{a}_{il} = \rho\delta_{kl} \,,$$

which gives

(26)
$$n\rho = \sum_{i,k} | a_{ik} |^2 \geq 0 \,.$$

From now on suppose $n \geq 2$. We wish to show that ρ is constant. In fact, we have by (6),

$$d(\sum_i \theta_{i,n+1} \wedge \bar{\theta}_{i,n+1}) = 0 \,,$$

so that it follows by exterior differentiation of (24) that

$$d\rho \wedge (\sum_i \theta_i \wedge \bar{\theta}_i) = 0 \,.$$

Put

$$d\rho = \sum_k (\rho_k \theta_k + \bar{\rho}_k \bar{\theta}_k)$$

and substitute into the above; we get immediately

$$d\rho = 0 .$$

If $\rho = 0$, we have by (26), $a_{ik} = 0$ and M is totally geodesic. From now on suppose that ρ *is a positive constant.*

We take the exterior derivative of the equation (21) and make use of (4) and (6). This gives

$$(27) \qquad \sum_k (da_{ik} - \sum_j a_{ij}\theta_{kj} - \sum_j a_{jk}\theta_{ij} + a_{ik}\theta_{n+1,n+1}) \wedge \theta_k = 0.$$

It follows that we can put

$$da_{ik} - \sum_j a_{ij}\theta_{kj} - \sum_j a_{jk}\theta_{ij} + a_{ik}\theta_{n+1,n+1} = \sum_j a_{ikj}\theta_j ,$$

where a_{ikj} are symmetric in all its indices. The complex conjugate of this equation will give a formula for $d\bar{a}_{ik}$. Differentiating (25) and substituting these expressions for da_{ik}, $d\bar{a}_{ik}$, we get

$$\sum_{i,j} (\bar{a}_{il} a_{ikj}\theta_j + a_{ik}\bar{a}_{ilj}\bar{\theta}_j) = 0 ,$$

from which it follows that

$$\sum_i \bar{a}_{il} a_{ikj} = 0 .$$

Since $\rho > 0$, we get from the last equation

$$a_{ikj} = 0 .$$

We have therefore the equation

$$(28) \qquad da_{ik} - \sum_j a_{ij}\theta_{kj} - \sum_j a_{jk}\theta_{ij} + a_{ik}\theta_{n+1,n+1} = 0 .$$

Equation (28) is valid for a holomorphically immersed hypersurface of dimension $\geqq 2$ in a kählerian manifold of constant holomorphic sectional curvature such that its induced metric is einsteinian. Notice that (28) is still valid if M is totally geodesic, for then $a_{ik} = 0$.

We now take the exterior derivative of (28). This gives, after simplification,

$$(29) \qquad (\rho - \frac{1}{4}K)(\delta_{ij}a_{kl} + \delta_{kj}a_{il} + \delta_{jl}a_{ik}) = 0 .$$

If $\rho - \dfrac{1}{4}K \neq 0$, we have

$$\delta_{ij}a_{kl} + \delta_{kj}a_{il} + \delta_{jl}a_{ik} = 0 .$$

Putting $j = l$ and summing, we get $(n + 2)a_{ik} = 0$, so that $a_{ik} = 0$ and M is totally geodesic. Hence M is not totally geodesic only if $\rho = K/4$, which implies $K > 0$ since $\rho > 0$. We have thus proved the first half of the theorem stated in the Introduction. Our next problem is to study the hypersurfaces satisfying the condition $\rho = \dfrac{K}{4} > 0$.

3. The complex projective space

For $K > 0$, V can be realized locally as the complex projective space P_{n+1} of dimension $n + 1$ with the Study-Fubini metric. We proceed to give a description of this metric.

Let V_{n+2} be the complex vector space of dimension $n + 2$, whose points are the ordered ennuples of complex numbers: $Z = (z_0, \cdots, z_{n+1})$. In V_{n+2} we introduce the hermitian scalar product

$$(W, Z) = \overline{(W, Z)} = \sum_A z_A \overline{w}_A , \quad W = (w_0, \cdots, w_{n+1}) . \tag{30}$$

The unitary group $U(n + 2)$ in $n + 2$ variables is the group of all linear homogeneous transformations on z_A leaving the scalar product (30) invariant. Let V^*_{n+2} be the subset of V_{n+2} obtained by the deletion of the zero vector. Then P_{n+1} is the orbit space of V^*_{n+2} under the action of the group $Z \to \lambda Z$, λ being a complex number $\neq 0$. We have thus the projection $\pi : V^*_{n+2} \to P_{n+1}$. To a point $p \in P_{n+1}$ a vector $Z \in \pi^{-1}(p)$ is called a homogeneous coordinate vector of p, and we will frequently identify p with Z. We put

$$Z_0 = Z/(Z, Z)^{\frac{1}{2}} . \tag{31}$$

so that $(Z_0, Z_0) = 1$. Then the Study-Fubini metric is given by

$$ds^2 = (dZ_0, dZ_0) - (dZ_0, Z_0)(Z_0, dZ_0) . \tag{32}$$

To study this metric let Z_A be a unitary frame in V_{n+2}, so that

$$(Z_A, Z_B) = \delta_{AB} . \tag{33}$$

In the space of all unitary frames in V_{n+2} let ω_{AB} be defined by

$$dZ_A = \sum_B \omega_{AB} Z_B , \tag{34}$$

so that we have

$$\omega_{AB} = - \overline{\omega}_{BA} = (dZ_A, Z_B) . \tag{35}$$

Then ω_{AB} are the Maurer-Cartan forms of $U(n + 2)$ and satisfy the structure equations

(36) $$d\omega_{AB} = \sum_C \omega_{AC} \wedge \omega_{CB}\,.$$

The same equations remain valid if we restrict ourselves to a frame field defined over a submanifold of V_{n+2}. The metric (32) can then be written

(37) $$ds^2 = \sum_\alpha \omega_{0\alpha}\bar{\omega}_{0\alpha}\,.$$

It is of the form (2) if we set

(38) $$\theta_\alpha = \omega_{0\alpha}\,.$$

Equations (3) and (4) will be satisfied, provided that we choose

(39) $$\theta_{\beta\alpha} = \omega_{\beta\alpha} - \delta_{\beta\alpha}\omega_{00}\,.$$

These are therefore the connection forms of the metric (32). By (36) we find the curvature forms of this metric to be

(40) $$\Theta_{\alpha\beta} = \theta_\beta \wedge \bar{\theta}_\alpha + \delta_{\alpha\beta} \sum_\gamma \theta_\gamma \wedge \bar{\theta}_\gamma\,.$$

Comparing with (16), we see that the metric (32) has constant holomorphic sectional curvature equal to 4. From the definition of the metric it is clear that $U(n+2)$ acts on P_{n+1} as a group of isometries.

Consider in P_{n+1} a hyperquadric defined by the equation

(41) $$\sum_{A,B} b_{AB} z_A z_B = 0\,, \quad b_{AB} = b_{BA}\,.$$

Under a unitary transformation

(42) $$z_A = \sum_B u_{AB} z_B^*$$

this goes into the hyperquadric

$$\sum_{A,B} b_{AB}^* z_A^* z_B^* = 0\,.$$

By introducing the matrices

(43) $$B = {}^tB = (b_{AB}), \quad B^* = {}^tB^* = (b_{AB}^*), \quad U = {}^t\bar{U}^{-1} = (u_{AB})\,,$$

we can express the relation between the coefficients b_{AB} and b_{AB}^* by the matrix equation

(44) $$B^* = {}^tUBU\,.$$

It follows that

$$B^*\bar{B}^* = {}^tUB\bar{B}\bar{U}\,.$$

Thus the eigenvalues of $B\bar{B}$ are invariant under the unitary transformation. In particular, the invariance of the trace of $B\bar{B}$ gives

$$\sum_{A,B} | b_{AB} |^2 = \sum_{A,B} | b^*_{AB} |^2 . \tag{45}$$

We will establish the following lemma:

Given a symmetric matrix B with complex elements, there exists a unitary matrix U such that ^tUBU is diagonal.

For $n = 0$, i.e., for a (2×2)-matrix B this can be verified by an elementary calculation, and we suppose the lemma true in this case. Let

$$\varphi(B) = \sum_{A \neq B} | b_{AB} |^2 , \tag{46}$$

i.e., $\varphi(B)$ is the sum of the squares of the absolute values of the non-diagonal elements of B. Suppose $| b_{01} | \geq | b_{AB} |$, $A \neq B$; this can always be achieved by interchanging the rows and columns when necessary. Let U_0 be a (2×2)-unitary matrix such that $^tU_0B^1U_0$ is diagonal, where $B^1 = \begin{pmatrix} b_{00} & b_{01} \\ b_{10} & b_{11} \end{pmatrix}$ and let

$$U = \begin{pmatrix} U_0 & 0 \\ 0 & I \end{pmatrix},$$

where I is the $(n \times n)$-unit matrix. Then we have

$$\sum_A | b^*_{AA} |^2 = \sum_A | b_{AA} |^2 + 2 | b_{01} |^2$$

and

$$\varphi(B^*) = \varphi(B) - 2 | b_{01} |^2 .$$

Under the assumption that $| b_{01} | \geq | b_{AB} |$, $A \neq B$, we have

$$\varphi(B) \leq (n+1)(n+2) | b_{01} |^2 ,$$

from which it follows that

$$\varphi(B^*) \leq \frac{n(n+3)}{(n+1)(n+2)} \varphi(B) . \tag{47}$$

Notice that the factor at the right-hand side before $\varphi(B)$ is < 1. We can therefore find a sequence of unitary matrices $U_1, \cdots, U_\nu, \cdots$, such that $\varphi(^tU_\nu BU_\nu)$ is strictly monotone decreasing and tends to zero. Since the unitary group is compact, there exists a unitary matrix U_∞ such that $\varphi(^tU_\infty BU_\infty) = 0$ and $^tU_\infty BU_\infty$ is diagonal. This proves the lemma.

It follows that the equation of the hyperquadric can be by unitary transformations brought to the normal form

$$(48) \qquad b_0 z_0^2 + \cdots + b_{n+1} z_{n+1}^2 = 0\,.$$

We can further suppose that b_A are real and ≥ 0. The ratios of b_A are invariants of the hyperquadric under $U(n+2)$. In particular, the hyperquadric

$$(49) \qquad z_0^2 + \cdots + z_{n+1}^2 = 0$$

will be called a *hypersphere*.

A one-one mapping $T\colon V_{n+2} \to V_{n+2}$ is called *antilinear*, if

$$(50) \qquad \begin{aligned} T(Z_1 + Z_2) &= T(Z_1) + T(Z_2), \\ T(\lambda Z) &= \bar{\lambda} T(Z)\,, \quad Z, Z_1, Z_2 \in V_{n+2}\,, \end{aligned}$$

λ being a complex number. It induces a one-one mapping in P_{n+1}. By the properties (50) an anti-linear mapping is completely determined by its effect on a frame.

4. Completion of the proof of the theorem

We wish to prove the second part of the theorem stated in the Introduction by showing that a hypersurface in P_{n+1} (with the Study-Fubini metric) whose induced metric is einsteinian and which is not totally geodesic is necessarily a hypersphere.

Continuing the proof of §2, we have $K = 4$ and $\rho = 1$. We apply a change of the frame field as defined by (17) with

$$(51) \qquad u_{i,n+1} = 0\,, \quad u_{n+1,n+1} = 1\,,$$

so that the normal vector e_{n+1} to M remains unchanged. By (18) and (19) we have respectively

$$\begin{aligned} \theta_i^* &= \sum_j \bar{u}_{ij} \theta_j\,, \\ \theta_{i,n+1}^* &= \sum_j u_{ij} \theta_{j,n+1}\,. \end{aligned}$$

If we set

$$(52) \qquad \theta_{i,n+1}^* = \sum_k a_{ik}^* \theta_k^*\,,$$

we get

$$(53) \qquad a_{ik}^* = \sum_{l,j} u_{il} a_{lj} u_{kj}\,.$$

Our lemma in §3 implies that unitary matrices (u_{ik}) can be so chosen that the matrix (a_{ik}^*) is diagonal. Moreover, since $\rho = 1$, we can even make it the unit matrix.

Suppose such a change of the frame field be already carried out. By dropping the asterisks, we have $a_{ik} = \delta_{ik}$ and

$$\theta_{i,n+1} = \theta_i \,. \tag{54}$$

Equation (28) becomes

$$\theta_{ik} + \theta_{ki} - \delta_{ik}\theta_{n+1,n+1} = 0 \,. \tag{55}$$

By (39) this gives

$$\omega_{ik} + \omega_{ki} - \delta_{ik}(\omega_{00} + \omega_{n+1,n+1}) = 0 \,. \tag{56}$$

We now modify Z_{n+1} by setting

$$Z^*_{n+1} = e^{i\phi} Z_{n+1} \,, \quad \phi \text{ real.}$$

Then

$$\omega^*_{n+1,n+1} = (dZ^*_{n+1}, Z^*_{n+1}) = i d\phi + \omega_{n+1,n+1}$$

and we have

$$\omega_{00} + \omega^*_{n+1,n+1} = i d\phi + \omega_{00} + \omega_{n+1,n+1} \,.$$

Since $i(\omega_{00} + \omega_{n+1,n+1})$ is real-valued and

$$d(i\omega_{00} + i\omega_{n+1,n+1}) = 0 \,,$$

we can determine ϕ so that

$$\omega_{00} + \omega^*_{n+1,n+1} = 0 \,.$$

Dropping the asterisks again, we have

$$\omega_{00} + \omega_{n+1,n+1} = 0 \tag{57}$$

and (56) gives

$$\omega_{ik} + \omega_{ki} = 0 \,. \tag{58}$$

Let T be an anti-linear transformation in P_{n+1}, so that

$$d(T(Z_A)) = \sum_B \bar{\omega}_{AB} T(Z_B) \,. \tag{59}$$

By using (57), (58) and (59), we find

$$\begin{aligned} d(Z_0 + T(Z_{n+1})) &= \omega_{00}(Z_0 + T(Z_{n+1})) + \sum_i \omega_{0i}(Z_i - T(Z_i)), \\ d(T(Z_0) + Z_{n+1}) &= -\,\omega_{00}(T(Z_0) + Z_{n+1}) - \sum_i \bar{\omega}_{0i}(Z_i - T(Z_i)), \\ d(Z_i - T(Z_i)) &= \omega_{0i}(T(Z_0) + Z_{n+1}) - \bar{\omega}_{0i}(Z_0 + T(Z_{n+1})) \\ &\quad + \sum_k \omega_{ik}(Z_k - T(Z_k)) \,. \end{aligned} \tag{60}$$

This is a differential system which is linear and homogeneous in the vectors $Z_0 + T(Z_{n+1})$, $T(Z_0) + Z_{n+1}$, $Z_i - T(Z_i)$. It follows that if these vectors are zero at a point of M, they are identically zero. We choose the anti-linear transformation T so that they are zero at $p_0 \in M$ and we have

$$(61) \qquad T(Z_0) = -Z_{n+1}, \quad T(Z_{n+1}) = -Z_0, \quad T(Z_i) = Z_i$$

everywhere on M. As a consequence we get

$$(62) \qquad (Z_0, T(Z_0)) = 0 .$$

We put

$$(63) \qquad \begin{aligned} &a_0 = (Z_0(p_0) - Z_{n+1}(p_0))/\sqrt{2}\,, \; a_{n+1} = i(Z_0(p_0) + Z_{n+1}(p_0))/\sqrt{2}\,, \\ &a_j = Z_j(p_0) . \end{aligned}$$

Then a_A is a unitary frame having the property

$$(64) \qquad Ta_A = a_A .$$

Let

$$Z_0(p) = \sum_A z_A a_A, \qquad p \in M .$$

Then

$$T(Z_0(p)) = \sum_A \bar{z}_A a_A ,$$

and equation (62) can be written

$$\sum_A z_A^2 = 0 .$$

This proves that M is a hypersphere.

Bibliography

[1] S. Chern, *Complex manifolds without potential theory,* to appear in van Nostrand notes series.

[2] Brian Smyth, *Differential geometry of complex hypersurfaces,* Ann. of Math. **85** (1967) 246-266.

UNIVERSITY OF CALIFORNIA, BERKELEY

Actes, Congrès intern. math., 1970. Tome 1, p. 41 à 53.

DIFFERENTIAL GEOMETRY; ITS PAST AND ITS FUTURE

by SHIING-SHEN CHERN (*)

A. Introduction.

It was almost a century ago, in 1872, that Felix Klein formulated his Erlanger Program. The idea of unifying the geometries under the group concept is simple and attractive and its applications in the derivation of different geometrical theorems from the same group-theoretic argument are usually of great elegance. This leads to the development of differential geometries of submanifolds in homogeneous (or Klein) spaces: conformal, affine, and projective differential geometries. The latter had in particular an energetic development in the twenties.

It was also about a century ago that the greatest modern differential geometer Elie Cartan was born (on April 9, 1869). Among his contributions of a basic nature are his systematic use of the exterior calculus and his clarification of the global theory of Lie groups. Fiber spaces also find their origin in Cartan's work.

Differential geometry is the study of geometry by the methods of infinitesimal calculus or analysis. Among mathematical disciplines it is probably the least understood ([1]). Many mathematicians feel there is no geometry beyond two and three dimensions. The advent into higher and even infinitely many dimensions does make the intuition unreliable and the dependence on algebra and analysis mandatory. The basis of algebra is the algebraic operations and the basis of analysis is the topological structure. I would like to surmise that the core of differential geometry is the Riemannian structure (in its broad sense).

The main object of study in differential geometry is, at least for the moment, the differentiable manifolds, structures on the manifolds (Riemannian, complex, or other), and their admissible mappings. On a manifold the coordinates are valid only locally and do not have a geometrical meaning themselves. Historically the difficulty in achieving a proper understanding of this situation must have been tremendous (I wonder whether this was part of the reason which caused Hadamard to admit his

(*) This paper was written when the author held a Research Professorship of the Miller Institute and was under partial support of NSF grant GP 20096.

([1]) G. D. BIRKHOFF, « The second is a disturbing secret fear that geometry may ultimately turn out to be no more than the glittering intuitional trappings of analysis ». Fifty years of American mathematics, *Semicentennial Addresses of Amer. Math. Soc.* (1938), p. 307.

G. W. MACKEY, « Geometrical intuition, while very helpful, is not reliable and cannot be depended upon for rigorous arguments », *Lectures on the Theory of Functions of a Complex Variable*, p. 21, Van Nostrand, notes.

psychological difficulty in the mastery of Lie groups) (²). For technical purposes the Ricci calculus was a powerful tool, but it is inadequate for global problems. Global differential geometry, with the exception of a few isolated results, had to wait till algebraic topology and Lie groups have paved the way.

Global differential geometry must be considered a young field. The notion of a differentiable manifold should have been in the minds of many mathematicians, but it was H. Whitney who found in 1936 a theorem to be proved: the imbedding theorem. In the case of the richer complex structure a definition of a Riemann surface by overlapping neighborhoods was given and the theory rigorously treated by H. Weyl in his famous book " Idee der Riemannschen Fläche, Göttingen, 1913 " (³), following which Caratheodory gave the first definition of a high-dimensional complex manifold. More general pseudo-group structures were treated by Veblen and J. H. C. Whitehead in 1932 [34]. Only special cases of the general theory, such as Riemannian, conformal, complex, foliated structures, etc. have been found significant.

B. The development of some fundamental notions and tools.

Perhaps the most far-reaching achievement in differential geometry in the last thirty years lies in its foundation. Not only are the notions clearly defined, but notations are devised to treat manifolds which could be infinite-dimensional. The notations are up to now on the diversive side and are thus at an experimental stage. We believe in the survival of the fittest. Important as these foundational works are, no mathematical discipline can prosper without deeper study and simple challenging problems. We will comment briefly on a few fundamental developments in differential geometry and its related subjects, without endeavoring to make the list complete.

(1) *Lie Groups.* — It is one of the happiest incidents in the history of mathematics that the structure of Lie groups can be so thoroughly analyzed. The existence of the five exceptional simple Lie groups makes a deep study necessary and leads to a better understanding. Even so the subject has unity and is so much simpler than (say) finite groups. The quotient spaces of Lie groups give a multitude of examples of manifolds which are easy to describe. They include the classically important spaces and form a reservoir on which new conjectures can be tested.

(2) *Fiber Spaces.* — When a manifold has a differentiable structure, it can be locally linearized, giving rise to the tangent bundle and the associated tensor bundles. The first idea of a connection in a fiber bundle with a Lie group can be found in Cartan's " espaces généralisés ". It was algebraic topology which focused on the simplest problems, e. g., the problem of introducing invariants which serve to distinguish a general

(²) J. HADAMARD, *Psychology of Invention in the Mathematical Field*, Princeton (1949), p. 115.

E. CARTAN, in his classical « Leçons sur la géométrie des espaces de Riemann » says, « La notion générale de variété est assez difficile à définir avec précision », p. 58.

(³) Weyl's book was dedicated to Felix KLEIN, to whom he acknowledged for the fundamental ideas. Weyl's definition of a Riemann surface and Hausdorff's introduction of his axioms in 1914 must have made it superfluous to give formally a definition of a differentiable manifold. Chevalley's book on Lie groups (1946) exerted a great influence in the clarification of many concepts attached to it.

fiber bundle from a product bundle. Among them are the characteristic classes. Characteristic classes with real coefficients can be represented by the curvature of a connection, the simplest example being the Gauss-Bonnet formula. The bundle structure is now an integral part of differential geometry.

(3) *Variational Methods.* — The importance of the notion of measure (length, area, volume, curvature, etc.) makes the variational method a powerful and indispensable tool. The study of geodesics on a Riemannian manifold is a brilliant chapter of mathematics. It led to Morse's creation of the critical point theory whose scope extends far beyond differential geometry. Another example is the Dirichlet problem and its application to elliptic operators. Multiple integral variational problems open a vista whose terrain is still rocky. It promises, however, a fertile field of work. When a geomatrical problem involves a function, either over the given manifold or in some related functional space, it always pays to look at its critical values and the second variation at them. Much of differential geometry utilizes this idea, in its various ramifications. The importance of variational method in differential geometry can hardly be over-emphasized.

(4) *Elliptic Differential Systems.* — The geometrical properties of differential geometry are generally expressed by differential equations or inequalities. Contrary to analysis special systems with their special properties received more attention. While analysis is the main tool, geometry furnishes the variety. Differential systems on manifolds with or without boundary are the prime objects of study.

Elliptic systems occupy a central position because of their rich properties, which follow from the severe restrictions on the set of solutions. Hodge's harmonic differential forms, with their applications to Kahlerian manifolds, will remain a crucial landmark. A simple idea of Bochner relates them to curvature and leads to vanishing theorems when the curvature satisfies proper " positivity " conditions. This has remained a standard method in the establishment of such theorems, which in turn give rise to existence theorems. The indices of linear elliptic operators on a compact manifold include some of the deepest invariants of manifolds (Atiyah, Bott, Singer).

In the study of mappings an important problem consists in the analysis of the singularities. Important progress has been made recently on the singularities of differentiable mappings (Whitney, Thom, Malgrange, Mather). If the mappings are defined by elliptic differential equations, cases are known where the singularities take relatively simple form. Singularities in differential geometry remain a relatively untouched subject.

C. **Formulation of some problems with discussion of related results.**

We will attempt to discuss some areas where it is believed that fruitful researches can be carried out. The limited time at my disposal and, above all, my own limitation make it impossible for the treatment to be even remotely exhaustive. Any subject left out carries no implication that it is considered less significant.

My object is to amuse you by stating some very simple problems which have so far defied the efforts of geometers. The danger in formulating such problems is that the line distinguishing them from mathematical puzzles is thin. Personally I think there is no such line except that the " serious " problems concern with a new domain where the phenomena are not well understood and the basic concepts not well developped. Geometry and analysis on manifolds are still at this stage and will remain so for years to come. When such problems are solved, similar ones will tend toward puzzles.

Many of the problems to be given below are known. It is hoped that its collection may attract mathematicians not engaged in this field and lead to further progress.

1. Riemannian manifolds whose sectional curvatures keep a constant sign

It was known to Riemann that the local properties of a Riemannian structure are completely determined by its sectional curvature. The latter is a function $R(\sigma)$ of a two-dimensional subspace σ of the tangent space at a point x, which is equal to the gaussian curvature of the surface generated by the geodesics tangent to σ at x. Manifolds for which $R(\sigma)$ keeps a constant sign for all σ have a simple geometrical meaning. For their global study it is important to require that they are not proper open subsets of larger manifolds and, following Hopf and Rinow, it is customary to impose the stronger completeness condition: every geodesic can be indefinitely extended. In fact, without the completeness requirement the sign of the sectional curvature imposes hardly any condition on the manifold, as Gromov [21] proved that there exists on any non-compact manifold a Riemannian metric for which the range of the values of $R(\sigma)$ is any open interval on the real line.

For *complete* Riemannian manifolds M for which $R(\sigma)$ keeps the same sign the two classical theorems are:

(1) Theorem of Hadamard-Cartan. — If $R(\sigma) \leqslant 0$, the universal covering manifold of M is diffeomorphic to R^n, $n = \dim M$.

(2) Theorem of Bonnet-Myers. — If $R(\sigma) \geqslant c\ (= \text{const}) > 0$, M has a diameter $\leqslant \pi/c^{1/2}$ and is therefore compact.

The case of positive curvature turns out to be more elusive. Cheeger and Gromoll [9] achieved what is essentially a structure theory of non-compact complete Riemannian manifolds M with $R(\sigma) \geqslant 0$ (all σ) by proving the following theorem. There is in M a compact totally geodesic and totally convex submanifold S_M (to be called the soul of M) without boundary such that M is diffeomorphic to the normal bundle of S_M. If the sectional curvature is strictly positive, then Gromoll and Meyer [20] proved that the soul is a point and M is diffeomorphic to R^n. In particular, M must be simply connected.

Compact Riemannian manifolds of positive curvature obviously satisfy the stronger condition $R(\sigma) \geqslant c > 0$ (all σ). By the Bonnet-Myers Theorem they are identical with the complete Riemannian manifolds with the same property. They are not necessarily simply connected, as the example of the non-euclidean elliptic space shows. So far the simply connected compact differentiable manifolds known to admit a Riemannian metric of positive curvature are the following [3]: (1) the n-sphere; (2) the complex projective space; (3) the quaternion projective space; (4) the Cayley plane; (5) two manifolds discovered by Berger, of dimensions 7 and 13 respectively.

It is very unlikely that there are no others, but nothing more is known. The following question was asked by H. Hopf:

PROBLEM I. — Does the product of two 2-dimensional spheres admit a Riemannian metric of strictly positive curvature?

More generally, it is not known whether the exotic 7-spheres, some of which are bundles of 3-spheres over 4-spheres, admit Riemannian metrics of positive curvature. The answer to the question in Problem I is probably negative. A supporting evidence is furnished by the following theorem of Berger [5]: Let M and N be compact Riemannian manifolds. Let $g(t)$ be a family of Riemannian structures on $M \times N$, such that $g(0)$ is the product structure and such that the following condition is satisfied:

$$\left.\frac{dR(\sigma)}{dt}\right|_{t=0} \geqslant 0$$

for all σ spanned at $x \in M \times N$ by a tangent vector to M and a tangent vector to N. Then

$$\left.\frac{dR(\sigma)}{dt}\right|_{t=0} = 0$$

for all such σ.

To get deeper topological properties of a manifold of positive curvature Rauch introduced the notion of *pinching*. M is said to be β-pinched if $0 < \beta \leqslant R(\sigma) \leqslant 1$ for all σ. After the pioneering work of Rauch the following are the main theorems on the topology of compact pinched Riemannian manifolds of positive curvature:

(1) (Berger-Klingenberg) [4, 25]. If a simply connected Riemannian manifold of positive curvature is β-pinched, $\beta > \frac{1}{4}$, it is homeomorphic to the n-sphere; if $\beta = \frac{1}{4}$ and it is not homeomorphic to the n-sphere, it is isometric to a symmetric space of rank 1.

(2) (Gromoll-Calabi) [19]. Let M be an n-dimensional compact simply connected Riemannian manifold of positive curvature. There exists a universal constant $\beta(n) < 1$, depending only on n, such that if M is $\beta(n)$-pinched, it is diffeomorphic to the standard n-sphere.

Similar problems can be studied on the global implications of curvature properties of complex Kählerian manifolds. A new feature is the notion of holomorphic sectional curvature, i. e., sectional curvature $R(\sigma)$, where σ is the two-dimensional real space underlying a complex line in the complex tangent space. A most attractive question is the following one formulated by Frankel:

PROBLEM II. — Let M be a compact Kählerian manifold of positive sectional curvature. Is M biholomorphically equivalent to the complex projective space?

Andreotti and Frankel [17] proved that the answer is affirmative if M is of dimension 2. The proof makes use of the classification of algebraic surfaces. Partial results were recently obtained by Kobayashi and Ochiai [26] for 3 dimensions.

2. EULER-POINCARÉ CHARACTERISTIC

Among the important topological invariants of a manifold is the Euler-Poincaré characteristic. Its role is well-known on problems such as the Lefschetz fixed-point theorem, singularities of vector fields, and indices of some elliptic operators. Geometrically it is closely related to the total curvature (*curvatura integra*) as expressed by the Gauss-Bonnet formula

$$\chi(M) = \frac{(-1)^m}{2^{3m}\pi^m m} \int_M (\sum_{i,j} \varepsilon_{i_1 \ldots i_{2m}} \varepsilon_{j_1 \ldots j_{2m}} R_{i_1 i_2 j_1 j_2} \cdots R_{i_{2m-1} i_{2m} j_{2m-1} j_{2m}}) dv \tag{1}$$

Here M is a compact orientable Riemannian manifold of even dimension $n = 2m$, $\chi(M)$ is its Euler-Poincaré characteristic, dv is the volume element, and R_{ijkl} are the components of the curvature tensor relative to ortho-normal frames. The $\varepsilon_{i_1 \ldots i_{2m}}$ is the Kronecker symbol and is zero if $i_1, \ldots, i_{2m}$ do not form a permutation of $1, \ldots, 2m$ and is equal to $+1$ or -1 according as the permutation is even or odd.

In spite of the explicit expression for $\chi(M)$ the following has not been established:

PROBLEM III AND CONJECTURE. — If M has sectional curvatures $\geqslant 0$, then $\chi(M) \geqslant 0$. If M has sectional curvatures $\leqslant 0$, then $\chi(M) \geqslant 0$ or $\leqslant 0$, according as $n \equiv 0$ or $2 \bmod 4$.

The above statement has been proved for $n = 4$ [10] and for the case that M has constant sectional curvature. A first approach would be to study the sign of the integrand in the Gauss-Bonnet formula, a purely algebraic problem. Even this algebraic problem seems to be of great interest [33].

As with the classical Gauss-Bonnet formula the relationship is more useful for compact manifolds with boundary (in which case a boundary integral should be added to make the formula (1) valid) and the problem is more interesting for non-compact manifolds, because a deeper study of the geometry will then be necessary. We will denote by $C(M)$ the right-hand side of (1) and we shall formulate the problem:

PROBLEM IV. — Let M be a complete Riemannian manifold of even dimension. Suppose $\chi(M)$ and $C(M)$ both exist, the latter meaning that the corresponding integral converges. Find a geometrical interpretation of the difference

$$\delta(M) = \chi(M) - C(M).$$

Of course $\delta(M) = 0$ if M is compact. In two dimensions Cohn-Vossen's classical inequality says that $\delta(M) \geqslant 0$. For a class of two-dimensional manifolds Finn and A. Huber [16, 23] obtained a geometrical interpretation of $\delta(M)$, which implies that it is non-negative. Partial results on Problem IV have been obtained by E. Portnoy [30]. Perhaps the case of Kählerian manifolds has a simpler answer and should be studied first.

In a different direction Satake [31] obtained a Gauss-Bonnet formula for his V-manifolds and applied it to automorphic functions and number theory. *V*-manifolds are essentially manifolds with singularities of a relatively simple type.

Another problem on the Euler-Poincaré characteristic concerns compact affinely connected manifolds which are locally flat. These can be described as manifolds with a linear structure, i. e., having a covering by coordinate neighborhoods such that the coordinate transformation in overlapping neighborhoods is linear.

PROBLEM V. — Let M be a compact manifold with an affine connection which is locally flat. Is its Euler-Poincaré characteristic equal to zero?

Bensecri proved that the answer is affirmative if M is of two dimensions (For proof and generalization cf. Milnor [27]). The high-dimensional case has been investigated by L. Auslander who proved the theorem [1]: suppose the affine connection be complete and suppose that the homomorphism $h: \pi_1(M) \to GL(n, R)$ defined by the holonomy group is not an isomorphism of the fundamental group $\pi_1(M)$ onto a discrete subgroup of $GL(n, R)$. Then $\chi(M) = 0$.

It is not known whether h can imbed $\pi_1(M)$ as a discrete subgroup of $GL(n, R)$.

In spite of great developments in algebraic topology there are simple problems on the Euler-Poincaré characteristic which remain unanswered.

3. Minimal submanifolds

A minimal submanifold is an immersion $x: M^n \to X^N$ of an n-dimensional differentiable manifold M^n (or simply M) into a Riemannian manifold X^N of dimension N, which *locally* solves the Plateau problem: Every point $x \in M$ has a neighborhood U such that U is of smallest n-dimensional area compared with other n-dimensional submanifolds having the same boundary ∂U. Analytically the condition can be expressed as follows: Let D^2x be the second differential on M in the sense of Levi-Civita. Then (D^2x, ξ), where ξ is a normal vector to M at x, is a quadratic differential form, the second fundamental form relative to ξ. The differential equation to be satisfied by M is

$$\operatorname{Tr}(D^2x, \xi) = 0, \qquad \text{all } \xi. \tag{2}$$

It is a system of non-linear elliptic partial differential equations of the second order, whose number is equal to the codimension $N - n$. A minimal submanifold of dimension one is a geodesic.

We wish to study the properties of complete minimal submanifolds in a given Riemannian manifold X^N (cf. [12]). Except for geodesics the interest has so far been restricted to the case when the ambient space X^N is either the Euclidean space E^N or the unit sphere $S^N(1)$ imbedded in E^{N+1}.

For a minimal submanifold $x: M^n \to E^N$ in the Euclidean space a condition equivalent to (2) is that the coordinate functions are harmonic (relative to the induced metric). It follows that for $n > 0$ a complete minimal submanifold in E^N is non-compact.

For various reasons the case of codimension one (i. e., the minimal hypersurfaces) is the most important. Let $x_1, \ldots, x_n, z$ be the coordinates in E^{n+1}. Consider minimal hypersurfaces defined by the equation

$$z = F(x_1, \ldots, x_n) \tag{3}$$

for all $x_1, \ldots, x_n$. The following fundamental theorem generalizes the classical theorem of Bernstein and was the combined effort of de Giorgi ($n = 3$), Almgren ($n = 4$), Simons ($n \leqslant 7$), Bombieri, de Giorgi, Giusti ($n \geqslant 8$) [6, 32]. The minimal hypersurface defined by (3) must be a hyperplane for $n \leqslant 7$ and is not always a hyperplane for $n \geqslant 8$.

The main reason for this difference is the existence of absolute minimum cones in high-dimensional Euclidean space, which in turn depends on properties of compact minimal hypersurfaces in $S^n(1)$. From a general viewpoint the study of compact minimal submanifolds in $S^N(1)$ is attractive for its own sake. The first uniqueness theorem is the theorem of Almgren-Calabi [11]. If a two-sphere is immersed as a minimal surface in $S^3(1)$, it must be the equator.

By a counter-example of Hsiang [22] this theorem is not true for the next dimension. However, the following question, which can be designated as the " spherical Bernstein problem ", is unanswered:

Problem VI. — Let the n-sphere be *imbedded* as a minimal hypersurface in $S^{n+1}(1)$. Is it an equator?

Two-dimensional minimal surfaces in E^N and in $S^N(1)$ have been more thoroughly studied, because of the application of complex function theory. If the surface is itself a two-sphere (hence in $S^N(1)$), severe restriction is imposed for global reason and we have the following theorem (Boruvka, do Carmo, Wallach, Chern, but mainly Calabi [8, 14]). Let the two-sphere be immersed in $S^N(1)$ as a minimal surface, such that it does not belong to an equator. Then we have: (1) N is even; (2) The total area of the surface is an integral multiple of 2π; (3) If the induced metric is of constant Gaussian curvature, it is completely determined up to motions in $S^N(1)$ and the Gaussian curvature has the value

$$K = \frac{2}{m(m+1)}, \qquad N = 2m. \tag{4}$$

(4) There are minimal two-spheres in $S^N(1)$ of non-constant Gaussian curvature; all these with a given area form a finite-dimensional space.

The immersion of the n-sphere as a minimal submanifold of $S^N(1)$ is a fascinating problem. If the induced metric has constant curvature, the immersion is given by the spherical harmonics (Takahashi). For $n > 2$ two isometric minimal immersions $S^n(a) \to S^N(1)$ are not necessarily equivalent under the motions of the ambient space (do Carmo, Wallach [15]). In view of the precise results on the two-sphere we wish to propose the following problem:

Problem VII. — Consider minimal immersions $S^n \to S^N(1)$ with total area $\leqslant A$ ($=$ const) and identify those which differ by a motion of the ambient space. Is the resulting set a finite-dimensional space with some natural topology?

4. Isometric mappings

A differentiable mapping f: M $\to V$ of Riemannian manifolds is called *isometric* if it preserves the lengths of tangent vectors. It is therefore necessarily an immersion, and $\dim M \leqslant \dim V$. Classical differential geometry deals almost exclusively with the case that V is the Euclidean space E^N of dimension N. We believe this is the most interesting case and we will adopt this restriction in our discussion.

The first problem is that of existence. Since the fundamental tensor on a Riemannian manifold of dimension n involves $n(n+1)/2$ components, Schläfli conjectured in 1871 that every Riemannian manifold of dimension n can be locally imbedded in E^N, with $N = \frac{1}{2}n(n+1)$. This was proved by Elie Cartan in 1927 for the real analytic case. For smooth non-analytic manifolds this local isometric imbedding problem is unsolved, even for $n = 2$, unless some restriction on the metric is imposed such as the Gaussian curvature keeping a constant sign. In other words, it is not known whether any smooth two-dimensional Riemannian manifold can be locally isometrically imbedded in E^3. The answer is probably negative.

The two important global imbedding theorems are:

(1) (Weyl's Problem). A compact two-dimensional Riemannian manifold of positive Gaussian curvature can be isometrically imbedded in E^3 (as a convex surface).

(2) (Nash's Theorem [18, 28]). A compact (resp. non-compact) C^∞ Riemannian manifold of dimension n can be isometrically imbedded in E^N,

$$N = \frac{1}{2}n(3n + 11) \quad (\text{resp. } N = 2(2n + 1)(3n + 7)) \ (^4)$$

The second problem is the uniqueness of the isometric imbedding, also called rigidity, which is the problem whether an isometric immersion is determined up to a rigid motion of the ambient space E^N. Most interesting is the classical case of surfaces in E^3. Cohn-Vossen proved the rigidity of compact surfaces with Gaussian curvature $K > 0$ and the theorem was extended by Voss [35] to the case $K \geqslant 0$. Even before Cohn-Vossen, Liebmann proved that a smooth family of isometric compact convex surfaces (i. e., $K > 0$) is trivial, i. e., it consists of the surfaces obtained by the rigid motion of one member of the family. It is not known whether the same is true when the curvature condition is dropped and we believe the following problem is fundamental:

PROBLEM VIII. — Let M be a compact surface and I be the interval $-1 < t < 1$. Let $f: M \times I \to E^3$ be a differentiable mapping such that $f_t : M \to E^3$ defined by $f_t(x) = f(x, t)$, $x \in M$, $t \in I$, is an immersion for each t. Suppose that the metric ds_t^2 induced by f_t on M is independent of t. Does there exist a rigid motion $g(t)$ such that

$$f_t(x) = g(t) f_0(x), \qquad x \in M, \tag{5}$$

where the right-hand side denotes the action on f_0 by $g(t)$?

The following remarks may be relevant to the problem. Cohn-Vossen [13] proved the existence of an unstable family of compact surfaces of revolution, i. e., that the above conclusion is not true if the hypothesis that ds_t^2 is independent of t is replaced by

$$\frac{\partial}{\partial t} ds_t^2 \,|_{t=0} = \frac{\partial^2}{\partial t^2} ds_t^2 \,|_{t=0} = 0 \tag{6}$$

There are well-known examples showing that Cohn-Vossen's rigidity theorem is not true without the convexity condition $K \geqslant 0$. A generalization of the latter condition to surfaces of higher genus is the notion of *tightness*. Let $f: M \to E^3$ be an immersed surface. The tangent plane at a point x is a local (resp. global) support plane if a neighborhood of the surface at x (resp. the whole surface $f(M)$) lies at one side of it. The surface is called tightly immersed if every local support plane is a global

(^{4}) The value for N in the case of non-compact manifolds is an improvement of NASH's value by GREENE [18].

support plane. A. D. Alexandrow proved that a real analytic tightly imbedded surface of genus one is rigid and Nirenberg [29] replaced the analyticity condition by some other conditions.

On the other hand, the notion of tightness has a meaning for polyhedral surfaces. In this case the rigidity problem asks whether the congruence of corresponding faces of two tightly imbedded polyhedral surfaces implies that they differ by a rigid motion. Cauchy's classical theorem says that this is true if the surfaces are of genus zero. But Banchoff [2] has constructed examples showing that this is untrue for surfaces of genus one. From these remarks it is anybody's guess whether the answer to the question in Problem VIII is affirmative or negative.

When M is of dimension greater than two, isometry is a strong condition and there are local rigidity theorems.

5. Holomorphic mappings

A holomorphic mapping $f: M \to V$ of complex manifolds is a continuous mapping which is locally defined by expressing the coordinates of the image point as holomorphic functions of those of the original point. The most significant example is the case when M is the complex line C and V is the complex projective line $P_1(C)$ (or the Riemann sphere), in which case the mapping is known as a meromorphic function. Much recent progress has been made in extending classical geometrical function theory to the study of holomorphic mappings.

A holomorphic mapping is called non-denegerate if the Jacobian matrix is of maximum rank at some point. For given M, V there may not exist a non-degenerate holomorphic mapping. Let B be a closed subset of V. Classically the following problem has been much studied.

Intersection or non-existence problem. Find B such that there is no non-degenerate holomorphic mapping $M \to V - B$, i. e., every non-degenerate holomorphic mapping $f: M \to V$ has the property $f(M) \cap B \neq \emptyset$.

The Picard theorem concerns the case $M = C$, $V = P_1(C)$, and B is the set of three distinct points. Clearly if the property holds for B, it holds for a subset containing B, so that a stronger theorem results from a smaller subset B. In view of the extreme importance and elegance of the Picard theorem, we wish to state the following conjecrure of Wu:

Problem and conjecture IX. — Let C_n be the n-dimensional complex number space and $P_n(C)$ the n-dimensional complex projective space. Let B be the set of $n + 2$ hyperplanes of $P_n(C)$ in general position (i. e., any $n + 1$ of them are the faces of a non-degenerate n-simplex). Then there is no non-degenerate holomorphic mapping $C_n \to P_n(C) - B$.

The Picard theorem says that this is true for $n = 1$. Wu has established this for $n \leqslant 4$. Moreover, if we set

$$\rho(n) = \begin{cases} \left(\dfrac{n}{2} + 1\right)^2 + 1, & n \text{ even} \\ \left(\dfrac{n+1}{2}\right)\left(\dfrac{n+3}{2}\right) + 1, & n \text{ odd}, \end{cases}$$

and let B' be the set of $\rho(n)$ hyperplanes in general position in $P_n(C)$, then Wu [36] proved that every holomorphic mapping $f: C_n \to P_n(C) - B'$ must reduce to a constant.

A far-reaching generalization of the Picard theory is the equi-distribution theory of Nevanlinna, which studies the frequency that a non-constant meromorphic function takes given values. In terms of vector bundles the problem can be generalized as follows [7]. Let M be a complex manifold and $p: E \to M$ a holomorphic vector bundle over M. A holomorphic mapping $s: M \to E$ is called a section if $p \cdot s =$ identity. Let W be a finite-dimensional vector space of holomorphic sections. Suppose the manifold and the bundle fulfill some convexity conditions (which are automatically satisfied in the classical case). Then we can define, to each $s(\neq 0) \in W$, a defect $\delta(s)$ satisfying the conditions: (1) $0 \leqslant \delta(s) \leqslant 1$; (2) $\delta(\lambda s) = \delta(s)$, $\lambda \in C - \{0\}$; (3) $\delta(s) = 1$ if s has no zero. The equi-distribution problem is to find an upper bound of an average of $\delta(s)$ (a sum in the case of a finite number of sections and an integral in the case of an infinite set). The problem has been studied recently by several authors.

Dual to the intersection problem is the extension problem: Given complex manifolds M, V and a closed subset $A \subset M$. When is a holomorphic mapping $M - A \to V$ the restriction of a holomorphic mapping $M \to V$?

Many extension theorems are known. In several complex variables the most famous are the Hartogs and Riemann extension theorems, which concern with the case that V is either the complex line or a bounded set of it. We wish to formulate the following problem of Hartogs type where the curvature of the image manifold enters into play:

PROBLEM X. — Let Δ be an n-ball in C_n, $n \geqslant 2$, and let V be a complete hermitian manifold of holomorphic sectional curvature $\leqslant 0$. Is it true that every holomorphic mapping of a neighborhood of the boundary $\partial\Delta$ of Δ into V extends into a holomorphic mapping of Δ into V?

It *is* known that without the curvature condition on V the assertion is not true [24]. The problem belongs to an area which might be described as " hyperbolic complex analysis ". The philosophy is that negative curvature of the receiving space limits the holomorphic mappings and allows strong theorems. In fact, a bounded holomorphic function is a mapping into a ball which has the non-euclidean hyperbolic metric.

REFERENCES

[1] L. AUSLANDER. — The structure of complete locally affine manifolds, *Topology*, 3 (1964), suppl. 1, pp. 131-139.

[2] T. BANCHOFF. — Non-rigidity theorems for tight polyhedra, to appear in *Archiv d. Math.*

[3] M. BERGER. — Les variétés riemanniennes homogènes normales simplement connexes a courbure strictement positive, *Ann. Scuola Norm. Sup. Pisa*, 15 (1961), pp. 179-246.

[4] —. — Sur quelques variétés riemanniennes suffisamment pincées, *Bull. Soc. Math. France*, 88 (1960), pp. 57-71.

[5] —. — Trois remarques sur les variétés riemanniennes à courbure positive, *Comptes Rendus* (Paris), 263 (1966), A 76-A 78.

[6] Bombieri, E. de Giorgi and E. Giusti. — Minimal cones and the Bernstein problem, *Inventiones Math.*, 7 (1969), pp. 243-268.

[7] R. Bott and S. S. Chern. — Hermitian vector bundles and the equidistribution of the zeroes of their holomorphic sections, *Acta Math.*, 114 (1965), pp. 71-112.

[8] E. Calabi. — Minimal immersions of surfaces in euclidean spheres, *J. Diff. Geom.*, 1 (1967), pp. 111-125.

[9] J. Cheeger and D. Gromoll. — The structure of complete manifolds of non-negative curvature, *Bull. Amer. Math. Soc.*, 74 (1968), pp. 1147-1150.

[10] S. Chern. — On curvature and characteristic classes of a Riemann manifold, *Abh. Math. Sem. Univ. Hamburg*, 20 (1955), pp. 117-126.

[11] —. — Simple proofs of two theorems on minimal surfaces, *L'Enseig. Math.*, 15 (1969), pp. 53-61.

[12] —. — Brief survey of minimal submanifolds, *Tagungsberichte, Oberwolfach* (1969).

[13] S. Cohn-Vossen. — Unstarre geschlossene Flächen, *Math. Ann.*, 102 (1929), pp. 10-29.

[14] M. do Carmo and N. Wallach. — Representations of compact groups and minimal immersions into spheres, to appear in *J. Diff. Geom.*

[15] —. — Minimal immersions of spheres into spheres, *Proc. Nat. Acad. Sci.* (USA), 63 (1969), pp. 640-642.

[16] R. Finn. — On a class of conformal metrics, with application to differential geometry in the large, *Comm. Math. Helv.*, 40 (1965), pp. 1-30.

[17] T. Frankel. — Manifolds with positive curvature, *Pacific J. Math.*, 11 (1961), pp. 165-174.

[18] Robert E. Greene. — Isometric embeddings of riemannian and pseudo-riemannian manifolds, *Memoirs Amer. Math. Soc.*, No. 97 (1970).

[19] D. Gromoll. — Differenzierbare Strukturen und Metriken positiver Krummung auf Spharen, *Math. Annalen*, 164 (1966), pp. 353-371.

[20] — and W. Meyer. — On complete open manifolds of positive curvature, *Annals of Math.*, 90 (1969), pp. 75-90.

[21] M. L. Gromov. — Stable mappings of foliations into manifolds, *Izvestia Akad. Nauk SSSR, Ser. Mat.*, 33 (1969), pp. 707-734.

[22] W. Y. Hsiang. — Remarks on closed minimal submanifolds in the standard riemannian *m*-sphere, *J. Diff. Geom.*, 1 (1967), pp. 257-267.

[23] A. Huber. — On subharmonic functions and differential geometry in the large, *Comm. Math. Helv.*, 32 (1957), pp. 13-72.

[24] H. Kerner. — Über die Fortsetzung holomorpher Abbildungen, *Archiv d. Math.*, 11 (1960), pp. 44-47.

[25] W. Klingenberg. — Über Riemannsche Mannigfaltigkeiten mit nach oben beschränkter Krümmung, *Annali di Mat.*, 60 (1962), pp. 49-59.

[26] S. Kobayashi and T. Ochiai. — On complex manifolds with positive tangent bundles, to appear in *J. Math. Soc. Japan*.

[27] J. Milnor. — On the existence of a connection with curvature zero, *Comm. Math. Helv.*, 32 (1958), pp. 215-223.

[28] J. Nash. — The imbedding problem for riemannian manifolds, *Ann. of Math.*, 63 (1956), pp. 20-64.

[29] L. Nirenberg. — Rigidity of a class of closed surfaces, Nonlinear Problems, R. E. Langer (Editor) (1963), pp. 177-193.

[30] E. Portnoy. — Toward a generalized Gauss-Bonnet formula for complete open manifolds, Stanford thesis (1969).

[31] I. Satake. — The Gauss-Bonnet theorem for V-manifolds, *J. Math. Soc. Japan*, 9 (1957), pp. 464-492.

[32] J. Simons. — Minimal varieties in riemannian manifolds, *Ann. of Math.*, 88 (1968), pp. 62-105.

[33] J. Thorpe. — The zeroes of non-negative curvature operators, to appear.

[34] O. VEBLEN and J. H. C. WHITEHEAD. — *Foundations of differential geometry*, Cambridge Univ. Press, New York (1932).
[35] K. VOSS. — Differentialgeometrie geschlossener Flächen im euklidischen Raum, *Jahresberichte deutscher Math. Ver.*, 63 (1960-1961), pp. 117-136.
[36] H. WU. — An n-dimensional extension of Picard's theorem, *Bull. Amer. Math. Soc.*, 75 (1969), pp. 1357-1361.

Department of Mathematics,
University of California
Berkeley, California 94720
(U. S. A.)

Added during proof, March 12, 1971: Problem X has been solved independently by P. Griffiths and B. Schiffman.

On the Minimal Immersions of the Two-sphere in a Space of Constant Curvature

SHIING-SHEN CHERN[1]

1. Introduction

This study arose from reading Calabi's interesting paper [2] on the minimal immersions of the two-sphere S^2 into an N-dimensional sphere S^N. Using an idea of H. Hopf, Calabi observed the strong implications of the fact that the submanifold is a two-sphere. His treatment made essential use of the global coordinates on S^N. We shall give a different approach by supposing only that the ambient space is a Riemannian manifold of constant sectional curvature. Some general remarks on the higher osculating spaces are intended to make the development more natural and to prepare the way for future applications to the case of submanifolds of higher dimensions.

2. Higher osculating spaces of a submanifold and fundamental forms

Let X be a Riemannian manifold of dimension N and

$$x: M \rightarrow X \tag{1}$$

be an immersion of a differentiable manifold M of dimension n into X. In this section we will agree on the following ranges of indices:

$$\begin{gathered} 1 \leqq A, B, C, D \leqq N, \\ 1 \leqq i, j \leqq n, \\ n+1 \leqq \alpha, \beta \leqq N. \end{gathered} \tag{2}$$

[1] Work done under partial support of NSF grant GP-8623.

Let e_A be a local orthonormal frame field. The Levi–Civita connection defines the covariant differentials

$$De_A = \sum_B \omega_{AB} e_B, \tag{3}$$

where

$$\omega_{AB} + \omega_{BA} = 0. \tag{4}$$

(See [3] for details.) If ω_B is a coframe field dual to e_A, so that we have

$$ds^2 = \sum_A \omega_A^2, \tag{5}$$

the structure equations of the space are

$$\begin{aligned} d\omega_A &= \sum_B \omega_B \wedge \omega_{BA}, \\ d\omega_{AB} &= \sum_C \omega_{AC} \wedge \omega_{CB} + \Omega_{AB}, \end{aligned} \tag{6}$$

where

$$\Omega_{AB} = -\tfrac{1}{2} \sum_{C,D} R_{ABCD} \omega_C \wedge \omega_D. \tag{7}$$

The space X is said to be of constant curvature c, if and only if

$$\Omega_{AB} = -c\omega_A \wedge \omega_B. \tag{8}$$

Throughout this paper we will suppose that X is of constant curvature.

To study the geometry of the immersed submanifold M we restrict ourselves (locally) to orthonormal frame fields over M such that e_i are tangent vectors at $x \in M$, so that e_α are normal vectors to M. Then we have

$$\omega_\alpha = 0. \tag{9}$$

By the first structure equation in (6) and Cartan's lemma, we can put

$$\omega_{i\alpha} = \sum_j h_{\alpha ij} \omega_j, \qquad h_{\alpha ij} = h_{\alpha ji}. \tag{10}$$

A curve C through x is a smooth function $x(s)$, $|s| < L$, with $x(0) = x$. We can suppose the parameter s to be its length. By covariant differentiations along C we get the vector fields

$$\frac{Dx}{ds}, \frac{D^2x}{ds^2}, \frac{D^3x}{ds^3}, \ldots \tag{11}$$

At $s = 0$ the vectors

$$\frac{Dx}{ds}, \frac{D^2x}{ds^2}, \ldots, \frac{D^mx}{ds^m}$$

are said to span the *osculating space of order m*, the osculating space of order one being the tangent line. The *osculating space $T_x^{(m)}$ of order m of M at $x \in M$* is defined to be the space spanned by all the osculating spaces of order m of the curves through x and lying on M. We then have

$$T_x^{(1)} \subset T_x^{(2)} \subset \cdots \subset T_x^{(m)} \subset \cdots, \tag{12}$$

where $T_x^{(1)}$ is the tangent space to M at x and we shall write $T_x^{(1)} = T_x$. Let

$$\dim T_x^{(m)} = n + p_1(x) + \cdots + p_{m-1}(x), \qquad m = 1, 2, \ldots. \tag{13}$$

A point $x \in M$ is called a *regular point of order m*, if

$$p_a(x) = \text{const}, \qquad a = 1, \ldots, m-1,$$

in a neighborhood of x.

We can write

$$Dx = \sum_i \omega_i e_i, \tag{14}$$

so that, by (3) and (10),

$$\begin{aligned} D^2x &= \sum_{i,\alpha} \omega_i \omega_{i\alpha} e_\alpha + \text{terms in } e_i \\ &= \sum_{i,j,\alpha} h_{\alpha ij} \omega_i \omega_j e_\alpha + \text{terms in } e_i. \end{aligned} \tag{15}$$

The quadratic differential forms

$$(e_\alpha, D^2x) = \sum_{i,j} h_{\alpha ij} \omega_i \omega_j \tag{16}$$

are called the *second fundamental forms* of M. The integer $p_1(x)$ defined previously is equal to the number of linearly independent vectors among $\sum_\alpha h_{\alpha ij} e_\alpha$, $i \leqq j$, and is also the number of linearly independent second fundamental forms. In the generic case we have $p_1(x) = n(n+1)/2$.

Suppose now that x is a regular point of order $m - 1 \geqq 2$, so that $p_1(y), \ldots, p_{m-2}(y)$ are constants when $y \in M$ is in a neighborhood of x. We shall use the following ranges of indices:

$$\begin{aligned} 1 &\leqq \lambda_0 = i \leqq n, \\ n + 1 &\leqq \lambda_1 \leqq n + p_1, \\ n + p_1 + 1 &\leqq \lambda_2 \leqq n + p_1 + p_2, \\ &\cdots \\ n + p_1 + \cdots + p_{m-2} + 1 &\leqq \lambda_{m-1} \leqq N. \end{aligned} \tag{17}$$

Let e_A be a local orthonormal frame field, such that $e_{\lambda_0}, e_{\lambda_1}, \ldots, e_{\lambda_b}$ span $T_x^{(b+1)}$, $b = 0, 1, \ldots, m-2$. We then have

$$\omega_{\lambda_{b-1}\lambda_{b+1}} = 0, \qquad b = 1, \ldots, m-2. \tag{18}$$

By exterior differentiation of (18) and making use of the second equation in (6) and the equation (8), we obtain

$$\sum_{\lambda_b} \omega_{\lambda_{b-1}\lambda_b} \wedge \omega_{\lambda_b\lambda_{b+1}} = 0, \qquad b = 1, \ldots, m-2. \tag{19}$$

This allows us to introduce by recurrence the quantities $h_{\lambda_b i_1 \cdots i_{b+1}}$ defined by the equations

$$\sum_{\lambda_b} h_{\lambda_b i_1 \cdots i_{b+1}} \omega_{\lambda_b \lambda_{b+1}} = \sum_{i_{b+1}} h_{\lambda_{b+1} i_1 \cdots i_{b+2}} \omega_{i_{b+2}}, \tag{20}$$

and the $h_{\lambda_b i_1 \cdots i_{b+1}}$ are symmetric in the Latin indices $i_1, \ldots, i_{b+1}$. We find

$$\begin{aligned} (e_{\lambda_{m-1}}, D^m x) &= \sum_{i,\lambda_1,\ldots,\lambda_{m-2}} \omega_i \omega_{i\lambda_1} \omega_{\lambda_1\lambda_2} \cdots \omega_{\lambda_{m-2}\lambda_{m-1}} \\ &= \sum_{i_1,\ldots,i_m} h_{\lambda_{m-1} i_1 \cdots i_m} \omega_{i_1} \cdots \omega_{i_m}, \end{aligned} \tag{21}$$

which are differential forms of degree m and are to be called the mth *fundamental forms* of M in X. By (20) it follows that, if the mth fundamental forms vanish at a point, those of higher order vanish at the same point.

For later applications it will be convenient to write the structure equations (6) in relation with the equations of parallel displacement (3). It suffices to take the second covariant differential in the sense of the exterior calculus. We then have

$$\hat{D}^2 e_A = \sum_B \Omega_{AB} e_B = -c\omega_A \wedge \sum_B \omega_B e_B, \tag{22}$$

where the notation $\hat{D}^2$ is used to distinguish it from the D^2 used previously. It follows that for our frame field attached to a submanifold M we have

$$\begin{aligned} \hat{D}^2 e_i &= -c\omega_i \wedge \sum_j \omega_j e_j, \\ \hat{D}^2 e_\alpha &= 0. \end{aligned} \tag{23}$$

3. Minimal surfaces

We consider the special case that M is a two-dimensional manifold immersed as a minimal surface in X (so that $n = 2$ in the formulas of the last section). The latter condition is expressed analytically by

$$\sum_i h_{\alpha ii} = 0. \tag{24}$$

From (20) it follows that

$$(25) \qquad \sum_j h_{\lambda_b jji_3 \cdots i_{b+1}} = 0, \qquad b = 1, 2, \ldots.$$

Because of the symmetry of $h_{\lambda_b i_1 \cdots i_{b+1}}$ in its Latin indices, the same relation holds when contracted with respect to any two Latin indices. At a generic point we have, therefore,

$$(26) \qquad p_1(x) = p_2(x) = \cdots = 2.$$

It will be convenient in this case to make use of the complex notation and the isothermal coordinates on M. We put

$$(27) \qquad \varphi = \omega_1 + i\omega_2,$$

so that the metric on M is

$$(28) \qquad ds^2 = \varphi\bar{\varphi}.$$

From the first equation of (6) we find

$$(29) \qquad d\varphi = -i\omega_{12} \wedge \varphi.$$

The Gaussian curvature K of M is defined by the equation

$$(30) \qquad d\omega_{12} = -\frac{i}{2} K\varphi \wedge \bar{\varphi}.$$

The existence of local isothermal coordinates means that we can write, locally,

$$(31) \qquad \varphi = \lambda dz.$$

Substituting into (29), we obtain

$$(32) \qquad (d\lambda + i\omega_{12}\lambda) \wedge dz = 0,$$

so that $d\lambda + i\omega_{12}\lambda$ is a multiple of dz. This remark will be useful later on.

From (21) and (25) we see that the mth fundamental forms at a generic point of a minimal surface can be written

$$(33) \qquad (e_\lambda, D^m x) = \sum_{i_1, \ldots, i_m} h_{\lambda i_1 \cdots i_m} \omega_{i_1} \cdots \omega_{i_m} = \operatorname{Re}(\overline{H}_\lambda^{(m)} \varphi^m), \qquad 2m - 1 \leqq \lambda \leqq N,$$

where

$$(34) \qquad H_\lambda^{(m)} = h_{\lambda \underbrace{1 \cdots 1}_{m}} + i h_{\lambda \underbrace{1 \cdots 1}_{m-1} 2}$$

4. A theorem on an elliptic differential system

We wish to prove the following theorem:

THEOREM. *Let $w_\alpha(z)$ be complex-valued functions which satisfy the differential system*

$$\frac{\partial w_\alpha}{\partial \bar{z}} = \sum_\beta a_{\alpha\beta} w_\beta, \qquad 1 \leqq \alpha, \beta \leqq p, \tag{35}$$

in a neighborhood of $z = 0$, where $a_{\alpha\beta}$ are complex-valued C^1-functions. Suppose the w_α do not all vanish identically in a neighborhood of $z = 0$. Then:

(1) *the common zeroes of w_α are isolated;*
(2) *at a common zero of w_α the ratios $w_1 : \cdots : w_p$ tend to a limit.*

LEMMA 1. *Under the hypotheses of the theorem let $w_\alpha = o(|z|^{r-1})$ at $z = 0$, $r \geqq 1$. Then $\lim_{z\to 0} w_\alpha(z) z^{-r}$ exists.*

PROOF. Let D be a disk of radius R about $z = 0$, and let $\zeta \in D$, $\zeta \neq 0$. Write

$$z = x + iy, \qquad \zeta = \xi + i\eta.$$

Let $\rho(z) = \max_\alpha |w_\alpha|$. From (35) we obtain

$$d\left\{\frac{w_\alpha\, dz}{z^r(z - \zeta)}\right\} = \frac{1}{z^r(z-\zeta)} \sum_\beta a_{\alpha\beta} w_\beta\, d\bar{z} \wedge dz.$$

By Stokes' Theorem it follows that

$$-2\pi i w_\alpha(\zeta)\zeta^{-r} + \int_{\partial D} \frac{w_\alpha\, dz}{z^r(z-\zeta)} = \iint_D \frac{\sum_\beta a_{\alpha\beta} w_\beta}{z^r(z-\zeta)}\, d\bar{z} \wedge dz. \tag{36}$$

Taking absolute value, we obtain

$$2\pi\rho(\zeta)|\zeta^{-r}| \leqq \int_{\partial D} \frac{\rho(z)|dz|}{|z^r(z-\zeta)|} + A \iint_D \frac{\rho(z)\, dx\, dy}{|z^r(z-\zeta)|}, \tag{37}$$

where $A > 0$ is a constant depending only on the system (35). With $z_0 \in D$, $z_0 \neq 0$, we integrate this inequality with respect to $d\xi\, d\eta / |\zeta - z_0|$ over D. Remarking that

$$\frac{1}{|(z-\zeta)(z_0 - \zeta)|} = \frac{1}{|z - z_0|}\left|\frac{1}{z-\zeta} + \frac{1}{\zeta - z_0}\right|,$$

$$\iint_D \frac{dx\, dy}{|z - \zeta|} < 2R, \tag{38}$$

the integration gives

$$2\pi \iint_D \frac{\rho(\zeta)\, d\xi\, d\eta}{|\zeta^r(\zeta - z_0)|} \leqq 4R \int_{\partial D} \frac{\rho(z)|dz|}{|z^r(z - z_0)|} + 4AR \iint_D \frac{\rho(z)\, dx\, dy}{|z^r(z - z_0)|}$$

or

$$(2\pi - 4AR) \iint_D \frac{\rho(z)\, dx\, dy}{|z^r(z - z_0)|} \leqq 4R \int_{\partial D} \frac{\rho(z)|dz|}{|z^r(z - z_0)|}.$$

We choose R so small that $2\pi - 4AR > 0$. Then the integral at the left-hand side of the preceding equation is bounded as $z_0 \to 0$. It follows from (37) that $\rho(\zeta)|\zeta^{-r}|$, and hence $|w_\alpha(\zeta)\zeta^{-r}|$, is bounded as $\zeta \to 0$. By (36) we see that $\lim_{\zeta \to 0} w_\alpha(\zeta)\zeta^{-r}$ exists.

LEMMA 2. *Under the hypotheses of the theorem suppose* $w_\alpha = o(|z|^{r-1})$, *all r. Then* $w_\alpha \equiv 0$ *in a neighborhood of* $z = 0$.

PROOF. Suppose, to the contrary, that $\rho(z_0) \neq 0$, $|z_0| < R$. We multiply (37) by $d\xi\, d\eta$ and integrate over D. This gives

$$2\pi \iint_D \rho(\zeta)|\zeta^{-r}|\, d\xi\, d\eta \leqq 2R \int_{\partial D} \rho(z)|z^{-r}|\, |dz| + 2AR \iint_D \rho(z)|z^{-r}|\, dx\, dy,$$

or

$$(2\pi - 2AR) \iint_D \rho(z)|z^{-r}|\, dx\, dy \leqq 2R \int_{\partial D} \rho(z)|z^{-r}|\, |dz|.$$

There exist two positive constants a and b independent of r such that the left-hand side of the above inequality is $\geqq a|z_0|^{-r}$ and the right-hand side is $\leqq bR^{-r}$. Combining them, we obtain

$$\frac{a}{b} \leqq \left|\frac{z_0}{R}\right|^r.$$

But this leads to a contradiction as $r \to \infty$.

From the two lemmas the theorem follows immediately. In fact, suppose $w_\alpha(0) = 0$ and suppose $w_\alpha(z)$ do not all vanish in a neighborhood of $z = 0$. By Lemma 2 there is a largest integer $r \geqq 1$ such that $w_\alpha = o(|z|^{r-1})$ at $z = 0$. By Lemma 1, $\lim_{z \to 0} w_\alpha(z)z^{-r}$ exists. These limits cannot be all zero, as otherwise we would have $w_\alpha = o(|z|^r)$. This implies the two statements in the theorem.

5. Minimal immersions of the two-sphere

Consider now the minimal immersion

$$S^2 \to X, \tag{39}$$

where, as before, X is a Riemannian manifold of dimension N and of constant curvature c. It will be convenient to make use of complex vectors, and we put

$$E_1 = e_1 + ie_2. \tag{40}$$

By (23) we have

$$\hat{D}^2 E_1 = -\frac{c}{2}\varphi \wedge \bar{\varphi} E_1. \tag{41}$$

On the other hand, using (10) and the conditions (24), we can write

$$DE_1 = -i\omega_{12}E_1 + \bar{\varphi}\sum_\alpha H_\alpha^{(2)} e_\alpha, \tag{42}$$

where

$$H_\alpha^{(2)} = h_{\alpha 11} + ih_{\alpha 12}. \tag{43}$$

The surface S^2 being supposed to be oriented, the frame e_1, e_2 in the tangent plane is defined up to a rotation, and E_1 up to the change

$$E_1 \to E_1^* = e^{i\tau}E_1, \tag{44}$$

where τ is real. Under the change (44), φ and $H_\alpha^{(2)}$ are changed according to

$$\begin{aligned} \varphi \to \varphi^* &= e^{i\tau}\varphi, \\ H_\alpha^{(2)} \to H_\alpha^{(2)*} &= e^{2i\tau}H_\alpha^{(2)}, \end{aligned} \tag{45}$$

from which it follows that

$$\bar{H}_\alpha^{(2)}\varphi^2 = \bar{H}_\alpha^{(2)*}\varphi^{*2}.$$

Thus the form

$$\sum_\alpha (\bar{H}_\alpha^{(2)})^2\varphi^4 \tag{46}$$

is globally defined on the surface S^2, being independent of the choice of the frame field.

Taking the exterior derivative of (42) and collecting the terms in e_α, we obtain, in view of (41),

$$\left(dH_\alpha^{(2)} + \sum_\beta H_\beta^{(2)}\omega_{\beta\alpha} + 2i\omega_{12}H_\alpha^{(2)}\right) \wedge \bar{\varphi} = 0,$$

or

$$d\bar{H}_\alpha^{(2)} + \sum_\beta \bar{H}_\beta^{(2)}\omega_{\beta\alpha} - 2i\omega_{12}\bar{H}_\alpha^{(2)} \equiv 0, \quad \text{mod } \varphi. \tag{47}$$

From (47) we derive

$$d\sum_\alpha (\bar{H}_\alpha^{(2)})^2 - 4i\omega_{12}\left(\sum_\alpha (\bar{H}_\alpha^{(2)})^2\right) \equiv 0, \quad \text{mod } \varphi. \tag{48}$$

We use the local isothermal coordinate z introduced in (31) and write the form (46) as $f(z)\,dz^4$. It follows from (32) and (48) that $f(z)$ is holomorphic, so that $f(z)\,dz^4$ is an Abelian form of degree 4 on S^2. This is possible only when $f(z) = 0$ or

(49) $$\sum_\alpha (H_\alpha^{(2)})^2 = 0,$$

or, by (43),

(50) $$\sum_\alpha h_{\alpha 11}^2 = \sum_\alpha h_{\alpha 12}^2, \qquad \sum_\alpha h_{\alpha 11} h_{\alpha 12} = 0.$$

Geometrically this means that the vectors $\sum_\alpha h_{\alpha 11} e_\alpha$ and $\sum_\alpha h_{\alpha 12} e_\alpha$ in the osculating space of the second order are perpendicular to each other and are of the same length

$$k_1 = \left(\sum_\alpha h_{\alpha 11}^2\right)^{1/2} \geqq 0.$$

The point is singular if and only if $H_\alpha^{(2)} = 0$ or $k_1 = 0$. From (47) and the theorem in §4 we see that, if $H_\alpha^{(2)} \not\equiv 0$, the singular points are isolated and that at each such singular point the "osculating space of the second order" is well defined and varies continuously with the point.

Suppose $k_1 \not\equiv 0$. We can choose the vectors e_3, e_4 of the frame field so that, if

(51) $$E_2 = e_3 + ie_4,$$

equation (42) can be written

(52) $$DE_1 = -i\omega_{12} E_1 + k_1 \bar{\varphi} E_2.$$

We extend the scalar product over complex vectors, so that it will be complex bilinear in both arguments. Then

(53) $$\begin{aligned} (E_1, \bar{E}_1) &= (E_2, \bar{E}_2) = 2, \\ (E_1, E_1) &= (E_1, E_2) = (E_2, E_2) = 0. \end{aligned}$$

Since D preserves the scalar product, we obtain from (3), (52), and (53),

$$DE_2 = -k_1 \varphi E_1 - i\omega_{34} E_2 + \Phi$$

where Φ is a linear combination of the vectors e_μ, $5 \leqq \mu \leqq N$. Suppose the point under consideration to be a regular point, i.e., $k_1 > 0$. Consideration of the terms in e_μ in $\hat{D}^2 E_1$ gives

$$\bar{\varphi} \wedge \Phi = 0,$$

so that Φ is of the form

$$\Phi = \frac{1}{k_1}\bar{\varphi}\Big(\sum_\mu H_\mu^{(3)} e_\mu\Big).$$

It should be remarked that with this notation, Re $(\bar{H}_\mu^{(3)}\varphi^3)$ are the third fundamental forms of the surface.

The form of degree 6:

$$\sum_\mu (\bar{H}_\mu^{(3)})^2 \varphi^6 \tag{54}$$

is independent of the choice of the frame field and, defined to be zero at a singular point, is well defined over the whole surface S^2. With the local isothermal coordinate z it can be written as $g(z)\,dz.^6$ We shall show that $g(z)$ is a holomorphic function. This follows immediately from the structure equations. In fact, consideration of the term in E_2 in $\hat{D}^2 E_1$ gives

$$\{dk_1 + ik_1(2\omega_{12} - \omega_{34})\} \wedge \bar{\varphi} = 0.$$

We can therefore put

$$\omega_{34} = 2\omega_{12} + \theta_1 \tag{55}$$

where θ_1 is a real one-form. Substituting back, we obtain

$$(dk_1 - ik_1\theta_1) \wedge \bar{\varphi} = 0$$

or

$$\theta_1 \equiv -id \log k_1, \quad \text{mod } \bar{\varphi}. \tag{56}$$

Since $D^2 E_2 = 0$, the terms involving e_μ give, on using (55) and (56),

$$d\bar{H}_\mu^{(3)} + \sum_\nu \bar{H}_\nu^{(3)}\omega_{\nu\mu} - 3i\omega_{12}\bar{H}_\mu^{(3)} \equiv 0, \quad \text{mod } \varphi, \qquad 5 \leqq \nu \leqq N, \tag{57}$$

from which it follows that

$$d\sum_\mu (\bar{H}_\mu^{(3)})^2 - 6i\omega_{12}\Big(\sum_\mu (\bar{H}_\mu^{(3)})^2\Big) \equiv 0, \quad \text{mod } \varphi. \tag{58}$$

Using (32), we derive, as before, that $g(z)$ is holomorphic. Thus the form (54) is an Abelian form of degree 6 and it must be zero. We can therefore define

$$E_3 = e_5 + ie_6 \tag{59}$$

such that

$$DE_2 = -k_1\varphi E_1 - i\omega_{34}E_2 + k_2\bar{\varphi}E_3.$$

Continuing in this way, we can define a local frame field e_A at a regular point of S^2, such that, if

$$E_s = e_{2s-1} + ie_{2s}, \qquad 1 \leqq s \leqq m, \tag{60}$$

we have

$$DE_s = -k_{s-1}\varphi E_{s-1} - i\omega_{2s-1,2s}E_s + k_s\bar{\varphi}E_{s+1}, \qquad 1 \leqq s \leqq m, \; k_0 = k_m = 0. \tag{61}$$

The integer m is the smallest integer such that DE_m is a linear combination of E_{m-1} and E_m. Equations (61) will be called the *Frenet–Boruvka Formulas* for the minimal immersion of a two-sphere in X. Under the change (44), E_s will be changed according to

$$E_s \to E_s^* = e^{sit}E_s. \tag{62}$$

Generalizing (55) we set

$$\omega_{2s-1,2s} = s\omega_{12} + \theta_{s-1}, \qquad 2 \leqq s \leqq m. \tag{63}$$

Then the one-forms θ_{s-1}, $2 \leqq s \leqq m$, remain invariant under a change of the frame field in the tangent plane.

Computing $\hat{D}^2E_s$ and using (41) and

$$\hat{D}^2E_s = 0, \qquad s > 1, \tag{64}$$

we obtain

$$k_1^2 = \tfrac{1}{2}(c - K) \geqq 0 \tag{65}$$

and

$$\begin{aligned} &\{dk_s + ik_s(\theta_s - \theta_{s-1})\} \wedge \varphi = 0, && \theta_0 = 0, \\ &d\theta_s + i\{k_s^2 - k_{s+1}^2 - \tfrac{1}{2}(s+1)K\}\varphi \wedge \bar{\varphi} = 0, && 1 \leqq s \leqq m - 1. \end{aligned} \tag{66}$$

These are the integrability conditions of (61), and they can be simplified. In fact, relative to the complex structure on S^2 we use the operators ∂, $\bar{\partial}$, and

$$d^c = i(\bar{\partial} - \partial). \tag{67}$$

From the fact that $d \log k_s$ and $\theta_s - \theta_{s-1}$ are real one-forms, we obtain from the first equation of (66),

$$\theta_s - \theta_{s-1} = d^c \log k_s, \qquad 1 \leqq s \leqq m - 1,$$

which gives

$$\theta_s = d^c \log (k_1 \cdots k_s), \qquad 1 \leqq s \leqq m - 1. \tag{68}$$

To a real-valued smooth function u let its Laplacian Δu be defined by

$$dd^c u = \frac{i}{2}\Delta u \varphi \wedge \bar{\varphi}. \tag{69}$$

Then, in view of (68), the second equation of (66) gives

$$\tfrac{1}{2}\Delta \log (k_1 \cdots k_s) + k_s^2 - k_{s+1}^2 - \tfrac{1}{2}(s+1)K = 0 \tag{70}$$

or

$$\tfrac{1}{2}\Delta \log k_s - k_{s-1}^2 + 2k_s^2 - k_{s+1}^2 - \tfrac{1}{2}K = 0. \tag{71}$$

We summarize our results in the following theorem:

THEOREM. *Let $S^2 \to X$ be a minimal immersion of the two-sphere into a Riemannian manifold of constant curvature. There is an integer m such that the osculating spaces of order m (and dimension $2m$) are parallel along the surface. The singular points are isolated. At a regular point a complete system of local invariants is given by the quantities $k_s > 0$, $1 \leqq s \leqq m-1$, which satisfy the conditions* (71).

The simplest case is when the Gaussian curvature K is constant. If $K = c$, the tangent plane is parallel along the surface and the surface is totally geodesic. If $K < c$, it follows from (65) and then recursively from (71) that k_s are all constants. The same relations give

$$K = \frac{2}{m(m+1)}c, \tag{72}$$

and all the k_s can be expressed in terms of c and m. The surface is thus determined up to an isometry in the space.

6. The case when the ambient space is the *N*-sphere

Consider the special case, studied by Calabi, of the minimal immersion

$$S^2 \to S^N, \tag{73}$$

where S^N is the sphere of radius 1 in R^{N+1} (so that $c = 1$). The preceding theorem has the consequence that the surface must lie on an even-dimensional great sphere S^{2m}. Without loss of generality we suppose $N = 2m$. In this case the tangent vectors of S^{2m} can be realized in R^{2m+1}, and the vectors E_s by vectors in the complex number space C^{2m+1}. If x is a point on the surface, we can write

$$dx = \tfrac{1}{2}\bar{\varphi}E_1 + \tfrac{1}{2}\varphi\bar{E}_1. \tag{74}$$

Moreover, the Levi–Civita connection is defined by orthogonal projection into the tangent spaces of S^{2m}, and we have

$$DE_1 = -\varphi x + dE_1,$$
$$DE_s = dE_s, \qquad 1 < s. \tag{75}$$

We interpret E_s as homogeneous coordinate vectors in the complex projective space $P_{2m}(C)$. Equation (61) with $s = m$ shows that E_m is an algebraic curve. It will be called the *generating curve* of the minimal surface. From the other equations of (61) we see that E_mE_{m-1} is the line tangent to the generating curve at E_m, etc., and $E_mE_{m-1}\cdots E_1$ is its osculating space of order $m - 1$. Since

$$(E_r, E_s) = 0, \qquad 1 \leqq r, s \leqq m,$$

the generating curve has the property that its osculating spaces of order $m - 1$ belong completely to the nonsingular hyperquadric Q_{2m-1} in $P_{2m}(C)$, the equation of Q_{2m-1} being

$$z_0^2 + z_1^2 + \cdots + z_m^2 = 0, \tag{76}$$

where $z_0, z_1, \ldots, z_m$ are the homogeneous coordinates in $P_{2m}(C)$. It is a classical result (see [4], p. 235) that the linear subspaces of dimension $m - 1$ on Q_{2m-1} form an irreducible algebraic variety of (complex) dimension $m(m + 1)/2$.

To construct the minimal surface from the generating curve, we take its conjugate complex curve $\bar{E}_m$. The corresponding osculating spaces of order m are conjugate complex and therefore meet in a real point x. The minimal surface is the locus of such points of intersection. Boruvka showed that if the generating curve is a normal curve, the resulting minimal surface has constant Gaussian curvature and is given by the spherical harmonics (see [1]). In general, the surface described by x may have isolated points without tangent plane, so that we obtain only a generalized surface.

By (65) we can relate the area of a minimal S^2 with a projective invariant of the generating curve, thus giving a geometrical interpretation of Calabi's result that this area is an integral multiple of 2π.

UNIVERSITY OF CALIFORNIA
BERKELEY

REFERENCES

1. BORUVKA, O., "Sur les surfaces représentées par les fonctions sphèriques de première espèce," *J. de Math. Pures et Appl.*, 1933, pp. 337–383.

2. CALABI, E., "Minimal immersions of surfaces in Euclidean spheres," *J. Diff. Geom.*, *1* (1967), pp. 111–125.
3. CHERN, S., *Minimal Submanifolds in a Riemannian Manifold.* Mimeographed lecture notes, University of Kansas, 1968.
4. HODGE, W. V. D., and D. PEDOE, *Methods of Algebraic Geometry*, Vol. II. New York: Cambridge University Press, 1952.

Reprinted from
Problems in Analysis, Princeton Univ. Press, (1970) 27–40

Reprinted from *Differentialgeometrie im Grossen.*
W. Klingenberg (ed.) **4** (1971) 43–60

BRIEF SURVEY OF MINIMAL SUBMANIFOLDS

Shiing-shen Chern in Berkeley

1. INTRODUCTION

The minimal surface equation in euclidean 3-space was first given by Lagrange in 1760. While remarking that there are solutions other than the plane, he did not give any explicitly. It was until 1776 that Meusnier found the minimal surfaces which are the helicoids and the catenoids. Meusnier's idea is instructive: he found them by determining: a) all minimal surfaces generated by lines parallel to a plane; b) all minimal surfaces of revolution.

The subsequent developments were dominated by the Plateau problem, till it was solved by T. Radò and J. Douglas in 1931. The following existence theorem was established: Given a closed rectifiable Jordan curve

C in space, there exists a generalized minimal surface with C as its boundary.

The minimal surface in this theorem was allowed to have isolated branch points where it fails to be an immersion. Recently Osserman proved that the solution is actually everywhere regular, i.e., free from branch points.

The problem can be generalized to an arbitrary riemannian manifold. Let X^N be a riemannian manifold of dimension N. An n-dimensional submanifold $M^n \to X^N$ is called minimal, if locally it is a submanifold of smallest n-dimensional area with a given boundary $C = \partial M^n$. The Plateau problem in this general case has been studied by E. R. Reifenberg, H. Federer, W. Fleming, E. de Giorgi, C. Morrey, etc., and existence theorems have been proved when "generalized solutions" are admitted.

We will be concerned only with smooth minimal submanifolds, defined by the condition that the first variation of the n-dimensional area is zero when its boundary ∂M^n is fixed. To express this condition we define the mean curvature vector as follows: Let D^2x be the second differential on M^n 1). If ξ is a normal vector to M^n at x, the scalar product

$$\text{(1)} \qquad \mathrm{II}_\xi = (D^2x, \xi)$$

is a quadratic differential form on M^n, which depends linearly on ξ; it is called the <u>second fundamental form</u>. Let $e_{n+1},\ldots,e_N$ be (locally) an orthonormal field of normal vectors to M^n. Then

$$\text{(2)} \qquad H = \frac{1}{n} \sum_{n+1\leq\alpha\leq N} \mathrm{Tr}(D^2x, e_\alpha)e_\alpha$$

is a field of normal vectors to M^n, independent of the choice of e_α; H is called the <u>mean curvature vector</u>. M^n is a minimal submanifold, if and only if $H = 0$; it is said to be of <u>constant mean curvature</u>, if the absolute differential $DH = 0$.

In the special case when X^N is the euclidean $(n+1)$-space E^{n+1} with the coordinates $x_1,\ldots,x_n$, z, and M^n is defined by

$$\text{(3)} \qquad z = z(x_1,\ldots,x_n),$$

the condition $H = 0$ becomes

$$\text{(4)} \qquad \sum_i \frac{\partial}{\partial x_i}\left(\frac{\frac{\partial z}{\partial x_i}}{\sqrt{1+\sum_k \left(\frac{\partial z}{\partial x_k}\right)^2}} \right) = 0, \qquad 1 \leq i,\ k \leq n.$$

In the general case a minimal submanifold M^n satisfies a non-linear elliptic system of differential equations of the second order with a number of equations equal to the codimension $N - n$. Our object is to study the properties of complete minimal submanifolds.

2. MINIMAL SUBMANIFOLDS ON A RIEMANNIAN MANIFOLD

A minimal submanifold of dimension one is a geodesic, about which extensive researches have been carried out. To my knowledge there are very few results on minimal submanifolds of dimension > 1 on an arbitrary riemannian manifold. I will mention the two following:

WIRTINGER'S THEOREM. A complex analytic submanifold on a Kaehler manifold is minimal.

HSIANG'S THEOREM. When a closed group of isometries acts on a compact riemannian manifold, an extremal orbit is a minimal submanifold.

Among consequences of these theorems are examples of minimal submanifolds. In analogy to results on geodesics the following problem is a natural one: Does a compact riemannian manifold always have a closed minimal surface (of dimension two)?

The answer in the general case is probably negative. But it might be affirmative when one imposes conditions on the manifold, such as a) being simply connected; b) having a constant sign for the sectional curvature c) having a metric which is a perturbation of a metric of constant curvature.

So far most of the studies have been restricted to the case where the ambient space X^N is either the euclidean space E^N or the sphere S^N with the metrics of constant curvature.

3. Examples

It is important to know some examples, not only because they are hard to get, even locally, but also because they serve as models with which other minimal submanifolds can be compared and results of a general nature can be derived.

In many ways the most notable minimal surface in E^3 is the Enneper surface. This has a generalization in E^N [6]: There is a simply connected, irreducible, rational surface of order N^2 in E^N, which is a minimal surface. It will be interesting to construct further examples of minimal submanifolds in E^N.

We denote by $S^N(a)$ the sphere of radius a in E^{N+1} with the induced metric. Its simplest minimal submanifolds are the great spheres, which are moreover totally geodesic (i.e., the second fundamental forms vanish identically). Of great interest are the closed minimal submanifolds on $S^N(1)$. We will state some theorems which lead to such examples.

THEOREM 1. *Let* O *be the center of the sphere* $S^N(1)$. *To a submanifold* $x: M^n \to S^N(1)$ *let* $\hat{x}: M^n \times I_\epsilon \to E^{N+1}$, *where* $I_\epsilon = \{t \mid \epsilon < t \leq 1, \epsilon > 0\}$, *be the cone over* M^n *defined by* $\hat{x}(p,t) = t\,x(p)$, $p \in M^n$, $t \in I_\epsilon$. *Then* $x(M^n)$ *is minimal in* $S^N(1)$ *if and only if* $\hat{x}(M^n \times I_\epsilon)$ is a minimal cone in E^{N+1}.

Combined with Wirtinger's theorem, it follows that the intersection of a complex analytic cone in the complex number space C_ν by its unit sphere $S^{2\nu-1}(1)$ of real dimension $2\nu - 1$ gives a minimal submanifold of codimension 2 on $S^{2\nu-1}(1)$.

THEOREM 2. *Let* $M^{n_k} \to S^{N_k}(a_k)$, $k = 1,2$, *be minimal submanifolds*. Suppose $a_1^2 + a_2^2 = 1$. *Then the immersion*

$$M^{n_1} \times M^{n_2} \to S^{N_1}(a_1) \times S^{N_2}(a_2) \to S^{N_1+N_2+1}(1)$$

(defined in an obvious way) *is minimal, if*

$$a_k^2 = \frac{n_k}{n_1+n_2}, \quad k = 1,2.$$

In particular, since $S^{N_k}(a_k)$ itself is a minimal submanifold of codimension zero on $S^{N_k}(a_k)$, we have the minimal submanifold

$$S^{N_1}\left(\sqrt{\frac{N_1}{N_1+N_2}}\right) \times S^{N_2}\left(\sqrt{\frac{N_2}{N_1+N_2}}\right) \to S^{N_1+N_2+1}(1).$$

We will call this the Clifford minimal hypersurface, the classical clifford surface being the case $N_1 = N_2 = 1$. Next to the great spheres these are perhaps the simplest closed minimal submanifolds on $S^N(1)$.

THEOREM 3. (Takahashi) Let $x: M^n \to E^{N+1}$, $|x| = 1$, be a submanifold in $S^N(1)$, and let H be its mean curvature vector and Δ its Laplacian relative to the induced riemannian metric. Then

$$(5) \qquad \Delta x = n(H-x).$$

In particular, the submanifold is minimal, if and only if x satisfies the differential equation

$$(6) \qquad \Delta x + nx = 0.$$

This theorem gives minimal submanifolds $x: S^n(a) \to S^N(1)$ by making use of the spherical harmonics of $S^n(a)$. These examples will be discussed further in a later section.

All the theorems 1, 2, and 3 are not difficult to prove. A much deeper theorem which gives the extent of the closed minimal submanifolds on $S^N(1)$ is the following:

THEOREM 4. (Lawson) Every compact surface but the projective plane can be minimally immersed into $S^3(1)$. Every compact orientable surface can be minimally imbedded in $S^3(1)$.

4. COMPLETE MINIMAL SUBMANIFOLDS IN E^N

Analogous to Theorem 3 in §3 it can be shown that the coordinate functions of a minimal submanifold in E^N are harmonic functions relative to the induced metric. This has the consequence that there is no compact minimal submanifold without boundary in E^N (except points) or, in other words, that every complete minimal submanifold of dimension > 0 in E^N is necessarily non-compact.

While the Plateau problem is one of existence, much of the geometry of minimal submanifolds deals with uniqueness problems. Most notable among them is the theorem of S. Bernstein (see [4] for a simple proof):

<u>Every minimal surface in</u> E^3 <u>with the coordinates</u> x, y, z, <u>which is representable in the form</u>

$$z = f(x,y)$$

<u>for all</u> x,y <u>is a plane</u>.

This theorem leads to the so-called Bernstein problem: Let

$$z = f(x_1,\dots,x_n), \quad \text{all } x_1,\dots,x_n,$$

be a minimal hypersurface in E^{n+1} with the coordinates $x_1,\dots,x_n,z$. Is the function f necessarily linear? The answer is affirmative for $n = 3$ (de Giorgi 1965), $n = 4$ (Almgren 1966), $n \leq 7$ (Simons 1967). Very recently Bombieri, de Giorgi, and Giusti proved that the answer is negative for $n \geq 8$. This must be one of the most exciting results on minimal submanifolds.

The case of minimal surfaces ($n = 2$) occupies a special position, because the problem is then closely related to the theory of holomorphic curves in a complex manifold. The connecting link is the Gauss map, defined as follows: Let $x: M^2 \to E^N$ be an oriented minimal surface and let $Gr(2,N)$ be the Grassman manifold of all oriented planes through a fixed point O of E^N. The Gauss map g:

$M^2 \to Gr(2,N)$ assigns to a point $p \in M^2$ the oriented plane through o parallel to the tangent plane to $x(M^2)$ at $x(p)$.

The surface M^2, through the induced metric, has an underlying conformal structure and is a complex curve (or Riemann surface). On the other hand,

$Gr(2,N) = \frac{SO(N)}{SO(2) \times SO(N-2)}$ has a complex structure invariant under the action of $SO(N)$. It makes $Gr(2,N)$ into a symmetric hermitian manifold, which is complex analytically equivalent to the non-degenerate hyperquadric in the complex projective space $P_{N-1}(C)$ of dimension $N-1$. We have the theorem: M^2 is a minimal surface in E^N if and only if the Gauss map g is anti-holomorphic.

The following is a geometrical generalization of the classical Bernstein theorem:

THEOREM. (DENSITY THEOREM). Let $x: M^2 \to E^N$ be a complete oriented minimal surface which is not a plane, and let $g: M^2 \to Gr(2,N) \subset P_{N-1}(C)$ be its Gauss map. The hyperplanes of $P_{N-1}(C)$ which meet the image $g(M^2)$ form a dense subset in the space P^*_{N-1} of all hyperplanes of $P_{N-1}(C)$. In the case $N = 3$ the image $g(M^2)$ is dense in $Gr(2,N)$.

The case $N = 3$ of the theorem is due to R. Osserman. It generalizes Bernstein's theorem, because $Gr(2,3)$ can be identified with the unit sphere with center 0 in E^3.

5. CLOSED MINIMAL SUBMANIFOLDS IN $S^N(1)$

We have given several examples of such submanifolds in §3. The simplest uniqueness theorem is the following:

THEOREM (ALMGREN-CALABI). A minimal surface $x: S^2 \to S^3(1)$ is necessarily a great sphere.

It is tempting to ask whether an immersed minimal hypersurface $x: S^{N-1} \to S^N(1)$ is a great hypersphere for $N \geqq 4$. That the answer to this question is negative is a result of the following example of Hsiang: In E^5 with the coordinates (x,y,z,u,v) take the intersection M^3 of the cubic cone

$$\begin{vmatrix} u & x & y \\ x & v & z \\ y & z & -u-v \end{vmatrix} = 0 \tag{7}$$

With the unit sphere $S^4(1)$. Then M^3 is a minimal hypersurface on $S^4(1)$ and is covered by S^3, so that this gives an immersion of S^3 in $S^4(1)$ as a minimal hypersurface without being a great hypersphere.

On the other hand, the following spherical Bernstein problem is unsolved: Is an imbedded minimal hypersurface $x: S^{N-1} \to S^N(1)$ a great hypersphere for $N \geq 4$?

In general, the question of immersing S^n as a minimal submanifold of $S^N(1)$ is a fascinating one. While there are numerous possibilities, an important case is when the induced metric is of constant sectional curvature, i.e., the case of isometric minimal immersions $x: S^n(a) \to S^N(1)$, where $S^n(a)$ has constant curvature $\frac{1}{a^2}$. By Takahashi's theorem the immersion will be given by the solutions of the equation (6). More precisely, we have the theorem:

Theorem. Let $x: M^n \to E^{N+1}$ be an isometric immersion of a riemannian manifold M^n, $x(p) = (x_o(p),\ldots,x_N(p))$, $p \in M^n$, where x_A, $0 \leq A \leq N$, satisfies the differential equation

$$\Delta x_A + \lambda x_A = 0, \qquad \lambda > 0. \tag{8}$$

Then $x(M^n)$ is contained in the sphere $S^N(\sqrt{\frac{n}{\lambda}})$ as a minimal submanifold.

In the particular case $M^n = S^n(a)$ the solutions of (8) are the spherical harmonics on $S^n(a)$. These lead to a family of isometric minimal immersions $x_s: S^n(a) \to S^N(1)$ first described by M. do Carmo and N. R. Wallach:

(9) $$N(s) = (2s+n-1)\frac{(s+n-2)!}{s!(n-1)!} - 1,$$

$$a^2 = \frac{s(s+n-1)}{n},$$

when s is a positive integer, being in fact the degree of the spherical harmonics in question ($N(s) + 1$ is the maximum number of linearly independent spherical harmonics of degree s in $n + 1$ variables.) The general problem of isometric minimal immersions $S^n(a) \to S^N(1)$ will be discussed in the paper of M. do Carmo.

There are minimal surfaces $S^2 \to S^N(1)$, whose induced metric is not of constant curvature.

6. INEQUALITIES

Among the most important and interesting results on minimal submanifolds are the inequalities which relate its various geometrical quantities. We will state three of the most important ones.

THEOREM 1. (E. Heinz and E. Hopf). Let

$$z = z(x,y) \quad , \quad x^2 + y^2 < R^2$$

be a minimal surface in E^3. Then at $x = y = 0$, the Gaussian curvature K satisfies the inequality

(10) $$|K_o| \leq \frac{A}{(1+(\text{grad } z)_o^2)R^2} \leq \frac{A}{R^2}$$

when A *is a universal constant.*

This inequality has been generalized by Osserman and Chern to parametric minimal surfaces in E^N. By letting $R \to \infty$, one derives the Bernstein theorem from (10).

The inequality (10) means geometrically that when a piece of minimal surface is larger, it is flater at the center. A similar bound on $(\text{grad } z)_o$ does not exist, because any plane is a minimal surface. However, if we impose the further condition $z > 0$, such an inequality was obtained by Finn and Serrin and extended to higher dimensions by Bombieri, de Giorgi, and M. Miranda:

THEOREM 2. *Let*

$$z = z(x_1,\ldots,x_n) > 0 \quad , \quad x_1^2+\ldots+x_n^2 < R^2$$

be a minimal hypersurface in E^{n+1} *with the coordinates* $(x_1,\ldots,x_n, z)$. *Then*

(11) $$|(\text{grad } z)_o| \leq c_1 \exp\left(c_2 \frac{z(0)}{R}\right) ,$$

when c_1, c_2 *are constants depending only on* n.

This theorem is one of the deepest results on minimal hypersurfaces. Its proof is difficult.

Our last inequality pertains to a closed minimal submanifold $M^n \to S^N(1)$. We denote by $\sigma(p) \geq 0$, $p \in M^n$, the square of the norm of the second fundamental form; it is an intrinsic invariant on M^n and is related in a simple way to the scalar curvature of the induced metric on M^n. Then we have

THEOREM 3. (Simons) Let $x: M^n \to S^N(1)$ be a closed minimal submanifold. Then

$$(12) \qquad \int_{M^n} (\sigma - \frac{n}{q})\, \sigma * 1 \geq 0 \quad , \quad q = 2 - \frac{1}{N-n} ,$$

Where $*1$ is the volume element of M^n.

It follows that either $\sigma = 0$, in which case the submanifold is totally geodesic or $\sigma \equiv \frac{n}{q}$ or there is a point $p \in M^n$ with $\sigma(p) > \frac{n}{q}$. It has been proved [5] that a minimal submanifold with $\sigma \equiv \frac{n}{q}$ is either a Clifford hypersurface or a Veronese surface on $S^4(1)$, which is the example in (9) with $n = 2$, $s = 2$.

7. PROBLEMS

In addition to the problems mentioned in the text we wish to add a few more:

PROBLEM 1. Give an elementary proof of the Bernstein-Simons theorem without the use of minimal currents. Is a Heinz-type inequality true for these dimensions?

PROBLEM 2. Let $S^3(a) \to S^7(1)$ be an isometric minimal immersion. Is it totally geodesic?

PROBLEM 3. Let $M^n \to S^N(1)$ be a closed minimal submanifold with $\sigma = \text{const.}$, $\sigma \neq 0, \frac{n}{q}$. What is the infimum of these values of σ?

Work done under partial support of National Science Foundation grant no. GP 13348.

* *

*

1)
$\frac{Dx}{ds}$ is the unit tangent vector to a curve on M^n and $\frac{D^2x}{ds^2}$ is its absolute derivative in the sense of Levi-Civita.

BIBLIOGRAPHY

A. Books and Survey Articles

1. Chern, S., Minimal submanifolds in a Riemannian manifold, University of Kansas, Department of Mathematics, Technical Report 19, 1968.

2. Federer, H., Geometric Measure Theory, Springer Verlag 1969.

3. Morrey, C. B., Multiple Integrals in the Calculus of Variations, Springer Verlag 1966.

4. Nitsche, J. C. C., On new results in the theory of minimal surfaces, Bull. Amer. Math. Soc. 71(1965), 195-270.

5. Osserman, R., A Survey of Minimal Surfaces, van Nostrand Reinhold, New York 1969.

6. Osserman, R., Minimal Varieties, to appear.

B. Papers

1. E. Bombieri, E. de Giorgi, and E. Giusti, Minimal cones and the Bernstein problem, Inventiones Math. 7, 243-268 (1969).

2. E. Bombieri, E. de Giorgi, and M. Miranda, Una maggiorazione a priori relativa alle ipersuperfici minimali non parametriche, Archive for Rat. Mech. and Analysis, 32, 255-267 (1968).

3. E. Calabi, Minimal immersions of surfaces in euclidean spheres, J. Diff. Geom. 1, 111-125 (1967).

4. S. S. Chern, Simple proofs of two theorems on minimal surfaces, L'Enseig. Math. 15, 53-61 (1969).

5. S. S. Chern, M. do Carmo, and S. Kobayashi, Minimal submanifolds of a sphere with second fundamental form of constant length, to appear in "Functional analysis and related fields, volume in honor of M. H. Stone".

6. S. S. Chern and R. Osserman, Complete minimal surfaces in euclidean n-space, J. d'Analyse Math. 19, 15-34 (1967).

7. M. do Carmo and N. Wallach, Minimal immersions of spheres into spheres, Proc. Nat. Acad. Sci. (USA) 1969.

8. W. Y. Hsiang, On the compact homogeneous minimal submanifolds, Proc. Nat. Acad. Sci. (USA), 56, 5-6 (1966).

9. H. B. Lawson, Jr., Compact minimal surfaces in S^3, Proc. Symp. in Pure Math.; Global Analysis, Amer. Math. Soc., 1969.

10. R. Osserman, A proof of the regularity everywhere of the classical solution to Plateau's problem, to appear.

11. J. Simons, Minimal varieties in Riemannian manifolds, Ann. of Math. 88, 62-105 (1968).

12. T. Takahashi, Minimal immersions of Riemannian manifolds, J. Math. Soc. Japan, 18, 380-385 (1966).

Differential Geometry, in honor of K. Yano
Kinokuniya, Tokyo, 1972, 73–94.

HOLOMORPHIC CURVES IN THE PLANE

SHIING-SHEN CHERN[1)]

1. Introduction

The theory of holomorphic (or meromorphic) curves was initiated by H. and J. Weyl in 1938 [3]. Its main problem, the proof of the so-called defect relations, was solved by Ahlfors [1]. It a natural generalization of the Picard-Borel-Nevanlinna theory and should be considered a beautiful chapter in complex differential geometry. A modern detailed treatment was given recently by H. Wu [4].

Further development of the theory is still hampered by the analytical complexity of the problem. By restricting to the special case of plane curves, this paper attempts at an exposition which will give the main ideas more clearly. However, with the exception of the last section on defect relations, we shall present our results in n dimensions, as no extra work will be involved. It is our contention that the fundamental relation is an integral-geometric inequality of Ahlfors. Such integral inequalities should be the just aim in the quantitative study of high-dimensional holomorphic mappings.

It will perhaps be of interest to state at the beginning our main result: Let $f: C \to P_n$ be a holomorphic mapping, where C is the complex line and P_n is the complex projective space of n dimensions, such that the image does not belong to a hyperplane. To a hyperplane α of P_n we can define a *defect* $\delta(\alpha)$ (see § 6 below), having the properties: 1) $0 \leqq \delta(\alpha) \leqq 1$; 2) $\delta(\alpha) = 1$, if $\alpha \cap f(C) = \emptyset$. Then, for a set of hyperplanes α_j, $1 \leqq j \leqq q$, in general position, we have

$$(1) \qquad \sum_{1 \leqq j \leqq q} \delta(\alpha_j) \leqq n + 1 .$$

We will give a proof of (1) in the case $n = 2$. As a corollary it follows that of four lines in general position in the plane, a plane curve must meet one of them.

---

1) Work done under partial support of NSF grant GP 20096.

2. Gauss-Bonnet formula in a semi-hermitian line bundle

We will review some basic facts on hermitian line bundles. For details cf. [2].

Let M be a complex manifold of dimension n. A holomorphic line bundle $p: E \to M$ is an element of the cohomology group $H^1(M, O^*)$, where O^* is the multiplicative sheaf of germs of non-vanishing holomorphic functions on M. Relative to an open covering $\mathscr{U} = \{U, V, W, \cdots\}$ of M the one-dimensional cohomology class is represented by the transition functions $g_{UV}: U \cap V \to C - \{0\}$, which are holomorphic and non-zero in $U \cap V$. The bundle $p^{-1}(U)$ over U has the local coordinates (x, ξ_U), $x \in U, \xi_U \in C$, such that in $U \cap V$ the local coordinates are related by

$$\xi_U g_{UV}(x) = \xi_V , \qquad x \in U \cap V . \tag{2}$$

The bundle is called hermitian if a positive definite hermitian norm is given in the fibers, which is C^∞ over M. That is, over each U of the covering $\mathscr{U}$ there is a C^∞-function $h_U > 0$, such that

$$\|\xi\|^2 = h_U |\xi_U|^2 = h_V |\xi_V|^2 \qquad \text{in } U \cap V . \tag{3}$$

Equation (3) is equivalent to

$$h_U = h_V |g_{UV}|^2 \qquad \text{in } U \cap V . \tag{4}$$

From this it follows that

$$\Omega = \frac{i}{2\pi} \partial\bar{\partial} \log h_U \tag{5}$$

is a globally defined form in M of bidegree $(1, 1)$, called the curvature form of the hermitian bundle. It is obviously a closed form and the cohomology class to which it belongs in the sense of de Rham's theorem is the characteristic class of the bundle E. We can write Ω as

$$\Omega = \frac{1}{4\pi i} d(\partial - \bar{\partial}) \log \|\xi\|^2 . \tag{6}$$

From now on let M be of dimension 1. We say that the functions $\{h_U\}$ satisfying (4) define a *semi-hermitian* structure if $h_U \geqq 0$ and if the zeroes of h_U are isolated such that at a point with the local coordinate ζ,

$$h_U = |\zeta|^{2m} h'_U , \qquad h'_U > 0 , \tag{7}$$

where m is an integer, called the *order* of the zero. We shall give the Gauss-Bonnet formula for a semi-hermitian holomorphic line bundle:

Let $p: E \to M$ be a semi-hermitian holomorphic line bundle over a one-dimensional complex manifold M. Let D be a compact domain of M with smooth boundary ∂D and $s: D \to E$ be a holomorphic section such that: 1) ∂D contains no zero of the hermitian structure; 2) $s(\partial D)$ does not meet the zero section of the bundle. Then

$$n(s) + n(h) = -\int_D \Omega + \frac{1}{2\pi i}\int_{\partial D} (\partial - \bar{\partial}) \log \|s\| \tag{8}$$

where $n(s)$ is the number of zeroes of the section s, $n(h)$ is the number of zeroes of the semi-hermitian structure in D, and $\|s\|$ is the hermitian norm of the section on the boundary.

This theorem follows immediately from the application of Stokes Theorem to (6). It suffices to observe that if

$$w(\zeta) = \zeta^k + \cdots$$

is a meromorphic function at $\zeta = 0$, then

$$k = \frac{1}{2\pi i} \lim_{\epsilon \to 0} \int_{\gamma_\epsilon} (\partial - \bar{\partial}) \log |w| ,$$

where γ_ϵ is a small circle about $\zeta = 0$.

3. Frenet formulas of curves in projective space

Let P_n be the complex projective space of n dimensions. It can be considered as the base space of the holomorphic line bundle

$$p: C_{n+1} - \{0\} \to P_n , \tag{9}$$

where C_{n+1} is the $(n + 1)$-dimensional number space. The inverse image of a point of P_n is called its homogeneous coordinate vector. It will be convenient not to distinguish between them.

If

$$Z = (z_0, \cdots, z_n) , \qquad W = (w_0, \cdots, w_n)$$

are vectors in C_{n+1}, we introduce the hermitian scalar product

$$(Z, W) = z_0\bar{w}_0 + \cdots + z_n\bar{w}_n . \tag{10}$$

The bundle (9) has an hermitian structure defined by the norm $|Z| = (Z, Z)^{1/2}$. By (5) its curvature form is

$$\Omega = \frac{i}{\pi}\partial\bar{\partial}\log|Z|\,. \tag{11}$$

To obtain further properties of this form Ω we consider the unitary group and its coset spaces, as follows:

$$U(n+1) \to U(n+1)/U(n) \to U(n+1)/U(1)\times U(n) = P_n\,, \tag{12}$$

where the second projection is the Hopf fibering. We identify the unitary group $U(n+1)$ with the space of all unitary frames $Z_\alpha \in C_{n+1} - \{0\}$, satisfying

$$(Z_\alpha, Z_\beta) = \delta_{\alpha\beta}\,, \qquad 0 \leqq \alpha, \beta, \gamma \leqq n\,. \tag{13}$$

Under this identification the first projection in (12) is defined by assigning to the frame Z_α its first vector Z_0.

The Maurer-Cartan forms $\theta_{\alpha\beta}$ of the unitary group $U(n+1)$ are defined by

$$dZ_\alpha = \sum_\beta \theta_{\alpha\beta} Z_\beta\,, \tag{14}$$

with

$$\theta_{\alpha\beta} + \bar{\theta}_{\beta\alpha} = 0\,. \tag{15}$$

They satisfy the Maurer-Cartan equations

$$d\theta_{\alpha\beta} = \sum_\gamma \theta_{\alpha\gamma}\wedge\theta_{\gamma\beta}\,. \tag{16}$$

The Fubini-Study metric on P_n is given by

$$ds^2 = \sum_j \theta_{0j}\bar{\theta}_{0j}\,, \qquad 1 \leqq j \leqq n\,, \tag{17}$$

and its associated Kähler form is

$$\hat{H} = \frac{i}{2}\sum_j \theta_{0j}\wedge\bar{\theta}_{0j}\,. \tag{18}$$

By (15) and (16) we can write

$$\hat{H} = \frac{1}{2i}\sum_j \theta_{0j}\wedge\theta_{j0} = \frac{1}{2i}d\theta_{00}\,. \tag{19}$$

$\hat{H}$ and Ω are related in a simple way; the relation can be derived as follows: We have

(20) $$Z_0 = Z/|Z| \, .$$

Differentiating

$$|Z|^2 = (Z, Z) \, ,$$

we get

$$2|Z| d|Z| = (dZ, Z) + (Z, dZ) \, .$$

It follows that

$$\begin{aligned}\theta_{00} &= (dZ_0, Z_0) = \frac{1}{|Z|^2}(dZ, Z) - \frac{1}{|Z|}d|Z| \\ &= \frac{1}{2|Z|^2}\{(dZ, Z) - (Z, dZ)\} = (\partial - \bar{\partial}) \log |Z| \, ,\end{aligned}$$

and hence that

(21) $$\Omega = \frac{1}{\pi}\hat{H} \, .$$

Thus Ω is positive definite. In fact, it is so normalized that it represents the generator of $H^2(P_n; Z)$; in other words, we have

(22) $$\int_{P_1} \Omega = 1 \, .$$

We will take Ω^n to be the volume element of P_n, so that P_n has the total volume 1.

With these preparations on the geometry in P_n consider a complex manifold M of dimension 1 and a holomorphic mapping $f: M \to P_n$. We will call $f(M)$ a holomorphic curve, not necessarily immersed. If ζ is a local coordinate on M, the curve can be locally represented by $Z(\zeta)$, expressing the homogeneous coordinate vector Z as a holomorphic function in ζ. We suppose that $f(M)$ does not belong to a hyperplane, i.e.,

(23) $$Z(\zeta) \wedge Z'(\zeta) \wedge \cdots \wedge Z^{(n)}(\zeta) \not\equiv 0;$$

such a curve will be called *non-degenerate*.

From (23) it follows that

(24) $$\Lambda_k \underset{\text{def}}{=} Z(\zeta) \wedge Z'(\zeta) \wedge \cdots \wedge Z^{(k)}(\zeta) \not\equiv 0 \, , \qquad k \leqq n \, .$$

The left-hand side of (24) defines the kth osculating space at ζ, the first osculating space being the tangent line. We can define at each point ζ a unitary frame Z_α, $0 \leqq \alpha \leqq n$, such that $Z_0, \cdots, Z_k$ span the kth osculating space. Each Z_α is defined up to the multiplication by a complex number of norm 1:

$$(25) \qquad Z_\alpha \to Z_\alpha^* = \exp(i\tau_\alpha)Z_\alpha\,, \qquad \tau_\alpha \text{ real.}$$

Such frames will be called the *Frenet frames* of the curve.

For a family of Frenet frames, equations (14) simplify to

$$(26) \qquad dZ_\alpha = \theta_{\alpha,\alpha-1}Z_{\alpha-1} + \theta_{\alpha\alpha}Z_\alpha + \theta_{\alpha,\alpha+1}Z_{\alpha+1}\,, \qquad 0 \leqq \alpha \leqq n\,.$$

These are the *Frenet formulas* of a holomorphic curve.

From (26) we find

$$(27) \qquad \begin{aligned} d(Z_0 \wedge \cdots \wedge Z_k) &= (\theta_{00} + \cdots + \theta_{kk})(Z_0 \wedge \cdots \wedge Z_k) \\ &\quad + \theta_{k,k+1}(Z_0 \wedge \cdots \wedge Z_{k-1} \wedge Z_{k+1})\,. \end{aligned}$$

Since $Z_0 \wedge \cdots \wedge Z_k$ and Λ_k both define the kth osculating space, we have

$$(28) \qquad Z_0 \wedge \cdots \wedge Z_k = \Lambda_k / |\Lambda_k|\,.$$

Comparing (27) and (28), we see that $\theta_{k,k+1}$ is of bidegree (1, 0) and hence that $\theta_{k+1,k}$ is of bidegree (0, 1).

The multi-vector $\Lambda_k(\zeta)$ defines a holomorphic curve in the Grassmann manifold of all k-dimensional linear spaces in P_n; it is called the *kth associated curve*. Its induced metric is given by

$$(29) \qquad \begin{aligned} ds_k^2 &= \{|\Lambda_k|^2(d\Lambda_k, d\Lambda_k) - (\Lambda_k, d\Lambda_k)(d\Lambda_k, \Lambda_k)\}/|\Lambda_k|^4 \\ &= \frac{|\Lambda_{k-1}|^2|\Lambda_{k+1}|^2}{|\Lambda_k|^4} d\zeta d\bar{\zeta} = h_k^2 d\zeta d\bar{\zeta}\,, \quad \text{say.} \end{aligned}$$

For $k < n$, ds_k^2 is semi-hermitian, i.e., the zeroes are isolated. The latter are called the *stationary points of order k*. By (27) we also have

$$(30) \qquad ds_k^2 = \theta_{k,k+1}\bar{\theta}_{k,k+1}\,.$$

We denote its Kähler form, divided by the factor π, by

$$(31) \qquad \Omega_k = \frac{i}{2\pi}\theta_{k,k+1} \wedge \bar{\theta}_{k,k+1} = \frac{i}{2\pi} h_k^2 d\zeta \wedge d\bar{\zeta}\,,$$

so that $\Omega_0 = \Omega$.

By the same computation which leads to formula (21), we find

$$\theta_{00} + \cdots + \theta_{kk} = (\partial - \bar{\partial}) \log |\Lambda_k| . \tag{32}$$

Taking its exterior derivative and using the structure equations (16), we get

$$\Omega_k = \frac{i}{\pi} \partial\bar{\partial} \log |\Lambda_k| , \qquad 0 \leqq k < n . \tag{33}$$

These n forms $\Omega_k, 0 \leqq k \leqq n-1$, are the fundamental local invariants of a holomorphic curve.

4. Plücker formulas

By (5), (29) and (33), the curvature form of the tangent bundle of M with the semi-hermitian metric ds_k^2 is $\Omega_{k-1} + \Omega_{k+1} - 2\Omega_k$.

Consider first the special case that M is a compact one-dimensional manifold without boundary. Then

$$v_k = \int_M \Omega_k , \qquad 0 \leqq k \leqq n-1 , \tag{34}$$

are integers. The Gauss-Bonnet formula in § 2 gives

$$-w_k - v_{k-1} + 2v_k - v_{k+1} = 2 - 2g , \qquad 0 \leqq k \leqq n-1 , \tag{35}$$

where w_k is the number of stationary points of order k and g is the genus of M. Formula (35) contains the classical Plücker formulas for algebraic curves.

More generally, let D be a compact domain of M with boundary ∂D, which contains no stationary point of order k. We denote by $w_k(D)$ the number of stationary points of order k in the interior of D and $v_k(D)$ the integral of Ω_k over D. The application of the Gauss-Bonnet formula then gives

$$\begin{aligned} &-w_k(D) - v_{k-1}(D) + 2v_k(D) - v_{k+1}(D) \\ &= \chi(D) + \frac{i}{2\pi} \int_{\partial D} (\partial - \bar{\partial}) \log \|s\| , \end{aligned} \tag{36}$$

where s is a vector field over ∂D pointing toward the interior of D, $\|s\|$ is the length of s, and $\chi(D)$ is the Euler characteristic of D.

We will consider the case that M is non-compact and, for simplicity, that $M = C$, the complex line. The argument applies without essential change to the case that M is a compact Riemann surface with a finite number of points deleted.

One advantage for C, however, is that the coordinate $\zeta = re^{i\theta}$ is now global. We set $u = \log r$ and take D to be the disk of radius $\exp u$. Then the numbers at the left-hand side of (36) are functions of u, and we get

$$\begin{aligned} &-w_k(u) - v_{k-1}(u) + 2v_k(u) - v_{k+1}(u) \\ &= \frac{i}{2\pi} \int_{r=\exp u} (\partial - \bar{\partial}) \log \frac{|\Lambda_{k-1}||\Lambda_{k+1}|}{|\Lambda_k|^2} . \end{aligned} \tag{37}$$

This follows, because we take s to be the radial vector field on the boundary ∂D and its integral cancels with the Euler characteristic $\chi(D)$, which is 1. Formula (37) can also be derived directly, and more simply, by Stokes Theorem.

5. Inequalities of Ahlfors

We will first consider an inversion problem in integral geometry: Let

$$dP_n = \Omega^n \tag{38}$$

be the volume element of P_n. Let $A \in C_{n+1}$, $|A| = 1$, be a fixed point of P_n. To an integrable function $g(t) \geqq 0$ consider the integral

$$\psi(t) = \int g(t\,|Z_0, A|^2) dP_n,\; Z_0 \in P_n, |Z_0| = 1 . \tag{39}$$

Since dP_n is invariant under the action of $U(n+1)$, $\psi(t)$ is independent of A. Our problem is to express $g(t)$ in terms of $\psi(t)$.

By (18), (21) and (38), we have

$$\begin{aligned} dP_n &= \left(\frac{i}{2\pi}\right)^n n!\, \theta_{01} \wedge \bar{\theta}_{01} \wedge \cdots \wedge \theta_{0n} \wedge \bar{\theta}_{0n} \\ &= \left(\frac{i}{2\pi}\right)^n n!(dZ_0, Z_1) \wedge \overline{(dZ_0, Z_1)} \wedge \cdots \\ &\qquad \wedge (dZ_0, Z_n) \wedge \overline{(dZ_0, Z_n)} . \end{aligned} \tag{40}$$

This expression is invariant under a unitary transformation of $Z_1, \cdots, Z_n$. With fixed A we write

$$(1 + r^2)^{1/2} Z_0 = A + rY_1 , \tag{41}$$

where

$$(A, Y_1) = 0 , \qquad |Y_1| = 1 , \tag{42}$$

and $0 \leqq r \leqq \infty$. Thus Y_1 describes the unit sphere in C_{n+1} orthogonal to A. To write dP_n in the new coordinate system r, Y_1 we put

$$(1 + r^2)^{1/2} Z_1 = -rA + Y_1 , \tag{43}$$

and choose $Z_2, \cdots, Z_n$, so that $Z_0, \cdots, Z_n$, and hence also $A, Y_1, Z_2, \cdots, Z_n$, are unitary frames. It follows that

$$\begin{aligned} (1 + r^2)(dZ_0, Z_1) &= dr + r(dY_1, Y_1) , \\ (1 + r^2)^{1/2}(dZ_0, Z_\lambda) &= r(dY_1, Z_\lambda) , \qquad 2 \leqq \lambda \leqq n . \end{aligned}$$

We have therefore, on putting $\sigma = r^2$,

$$\begin{aligned} dP_n = -\left(\frac{i}{2\pi}\right)^n n! \frac{\sigma^{n-1}}{(1+\sigma)^{n+1}} d\sigma \wedge (dY_1, Y_1) \wedge (dY_1, Z_2) \\ \wedge \overline{(dY_1, Z_2)} \wedge \cdots \wedge (dY_1, Z_n) \wedge \overline{(dY_1, Z_n)} . \end{aligned}$$

Let

$$\begin{aligned} dP_{n-1} = \left(\frac{i}{2\pi}\right)^{n-1} (n-1)! \, (dY_1, Z_2) \wedge \overline{(dY_1, Z_2)} \\ \wedge \cdots \wedge (dY_1, Z_n) \wedge \overline{(dY_1, Z_n)} \end{aligned} \tag{44}$$

be the volume element in $A^\perp$, the hyperplane in P_n orthogonal to A. Then we can write

$$dP_n = -\frac{ni}{2\pi} \frac{\sigma^{n-1}}{(1+\sigma)^{n+1}} d\sigma \wedge (dY_1, Y_1) \wedge dP_{n-1} . \tag{45}$$

The unit sphere described by Y_1 is a circle bundle over P_{n-1}. If we denote its fiber coordinate by α, we have

$$(dY_1, Y_1) = i\, d\alpha + \text{form in base.}$$

Hence

$$dP_n = \frac{n}{2\pi} \frac{\sigma^{n-1}}{(1+\sigma)^{n+1}} d\sigma \wedge d\alpha \wedge dP_{n-1} . \tag{46}$$

From (41) we have

$$|Z_0, A| = \frac{1}{(1+\sigma)^{1/2}} . \tag{47}$$

As $\int dP_{n-1} = 1$ and α runs from 0 to 2π, the integration of (39) gives

$$\psi(t) = n\int_0^\infty g\left(\frac{t}{1+\sigma}\right)\frac{\sigma^{n-1}}{(1+\sigma)^{n+1}}d\sigma .$$

By the change of variables

$$\tau = \frac{1}{1+\sigma},$$

we can write

$$\psi(t) = n\int_0^1 (1-\tau)^{n-1}g(t\tau)d\tau . \tag{48}$$

From this the inversion can be achieved by putting

$$u = t\tau .$$

Then

$$t^n\psi(t) = n\int_0^t (t-u)^{n-1}g(u)du ,$$

and we get

$$g(t) = \frac{1}{n!}\,\frac{d^n}{dt^n}(t^n\psi(t)) . \tag{49}$$

Thus the inverse operator of the integral operator (39) is a differential operator.

The density Ω^n in P_n leads to a density in the space P_n^* of all hyperplanes ξ of P_n, by the definition

$$d\xi = |d\xi^\perp| = |dP_n| , \tag{50}$$

where $\xi^\perp$ is the *pole* of ξ, i.e., the point orthogonal to all points of ξ. Let $f: M \to P_n$ be a holomorphic curve and D be a compact domain of M. We shall prove the *Crofton formula*

$$\int n(f(D) \cap \xi)d\xi = v_0(D) , \tag{51}$$

where $n(f(D) \cap \xi)$ denotes the number of common points of $f(D)$ and ξ, and the integration is over P_n^*.

Suppose the curve be given by $Z_0(\zeta)$, $\zeta \in M$, $|Z_0| = 1$ and suppose

$$dZ_0 = \theta_{00}Z_0 + \theta_{01}Z_1 , \quad (Z_0, Z_1) = 0 , \quad |Z_1| = 1 ,$$

so that Z_0Z_1 is the tangent line at Z_0. We will coordinatize ξ by the point Z_0 where it meets the curve and the point $\xi^\perp \in Z_0^\perp$, and express $d\xi$ in terms of this coordinate system. For this purpose let $Y_2, \cdots, Y_n$ be such that $\xi^\perp, Z_0, Y_2, \cdots, Y_n$ form a unitary frame. Then, by (40),

$$d\xi = \pm \frac{i}{2\pi} n(d\xi^\perp, Z_0) \wedge \overline{(d\xi^\perp, Z_0)} \wedge d\xi^\perp(Z_0^\perp) ,$$

where

$$\begin{aligned} d\xi^\perp(Z_0^\perp) = \left(\frac{i}{2\pi}\right)^{n-1}(n-1)!\,(d\xi^\perp, Y_2) \wedge \overline{(d\xi^\perp, Y_2)} \\ \wedge \cdots \wedge (d\xi^\perp, Y_n) \wedge \overline{(d\xi^\perp, Y_n)} \end{aligned} \tag{52}$$

is the volume element of $\xi^\perp$ in the hyperplane $Z_0^\perp$.

Now we have

$$(\xi^\perp, Z_0) = 0 ,$$

from which follows

$$(d\xi^\perp, Z_0) = -(\xi^\perp, dZ_0) = -\bar{\theta}_{01}(\xi^\perp, Z_1) .$$

Using (31), we get

$$d\xi = n\,|\,\xi^\perp, Z_1|^2 \Omega_0 \wedge d\xi^\perp(Z_0^\perp) . \tag{53}$$

We now carry out the integration in (51) by holding the point $Z_0(\zeta)$ fixed. The integral

$$\int |\,\xi^\perp, Z_1|^2 d\xi^\perp(Z_0^\perp)$$

is a universal constant, since Z_1 is fixed. Hence we get

$$\int n(f(D) \cap \xi) d\xi = c v_0(D) ,$$

where c is a constant independent of the curve. If D is the projective line, we have $c = 1$. Thus (51) is proved.

Let α be a fixed hyperplane of P_n, $|\alpha^\perp| = 1$. Let $g(t) \geqq 0$, $t \geqq 0$, be an integrable function. Generalizing the Crofton formula, we wish to find a lower bound for the integral

$$\int n(f(D) \cap \xi) g(|\alpha, \xi|^2) d\xi ,$$

under the condition

$$\int_{P_n^*} g(|\alpha, \xi|^2) d\xi = 1 . \tag{54}$$

We write

$$\alpha^\perp = \nu Z_0 + \mu X , \tag{55}$$

with

$$(Z_0, X) = 0 , \qquad |X| = 1 . \tag{56}$$

Then

$$\nu = (\alpha^\perp, Z_0) , \qquad |\nu|^2 + |\mu|^2 = 1 , \tag{57}$$

and

$$|\alpha, \xi| = |\alpha^\perp, \xi^\perp| = |\mu||X, \xi^\perp| , \tag{58}$$

so that

$$g(|\alpha, \xi|^2) d\xi = n |\xi^\perp, Z_1|^2 g(|\mu|^2 |X, \xi^\perp|^2) \Omega_0 \wedge d\xi^\perp(Z_0^\perp) . \tag{59}$$

We again hold $Z_0(\zeta)$ fixed and integrate over the hyperplane $Z_0^\perp$. X and Z_1 being fixed points in $Z_0^\perp$, we write

$$Z_1 = yX + y_1 X_1 , \quad (X, X_1) = 0 , \quad |X_1| = 1 . \tag{60}$$

Then

$$y = (Z_1, X) , \qquad |y|^2 + |y_1|^2 = 1 , \tag{61}$$

and

$$|\xi^\perp, Z_1|^2 \geqq |y|^2 |\xi^\perp, X|^2 + y\bar{y}_1 (X, \xi^\perp)(\xi^\perp, X_1) + \bar{y} y_1 (\xi^\perp, X)(X_1, \xi^\perp) . \tag{62}$$

Since $(X, \xi^\perp)$, as $\xi^\perp$ runs over $Z_0^\perp$, takes values of opposite sign, the integral over $Z_0^\perp$ of the last two terms as the right-hand side, when multiplied by $g(|\mu|^2 |X, \xi^\perp|^2)$, is zero. Hence we get

$$\int n(f(D) \cap \xi) g(|\alpha, \xi|^2) d\xi \geqq n \int_D \frac{|Z_1, X|^2}{|\mu|^2} \psi(|\mu|^2) \Omega_0 , \tag{63}$$

where

(64) $$\psi(|\mu|^2) = \int_{\xi^\perp \in Z_0^\perp} g(|\mu|^2 |X, \xi^\perp|^2)|\mu|^2|X, \xi^\perp|^2 d\xi^\perp(Z_0^\perp) .$$

By our solution of the inversion problem at the beginning of this section, (64) is equivalent to

(65) $$(n-1)!\, tg(t) = \frac{d^{n-1}}{dt^{n-1}}(t^{n-1}\psi(t)) .$$

On the other hand, by (48), condition (54) can be written as

(66) $$n\int_0^1 (1-\tau)^{n-1} g(\tau)d\tau = 1 .$$

We shall transform (66) into a condition involving $\psi(t)$. In fact, (65) is equivalent to

$$t^{n-1}\psi(t) = (n-1)\int_0^t (t-u)^{n-2} ug(u)du .$$

It follows that

$$\begin{aligned}\int_0^1 \frac{\psi(t)}{t}dt &= (n-1)\int_0^1 t^{-n}dt \int_0^t (t-u)^{n-2}ug(u)du \\ &= (n-1)\int_0^1 g(u)du \int_u^1 (1-ut^{-1})^{n-2}ut^{-2}dt \\ &= \int_0^1 (1-u)^{n-1}g(u)du .\end{aligned}$$

Hence condition (66) can be written as

(67) $$n\int_0^1 t^{-1}\psi(t)dt = 1 .$$

With $g(t)$ and $\psi(t)$ related by (65) and $\psi(t)$ satisfying (67), we have proved the inequality

(68) $$\begin{aligned} n\int_D &\frac{|\alpha^\perp, Z_1|^2}{(1-|\alpha^\perp, Z_0|^2)^2}\psi(1-|\alpha^\perp, Z_0|^2)\Omega_0 \\ &\leqq \int n(f(D) \cap \xi)g(|\alpha, \xi|^2)d\xi .\end{aligned}$$

Different inequalities result from different permissible choices of $\psi(t)$. In particular, it is sufficient that $\psi(t), 0 \leqq t \leqq 1$, has non-negative derivatives of orders $\leqq n-1$ and satisfies (67).

The two simplest choices are

$$n\psi(t) = t\,, \tag{69}$$
$$n\psi(t) = (1-\lambda)t(1-t)^{-\lambda}\,, \qquad 0 < \lambda < 1\,.$$

They give respectively the inequalities

$$\int_D \frac{|\alpha^\perp, Z_1|^2}{1-|\alpha^\perp, Z_0|^2}\Omega_0 \leqq \int n(f(D) \cap \xi) g(|\alpha, \xi|^2) d\xi\,, \tag{70}$$

$$(1-\lambda)\int_D \frac{|\alpha^\perp, Z_1|^2}{(1-|\alpha^\perp, Z_0|^2)|\alpha^\perp, Z_0|^{2\lambda}}\Omega_0 \leqq \int n(f(D) \cap \xi) g(|\alpha, \xi|^2) d\xi\,. \tag{71}$$

Notice that the left-hand sides could be improper integrals, in fact, the integral in (70) at points $Z_0(\zeta) = \alpha^\perp$ and the integral in (71) at such points and at points where $(Z_0(\zeta), \alpha^\perp) = 0$.

6. First and second main theorems in integrated form

Let α be a fixed hyperplane in P_n, $|\alpha^\perp| = 1$. By § 3 we can write

$$\Omega_0 = \frac{1}{2\pi i} d(\partial - \bar\partial) \log \frac{|Z|\cdot|\alpha^\perp|}{|Z, \alpha^\perp|}\,, \tag{72}$$

because $\log|Z(\zeta), \alpha^\perp|$ is harmonic. This formula is valid in $P_n - \alpha$. From it follows immediately the *first main theorem*:

Let $f\colon M \to P_n$ be a non-degenerate holomorphic curve and $D \subset M$ a compact domain such that $f(\partial D) \cap \alpha = \emptyset$. The number of points in $f(D) \cap \alpha$, counted with multiplicities, is given by

$$n(D, \alpha) = v_0(D) + \frac{1}{2\pi}\int_{f(\partial D)} d^c \log \frac{|Z, \alpha^\perp|}{|Z|\cdot|\alpha^\perp|}\,, \tag{73}$$

where

$$d^c = i(\bar\partial - \partial)\,. \tag{74}$$

Let $\tau\colon M \to R^+$ be a harmonic exhaustion function on M. Let

$$D_u = \{\zeta \in M \mid \tau(\zeta) \leqq u\}\,. \tag{75}$$

For simplicity we write

$$(76) \qquad n(D_u, \alpha) = n(u, \alpha) , \qquad v_0(D_u) = v_0(u) .$$

Then (73) can be written as

$$(77) \qquad n(u, \alpha) - v_0(u) = \frac{1}{2\pi} \int_{f(\partial D_u)} d^c \log \frac{|Z, \alpha^\perp|}{|Z| \cdot |\alpha^\perp|} .$$

By a standard argument the integral operator and the differential operator d^c at the right-hand side of (77) can be interchanged, and we can integrate the equation (77) with respect to u. We put

$$(78) \qquad N(u, \alpha) = \int_0^u n(t, \alpha) dt , \quad T_k(u) = \int_0^u v_k(t) dt , \qquad 0 \leqq k \leqq n-1 ,$$

$$(79) \qquad m(u, \alpha) = \frac{1}{2\pi} \int_{\partial D_u} \log \frac{|Z(\zeta)| \cdot |\alpha^\perp|}{|Z(\zeta), \alpha^\perp|} d^c \tau \geqq 0 .$$

The integration of (77) then gives the *integrated form of the first main theorem*.

$$(80) \qquad N(u, \alpha) + m(u, \alpha) = T_0(u) + m(0, \alpha) .$$

As a corollary we have the fundamental inequality:

$$(81) \qquad N(u, \alpha) < T_0(u) + \text{const.}$$

Suppose M has an infinite harmonic exhaustion. The *defect* of α is defined by

$$(82) \qquad \delta(\alpha) = \liminf_{u \to \infty} \frac{m(u, \alpha)}{T_0(u)} = 1 - \limsup_{u \to \infty} \frac{N(u, \alpha)}{T_0(u)} .$$

By (79) and (80) we have

$$(83) \qquad 0 \leqq \delta(\alpha) \leqq 1 .$$

Also, if $f(M) \cap \alpha = \emptyset$, $\delta(\alpha) = 1$.

We wish to remark that the study of the asymptotic behavior of a non-compact non-degenerate holomorphic curve (with harmonic exhaustion) relative to a hyperplane α necessitates the introduction of quantities more general than $m(u, \alpha)$. This is perhaps best explained as follows: Let $Z_0(\zeta), \cdots, Z_n(\zeta)$ be the Frenet frame at $\zeta \in M$. Write

$$\alpha^{\perp} = w_0 Z_0(\zeta) + \cdots + w_n Z_n(\zeta) . \tag{84}$$

We put

$$m_k(u, \alpha) = \frac{1}{4\pi} \int_{\partial D_u} - \log (|w_0|^2 + \cdots + |w_k|^2) d^c\tau \geqq 0 , \qquad 0 \leqq k \leqq n-1 , \tag{85}$$

so that $m_0(u, \alpha) = m(u, \alpha)$. As to be expected, the Ahlfors inequalities lead to relations between the differences $m_k(u, \alpha) - m_{k-1}(u, \alpha)$.

Formula (37) was called the *second main theorem* by H. Weyl. If M has a harmonic exhaustion, it can also be integrated. For simplicity we again restrict ourselves to the case $M = C$, using the notations introduced at the end of § 4. We put

$$W_k(u) = \int_0^u w_k(t) dt \geqq 0 , \tag{86}$$

$$S_k(u) = \frac{1}{2\pi} \int_0^{2\pi} \log h_k d\theta , \tag{87}$$

(h_k was defined in (29)). Then we have the integrated form of the second main theorem

$$\begin{aligned} W_k(u) + T_{k-1}(u) - 2T_k(u) + T_{k+1}(u) &= S_k(u) - S_k(0) , \\ T_n(u) = 0 , \quad 0 \leqq k &\leqq n-1 . \end{aligned} \tag{88}$$

Between $S_k(u)$ and $T_k(u)$ there is an integral inequality, to be derived as follows: We have

$$T_k(u) = \frac{1}{\pi} \int_0^u dt \int_0^t \exp (2s) ds \int_{r = \exp s} h_k^2 d\theta . \tag{89}$$

By the concavity of the logarithm function we have

$$2S_k(u) \leqq \log \left\{ \frac{1}{2\pi} \int_0^{2\pi} h_k^2 d\theta \right\} . \tag{90}$$

It follows that

$$2 \int_0^u dt \int_0^t \exp (2s + 2S_k(s)) ds \leqq T_k(u) . \tag{91}$$

Inequality (91) will be the type of inequalities that are encountered in the study of value distributions. More generally, suppose

$$\int_{u_0}^{u} dt \int_{u_0}^{t} \exp(K\theta(s))ds < C\Phi(u) + C' , \tag{92}$$

where $\theta(u)$ and $\Phi(u)$ are continuous functions on $u_0 \leqq u < \infty$ and K, C, C' are positive constants. Following H. and J. Weyl, we express this relationship by writing

$$\theta = \mu(\Phi) . \tag{93}$$

By a well-known lemma, this implies

$$\| \quad \theta(u) < \kappa \log (C\Phi(u) + C') , \qquad \kappa > 1 , \tag{94}$$

where the sign $\|$ before the inequality means that it is valid in $[0, \infty] - I$, where I is a union of intervals satisfying $\int_I d \log u < \infty$. For details of these results, cf. [4], pp. 54–55, 131, 135.

The inequality (91) now implies

$$S_k(u) = \mu(T_k(u)) , \qquad 0 \leqq k \leqq n-1 . \tag{95}$$

7. Plane curves

Consider a non-degenerate holomorphic curve $f: C \to P_2$. Let ζ be the coordinate in C, and, as above,

$$\zeta = re^{i\theta} , \qquad u = \log r . \tag{96}$$

According to the general theory there are two order functions $T_0(u)$ and $T_1(u)$, such that

$$\begin{aligned} W_k + T_{k-1} - 2T_k + T_{k+1} &= \mu(T_k(u)) , \\ T_2(u) = 0 , \qquad k &= 0, 1 . \end{aligned} \tag{97}$$

Since $W_k \geqq 0$, this relation is still true when W_k is dropped. That is, we have

$$\begin{aligned} -2T_0 + T_1 &= \mu(T_0(u)) , \\ T_0 - 2T_1 &= \mu(T_1(u)) . \end{aligned} \tag{98}$$

It follows that

(99) $$T_0 < \beta T_1\,, \quad T_1 < \beta T_0\,, \quad \beta > 2\,.$$

Thus $T_0(u)$ and $T_1(u)$ have the same order, which in a sense sets the *transcendental level* of the curve. We will denote this order by $T(u)$, so that

(100) $$\mu(T_0(u)) = \mu(T_1(u)) = \mu(T(u))\,.$$

We shall derive the first defect relation from the inequality (71). We write (81) as

$$T_0(u) + \text{const} > \int_0^u n(t, \xi)dt$$

and integrate it with respect to the density $g(|\alpha, \xi|^2)d\xi$, with a fixed α. Using the w_ρ, $0 \leqq \rho \leqq n$, introduced in (84), this gives

$$\begin{aligned} T_0(u) + \text{const} &> (1-\lambda)\int_0^u dt \int_{D_t} \frac{|w_1|^2}{(1-|w_0|^2)|w_0|^{2\lambda}}\Omega_0 \\ &\geqq (1-\lambda)\int_0^u dt \int_{D_t} \frac{|w_1|^2}{|w_0|^{2\lambda}}\Omega_0\,, \qquad 0 < \lambda < 1\,. \end{aligned}$$

It follows that

(101) $$(1-\lambda)\int_0^u dt \int_{D_t} \frac{|w_0|^2 + |w_1|^2}{|w_0|^{2\lambda}}\Omega_0 < (2-\lambda)T_0(u) + \text{const.}$$

But

(102) $$\Omega_0 = \frac{i}{2\pi} h_0^2 d\zeta \wedge d\bar{\zeta} = \frac{1}{\pi}\exp(2u)h_0^2 du \wedge d\theta\,.$$

Emphasizing the dependence on α, we write (101) as

(103) $$\begin{aligned} &\frac{1-\lambda}{\pi}\int_0^u dt \int_0^t \exp(2s)ds \int_0^{2\pi} h_0^2 \frac{|Z_0, \alpha^\perp|^2 + |Z_1, \alpha^\perp|^2}{|Z_0, \alpha^\perp|^{2\lambda}} d\theta \\ &\qquad < (2-\lambda)T_0(u) + \text{const.} \end{aligned}$$

We put

(104) $$\Phi(\alpha) = (|Z_0, \alpha^\perp|^2 + |Z_1, \alpha^\perp|^2)/|Z_0, \alpha^\perp|^{2\lambda}\,.$$

Let $\alpha_1, \cdots, \alpha_q$ be a set of lines of P_2 in general position, with $|\alpha_i^\perp| = 1$. Every point of P_2 has a neighborhood which meets at most two of the α's.

Since P_2 is compact, there exists $L \geqq 1$ such that

$$(105)\qquad \begin{aligned} |Z_0, \alpha_i^{\perp}| &\leqq L^{-1} \quad \text{for at most two of the } \alpha\text{'s},\\ \Phi(\alpha_i) &< L^{2\lambda} \quad \text{for the remaining } \alpha\text{'s}. \end{aligned}$$

Let $\sum'$ be the sum over the α_i with $\Phi(\alpha_i) \geqq L^{2\lambda}$. Then, by the concavity of the logarithm,

$$\begin{aligned} \log \sum \Phi(\alpha_i) &\geqq \log \sum{}' \Phi(\alpha_i) \geqq \tfrac{1}{2} \sum{}' \log \Phi(\alpha_i)\\ &\geqq \tfrac{1}{2} \{\sum \Phi(\alpha_i) - 2q\lambda \log L\}\,. \end{aligned}$$

Also by the concavity of the logarithm and by (88), we have

$$\begin{aligned} \log \frac{1}{2\pi} \int_0^{2\pi} h_0^2 \sum \Phi(\alpha_i) d\theta &\geqq \frac{1}{2\pi} \int_0^{2\pi} \log \{h_0^2 \sum \Phi(\alpha_i)\} d\theta\\ = 2(W_0(u) - 2T_0(u) + T_1(u)) &= \frac{1}{2\pi} \int_0^{2\pi} \log \sum \Phi(\alpha_i) d\theta + \text{const}\\ \geqq 2(W_0(u) - 2T_0(u) + T_1(u)) &+ \frac{1}{4\pi} \sum_i \int_0^{2\pi} \log \Phi(\alpha_i) d\theta + \text{const.} \end{aligned}$$

Taking the inequality (103) for $\alpha = \alpha_i$ and summing with respect to i, we get

$$\frac{1-\lambda}{\pi} \int_0^u dt \int_0^t \exp(2s) ds \int_0^{2\pi} h_0^2 \sum_i \Phi(\alpha_i) d\theta < 2qT_0(u) + \text{const.}$$

We now choose λ so that $(1-\lambda)T_0(u) = 1$. We have then

$$\frac{1}{4\pi} \sum_i \int_0^{2\pi} \log \Phi(\alpha_i) d\theta \leqq 2(-W_0 + 2T_0 - T_1) + \mu(T(u)^2)\,.$$

By (85), the left-hand side is equal to

$$\begin{aligned} &\sum_i \{\lambda m_0(u, \alpha_i) - m_1(u, \alpha_i)\}\\ &= \sum_i \{m_0(u, \alpha_i) - m_1(u, \alpha_i)\} - (1-\lambda) \sum_i m_0(u, \alpha_i)\,. \end{aligned}$$

But

$$(1-\lambda) m_0(u, \alpha_i) \leqq (1-\lambda) T_0(u) + \text{const},$$

where the first term at the right-hand side is 1. It follows that

$$\sum_i \{m_0(u,\alpha_i) - m_1(u,\alpha_i)\} \leqq 2(-W_0 + 2T_0 - T_1) + \mu(T(u)^2) . \tag{106}$$

A weaker inequality is

$$\sum_i \{m_0(u,\alpha_i) - m_1(u,\alpha_i)\} \leqq 2(2T_0 - T_1) + \mu(T(u)^2) . \tag{107}$$

To prove the defect relation (1) for $n = 2$ we need an upper bound for $\sum_i m_1(u, \alpha_i)$. Unfortunately this does not follow from (70). It is, however, of interest to see how a more careful estimate gives, in the case $n = 2$, a better inequality. Instead of the inequality (62) we have

$$\begin{aligned} |\xi^\perp, Z_1|^2 = |y|^2|\xi^\perp, X|^2 + |y_1|^2|\xi^\perp, X_1|^2 + y\bar{y}_1(X, \xi^\perp)(\xi^\perp, X_1) \\ + \bar{y}y_1(\xi^\perp, X)(X_1, \xi^\perp) . \end{aligned} \tag{108}$$

Since $n = 2$, we have

$$|\xi^\perp, X|^2 + |\xi^\perp, X_1|^2 = 1 \tag{109}$$

for $\xi^\perp \in Z_0^\perp$. Hence

$$\begin{aligned} |\xi^\perp, Z_1|^2 = 1 - |Z_1, X|^2 + |\xi^\perp, X|^2(-1 + 2|Z_1, X|^2) \\ + y\bar{y}_1(X, \xi^\perp)(\xi^\perp, X_1) + \bar{y}y_1(\xi^\perp, X)(X_1, \xi^\perp) . \end{aligned}$$

When integrated with respect to $g(|\mu|^2|X, \xi^\perp|^2)d\xi^\perp(Z_0^\perp)$, the last two terms will give zero. We put

$$\int g(t|X, \xi^\perp|^2)d\xi^\perp(Z_0^\perp) = \varphi(t) , \tag{110}$$

$$\int g(t|X, \xi^\perp|^2)t|X, \xi^\perp|^2 d\xi^\perp(Z_0^\perp) = \psi(t) . \tag{111}$$

Then we have, for $n = 2$, instead of (63) the relation

$$\begin{aligned} &\tfrac{1}{2}\int n(f(D) \cap \xi)g(|\alpha, \xi|^2)d\xi \\ &= \int_D \Big\{(1 - |Z_1, X|^2)\varphi(|\mu|^2) \\ &\quad + \frac{1}{|\mu|^2}(-1 + 2|Z_1, X|^2)\psi(|\mu|^2)\Big\}\Omega_0 . \end{aligned} \tag{112}$$

By (65) and (67), $\psi(t)$ satisfies the conditions

$$tg(t) = t\psi' + \psi ,$$

$$\int_0^1 t^{-1}\psi(t)dt = \tfrac{1}{2} .$$

From (48) $\varphi(t)$ is given in terms of $g(t)$ by

$$\varphi(t) = \int_0^1 g(t\tau)d\tau \ .$$

For $n = 2$ the choice

$$\psi(t) = \tfrac{1}{2}(1-\lambda)t^{1-\lambda}\ , \qquad 0 < \lambda < 1\ , \tag{113}$$

is allowable. With this choice of $\psi(t)$ we have

$$\varphi(t) = \tfrac{1}{2}(2-\lambda)t^{-\lambda}\ . \tag{114}$$

The integral formula (112) then becomes

$$\begin{aligned}\int n(f(D)\cap\xi)g(|\alpha,\xi|^2)d\xi &= \int_D \frac{1-\lambda|Z_1,X|^2}{(1-|\alpha^\perp,Z_0|^2)^\lambda}\Omega_0 \\ &= \int_D \frac{\Omega_0}{(1-|\alpha^\perp,Z_0|^2)^\lambda} - \lambda\int_D \frac{|\alpha^\perp,Z_1|^2}{(1-|\alpha^\perp,Z_0|^2)^{1+\lambda}}\Omega_0\ .\end{aligned} \tag{115}$$

On the other hand, we have from (68),

$$(1-\lambda)\int_D \frac{|\alpha^\perp,Z_1|^2}{(1-|\alpha^\perp,Z_0|^2)^{1+\lambda}}\Omega_0 \leqq \int n(f(D)\cap\xi)g(|\alpha,\xi|^2)d\xi\ . \tag{116}$$

Integrating the inequality (81) as above, we get from (115) and (116) respectively,

$$\begin{aligned}T_0(u) + \text{const} > &\int_0^u dt\int_{D_t} \frac{\Omega_0}{(1-|\alpha^\perp,Z_0|^2)^\lambda} \\ &- \lambda\int_0^u dt\int_{D_t} \frac{|\alpha^\perp,Z_1|^2}{(1-|\alpha^\perp,Z_0|^2)^{1+\lambda}}\Omega_0\ ,\end{aligned} \tag{117}$$

$$T_0(u) + \text{const} > (1-\lambda)\int_0^u dt\int_{D_t} \frac{|\alpha^\perp,Z_1|^2}{(1-|\alpha^\perp,Z_0|^2)^{1+\lambda}}\Omega_0\ . \tag{118}$$

Combining (117) and (118), we have

$$(1-\lambda)\int_0^u dt\int_{D_t} \frac{\Omega_0}{(1-|\alpha^\perp,Z_0|^2)^\lambda} < T_0(u) + \text{const.} \tag{119}$$

Our plane curve has a dual curve described by its tangent lines. It can be considered as the curve $Z_2(\zeta)$. When applied to the curve $Z_2(\zeta)$, the inequality (119) becomes

$$(120) \qquad (1-\lambda)\int_0^u dt \int_{D_t} \frac{\Omega_1}{(1-|\alpha^\perp, Z_2|^2)^\lambda} < T_1(u) + \text{const.}$$

Applying to (120) the arguments used in deriving (106), we get

$$(121) \qquad \sum_i m_1(u,\alpha_i) \leqq -W_1 - T_0 + 2T_1 + \mu(T(u)^2)\,,$$

or the weaker inequality

$$(122) \qquad \sum_i m_1(u,\alpha_i) \leqq -T_0 + 2T_1 + \mu(T(u)^2)\,.$$

Combining (107) and (122), we have

$$(123) \qquad \sum_i m_0(u,\alpha_i) \leqq 3T_0 + \mu(T(u)^2)\,.$$

The defect relation (1) for $n = 2$ follows immediately from (123).

Remark. To derive the defect relation (1) for a holomorphic curve in P_n it is necessary to consider all the associated curves. The fundamental inequality to be proved is

$$(124) \qquad \frac{1-\lambda}{\pi}\int_0^u dt \int_0^t \exp(2s)ds \int_0^{2\pi} h_k^2 \frac{|Z_0,\alpha^\perp|^2 + \cdots + |Z_{k+1},\alpha^\perp|^2}{(|Z_0,\alpha^\perp|^2 + \cdots + |Z_k,\alpha^\perp|^2)^\lambda} d\theta$$
$$< CT_k(u) + C'\,, \quad 0 \leqq k \leqq n-1\,, \quad 0 < \lambda < 1\,,$$

where $Z_0, \cdots, Z_n$ are the Frenet frames of the curve and C, C' are positive constants. With our setup it should not be difficult to prove (124) by integral geometry in a Grassmann manifold.

References

[1] L. V. Ahlfors, *The theory of meromorphic curves,* Acta Soc. Sci. Fenn. Ser. A, vol. **3**, no. 4.

[2] S. S. Chern, *Complex manifolds without potential theory,* van Nostrand, 1967.

[3] H. Weyl & J. Weyl, *Meromorphic functions and analytic curves,* Princeton, 1943.

[4] H. Wu, *The equidistribution theory of holomorphic curves,* Annals of Math. Studies, Princeton, 1970.

UNIVERSITY OF CALIFORNIA
BERKELEY, CALIF. 94720
U.S.A.

FRENET FRAMES ALONG HOLOMORPHIC CURVES

Shiing-Shen Chern
Michael J. Cowen
Albert L. Vitter III

University of California at Berkeley
Princeton University
Tulane University

1. INTRODUCTION

The study of holomorphic curves has received in recent years a revival of interest and new impetus, mainly as a special case in the general framework of holomorphic mappings. The most refined results are for holomorphic curves in complex projective space. This theory was initiated by H. and J. Weyl in 1938 [7] and its most important results, the defect relations, are due to Ahlfors [1]. A principal tool in both the original and modern (see [5] and [8]) contributions to this subject has been the systematic use of Frenet frames along holomorphic curves in projective space. In this paper we give the concept of a Frenet frame a differential geometric definition in the context of holomorphic curves on general hermitian manifolds. We characterize those hermitian manifolds possessing a Frenet frame for every holomorphic curve by curvature conditions. In particular, we show that if such a manifold is Kähler then it has constant holomorphic sectional curvature. The classical definition of Frenet frames on projective spaces is then examined. Finally, we consider the application of Frenet frames to the study of holomorphic sections of line bundles. Here too, the classical case (hyperplane bundle) is examined in terms of the general theory.

2. REVIEW OF HERMITIAN GEOMETRY (cf. [2])

Let M be a hermitian manifold of dimension n. Locally its metric can be written

$$<,> = \sum_k \omega_k \wedge \overline{\omega}_k$$

where $\omega_1,\ldots\omega_n$, called the coframe, are forms of bidegree (1,0) and are determined up to unitary transformation (we will agree that all small Latin letters in this section have range $1,\ldots n.$). Let $e_1,\ldots e_n$ be the unitary frame dual to $\omega_1,\ldots\omega_n$. It is well-known that there is a unique connection

$$De_i = \sum_k \omega_{ik}\, e_k$$

on M characterized by the conditions

$$\omega_{ik} + \overline{\omega}_{ki} = 0 \tag{1}$$

$$T_i \equiv d\omega_i - \sum_k \omega_k \wedge \omega_{ki} \text{ of bidegree } (2,0). \tag{2}$$

The first condition means that the connection preserves the hermitian structure. $T_1,\ldots T_n$ are called the torsion forms. From (1) and (2) it follows that

$$d\omega_{ik} = \sum_j \omega_{ij} \wedge \omega_{jk} + \Omega_{ik} \tag{3}$$

where the curvature forms Ω_{ik} are of bidegree (1,1) and satisfy

$$\Omega_{ik} + \overline{\Omega}_{ki} = 0\,. \tag{4}$$

More explicitly we have

$$\Omega_{ij} = \sum_{k,\ell} R_{ijk\ell}\, \omega_k \wedge \overline{\omega}_\ell$$

where

$$R_{ijk\ell} = \overline{R}_{ji\ell k}\,. \tag{5}$$

For each $p \in M$, $X,Y \in TM_p$ the curvature forms define a linear transformation $\Omega_p(X,\overline{Y}):TM_p \to TM_p$ determined by

$$<\Omega_p(X,\overline{Y})e_i,e_j> = \Omega_{ij}(X,\overline{Y})$$

Let $L \xrightarrow{\pi} M$ be a holomorphic line bundle with hermitian norm. Relative to an open covering $\{U,V\ldots\}$ of M, the points of $\pi^{-1}(U)$ will have local coordinates (x,y_U), $x \in U$, $y_U \in \mathbb{C}$, which are related by the equations of transition

$$y_U g_{UV} = y_V \qquad \text{in } \pi^{-1}(U \cap V)$$

where g_{UV} is a holomorphic function without zero in $U \cap V$. The hermitian norm is given in terms of coordinates by

$$\|y\|^2 = h_U |y_U|^2$$

where $h_U > 0$ is a C^∞ function in U. In $U \cap V$ we have therefore

$$h_U |y_U|^2 = h_V |y_V|^2$$

or

$$h_U = h_V |g_{UV}|^2 .$$

A connection on L of bidegree (1,0), uniquely determined by the condition that it leaves the hermitian norm invariant is given by the form

$$\phi_U = \partial \log h_U .$$

Its curvature form is

$$\Phi = - \partial\bar{\partial} \log h_U .$$

For a section y_U its covariant differential is

$$Dy_U = dy_U + \phi_U y_U .$$

When y_U is holomorphic, this is a form of bidegree (1,0) with values in L.

3. FRENET FRAMES

A holomorphic curve in M is a holomorphic mapping $f: X \to M$, where X is a simply connected complex manifold of dimension one (the most important case is $X = \mathbb{C}$). We assume f is not the constant mapping. Let ζ and z_i, $i = 1$ to M, be the local coordinates of X and M respectively and consider a neighborhood of $\zeta = 0$. Then f is given by $z_i = z_i(\zeta)$ and its tangent vector $f'(0) \in TM_{z(0)}$ is given by

$$\sum_i \frac{\partial z_i}{\partial \zeta}(0) \frac{\partial}{\partial z_i} = \zeta^p V$$

where p is a non-negative integer and V a non-zero vector. The isolated points where $p > 0$ are called the *stationary points of* f *of order zero.*

By a *unitary frame for* M *along* f we mean C^∞ mappings $e_i: X \to TM$ such that $e_1(p), \ldots e_n(p)$ is a unitary basis for $TM_{f(p)}$ for each $p \in X$. The covariant differential of the field e_i is given by $De_i = \sum_j f^*(\omega_{ij})e_j$ which will be denoted simply by $De_i = \sum_j \omega_{ij} e_j$. A unitary frame for M along f is called a *Frenet frame* if it has the additional properties at each point of X:

(i) $e_1 = \|v\|^{-1} v$

(ii) $De_i = \omega_{ii-1} e_{i-1} + \omega_{ii} e_i + \omega_{ii+1} e_{i+1}$ for $i = 1$ to $n - 1$ where ω_{ii+1} is a $(1,0)$ form.

The points of X where ω_{ii+1} vanishes are called the *stationary points of* f *of order* i.

In attempting to construct a Frenet frame along f one begins by defining e_1 locally according to (i) and completing to a local unitary frame $e_1, \ldots e_n$. We have

$$\begin{aligned} \omega_r &= 0 \qquad (\text{i.e., } f^*(\omega_r) = 0)\, r = 2 \text{ to } n \\ \omega_1 &= h_1\, d\zeta . \end{aligned} \tag{6}$$

We then need the following result [3]: Let $w_\lambda(\zeta)$ be complex-valued functions which satisfy the differential system

$$\frac{\partial w_\lambda}{\partial \bar{\zeta}} = \sum_\mu a_{\lambda\mu}(\zeta) w_\mu \qquad 1 \le \lambda,\ \mu \le n \tag{7}$$

in a neighborhood of $\zeta = 0$, where $a_{\lambda\mu}$ are complex valued C^1-functions. Suppose the w_λ do not all vanish identically. Then w_λ are of the form

$$w = \zeta^p \tilde{w}_\lambda, \tag{8}$$

where p is an integer ≥ 0 and $\tilde{w}_\lambda(0)$ are not all zero.

Theorem 1: *Let* M *be a hermitian manifold. consider the following two curvature conditions:*

$$\Omega_p(X,\bar{X})X = a(X)X \quad \textit{where} \quad a(X) \in \mathbb{R}, \quad X \in TM_p \quad \forall p \in M \tag{9A}$$

$$\Omega_p(X,\bar{X})Y = b(X)Y \quad \textit{where} \quad b(X) \in \mathbb{R}, \quad \forall X,Y \in TM_p,\ X \perp Y,\ \forall p \in M \tag{9B}$$

M *has a Frenet frame for every holomorphic curve iff (i) dimension* $M = 2$ *or (ii) dimension* $M = 3$ *and* (9A) *holds or (iii) dimension* $M \geq 4$ *and both* (9A) *and* (9B) *hold.*

Proof: First, the existence of a Frenet frame along every holomorphic curve will be shown to follow from the curvature and dimension conditions. From (2) and (6) we have

$$\omega_1 \wedge \omega_{1r} = 0 \qquad r = 2 \text{ to } n. \tag{10}$$

This implies that at a non-stationary point of f, ω_{1r} is a multiple of ω_1, i.e.

$$\omega_{1r} = h_r d\zeta \qquad r = 2 \text{ to } n. \tag{11}$$

By continuity, this is also true at a stationary point and hence at all points. If $n = 2$, we are done. If $n \geq 3$, Ω_{1r} along f is

$$\Omega_{1r}(e_1,\bar{e}_1)\|f'\|^2 d\zeta \wedge d\bar{\zeta} = 0$$

by (9A) and so differentiation of (11) gives

$$(dh_r - h_r\omega_{11} + \sum_q h_q \omega_{qr}) \wedge d\zeta = 0 \qquad 2 \leq q, r \leq n.$$

The expression in parenthesis is therefore a multiple of $d\zeta$ which means that $h_r(\zeta)$, $r = 2$ to n, satisfy a differential system of type (7). Hence the conclusion (8) implies $De_1 = \omega_{11}e_1 + d\zeta \wedge \sum_r h_r e_r$ where either $h_r \equiv 0$ or $h_r(\zeta) = \zeta^p \tilde{h}(\zeta)$ such that the $\tilde{h}_r(0)$ are not all zero. If $h_r \equiv 0$ for all r, $e_1,\ldots e_n$ is trivially a Frenet frame; if not we can make a unitary change of $e_2,\ldots e_n$ so that $De_1 = \omega_{11}e_1 + \omega_{12}e_2$ where ω_{12} is a (1,0) form. Now since $\omega_{1r} = 0$ for $r \geq 3$ and $\Omega_{ij}(e_1,\bar{e}_1) = 0$ for $i \neq j$, (3) yields $0 = \bar{\omega}_{12} \wedge \omega_{23}$ and so ω_{23} is a (1,0) form. If $n = 3$ we are done. If $n \geq 4$, continuing the above procedure produces a Frenet frame locally on X. If $e_1,\ldots e_n$ is a Frenet frame on an open set $U \subset X$ then the e_j are defined up to $e_j \rightarrow \exp(i\tau_j)e_j$ where τ_j is a real valued C^∞ function on U. A standard patching arguement, using the simple connectivity of X, now yields a Frenet frame globally on X.

Now assume M has a Frenet frame along every holomorphic curve and that the dimension of M is ≥ 3. Using local coordinates one sees

$\forall p \in M$ and X,Y unit vectors in TM_p such that $X \perp Y$ there is a holomorphic curve on M through p whose Frenet frame at p is of the form $e_1 = X, e_2 = Y, e_3, \ldots e_n$. Then

$$\Omega(X,\overline{X})X = \sum_j \Omega_{ij}(X,\overline{X})\, e_j$$
$$= \Omega_{11}(X,\overline{X})X + \Omega_{12}(X,\overline{X})Y$$

($\Omega_{1r}(X,\overline{X}) = 0$ for $r \geq 3$ because of the Frenet frame conditon and (3)). Since the left hand side is independent of Y and since $n \geq 3$, $\Omega_{12}(X,\overline{X}) = 0$ and so $\Omega(X,\overline{X})X = \Omega_{11}(X,\overline{X})X$.

$$\Omega(X,\overline{X})Y = \Omega_{21}(X,\overline{X})X + \Omega_{22}(X,\overline{X})Y + \Omega_{23}(X,\overline{X})e_3$$

$\Omega_{21}(X,\overline{X}) = -\overline{\Omega}_{12}(X,\overline{X}) = 0$ and e_3 can be chosen to be an arbitrary unit vector in TM_p orthogonal to X and Y. Since the left hand side is independent of e_3, $n \geq 4$ implies that $\Omega_{23}(X,\overline{X}) = 0$ and so $\Omega(X,\overline{X})Y = \Omega_{22}(X,\overline{X})Y$.

Q.E.D.

The next theorem exploits equations (9A) and (9B) to give a precise description of the curvature of M.

Theorem 2: *A hermitian manifold* M *of dimension* ≥ 4 *has a Frenet frame along every holomorphic curve iff the curvature of* M *at each point has the following form:*

$$\Omega_{ij} = \rho\, \omega_j \wedge \overline{\omega}_i \qquad \forall i \neq j$$
$$\Omega_{ii} = \rho\, \omega_i \wedge \overline{\omega}_i + \sum_j b_j \omega_j \wedge \overline{\omega}_j + \sum_{j<k} [c_{jk}\omega_j \wedge \overline{\omega}_k + \overline{c}_{jk}\omega_k \wedge \overline{\omega}_j] \tag{12}$$

where $\rho \in \mathbb{R}$ *is independent of* i *and* j, $b_j \in \mathbb{R}$, *and* $c_{jk} \in \mathbb{C}$ *is independent of* i.

A hermitian manifold M *of dimension* $= 3$ *has a Frenet frame for every holomorphic curve iff the curvature of* M *at each point has the following form:*

$$\Omega_{ij} = \rho_{ij}\omega_j \wedge \overline{\omega}_i \qquad \forall i \neq j$$
$$\Omega_{ii} = (b_{ji} + \rho_{ij})\omega_i \wedge \overline{\omega}_i + \sum_{j|j\neq i} b_{ij}\omega_j \wedge \overline{\omega}_j + \sum_{j<k} [c_{jk}\omega_j \wedge \overline{\omega}_k + \overline{c}_{jk}\omega_k \wedge \overline{\omega}_j] \tag{13}$$

where $\rho_{ij} = \rho_{ji} \in \mathbb{R}$, $b_{ij} \in \mathbb{R}$, *and* $c_{jk} \in \mathbb{C}$ *is independent of* i.

Proof: Equation (9A) for the ith and jth components of $\Omega(X,\overline{X})X$ $(i \neq j)$ yields

$$a(X)\,X_i = \sum_{\mu,k,\ell} R_{\mu i k \ell}\, X_\mu\, X_k\, \overline{X}_\ell$$

$$a(X)\,X_j = \sum_{\mu,k,\ell} R_{\mu j k \ell}\, X_\mu\, X_k\, \overline{X}_\ell\,.$$

Multiplying the first equation by X_j and the second by X_i we get

$$X_j \sum_{\mu,k,\ell} R_{\mu i k \ell} X_\mu X_k \overline{X}_\ell = X_i \sum_{\mu,k,\ell} R_{\mu j k \ell} X_\mu X_k \overline{X}_\ell\,. \tag{14}$$

Comparing coefficients one sees that

$$R_{\mu i k \ell} = 0 \quad \text{unless} \quad k \text{ or } \mu = i \tag{15}$$

and so by (5)',

$$R_{i \mu \ell k} = 0 \quad \text{unless} \quad k \text{ or } \mu = i\,. \tag{16}$$

Considering the other terms in (14) with the aid of (15) and (16) yields

$$R_{iiii} = R_{jjii} + R_{ijji} \tag{17}$$

$$R_{iik\ell} \text{ is independent of } i \text{ for } k \neq \ell. \tag{18}$$

Applying (5) again we get

$$R_{ijji} = R_{jiij} \in \mathbb{R}. \tag{19}$$

Furthermore it is clear that (15), (17), and (18) are equivalent to (9A). This proves (13) in the case $n = 3$.

Now considering equation (9B) applied in the case $Y = e_j$, $X = e_i$, we get $b(e_i) = R_{jjii}$ and so

$$R_{jjii} \text{ is independent of } j \text{ for } j \neq i\,. \tag{20}$$

Equation (17) now implies that R_{ijji} is independent of $j \neq i$ and then (19) gives

$$R_{ijji} \quad i \neq j \text{ is independent of } i \text{ and } j\,. \tag{21}$$

Thus (12) is proven in the case $n \geq 4$.

There are no more relations among the components of the curvature for the following reason. If ϕ is a C^∞ real valued function on M, make the conformal change of metric $\widetilde{\langle,\rangle} = e^\phi \langle,\rangle$. The curvature form of $\widetilde{\langle,\rangle}$

is $\tilde{\Omega} = \Omega - \partial\bar{\partial}\phi I$ where I is the $n \times n$ identity matrix. Therefore $\tilde{\Omega}$ satisfies the conditions of Theorem 1 iff Ω does. For an arbitrary $p \in M$ and hermitian symmetric $n \times n$ matrix (C_{jk}), ϕ can be chosen so that $(\partial^2\phi/\partial z_j \partial \bar{z}_k)|p = (C_{jk})$.

Q.E.D.

Theorem 3: *A Kähler manifold* M *of dimension* ≥ 3 *has a Frenet frame for every holomorphic curve iff it has constant holomorphic sectional curvature.*

Proof: The Kähler identities $R_{kji\ell} = R_{ijk\ell} = R_{i\ell kj}$ applied to (15) through (18) yield

$$R_{ijji} = R_{jjii} = R_{jiij} = R_{iijj} \tag{22}$$

$$R_{iik\ell} = R_{kii\ell} = 0 \quad k \neq \ell \tag{23}$$

$$R_{iiii} = 2R_{iijj} \quad j \neq i \tag{24}$$

and so the components in (22) and (24) are independent of i and j. Therefore the curvature of M has the form

$$\Omega_{ij} = \rho[\omega_j \wedge \omega_i + \delta^i_j \sum_K \omega_k \wedge \bar{\omega}_k].$$

This is precisely the curvature form of a Kähler manifold of constant holomorphic sectional curvature 2ρ.

4. HOLOMORPHIC CURVES IN $\mathbb{P}^n$

The classical case of holomorphic curves in complex projective case will now be examined. Let $Z = (z_o, \ldots z_n) \in \mathbb{C}^{n+1} - \{0\}$ be a set of homogeneous coordinates for a point $[Z]$ in $\mathbb{P}^n$. Introduce the usual scalar product on $\mathbb{C}^{n+1}$, $\langle Z,W\rangle = \sum_{i=0}^{n} z_i \bar{w}_i$, and identify the unitary group $U(n+1)$ with the space of all unitary frames for $\mathbb{C}^{n+1}$. $\mathbb{P}^n$ can be regarded as the base space of the bundle

$$U(n+1) \xrightarrow{p_1} U(n+1)/U(n) \xrightarrow{p_2} U(n+1)/U(1) \times U(n) = \mathbb{P}^n$$

where p_1 takes a frame $Z_o, \ldots Z_n$ to its first vector Z_o and p_2 is the Hopf map. The Maurer-Cartan forms of the unitary group are defined by

$$dZ_\alpha = \sum_\alpha \theta_{\alpha\beta} Z_\beta \qquad \alpha,\beta = 0 \text{ to } n \tag{25}$$

with

$$\Theta_{\alpha\beta} + \overline{\Theta}_{\beta\alpha} = 0 \tag{26}$$

and they satisfy the Maurer-Cartan equations

$$d\Theta_{\alpha\beta} = \sum_{\gamma} \Theta_{\alpha\gamma} \wedge \Theta_{\gamma\beta} . \tag{27}$$

In this setup, $T\mathbb{P}^n_{[Z]}$ is identified with $\{W \in \mathbb{C}^{n+1} : \langle W,Z \rangle = 0\}$ and so the fiber of the map P_1 over Z becomes the space of unitary frames for $T\mathbb{P}^n_{[Z]}$ where the hermitian metric on $\mathbb{P}^n$ is the Fubini-Study metric

$$ds^2 = \sum_{j=1}^{n} \Theta_{oi} \wedge \overline{\Theta}_{oi}.$$

The connection forms can now be read off from (27):

$$\omega_{ij} = \Theta_{ij} - \delta^i_j \Theta_{oo} \qquad 1 \le i,j \le n . \tag{28}$$

The curvature forms can be read off from the Maurer-Cartan equations:

$$\Omega_{ij} = \Theta_{oj} \wedge \overline{\Theta}_{oi} + \delta^i_j \sum_{k=1}^{n} \Theta_{ok} \wedge \overline{\Theta}_{ok} . \tag{29}$$

Therefore $\mathbb{P}^n$ has constant holomorphic sectional curvature equal to 2.

Let $f:X \to \mathbb{P}^n$ be a holomorphic curve. Using a local coordinate ζ on X and homogeneous coordinates on $\mathbb{P}^n$, f is given by an $n + 1$ dimensional holomorphic vector function $Z(\zeta) = (z_o(\zeta),\ldots,z_n(\zeta))$. We assume that $f(X)$ does not belong to any hyperplane, i.e., $Z(\zeta) \wedge Z'(\zeta) \wedge \ldots \wedge Z^{(n)}(\zeta) \not\equiv 0$. The classical way to define a unitary frame along f is by requiring that $Z_o(\zeta) \equiv \|Z(\zeta)\|^{-1} Z(\zeta)$, $Z_1(\zeta),\ldots Z_k(\zeta)$ form a unitary basis for the vector space spanned by $Z(\zeta)$, $Z'(\zeta),\ldots Z^{(k)}(\zeta)$ (the kth osculating space at ζ) for $k = 1$ to n and for all $\zeta \in X$. We will show that this is the Frenet frame defined in section 3. Equations (25) restricted to f simplify to give

$$dZ_i = \Theta_{ii-1} Z_{i-1} + \Theta_{ii} Z_i + \Theta_{ii+1} Z_{i+1} \tag{30}$$

for Θ_{ii+1} a multiple of $d\zeta$. This along with (28) implies that $\omega_{ij} = 0$ unless $|i - j| \le 1$ and that ω_{ii+1} is a multiple of $d\zeta$. Therefore

$$DZ_i = \omega_{ii-1} Z_{i-1} + \omega_{ii} Z_i + \omega_{ii+1} Z_{i+1}$$

and so $Z_1,\ldots Z_n$ is a Frenet frame.

5. COVARIANT DERIVATIVES OF A HOLOMORPHIC SECTION ALONG A CURVE

Let $L \to M$ be a holomorphic line bundle and s a holomorphic section. Assume that the holomorphic curve $f: X \to M$ has a Frenet frame. In value distribution theory one is often interested in how the image of f intersects the zero set of s in M. Thus it is natural to study the section s restricted to f (or, more precisely, the holomorphic section $s \circ f$ of the bundle $f^*(L) \to X$). We will define successive covariant derivatives of s along f. The definition will not work for all sections because the existence of the covariant derivative will imply conditions on s. We will then show how our definition applies to the classical case of the hyperplane bundle H over $\mathbb{P}^n$ and examine its failure in the case of higher powers (≥ 2) of H.

Let $E_k \to X$ be the holomorphic sub-bundle of f^*TM consisting of the kth osculating spaces of f, i.e., the spaces spanned by $e_1, \ldots e_k$, and let $F_k \to X$ be the holomorphic line bundle defined by $F_k \equiv E_k^*/E_{k-1}^*$ $k = 1$ to n (E_o defined to be trivial). The fact that E_k is a holomorphic bundle is equivalent to the existence of a Frenet frame along X. The hermitian metric and connection on TM induce those on F_k; the connection form is $-\omega_{kk}$. The connection form on $f^*L \otimes F_k$ is therefore $\phi - \omega_{kk}$. The kth covariant derivative s_k of s is a C^∞ section of $f^*L \otimes F_k$ defined recursively by

$$Ds_k = ds_k + s_k(\phi - \omega_{kk}) = s_{k+1}\omega_{kk+1} + t_{k+1}d\bar{\zeta} \tag{31}$$

$$0 \leq k \leq n-1, \; s_o \equiv s, \; t_1 = 0,$$

$$\omega_{o1} \equiv \omega_1 \text{ and } \omega_{oo} \equiv 0.$$

Since ω_{kk+1} may have isolated zeroes (we consider only curves for which none of the ω_{kk+1} vanish identically) the existence of s_{k+1} implies an assumption of a high order zero of the $d\zeta$ component of Ds_k at all the stationary points of f of order k.

Now consider the hyperplane bundle $H \to \mathbb{P}^n$. The connection form on H is $\phi = -2\,\partial \log \|z\|$ and from [4], p. 77, we have

$$\Theta_{oo} = (\partial - \bar{\partial}) \log \|z\|.$$

A holomorphic section of H restricted to f has the form $s(\zeta) = \langle Z(\zeta), A\rangle$ (the zero set of s is the hyperplane in $\mathbb{P}^n$ given by

$0 = \sum_{\alpha=0}^{n} \bar{A}_\alpha z_\alpha)$. Define the functions $w_\alpha(\zeta), \alpha = 0$ to n, by $A = \sum_{\alpha=0}^{n} w_\alpha(\zeta) Z_\alpha(\zeta)$. Since A is constant we get

$$d\bar{w}_\alpha = \sum_\beta \Theta_{\alpha\beta} \bar{w}_\beta . \tag{32}$$

We will show that the covariant derivative s_k is related to w_k by

$$s_k = \|z\| \bar{w}_k . \tag{33}$$

Let $\tilde{s}_k \equiv \|z\|\bar{w}_k$ $k = 0$ to n. Then

$$d \log \tilde{s}_k = \partial \log \|z\| + \bar{\partial} \log \|z\| + d \log \bar{w}_k$$

$$= \partial \log \|z\| + \bar{\partial} \log \|z\| + \frac{\|z\|}{\tilde{s}_k} \sum_{\mu=0}^{n} \Theta_{k\mu} \bar{w}_\mu$$

$$d\tilde{s}_k = - \Theta_{oo} \tilde{s}_k - \phi \tilde{s}_k + \sum_{\mu=0}^{n} \Theta_{k\mu} \tilde{s}_\mu$$

$$d\tilde{s}_k + (\phi - \omega_{kk})\tilde{s}_k = \tilde{s}_{k+1} \omega_{kk+1} + t_{k+1} d\bar{\zeta} .$$

This is exactly the equation (31) defining s_k.

The functions w_k play an important role in the value distribution theory of holomorphic curves in $\mathbb{P}^n$. Formula (33) gives them a new geometric interpretation. In the classical theory introduce

$$\phi_k \equiv \sum_{\nu=0}^{k} |w_\nu|^2 \qquad 0 \le k \le n.$$

The fundamental inequalities of Ahlfors are integral inequalities involving the ϕ_k. Their original proof by Ahlfors made use of integral geometry, cf [1]. Recently Cowen and Griffiths [5] gave a proof based on the "method of negative curvature" where the basic analytic step is to find formulas for $\partial\bar{\partial} \log \phi_k$ and $\partial\bar{\partial} \log \log \phi_k$.

We will now examine the higher tensor powers of the hyperplane bundle. The line bundle $H^m \to \mathbb{P}^n$ $(m \ge 2)$ restricted to the holomorphic curve f has connection form $\phi = - m \,\partial \log \|z\|^2$ and a holomorphic section of H^m along f is given by $s(\zeta) = P(Z(\zeta))$, for P a homogeneous polynomial of degree m. Let $Z_o \equiv \|z\|^{-1} Z$ and $Z_1, \ldots Z_n$ be the Frenet frame along f. Replacing the constant vector $\bar{A}$ in the $m = 1$ case we have the gradient of P, denoted $\frac{\partial P}{\partial z} = \left(\frac{\partial P}{\partial z_o}, \ldots \frac{\partial P}{\partial z_n}\right)$. Define the functions $w_k(\zeta)$ by $\overline{\frac{\partial P}{\partial z}}(Z(\zeta)) = \sum_{k=0}^{n} w_k(\zeta) Z_k(\zeta)$.

Define the hessian of P to be the n + 1 by n + 1 matrix given by

$$\text{Hess } P \equiv \left(\frac{\partial^2 P}{\partial z_i \partial z_j}\right) .$$

Using Euler's formula we get

$$s = \frac{1}{m} < Z, \overline{\frac{\partial P}{\partial z}} > \tag{34}$$

$$\frac{\partial P}{\partial z} = \frac{1}{m-1} Z \cdot \text{Hess } P = \frac{\|Z\|}{m-1} Z_o \cdot \text{Hess } P \tag{35}$$

where $Z \cdot \text{Hess } P$ means the product of the 1 by n + 1 matrix Z with the n + 1 by n + 1 matrix Hess P to produce a 1 by n + 1 matrix. We will also use the formula

$$dZ = d(\|Z\|Z_o) = d(\|Z\|)Z_o + \|Z\|(\Theta_{oo}Z_o + \Theta_{o1}Z_1). \tag{36}$$

The first two covariant derivatives of s will now be computed.

$$\begin{aligned}
Ds &= ds + \phi s \\
&= \langle dZ, \overline{\frac{\partial P}{\partial z}}\rangle - m \frac{<dZ,Z>}{\|Z\|^2} \frac{1}{m} < Z, \overline{\frac{\partial P}{\partial z}} > \\
&= d(\|Z\|)\langle Z_o, \overline{\frac{\partial P}{\partial z}} \rangle \\
&\quad + \|Z\| \,\Theta_{oo} \langle Z_o, \overline{\frac{\partial P}{\partial z}}\rangle + \|Z\| \,\Theta_{o1} \langle Z_1, \overline{\frac{\partial P}{\partial z}} \rangle \\
&\quad - (d(\|Z\|) + \|Z\| \,\Theta_{oo}) \langle Z_o, \overline{\frac{\partial P}{\partial z}} \rangle \\
&= \|Z\| \,\overline{w}_1 \,\Theta_{o1}
\end{aligned}$$

so that $s_1 = \|Z\| \overline{w}_1 = \|Z\| \langle Z_1, \overline{\frac{\partial P}{\partial z}} \rangle$ as in the m = 1 case.

$$\begin{aligned}
Ds_1 &= ds_1 + (\phi - \omega_{11})s_1 \\
&= d(\|Z\|) \langle Z_1, \overline{\frac{\partial P}{\partial z}} \rangle \\
&\quad + \|Z\| \langle \Theta_{1o}Z_o + \Theta_{11} Z_1 + \Theta_{12} Z_2, \overline{\frac{\partial P}{\partial z}} \rangle \\
&\quad + \|Z\| \langle Z_1, \overline{dZ \cdot \text{Hess } P} \rangle \\
&\quad + \|Z\| \langle Z_1, \overline{\frac{\partial P}{\partial z}} \rangle \left(- m \frac{\langle dZ, Z \rangle}{\|Z\|^2} - \Theta_{11} + \Theta_{oo}\right)
\end{aligned}$$

which, using (34), (35), and (36), reduces to

$$Ds_1 = - mP(Z) \,\overline{\Theta}_{o1} + \|Z\|w_2 \,\Theta_{12} + \langle Z_1 \cdot \text{Hess } P, \overline{Z}_1 \rangle \,\Theta_{o1} .$$

Therefore the $d\zeta$ component of Ds_1 is not a multiple of Θ_{12} and so s_2 cannot be defined according to the procedure of Section 4. The coefficient of the extra term, $\langle Z_1 \cdot \text{Hess } P, \bar{Z}_1 \rangle$, has a geometric significance: at points where f is tangent to the projective hypersurface $V(P)$ defined by P, i.e., $P(Z) = 0$ and $\langle Z_1, \overline{\frac{\partial P}{\partial z}} \rangle = 0$, the second fundamental form of $V(P) \subset \mathbb{P}^n$ in the direction Z_1 is given by (cf. [6])

$$\frac{\langle Z_1 \cdot \text{Hess } P, \bar{Z}_1 \rangle}{2\|\frac{\partial P}{\partial z}\|}.$$

REFERENCES

1. Ahlfors, L. V., *The theory of meromorphic curves*, Finska Vetenskaps-Societeten, Helsingfors: Acta Soc. Sci. Fenn. Ser. A, vol. 3, no. 4 (1941), 3-31.

2. Chern, S.-S., *Complex Manifolds Without Potential Theory*, van Nostrand, 1967.

3. ________, *On the minimal immersions of the two sphere in a space of constant curvature*, Problems in Analysis, in Honor of S. Bochner, Princeton 1970, 27-40.

4. ________, *Holomorphic curves in the plane*, Differential Geometry in Honor of K. Yano, Tokyo 1972, 73-94.

5. Cowen, M., and P. Griffiths, *Holomorphic curves and metrics of negative curvature*, to appear.

6. Vitter, A., *On the curvature of complex hypersurfaces*, to appear in Indiana University Mathematics Journal.

7. Weyl, H., and J. Weyl, *Meromorphic Functions and Analytic Curves*, Princeton, 1943.

8. Wu, H., *The equidistribution theory of holomorphic curves*, Annals of Math. Studies, Princeton, 1970.

Reprinted from *Proc. of Conf. on Value Distribution Theory*, Tulane Univ. (1974) 191–203

REAL HYPERSURFACES IN COMPLEX MANIFOLDS

BY

S. S. CHERN and J. K. MOSER

University of California
Berkeley, Cal., USA

New York University
New York, N. Y., USA

Introduction

Whether one studies the geometry or analysis in the complex number space C_{n+1}, or more generally, in a complex manifold, one will have to deal with domains. Their boundaries are real hypersurfaces of real codimension one. In 1907, Poincaré showed by a heuristic argument that a real hypersurface in C_2 has local invariants under biholomorphic transformations [6]. He also recognized the importance of the special unitary group which acts on the real hyperquadrics (cf. § 1). Following a remark by B. Segre, Elie Cartan took, up again the problem. In two profound papers [1], he gave, among other results, a complete solution of the equivalence problem, that is, the problem of finding a complete system of analytic invariants for two real analytic real hypersurfaces in C_2 to be locally equivalent under biholomorphic transformations.

Let $z^1, \ldots, z^{n+1}$ be the coordinates in C_{n+1}. We study a real hypersurface M at the origin 0 defined by the equation

$$r(z^1, \ldots, z^{n+1}, \bar{z}^1, \ldots, \bar{z}^{n+1}) = 0, \tag{0.1}$$

where r is a real analytic function vanishing at 0 such that not all its first partial derivatives are zero at 0. We set

$$z = (z^1, \ldots, z^n), \quad z^{n+1} = w = u + iv. \tag{0.2}$$

After an appropriate linear coordinate change the equation of M can be written as

$$v = F(z, \bar{z}, u), \tag{0.3}$$

where F is real analytic and vanishes with its first partial derivatives at 0. Our basic assumption on M is that it be nondegenerate, that is, the Levi form

This work was partially supported by the National Science Foundation, Grants GP-20096 and GP-34785X. We wish to thank the Rockefeller University for their hospitality where the first author was a visitor in the Spring of 1973.

15 – 742902 *Acta mathematica* 133. Imprimé le 18 Février 1975

$$\langle z, z\rangle = \sum_{1\leqslant\alpha,\beta\leqslant n} g_{\alpha\bar\beta} z^\alpha \bar z^\beta, \quad g_{\alpha\bar\beta} = \left(\frac{\partial^2 F}{\partial z^\alpha \partial \bar z^\beta}\right)_0 \tag{0.4}$$

is nondegenerate at 0. In § 2, 3 we study the problem of reducing the equation to a normal form by biholomorphic transformations of z, w. This is first studied in terms of formal power series in § 2 and their convergence to a holomorphic mapping is established in § 3. The results are stated in Theorems 2.2 and 3.5. It is worth noting that the convergence or existence proof is reduced to that of ordinary differential equations.

The normal form is found by fitting the holomorphic image of a hyperquadric closely to the given manifold. For $n=1$ this leads to 5th order osculation of the holomorphic image of a sphere at the point in question, while for $n\geqslant 2$ the approximation is more complicated. In both cases, however, the approximation takes place along a curve transversal to the complex tangent space. The family of the curves so obtained satisfies a system of second order differential equations which is holomorphically invariantly associated with the manifold. For a hyperquadric, or the sphere, these curves agree with the intersection of complex lines with the hyperquadric. For $n=1$ the differential equations can be derived from those of the sphere by constructing the osculating holomorphic image of the sphere, while for $n>1$ such a simple interpretation does not seem possible. This family of curves is clearly of basic importance for the equivalence problem. At first the differential equations for these curves are derived for real analytic hypersurfaces but they remain meaningful and invariant for five times continuously differentiable manifolds.

On the other hand, equation (0.1) implies

$$i\partial r = -i\bar\partial r, \tag{0.5}$$

which is therefore a real-valued one-form determined by M up to a non-zero factor; we will denote the common expression by θ. Let T_x and T_x^* be respectively the tangent and cotangent spaces at $x\in M$. As a basis of T_x^* we can take θ, Re (dz^α), Im (dz^α), $1\leqslant\alpha\leqslant n$. The annihilator $T_{x,C}=\theta^\perp$ in T has a complex structure and will be called the complex tangent space of M at x. Such a structure on M has been called a Cauchy-Riemann structure [8]. The assumption of the nondegeneracy of the Levi form defines a conformal hermitian structure in $T_{x,C}$. To these data we apply Cartan's method of equivalence, generalizing his work for $\mathbf{C}_2$. It turns out that a unique connection can be defined, which has the special unitary group as the structure group and which is characterized by suitable curvature conditions (Theorem 5.1). The successive covariant derivatives of the curvature of the connection give a complete system of analytic invariants of M under biholomorphic transformations. The result is, however, of wider validity. First, it suffices that the Cauchy-Riemann structure be defined abstractly on a real manifold of dimension $2n+1$. Secondly, the connection and the resulting invariants are also defined under weaker smoothness conditions, such as C^∞, although their identity will in general not insure equivalence without real analyticity. In this connection we mention the deep result of C. Fefferman [2] who showed that a biholomorphic mapping between two strictly pseudoconvex domains with smooth boundaries is smooth up to the boundary.

The equivalence problem was studied by N. Tanaka for real hypersurfaces in $\mathbf{C}_{n+1}$ called by him regular, which are hypersurfaces defined locally by the equation (0.3) where F does not involve u [7 I]. Later Tanaka stated the result in the general case [7 II], but the details, which are considerable, were to our knowledge never published.

One interesting feature of this study is the difference between the cases $\mathbf{C}_2$ and $\mathbf{C}_{n+1}$, $n \geqslant 2$. There is defined in general a tensor which depends on the partial derivatives of r up to order four inclusive and which vanishes identically when $n=1$. Thus there are invariants of order four in the general case, while for $n=1$ the lowest invariant occurs in order six. This distinction is also manifest from the normal forms.

The Cauchy-Riemann structure has another formulation which relates our study to systems of linear homogeneous partial differential equations of first order with complex coefficients. In fact, linear differential forms being covariant vector fields, the dual or annihilator of the space spanned by θ, dz^α will be spanned by the complex vector fields X_α, $1 \leqslant \alpha \leqslant n$, which are the same as complex linear homogeneous partial differential operators (cf. § 4). The question whether the differential system

$$X_\alpha w = 0, \quad 1 \leqslant \alpha \leqslant n, \tag{0.6}$$

has $n+1$ functionally independent solutions means exactly whether an abstractly given Cauchy-Riemann structure can be realized by one arising from a real hypersurface in $\mathbf{C}_{n+1}$. The answer is not necessarily affirmative. Recently, Nirenberg gave examples of linear differential operators X in three real variables such that the equation

$$Xw = 0 \tag{0.7}$$

does not have a nonconstant local solution [5].

It may be interesting to carry out this correspondence in an example. In $\mathbf{C}_2$ with the coordinates

$$z = x+yi, \quad w = u+vi, \tag{0.8}$$

consider the real hyperquadric M defined by

$$v = z\bar{z} = x^2+y^2. \tag{0.9}$$

On M we have

$$\theta = \tfrac{1}{2}dw - i\bar{z}dz = (\tfrac{1}{2}du + xdy - ydx), \quad dz = dx + idy.$$

Solving the equations

$$\theta = dz = 0,$$

we get

$$dx: dy: du = -i: 1: -2z.$$

The corresponding operator, defined up to a factor, is

$$L = -i\frac{\partial}{\partial x} + \frac{\partial}{\partial y} - 2z\frac{\partial}{\partial u} = -2i\left\{\frac{1}{2}\left(\frac{\partial}{\partial x} + i\frac{\partial}{\partial y}\right) - i(x+yi)\frac{\partial}{\partial u}\right\}, \tag{0.10}$$

which is the famous operator discovered by Hans Lewy.

The spirit of our study parallels that of classical surface theory. We list the corresponding concepts as follows:

Surfaces in euclidean 3-space	Real hypersurfaces in C_{n+1}
Group of motions	Pseudo-group of biholomorphic transformations
Immersed surface	Non-degenerate real hypersurface
Plane	Real hyperquadric
Induced riemannian structure	Induced CR-structure
Isometric imbedding	Existence of local solutions of certain systems of PDEs
Geodesics	Chains

Because of the special role played by the real hyperquadrics we will devote § 1 to a discussion of their various properties. Section 2 derives the normal form for formal power series and § 3 provides a proof that the resulting series converges to a biholomorphic mapping. These results were announced in [4]. In § 4 we solve the equivalence problem of the integrable G-structures in question in the sense of Elie Cartan. The solution is interpreted in § 5 as defining a connection in an appropriate bundle. Finally, the results of the two approaches, extrinsic and intrinsic respectively, are shown to agree with each other in § 6.

In the appendix we include results of S. Webster who derived some important consequences from the Bianchi identities.

1. The real hyperquadrics

Among the non-degenerate real hypersurfaces in C_{n+1} the simplest and most important are the real hyperquadrics. They form a prototype of the general non-degenerate real hypersurfaces which in turn derive their important geometrical properties from the "osculating" hyperquadrics. In fact, a main aim of this paper is to show how the geometry of a general non-degenerate real hypersurface can be considered as a generalization of that of real hyperquadrics. We shall therefore devote this section to a study of this special case.

Let z^α, z^{n+1} $(=w=u+iv)$, $1\leqslant\alpha\leqslant n$, be the coordinates in $\mathbf{C}_{n+1}$. A real hyperquadric is defined by the equation

$$v=h_{\alpha\bar\beta}z^\alpha z^{\bar\beta},\quad z^{\bar\beta}=\overline{z^\beta}, \tag{1.1}$$

where $h_{\alpha\bar\beta}$ are constants satisfying the conditions

$$h_{\alpha\bar\beta}=\bar h_{\beta\bar\alpha}=h_{\bar\beta\alpha},\quad \det\,(h_{\alpha\bar\beta})\neq 0. \tag{1.2}$$

Throughout this paper we will agree that small Greek indices run from 1 to n, unless otherwise specified, and we will use the summation convention. By the linear fractional transformation

$$Z^\alpha=\frac{2z^\alpha}{w+i},\quad W=\frac{w-i}{w+i}; \tag{1.3}$$

equation (1.1) goes into

$$h_{\alpha\bar\beta}Z^\alpha Z^{\bar\beta}+W\overline{W}=1. \tag{1.4}$$

This defines a hypersphere of dimension $2n+1$ when the matrix $(h_{\alpha\bar\beta})$ is positive definite. In general, we suppose $(h_{\alpha\bar\beta})$ to have p positive and q negative eigenvalues, $p+q=n$.

In order to describe a group which acts on the hyperquadric Q defined by (1.1), we introduce homogeneous coordinates ζ^A, $0\leqslant A\leqslant n+1$, by the equations

$$z^i=\zeta^i/\zeta^0,\quad 1\leqslant i\leqslant n+1. \tag{1.5}$$

$\mathbf{C}_{n+1}$ is thus imbedded as an open subset of the complex projective space $\mathbf{P}_{n+1}$ of dimension $n+1$. In homogeneous coordinates Q has the equation

$$h_{\alpha\bar\beta}\zeta^\alpha\zeta^{\bar\beta}+\frac{i}{2}(\bar\zeta^0\zeta^{n+1}-\zeta^0\bar\zeta^{n+1})=0. \tag{1.6}$$

For two vectors in $\mathbf{C}_{n+2}$:

$$Z=(\zeta^0,\zeta^1,\ldots,\zeta^{n+1}),\quad Z'=(\zeta'^0,\zeta'^1,\ldots,\zeta'^{n+1}), \tag{1.7}$$

we introduce the hermitian scalar product

$$(Z,Z')=h_{\alpha\bar\beta}\zeta^\alpha\bar\zeta'^\beta+\frac{i}{2}(\zeta^{n+1}\bar\zeta'^0-\zeta^0\bar\zeta'^{n+1}). \tag{1.8}$$

This product has the following properties:

(1) (Z, Z') is linear in Z and anti-linear in Z';

(2) $\overline{(Z, Z')} = (Z', Z)$;

(3) Q is defined by

$$(Z, Z) = 0. \tag{1.6a}$$

Let $SU(p+1, q+1)$ be the group of unimodular linear homogeneous transformations on ζ^A, which leave the form (Z, Z) invariant. Then Q is a homogeneous space with the group $SU(p+1, q+1)$ as its group of automorphisms. Its normal subgroup K of order $n+2$, consisting of the transformations

$$\zeta^{*A} = \varepsilon\zeta^A, \quad \varepsilon^{n+2} = 1, \quad 0 \leqslant A \leqslant n+1, \tag{1.9}$$

leaves Q pointwise fixed, while the quotient group $SU(p+1, q+1)/K$ acts on Q effectively.

By a Q-frame is meant an ordered set of $n+2$ vectors $Z_0, Z_1, \ldots, Z_{n+1}$ in $\mathbf{C}_{n+2}$ satisfying

$$(Z_\alpha, Z_\beta) = h_{\alpha\bar{\beta}}, \quad (Z_0, Z_{n+1}) = -(Z_{n+1}, Z_0) = -\frac{i}{2}, \tag{1.10}$$

while all other scalar products are zero, and

$$\det(Z_0, Z_1, \ldots, Z_{n+1}) = 1. \tag{1.11}$$

For later use it will be convenient to write (1.10) as

$$(Z_A, Z_B) = h_{A\bar{B}}, \quad 0 \leqslant A, B \leqslant n+1, \tag{1.10a}$$

where

$$h_{0,\overline{n+1}} = -h_{n+1,\bar{0}} = -\frac{i}{2}, \tag{1.10b}$$

while all other h's with an index 0 or $n+1$ are zero. There is exactly one transformation of $SU(p+1, q+1)$ which maps a given Q-frame into another. By taking one Q-frame as reference, the group $SU(p+1, q+1)$ can be identified with the space of all Q-frames. In fact, let Z_A, Z_A^* be two Q-frames and let

$$Z_A^* = a_A^B Z_B. \tag{1.12}$$

The linear homogeneous transformation on $\mathbf{C}_{n+2}$ which maps the frame Z_A to the frame Z_A^* maps the vector $\zeta^A Z_A$ to

$$\zeta^A Z_A^* = \zeta^A a_A^B Z_B. \tag{1.13}$$

If we denote the latter vector by $\zeta^{*B} Z_B$, we have

$$\zeta^{*B} = a_A^B \zeta^A, \tag{1.14}$$

which is the most general transformation of $SU(p+1, q+1)$ when Z_A^* runs over all Q-frames.

Let H be the isotropy subgroup of $SU(p+1, q+1)$, that is, its largest subgroup leaving a point Z_0 of Q fixed. The most general change of Q-frames leaving the point Z_0 fixed is

$$\left.\begin{aligned} Z_0^* &= tZ_0, \\ Z_\alpha^* &= t_\alpha Z_0 + t_\alpha{}^\beta Z_\beta, \\ Z_{n+1}^* &= \tau Z_0 + \tau^\beta Z_\beta + \bar{t}^{-1} Z_{n+1}, \end{aligned}\right\} \tag{1.15}$$

where

$$\left.\begin{aligned} t_\alpha &= -2\,itt_\alpha{}^\varrho \tau^{\bar\sigma} h_{\varrho\bar\sigma} = -2\,itt_\alpha{}^\varrho \tau_\varrho, \\ t\bar{t}^{-1} &\det(t_\alpha{}^\beta) = 1, \\ t_\alpha{}^\varrho \bar{t}_\beta{}^\sigma h_{\varrho\bar\sigma} &= h_{\alpha\bar\beta}, \\ h_{\varrho\bar\sigma} \tau^\varrho \tau^{\bar\sigma} + \frac{i}{2}&(\bar\tau \bar{t}^{-1} - \tau t^{-1}) = 0. \end{aligned}\right\} \tag{1.16}$$

In the first equation of (1.16) we have used $h_{\varrho\bar\sigma}$ to raise or lower indices. Observe that the last equation of (1.16) means that the point Z_{n+1}^* lies on Q, as does Z_{n+1}; the equation can also be written

$$\operatorname{Im}(\tau t^{-1}) = -h_{\varrho\bar\sigma}\tau^\varrho \tau^{\bar\sigma}. \tag{1.17}$$

H is therefore the group of all matrices

$$\begin{pmatrix} t & 0 & 0 \\ t_\alpha & t_\alpha{}^\beta & 0 \\ \tau & \tau^\beta & \bar{t}^{-1} \end{pmatrix} \tag{1.18}$$

with the conditions (1.16) satisfied. Its dimension is n^2+2n+2. By (1.14) the corresponding coordinate transformation is

$$\left.\begin{aligned} \zeta^{*0} &= t\zeta^0 + t_\alpha \zeta^\alpha + \tau \zeta^{n+1}, \\ \zeta^{*\beta} &= t_\alpha{}^\beta \zeta^\alpha + \tau^\beta \zeta^{n+1}, \\ \zeta^{*n+1} &= \bar{t}^{-1}\zeta^{n+1}, \end{aligned}\right\} \tag{1.19}$$

or, in terms of the non-homogeneous coordinates defined in (1.5),

$$\left.\begin{aligned} z^{*\beta} &= (t_\alpha{}^\beta z^\alpha + \tau^\beta w)\, t^{-1}\delta^{-1} \\ w^* &= |t|^{-2} w\delta^{-1}, \end{aligned}\right\} \tag{1.20}$$

where
$$\delta = 1 + t^{-1} t_\alpha z^\alpha + t^{-1}\tau w. \tag{1.21}$$

We put
$$C_\alpha{}^\beta = t^{-1} t_\alpha^\beta, \quad C_\alpha{}^\beta a^\alpha = t^{-1}\tau^\beta, \quad \varrho = |t|^{-2}. \tag{1.22}$$

Then (1.20) can be written

$$\left.\begin{aligned} z^{*\beta} &= C_\alpha{}^\beta (z^\alpha + a^\alpha w)\,\delta^{-1}, \\ w^* &= \varrho w \delta^{-1}. \end{aligned}\right\} \tag{1.23}$$

By (1.16) the coefficients in (1.23) satisfy the conditions

$$C_\alpha^{\ \lambda} C_{\bar\beta}^{\ \bar\sigma} h_{\lambda\bar\sigma} = \varrho h_{\alpha\bar\beta}, \tag{1.24}$$

and the coefficients in δ satisfy

$$\left.\begin{aligned} t^{-1} t_\alpha &= -2ia_\alpha = -2ih_{\alpha\bar\beta} a^{\bar\beta}, \\ \operatorname{Im}(t^{-1}\tau) &= -h_{\alpha\bar\beta} a^\alpha a^{\bar\beta}. \end{aligned}\right\} \tag{1.25}$$

Equations (1.23) give the transformations of the isotropy group H in non-homogeneous coordinates.

Incidentally, the hyperquadric Q can be viewed as a Lie group. To see this we consider the isotropy subgroup leaving Z_{n+1} fixed. The relevant formulae are obtained from (1.19) by the involution $\zeta^0 \to \zeta^{n+1}$, $\zeta^{n+1} \to -\zeta^0$, $\zeta^\alpha \to \zeta^\alpha$ $(\alpha = 1, 2, ..., n)$:

$$\left.\begin{aligned} \zeta^{*0} &= \bar t^{-1} \zeta^0 \\ \zeta^{*\beta} &= -\tau^\beta \zeta^0 + t_\alpha^{\ \beta} \zeta^\alpha \\ \zeta^{*n+1} &= -\tau\zeta^0 + t_\alpha \zeta^\alpha + t\zeta^{n+1} \end{aligned}\right\} \tag{1.26}$$

with the same restrictions (1.16) on the coefficients. We consider the subgroup obtained by choosing

$$t_\alpha^{\ \beta} = \delta_\alpha^{\ \beta}, \qquad t = 1 \tag{1.27}$$

and hence, by (1.16),

$$t_\alpha = -2ih_{\alpha\bar\beta}\overline{\tau^\beta}, \quad \operatorname{Im} \tau + h_{\alpha\bar\beta}\tau^\alpha \overline{\tau^\beta} = 0.$$

In non-homogeneous coordinates we obtain

$$\left.\begin{aligned} z^{*\alpha} &= a^\alpha + z^\alpha \\ w^* &= b + 2ih_{\alpha\bar\beta} z^\alpha \overline{a^\beta} + w \end{aligned}\right\} \tag{1.28}$$

where $$a^\alpha = -\tau^\alpha, \quad b = -\tau, \quad \operatorname{Im} b = h_{\alpha\bar\beta} a^\alpha \overline{a^\beta}.$$

Thus the point with the coordinates $(a^1, a^2, ..., a^n, b)$ can be viewed as a point on Q. If we take the point $(z^1, z^2, ..., z^n, w)$ also in Q then (1.28) defines a noncommutative group law on Q, making Q a Lie group. Moreover, the $(n+2)^2-1$ dimensional group $SU(p+1, q+1)/K$ is generated by the subgroup (1.26) satisfying (1.27) and the isotropy group H.

The Maurer-Cartan forms of $SU(p+1, q+1)$ are given by the equations

$$dZ_A = \pi_A^{\ B} Z_B. \tag{1.29}$$

They are connected by relations obtained from the diffentiation of (1.10a) which are

$$\pi_{A\bar B} + \pi_{\bar B A} = 0, \tag{1.30}$$

where the lowering of indices is relative to $h_{A\bar{B}}$. For the study of the geometry of Q it will be useful to write out these equations explicitly, and we have

$$\left.\begin{aligned} \pi_{\alpha\bar{\beta}} + \pi_{\bar{\beta}\alpha} &= 0, \\ \pi_0{}^{n+1} - \bar{\pi}_0{}^{n+1} = \pi_{n+1}{}^0 - \bar{\pi}_{n+1}{}^0 &= 0, \\ \bar{\pi}_0{}^0 + \pi_{n+1}{}^{n+1} &= 0, \\ \tfrac{1}{2} i\bar{\pi}_\alpha{}^0 + \pi_{n+1}{}^\beta h_{\beta\bar{\alpha}} &= 0, \\ \bar{\pi}_\alpha{}^{n+1} + 2i\pi_0{}^\beta h_{\beta\bar{\alpha}} &= 0. \end{aligned}\right\} \tag{1.30 a}$$

Another relation between the π's arises from the differentiation of (1.11). It is

$$\pi_A{}^A = 0, \tag{1.31}$$

or, by (1.30 a),

$$\pi_\alpha{}^\alpha + \pi_0{}^0 - \bar{\pi}_0{}^0 = 0. \tag{1.31 a}$$

The structure equations of $SU(p+1, q+1)$ are obtained by the exterior differentiation of (1.29) and are

$$d\pi_A{}^B = \pi_A{}^C \wedge \pi_C{}^B, \qquad 0 \leqslant A, B, C \leqslant n+1. \tag{1.32}$$

The linear space T_C spanned by $Z_0, Z_1, \ldots, Z_n$ is the complex tangent space of Q at Z_0. It is of complex dimension n, in contrast to the real tangent space of real dimension $2n+1$ of Q, which is defined in the tangent bundle of P_{n+1}, and not in P_{n+1} itself. The intersection of Q by a complex line transversal to T_C is called a *chain*. One easily verifies that a complex line intersecting T_C transversally at some point of Q is transversal to T_C at every other point of intersection with Q. Without loss of generality, suppose the complex line be spanned by Z_0, Z_{n+1}. The line Z_0, Z_{n+1} being fixed, it follows that along a chain dZ_0, dZ_{n+1} are linear combinations of Z_0, Z_{n+1}. Hence the chains are defined by the system of differential equations

$$\pi_0{}^\alpha = \pi_{n+1}{}^\alpha = 0. \tag{1.33}$$

Through every point of Q and any preassigned direction transversal to T_C there is a unique chain. Since the complex lines in P_{n+1} depend on $4n$ real parameters, the chains on Q depend on $4n$ real parameters. The notion of a chain generalizes to an arbitrary real hypersurface of $\mathbf{C}_{n+1}$.

§ 2. Construction of a normal form

(a) In this section we consider the equivalence problem from an extrinsic point of view. Let

$$r(z^1, z^2, \ldots, z^{n+1}, \overline{z^1}, \ldots, \overline{z^{n+1}}) = 0$$

denote the considered hypersurface M in $\mathbb{C}^{n+1}$, where r is a real analytic function whose first derivatives are not all zero at the point of reference. Taking this point to be the origin we subject M to transformations holomorphic near the origin and ask for a simple normal form. At first we will avoid convergence questions by considering merely formal power series postponing the relevant existence problem to the next section.

We single out the variables

$$z^{n+1} = w = u + iv, \quad \overline{z^{n+1}} = u - iv$$

and assume that we have

$$r_{z^\alpha} = 0, \quad \alpha = 1, \ldots, n$$

$$r_w = -r_{\bar{w}} \neq 0$$

at the origin. This can be achieved by a linear transformation. Solving the above equation for v we obtain

$$v = F(z, \bar{z}, u) \tag{2.1}$$

where F is a real analytic function in the $2n+1$ variables $z, \bar{z}, u$, which vanishes at the origin together with its first derivatives. This representation lacks the previous symmetry but has the advantage that F is uniquely determined by M.

We subject this hypersurface to a holomorphic transformation

$$z^* = f(z, w), \quad w^* = g(z, w), \tag{2.2}$$

where f is n-vector valued holomorphic, g a holomorphic scalar. Moreover, f, g are required to vanish at the origin and should preserve the complex tangent space (2.1) at the origin: $w = 0$. Thus we require

$$f = 0, \quad g = 0, \quad \frac{\partial g}{\partial z} = 0 \quad \text{at } z = w = 0. \tag{2.3}$$

The resulting hypersurface M^* will be written

$$v^* = F^*(z^*, \overline{z^*}, u^*).$$

Our aim is to choose (2.2) so as to simplify this representation of M^*.

From now on we drop the assumption that F is real analytic but consider it as a formal power series in $z^1, \ldots, z^n, \bar{z}^1, \ldots, \bar{z}^n$, and u with the reality condition

$$\overline{F(z, \bar{z}, u)} = F(\bar{z}, z, u).$$

Moreover, F is assumed to have no constant or linear terms. This linear space of formal power series will be denoted by $\mathfrak{F}$. Similarly, we consider transformations (2.2) given by formal power series f, g in $z^1, \ldots, z^n, w$ without constant term and—according to (2.3)—

no terms linear in z for g. These formal transformations constitute a group under composition which we call $\mathcal{G}$. Often we combine f and g to a single element h.

For the following it is useful to decompose an element $F \in \mathfrak{F}$ into semihomogeneous parts:

$$F = \sum_{\nu=2}^{\infty} F_\nu(z, \bar{z}, u)$$

where $F_\nu(tz, t\bar{z}, t^2u) = t^\nu F_\nu(z, \bar{z}, u)$ for any $t > 0$. Thus we assign u the "weight" 2 and z, $\bar{z}$ the "weight" 1. To simplify the terms of weight $\nu = 2$ we observe that they do not contain u—since F contains no linear terms—so that

$$F_2 = Q(z) + \overline{Q(z)} + H(z, z)$$

where Q is a quadratic form of z and H a hermitian form. The transformation

$$\begin{pmatrix} z \\ w \end{pmatrix} \mapsto \begin{pmatrix} z \\ w - 2iQ(z) \end{pmatrix}$$

removes the quadratic form, so that we can and will assume that $F_2 = H(z, z)$ is a hermitian form. This form, the Levi form, will be of fundamental importance in the following. In the sequel we will require that this form which we denote by

$$\langle z, z \rangle = F_2$$

is a *nondegenerate* hermitian form. If $\langle z, z \rangle$ is positive the hypersurface M is strictly pseudoconvex. With $\langle z_1, z_2 \rangle$ we denote the corresponding bilinear form, such that

$$\langle \lambda z_1, \mu z_2 \rangle = \lambda \bar{\mu} \langle z_1, z_2 \rangle$$

With this simplification M can be represented by

$$v = \langle z, z \rangle + F \tag{2.4}$$

where

$$F = \sum_{\nu=3}^{\infty} F_\nu$$

contains terms of weight $\nu \geq 3$ only. Now we have to restrict the transformation (2.2) by the additional requirement that $\partial^2 g / \partial z^\alpha \partial z^\beta$ vanishes at the origin.

(b) Normal forms. To determine a formal transformation in $\mathcal{G}$ simplifying M^* we write it in the form

$$z^* = z + \sum_{\nu=2}^{\infty} f_\nu, \quad w^* = w + \sum_{\nu=3}^{\infty} g_\nu, \tag{2.5}$$

where

$$f_\nu(tz, t^2w) = t^\nu f_\nu(z, w), \quad g_\nu(tz, t^2w) = t^\nu g_\nu(z, w),$$

and call ν the "weight" of these polynomials f_ν, g_ν. Inserting (2.5) into

$$v^* = \langle z^*, z^*\rangle + F^*$$

and restricting the variables z, w to the hypersurface (2.4) we get the transformation equations, in which z, $\bar{z}$, u are considered as independent variables. Collecting the terms of weight μ in the relation we get

$$F_\mu + \operatorname{Im} g_\mu(z, u+i\langle z, z\rangle) = 2 \operatorname{Re} \langle f_{\mu-1}, z\rangle + F^*_\mu + \ldots$$

where the dots indicate terms depending on $f_{\nu-1}$, g_ν, F_ν, F^*_ν with $\nu<\mu$. In F_μ the arguments are z, $w=u+i\langle z, z\rangle$. We introduce the linear operator L mapping $h=(f, g)$ into

$$Lh = \operatorname{Re}\{2\langle z, f\rangle + ig\}_{w=u+i\langle z, z\rangle} \tag{2.6}$$

and write the above relation as

$$Lh = F_\mu - F^*_\mu + \ldots \quad \text{for } h=(f_{\mu-1}, g_\mu) \tag{2.7}$$

and note that L maps $f_{\mu-1}$, g_μ into terms of weight μ.

In order to see how far one can simplify the power series F^*_μ one has to find a complement of the range of the operator L which is a matter of linear algebra. More precisely we will determine a linear subspace $\mathcal{N}$ of $\mathcal{F}$ such that $\mathcal{N}$ and the range of L span $\mathcal{F}$; i.e., if $\mathcal{V}$ denotes the space of $h=(f, g)$ with $f=\sum_{\nu=2}^{\infty} f_\nu$; $g=\sum_{\nu=3}^{\infty} g_\nu$, then we require that

$$\mathcal{F} = L\mathcal{V} + \mathcal{N} \text{ and } \mathcal{N} \cap L\mathcal{V} = (0). \tag{2.8}$$

Thus $\mathcal{N}$ represents a complement of the range of L.

Going back to equation (2.7) it is clear that we can require that F^*_μ belongs to $\mathcal{N}$ and solve the resulting equation for h. Using induction it follows that (2.5) can be determined such that the function F^* belongs to $\mathcal{N}$. We call such a hypersurface M^* with $F^* \in \mathcal{N}$ in "normalform". It is of equal importance to study how much freedom one has in transforming (2.4) into normal form which clearly depends on the null space of L. Thus we have reduced the problem of finding a transformation into normal form of M to the determination of a complement of the range $\mathcal{N}$ and the null space of the operator L. Our goal will be to choose $\mathcal{N}$ such that the elements N in $\mathcal{N}$ vanish to high order at the origin so that the hypersurface M^* can be approximated to high degree by the quadratic hypersurface $v=\langle z, z\rangle$.

(c) Clearly a transformation into a normal form can be unique only up to holomorphic mappings preserving the hyperquadric $v=\langle z, z\rangle$ as well as the origin. These mappings form the $(n+1)^2+1$ dimensional isotropic group H studied in § 1 and given by (1.23). We will make use of H to normalize the holomorphic mapping transforming M into normal form.

After the above preparation we may consider the group $\mathcal{G}_1$ of all formal transformations preserving the family of formal hypersurfaces

$$v = \langle z, z \rangle + \{\text{weight} \geqslant 3\},$$

as well as the origin. One verifies easily that the elements of $\mathcal{G}_1$ are of the form

$$z^* = Cz + \{\text{weight} \geqslant 2\}; \quad w^* = \varrho w + \langle \text{weight} \geqslant 3 \rangle,$$

where $\langle Cz, Cz \rangle = \varrho \langle z, z \rangle$. Using the form (1.23) one sees that any $\phi \in \mathcal{G}_1$ can be factored uniquely as

$$\phi = \psi \circ \phi_0$$

with $\phi_0 \in H$ and ψ a formal transformation of the form (2.5) with

$$f_2(0, w) = 0, \quad \operatorname{Re} \frac{\partial^2}{\partial w^2} g_4(0, w) = 0 \quad \textit{at } w = 0.$$

The first term can be normalized by choice of a^α $(\alpha = 1, ..., n)$ in (1.23) and the second by $\operatorname{Re}(t^{-1}\tau)$. We summarize the normalization conditions for ψ by requiring that the series

$$\left.\begin{array}{l} f, \dfrac{\partial}{\partial z^\alpha} f, \quad \dfrac{\partial}{\partial w} f \\[2ex] g, \dfrac{\partial}{\partial z^\alpha} g, \quad \dfrac{\partial}{\partial w} g, \quad \dfrac{\partial^2 g}{\partial z^\alpha \partial z^\beta}, \quad \operatorname{Re}\left(\dfrac{\partial^2 g}{\partial w^2}\right) \end{array}\right\} \tag{2.9}$$

all have no constant term.

From now on we may restrict ourselves to transformations (2.5) with the normalization (2.9). The submanifold of power series $h = (f, g)$ with the condition (2.9) will be called $\mathfrak{V}_0$. Similarly, we denote the restriction of the operator L to $\mathfrak{V}_0$ by L_0. We will see that L_0: $\mathfrak{V}_0 \to \mathfrak{F}$ is injective. This implies, in particular, that the most general formal power series mapping preserving $v = \langle z, z \rangle$ and the origin belongs to the isotropic group H.

(d) The operator L introduced above is of basic importance. To describe it more conceptually we interpret $h = (f, g)$ as a holomorphic vector field

$$X = \sum_\alpha f^\alpha \frac{\partial}{\partial z^\alpha} + g \frac{\partial}{\partial w} + \sum \bar{f}^\alpha \frac{\partial}{\partial \bar{z}^\alpha} + \bar{g} \frac{\partial}{\partial \bar{w}}$$

near the manifold, M. We describe the manifold $v = \langle z, z \rangle$ by

$$r(z, \bar{z}, w, \bar{w}) = \langle z, z \rangle - \frac{1}{2i}(w - \bar{w}) = 0.$$

Then
$$Lh = \mathfrak{L}_X r|_{r=0}$$

is the Lie-derivative of r along the holomorphic vector field X restricted to $r=0$. Of course, L is meaningful only up to a nonvanishing real factor.

For example, if we represent the manifold $r=0$ by

$$Q = \langle Z, Z\rangle + W\overline{W} = 1$$

we can associate with a holomorphic vector field

$$X = \sum_\alpha A^\alpha \frac{\partial}{\partial Z^\alpha} + \overline{A^\alpha}\frac{\partial}{\partial \overline{Z}^\alpha} + B\frac{\partial}{\partial W} + \overline{B}\frac{\partial}{\partial \overline{W}}$$

the Lie derivative of the above quadratic form

$$\mathfrak{L}_X Q = 2\,\mathrm{Re}\,\{\langle A, Z\rangle + B\overline{W}\}$$

restricted to $Q=1$.

For the following we will determine the kernel and a complement of the range for L in the original variables z, w. To formulate the result we order the elements F in terms of powers of z, $\bar{z}$ with coefficients being power series in u. Thus we write

$$F = \sum_{k,l\geq 0} F_{kl}$$

where
$$F_{kl}(\lambda z, \mu\bar{z}, u) = \lambda^k \mu^l F_{kl}(z, \bar{z}, u)$$

for all complex numbers λ, μ, and call (k, l) the "type" of F_{kl}.

The basic hermitian form will be written as

$$\langle z, z\rangle = \sum_{\alpha,\bar{\beta}} h_{\alpha\bar{\beta}} z^\alpha \bar{z}^{\beta}, \quad \bar{h}_{\alpha\bar{\beta}} = h_{\beta\bar{\alpha}}.$$

Using the notation of tensor calculus we define the contraction $\mathrm{tr}\,(F_{kl}) = G_{k-1,l-1}$ of

$$F_{kl} = \sum a_{\alpha_1\ldots\alpha_k\bar{\beta}_1\ldots\bar{\beta}_l} z^{\alpha_1}\ldots z^{\alpha_k}\bar{z}^{\beta_1}\ldots\bar{z}^{\beta_l}$$

where we assume that the coefficients $a_{\alpha_1\ldots\bar{\beta}_l}$ are unchanged under permutation of $\alpha_1, \ldots, \alpha_k$ as well as of $\bar{\beta}_1, \ldots, \bar{\beta}_l$. We define for $k, l \geq 1$

$$\mathrm{tr}\,(F_{kl}) = \sum b_{\alpha_1\ldots\alpha_{k-1}\bar{\beta}_1\ldots\bar{\beta}_{l-1}} z^{\alpha_1}\ldots z^{\alpha_{k-1}}\bar{z}^{\beta_1}\ldots\bar{z}^{\beta_{l-1}} \tag{2.10}$$

where
$$b_{\alpha_1\ldots\alpha_{k-1}\bar{\beta}_1\ldots\bar{\beta}_{l-1}} = \sum_{\alpha_k,\bar{\beta}_l} h^{\alpha_k\bar{\beta}_l} a_{\alpha_1\ldots\alpha_k\bar{\beta}_1\ldots\bar{\beta}_l}.$$

Here $h^{\alpha\bar{\beta}}$ is defined as usual by
$$h^{\alpha\bar{\beta}} h_{\gamma\bar{\beta}} = \delta^\alpha{}_\gamma$$
being the Kronecker symbol.

For the description of a complement of the range of L_0 we decompose the space $\mathfrak{F}$ of real formal power series as

$$\mathfrak{F}=\mathcal{R}+\mathcal{N}$$

where $\mathcal{R}$ consists of series of the type

$$R=\sum_{\min(k,l)\leqslant 1} R_{kl}+G_{11}\langle z,z\rangle+(G_{10}+G_{01})\langle z,z\rangle^2+G_{00}\langle z,z\rangle^3$$

G_{jm} being of type (j, m), and where

$$\mathcal{N}=\{N\in\mathfrak{F};\ N_{kl}=0\ \min(k,l)\leqslant 1;\quad \operatorname{tr} N_{22}=(\operatorname{tr})^2N_{32}=(\operatorname{tr})^3N_{33}=0\}. \tag{2.11}$$

This constitutes a decomposition of $\mathfrak{F}$, i.e. any F can uniquely be written as $F=R+N$ with $R\in\mathcal{R}$, $N\in\mathcal{N}$. Thus $PF=R$ defines a projection operator with range $\mathcal{R}$ and null space $\mathcal{N}$. One computes easily that

$$PF=\sum_{\min(k,l)\leqslant 1} F_{kl}+G_{11}\langle z,z\rangle+(G_{10}+G_{01})\langle z,z\rangle^2+G_{00}\langle z,z\rangle^3 \tag{2.12}$$

where

$$G_{11}=\frac{4}{n+2}\operatorname{tr}(F_{22})-\frac{2}{(n+1)(n+2)}(\operatorname{tr})^2(F_{22})\langle z,z\rangle$$

$$G_{10}=\frac{6}{(n+1)(n+2)}(\operatorname{tr})^2F_{32}$$

$$G_{00}=\frac{6}{n(n+1)(n+2)}(\operatorname{tr})^3F_{33}$$

In particular, for $n=1$

$$PF=\sum_{\min(k,l)\leqslant 1} F_{kl}+F_{22}+F_{23}+F_{32}+F_{33}. \tag{2.13}$$

While for $n>1$ it is a requirement that $\langle z,z\rangle^l$ divides F_{kl} $(k\geqslant l)$, this is automatically satisfied for $n=1$.

Evidently this decomposition is invariant under linear transformations of z which preserve the hermitian form $\langle z,z\rangle$.

The space $\mathcal{N}$ turns out to be an ideal in $\mathfrak{F}$ under multiplication with real formal power series. We will not use this fact, however, and turn to the main result about the kernel and corange of L_0:

LEMMA 2.1. *L_0 maps $\mathcal{V}_0$ one to one onto $\mathcal{R}_3=P\mathfrak{F}_3$, where $\mathfrak{F}_3$ denotes the space of those $F\in\mathfrak{F}$ containing terms of weight $\geqslant 3$ only, and $\mathcal{V}_0$ is the space of formal power series satisfying* (2.9).

Before proving this lemma we draw the crucial conclusion from it: For any $F \in \mathfrak{F}_3$ the equation

$$L_0 h = F \pmod{\mathfrak{N}}$$

can uniquely be solved for h in $\mathfrak{V}_0$, since this equation is equivalent to $PL_0 h = PF$. Thus $\mathfrak{N}$ represents a complement of the range of L_0 and applying our previous considerations on normal forms we obtain

THEOREM 2.2. *A formal hypersurface M can be transformed by a formal transformation*

$$z^* = z + f(z, w), \quad w^* = w + g(z, w)$$

normalized by (2.9) *into a normal form*

$$v^* = \langle z^*, z^* \rangle + N \quad \text{with } N \in \mathfrak{N}.$$

Moreover, this transformation is unique.

COROLLARY. *The only formal power series transformations which preserve $v = \langle z, z \rangle$ and the origin are given by the fractional linear transformations* (1.23) *constituting the group H.*

(e) Obviously it suffices to show that the equation

$$Lh = F \pmod{\mathfrak{N}}$$

possesses a unique solution $h \in \mathfrak{V}_0$. Here F is a formal power series containing terms of weight $\geqslant 3$ only. Collecting terms of equal type we have to solve the equations

$$(Lh)_{kl} = F_{kl} \quad \text{for } \min(k, l) \leqslant 1$$

$$(Lh)_{kl} \doteq F_{kl} \pmod{\mathfrak{N}} \quad \text{for } (k, l) = (2, 2), \ (3, 2), \ (3,3).$$

For this purpose we calculate $(Lh)_{kl}$ for the above types (k, l); because of the real character of F we may take $k \geqslant l$. We will use the identity

$$f(z, u + i\langle z, z \rangle) = \sum_{\nu=1}^{\infty} \left(\frac{\partial}{\partial w}\right)^{\nu} f(z, u) \frac{i^{\nu} \langle z, z \rangle^{\nu}}{\nu!}.$$

Expanding $f(z, w), g(z, w)$ in powers of $z, \bar{z}$ we write

$$f = \sum_{k=0}^{\infty} f_k, \qquad g = \sum_{k=0}^{\infty} g_k$$

where $$f_k(tz, w) = t^k f_k(z, w), \quad g_k(tz, w) = t^k g_k(z, w).$$

This notation should not be confused with the previous one which combined terms of equal weight, and which will no longer be needed.

We write Lh in the form

$$Lh = \mathrm{Re}\{2\langle f,z\rangle + ig\}_{w=u+i\langle z,z\rangle} = \langle f + f'i\langle z,z\rangle + \ldots, z\rangle + \frac{i}{2}(g + g'i\langle z,z\rangle + \ldots) + \text{complex conj.}$$

where the arguments of $f, f', \ldots, g, g', \ldots$ are z, u, and the prime indicates differentiation with respect to u. Now we collect terms of equal type (k, l). For example, if $k \geq 2$ the terms of type $(k, 0)$ and $(k+1, 1)$, respectively are

$$\frac{i}{2}g_k, \quad \langle f_{k+1}, z\rangle - \tfrac{1}{2}g_k'\langle z,z\rangle$$

so that we have

$$\left.\begin{aligned} ig_k &= 2F_{k0} \\ 2\langle f_{k+1}, z\rangle - g_k'\langle z,z\rangle &= 2F_{k+1,1} \end{aligned} \quad (k \geq 2).\right\} \tag{2.14 a}$$

For $k=1$ one gets additional terms and an easy calculation shows

$$\left.\begin{aligned} &ig_1 + 2\langle z, f_0\rangle = 2F_{10} \\ &-g_1'\langle z,z\rangle + 2\langle f_2, z\rangle - 2i\langle z, f_0'\rangle\langle z,z\rangle = 2F_{21} \\ &-\frac{i}{2}g_1''\langle z,z\rangle^2 + 2i\langle f_2', z\rangle\langle z,z\rangle - \langle z, f_0''\rangle\langle z,z\rangle^2 = 2F_{32} \quad (\mathrm{mod}\ \mathcal{N}). \end{aligned}\right\} \tag{2.14 b}$$

Finally, for $k=0$ one obtains four real equations,

$$\left.\begin{aligned} &-\mathrm{Im}\, g_0 = F_{00} \\ &\tfrac{1}{2}\mathrm{Im}\, g_0''\langle z,z\rangle^2 - 2\,\mathrm{Im}\langle f_1', z\rangle\langle z,z\rangle = F_{22} \quad (\mathrm{mod}\ \mathcal{N}) \\ &-\mathrm{Re}\, g_0'\langle z,z\rangle + 2\,\mathrm{Re}\langle f_1, z\rangle = F_{11} \\ &\tfrac{1}{6}\mathrm{Re}\, g_0'''\langle z,z\rangle^3 - \mathrm{Re}\langle f_1'', z\rangle\langle z,z\rangle^2 = F_{33} \quad (\mathrm{mod}\ \mathcal{N}). \end{aligned}\right\} \tag{2.14 c}$$

Thus we obtain three groups of decoupled systems of differential equations; actually the last system (2.14c) decouples into two groups.

The solution of these systems is elementary: Equations (2.14a) can be solved uniquely for f_{k+1}, g_k $(k \geq 2)$. Equations (2.14b) are equivalent to

$$\begin{aligned} &ig_1 + 2\langle z, f_0\rangle = 2F_{10} \\ &-g_1'\langle z,z\rangle + 2\langle f_2, z\rangle - 2i\langle z, f_0'\rangle\langle z,z\rangle = 2F_{21} \\ &-4\langle z, f_0''\rangle\langle z,z\rangle^2 = 2F_{32} - 2iF_{21}'\langle z,z\rangle - F_{10}''\langle z,z\rangle^2 \quad (\mathrm{mod}\ \mathcal{N}). \end{aligned}$$

Since the last equation has to be solved (mod $\mathcal{N}$) only we replace the right-hand side by its projection into $\mathcal{R}$, which we call $G_{10}\langle z,z\rangle^2$ so that

$$-4\langle z, f_0''\rangle = G_{10}.$$

With such a choice of f_0 one solves the first equation for g_1 and then the second for f_2. Here f_0 is fixed up to a linear function in w; but by our normalization (2.9), f_0 and hence g_1, f_2 are uniquely determined.

Finally we have to solve (2.14c): Since

$$F_{22} = G_{11}\langle z, z\rangle + N_{22}, \quad N_{22} \in \mathcal{N}$$

the second equation takes the form

$$\tfrac{1}{2}\operatorname{Im} g_0''\langle z, z\rangle - 2\operatorname{Im}\langle f_1', z\rangle = G_{11}$$

which can be solved with the first for $\operatorname{Im} g_0$ and $\operatorname{Im}\langle f_1', z\rangle = (d/du)\operatorname{Im}\langle f_1, z\rangle$. Since f_1 vanishes for $u=0$ we determine $\operatorname{Im} g_0$, $\operatorname{Im}\langle f_1, z\rangle$ uniquely in this way.

The last two equations of (2.14c) are equivalent to

$$\begin{aligned} -\operatorname{Re} g_0'\langle z, z\rangle + 2\operatorname{Re}\langle f_1, z\rangle &= F_{11} \\ -\tfrac{1}{3}\operatorname{Re} g_0''' &= G_{00} \end{aligned}$$

where we used that

$$F_{33} + \tfrac{1}{2}F_{11}''\langle z, z\rangle^2 = G_{00}\langle z, z\rangle^3 \pmod{\mathcal{N}}.$$

Clearly, the last equation can be solved for $\operatorname{Re} g_0'''$ and then the first for $\operatorname{Re}\langle f_1, z\rangle$. Thus g_0 is determined up to $aw + bw^2$, a, b real. But by our normalization both $a=0$ and $b=0$, and $\operatorname{Re} g_0$, $\operatorname{Re}\langle f_1, z\rangle$ are uniquely determined.

Thus, summarizing, all equations can be satisfied by f_k, g_k satisfying the normalization (2.9) and uniquely so. This concludes the proof of the Lemma 2.1 and hence of Theorem 2.2.

§ 3. Existence theorems

(a) So far we considered only formal series and now turn to the case of real analytic hypersurfaces M. We will show that the formal series transforming M into normal form are, in fact, convergent and represent holomorphic mappings. In the course of the proof we will obtain a geometrical interpretation of the condition

$$\operatorname{tr} N_{22} = 0, \quad (\operatorname{tr})^2 N_{32} = 0, \quad (\operatorname{tr})^3 N_{33} = 0$$

describing the normal form.

We begin with a transformation into a partial normal form: Let M be a real analytic hypersurface and γ a real analytic arc on M which is transversal to the complex tangent space of M. Moreover, we give a frame of linear independent vectors $e_\alpha \in T_{\mathbf{C}}$ ($\alpha = 1, \ldots, n$), also real analytic along the curve γ. All these data γ, e_α are given locally near a distinguished point p on γ.

Theorem 3.1. *Given a real analytic hypersurface M with the above data γ, e_α there exists a unique holomorphic mapping ϕ taking p into the origin $z=w=0$, γ into the curve $z=0$, $w=\xi$, where ξ is a real parameter ranging over an interval, and e_α into $\phi_*(e_\alpha)=\partial/\partial z^\alpha$ and the hypersurface into $\phi^*(M)$ given by*

$$v=F_{11}(z,\bar{z},u)+\sum_{\min(k,l)\geq 2}F_{kl}(z,\bar{z},u). \tag{3.1}$$

Proof. We may assume that the variables $z=(z^1,\dots,z^n)$ and w are so introduced that p is given by $z=0$, $w=0$ and the complex tangent space of M by $w=0$. If γ is given by

$$z=p(\xi),\quad w=q(\xi)$$

where $\xi=0$ corresponds to $z=0$, $w=0$ then $q'(0)\neq 0$. The transformation

$$z=p(w^*)+z^*,\quad w=q(w^*)$$

is holomorphic and takes the curve γ into $z^*=0$, $w^*=\xi$. Changing the notation and dropping the star we can assume that the hypersurface is given by

$$v=F(z,\bar{z},u)$$

and γ by $z=0$, $w=\xi$, so that $F(0,0,u)=0$.

The function $F(z,\bar{z},u)$ is given by convergent series and is real. In the variables x^α, y^α given by $z^\alpha=x^\alpha+iy^\alpha$, $\bar{z}^\alpha=x^\alpha-iy^\alpha$ the function $F(z,\bar{z},u)$ is real analytic. The space of these functions, real analytic in some neighborhood of the origin and vanishing at the origin will be denoted by $\mathfrak{F}^\omega$. In the following it will be a useful observation that z, $\bar{z}$ can be considered as independent variables for $F\in\mathfrak{F}^\omega$.

Lemma 3.2. *If $F\in\mathfrak{F}^\omega$ and $F(0,0,u)=0$ then there exists a unique holomorphic transformation*

$$z^*=z;\quad w^*=w+g(z,w);\quad g(0,w)=0$$

taking

$$v=F(z,\bar{z},u)$$

into

$$v^*=F^*(z^*,\bar{z}^*,u^*)$$

where

$$F^*_{k0}=F^*_{0k}=0\quad\text{for } k=1,2,\dots. \tag{3.2}$$

Proof. The conditions (3.2) can be expressed by

$$F^*(z^*,0,u)=0 \tag{3.3}$$

and a second equation which follows on account of the real character of F^*. The transformation formula gives

$$F^*(z^*, \bar{z}^*, u^*) = \frac{1}{2i}\left(g(z,w) - \overline{g(z,w)}\right) + F(z, \bar{z}, u)$$

where $$u^* = u + \tfrac{1}{2}\left(g(z,w) + \overline{g(z,w)}\right), \quad w = u + iF(z, \bar{z}, u).$$

Keeping in mind that $z, \bar{z}, u$ can be viewed as independent variables, we set $\bar{z}=0$ in the above equations. Observing that $\overline{g(z,w)}=0$ for $\bar{z}=0$, since $g(0,w)=0$, we obtain with (3.3)

$$0 = \frac{1}{2i}\, g(z, u + iF(z,0,u)) + F(z,0,u) \tag{3.4}$$

as condition for the function g. To solve this equation we set

$$s = u + iF(z, 0, u).$$

Since, by assumption, $F(z,0,u)$ vanishes for $z=0$ we can solve this equation for u:

$$u = s + G(z, s) \quad \text{where } G(0, s) = 0.$$

Equation (3.4) takes the form

$$0 = \frac{1}{2i}\, g(z,s) + \frac{1}{i}(s-u)$$

or $$u = s + \tfrac{1}{2} g(z,s).$$

Thus $g(z,w) = 2G(z,w)$ is the desired solution which vanishes for $z=0$. It is clear that the steps can be reversed, and Lemma 3.2 is proven.

Thus we may assume that M is of the form

$$v = F(z, \bar{z}, u) = \sum_{\min(k,l)\geq 1} F_{kl}(z, \bar{z}, u),$$

and the curve γ is given by $z=0$, $w=\xi$. Now we will require that $F_{11}(z,\bar{z},0)$ is a nondegenerate hermitian form.

LEMMA 3.3. *If* $F \in \mathfrak{F}^\omega$ *and*

$$F_{k0} = 0 = F_{0k} \quad \textit{for } k = 0, 1, \ldots$$

and $F_{11}(z,\bar{z},0)$ *nondegenerate then there exists a holomorphic transformation*

$$z^* = z + f(z,w); \quad w^* = w \tag{3.5}$$

with $f(0,w)=0$, $f_z(0,w)=0$ *and such that* $v = F(z,\bar{z},u)$ *is mapped into*

$$v^* = F^*_{11}(z^*, \bar{z}^*, u^*) + \sum_{\min(k,l)\geq 2} F^*_{kl}. \tag{3.6}$$

Proof. By $O_{\varkappa\lambda}$ we will denote a power series in $z, \bar{z}$ containing only terms of type (k, l) with $k \geqslant \varkappa$ and $l \geqslant \lambda$. Thus $F(z, \bar{z}, u)$ can be written as

$$F(z, \bar{z}, u) = F_{11}(z, \bar{z}, u) + \sum_{\alpha=1}^{n} z^\alpha A_\alpha(\bar{z}, u) + \sum_{\alpha=1}^{n} \overline{z^\alpha} \, \overline{A_\alpha(\bar{z}, u)} + O_{22}$$

where
$$A_\alpha(\bar{z}, u) = \frac{\partial}{\partial z^\alpha}(F - F_{11})\big|_{z=0} = O_{02}.$$

We restrict u to such a small interval in which the Levi form

$$F_{11}(z, \bar{z}, u) = \sum h_{\alpha\bar{\beta}}(u) z^\alpha \overline{z^\beta}$$

is nondegenerate. If $(h^{\alpha\bar{\beta}})$ is the inverse matrix of $(h_{\alpha\bar{\beta}})$ and the holomorphic vector function $f(z, w)$ is defined by

$$\overline{f^\beta(z, u)} = \sum_\alpha h^{\alpha\bar{\beta}}(u) A_\alpha(\bar{z}, u) \in O_{02} \tag{3.7}$$

then
$$\begin{aligned} F_{11}(z + f, \bar{z} + \bar{f}, u) &= F_{11}(z, \bar{z}, u) + \sum z^\alpha A_\alpha + \sum \overline{z^\alpha} \, \overline{A_\alpha} + O_{22} \\ &= F(z, \bar{z}, u) + O_{22} \end{aligned}$$

so that $v = F(z, \bar{z}, u)$ is transformed by (3.5), defined by (3.7), into

$$v^* = F_{11}(z^*, \bar{z}^*, u^*) + O_{22}.$$

Note also that, by (3.6), $f(z, u) \in O_{20}$ which finishes the proof.

With these two lemmas we see that the coordinates can be so chosen that γ is given by the u-axis: $z = 0$, $w = \xi$ and M given by (3.6). Actually the coordinates are not uniquely fixed by these requirements but the most general holomorphic transformation preserving the parametrized curve γ: $z = 0$, $w = \xi$ and the form (3.6) of M is given by

$$z^* = M(w) z, \quad w^* = w$$

where $M(w)$ is a nonsingular matrix depending holomorphically on w. This matrix can be used to transform the frame e_α into $\partial/\partial z^\alpha$ which, in turn, fixes $M(w)$ uniquely. This completes the proof of Theorem 3.1.

In order to make the hermitian form $F_{11}(z, \bar{z}, u)$ independent of u we perform a linear transformation

$$z^* = C(w) z, \quad w^* = w$$

and determine C such that

$$F_{11}(C(u) z, \overline{C(u) z}, 0) = F_{11}(z, \bar{z}, u).$$

The choice of $C(u)$ becomes unique if we require that $C(u)$ be hermitian with respect to the form

$$F_{11}(z, \bar{z}, 0) = \langle z, z\rangle$$

i.e.

$$F_{11}(Cz, \bar{z}, 0) = F_{11}(z, \bar{C}\bar{z}, 0)$$

Denoting the matrix $(h_{\alpha\beta}(u))$ by $H(u)$ these requirements amount to the two matrix equations

$$\left.\begin{aligned} C^*(u)H(0)C(u) &= H(u) \\ H(0)C(u) &= C^*(u)H(0). \end{aligned}\right\} \tag{3.8}$$

Eliminating $C^*(u)$ we obtain

$$C^2(u) = H(0)^{-1}H(u).$$

Since the right-hand side is close to the identity matrix for small u there exists a unique matrix $C(u)$ with $C(0)=I$. This solution depends analytically on u and, morever, satisfies automatically the relation (3.8). Indeed, if $C(u)$ is a solution so is $H^{-1}(0)C^*(u)H(0)$ which also reduces to the identity for $u=0$. By uniqueness it agrees with $C(u)$ yielding (3.8).

Thus we can assume that the hypersurface is represented by

$$v = \langle z, z\rangle + \sum_{\min(k,l)\geq 2} F_{kl}(z, \bar{z}, u) \tag{3.9}$$

and γ is given by $z=0$, $w=\xi$. The freedom in the change of variables preserving γ and the above form of M is given by linear map $z\to U(w)z$, $w\to w$ which preserve the form $\langle z, z\rangle$. In other words we can prescribe an analytic frame $e_\alpha(u)$ $(\alpha=1, \ldots, n)$ along the u-axis which is normalized by

$$\langle e_\alpha, e_\beta\rangle = h_{\alpha\bar{\beta}} \text{ where } \langle z, z\rangle = \sum h_{\alpha\bar{\beta}} z^\alpha \overline{z^\beta}.$$

The coefficients of $F_{kl}(z, \bar{z}, u)$ in (3.9) can be viewed as functionals depending on the curve γ: $z=p(\xi)$, $w=q(\xi)$. These are, of course, *local* functionals and more precisely we have

LEMMA 3.4. *The coefficients of F_{kl} in* (3.9) *depend analytically on p, q, $\bar{p}$, $\bar{q}$ and their derivatives of order $\leq k+l$. More precisely, these coefficients depend rationally on the derivatives p', $\bar{p}'$, q', etc.*

Proof. Let $v=G(z, \bar{z}, u)$ represent the given hypersurface containing the curve $z=p(\xi)$, $w=q(\xi)$ where $\operatorname{Re} q'(0) \neq 0$. The condition that this curve be transversal to the complex tangent space amounts to

$$\operatorname{Re}\{q' - 2iG_z p' - iG_u q'\} \neq 0 \tag{3.10}$$

which we require for $\xi=0$. First we subject the hypersurface to the transformation

$$z = p(w^*) + z^*, \quad w = q(w^*)$$

and study how the resulting hypersurface depends on p, q. This hypersurface is given implicitly by

$$\frac{1}{2i}\{q - \bar{q}\} - G(p + z^*, \bar{p} + \overline{z^*}, \tfrac{1}{2}(\bar{q} + \bar{q})) = 0 \tag{3.11}$$

where the arguments in p, q are w^*. Under the assumption (3.10) we can solve this equation for v^* to obtain the desired representation. Since the given curve was assumed to lie on the given hypersurface we have $v^* = 0$ as a solution of (3.11) if $z^* = 0$, $\overline{z^*} = 0$ Therefore the solution of (3.11)

$$v^* = F^*(z^*, \overline{z^*}, u^*) \tag{3.12}$$

vanishes for $z^* = 0, \overline{z^*} = 0$. We expand the terms in (3.11) in powers of $z^*, \overline{z^*}, v^*$ and investigate the dependence of the coefficients on $p(u^*)$, $q(u^*)$ and their derivatives.

To simplify the notation we drop the star and denote the left-hand side of (3.11) by

$$\Phi(z, \bar{z}, u, v) = \sum_{\zeta + \nu > 0} \Phi_{\zeta\nu}$$

where $\Phi_{\zeta\nu}$ is a polynomial in $z, \bar{z}, v$, homogeneous of degree ζ in $z, \bar{z}$ and of degree ν in v. The equation (3.11) takes the form

$$Av + \Phi_{10} + \sum_{\zeta + \nu \geqslant 2} \Phi_{\zeta\nu} = 0 \tag{3.13}$$

where
$$Av = \Phi_{01} = \mathrm{Re}\,\{q' - 2iG_z(p, \bar{p}, \tfrac{1}{2}(q + \bar{q}))p' - iG_u q'\}v$$

Thus A is an analytic function of $p, \bar{p}, q, \bar{q}$ and their derivatives, in fact, depending linearly on the latter. Moreover, by (3.10), we have $A \neq 0$ for small $|u|$.

Similarly, the coefficients of $\Phi_{\zeta\nu}$ are analytic functions of $p, \bar{p}, q, \bar{q}$ at $\xi = u$ and their derivatives of order $\leqslant \nu$. This becomes clear if one replaces $q(u + iv)$ by $q(u) + q'(u)iv + \ldots$ and similarly for $p(u + iv)$ in (3.11) and rewrites the resulting expressions as the series Φ in $z, \bar{z}, v$. In fact, the coefficients of Φ will depend polynomially on $p, \bar{p}', q'$ etc. Finally to obtain the same property for the coefficients of F^* in (3.12) we solve (3.13) for v as a power series in $z, \bar{z}$; let

$$v = V_1 + V_2 + \ldots$$

where V_ζ are homogeneous polynomials in $z, \bar{z}$ of degree ζ. We obtain V_ζ by comparison of coefficients in (3.13) in a standard fashion, which gives $A\ V_\zeta$ as a polynomial in $V_1, V_2, \ldots, V_{\zeta-1}$ with coefficients analytic in $p, q, \bar{p}, \bar{q}$ and their derivatives of order $\leqslant \zeta$; in dependence on the derivatives they are rational, the denominator being a power of A.

This proves the statement about the analytic behavior of the coefficients of F^* in (3.12). To complete the proof we have to subject this hypersurface to the holomorphic transformation of Lemma 3.2, 3.3 which preserve the curve $z=0$, $w=\xi$. From the proofs of these lemmas it is clear that the coefficients of the transformation as well as of the resulting hypersurface (3.6) have the stated analytic dependence on p, q. The same is true of the transformation $z \to C(w)z$, $w \to w$ which leads to (3.9).

(b) Returning to (3.9) it remains to satisfy the relations

$$\operatorname{tr} F_{22}=0, \quad (\operatorname{tr})^2 F_{32}=0, \quad (\operatorname{tr})^3 F_{33}=0$$

which give rise to a set of differential equations for the curve γ and for the associated frame.

We begin with the condition $(\operatorname{tr})^2 F_{32}=0$ which gives rise to a differential equation of second order for the curve γ, where the parametrization is ignored. For this purpose we assume that the parametrization is fixed, say by $\operatorname{Re} q(\xi)=\xi$ and study the dependence of F_{32} on $p(\xi)$. According to Lemma 3.4 the coefficients of F_{32} are analytic functions of $p, \bar{p}$ and their derivatives up to order 5. But if the hypersurface is in the form (3.9) then F_{32} depends on the derivatives of order $\leqslant 2$ and is of the form

$$F_{32}=\langle z, Bp''\rangle\langle z, z\rangle^2+K_{32}' \tag{3.14}$$

where K_{32}, B depend on p, $\bar{p}$, p', $\bar{p}'$ analytically, and B is a nonsingular matrix for small $|u|$.

To prove this statement we recall that (3.9) was obtained by a transformation

$$z \to p(w)+C(w)z+\dots, \quad w \to q(w)+\dots$$

We choose $\operatorname{Re} q(u)=u$ fixing the parametrization; $\operatorname{Im} q(u)$ is determined by $p, \bar{p}$. To study the dependence of F_{32} at $u=u_0$ we subject (3.9) to the transformation

$$z=s(w^*)+z^*+\dots, \quad w=q(w^*+u_0) \tag{3.15}$$

which amounts to replacing $p(u)$ by $p^*(u^*)=p(u_0+u)+C(u_0+u)s(u)$. Considering p and p' fixed at $u=u_0$ we require $s(0)=0$, $s'(0)=0$ and investigate the dependence of F_{32} on the germ of s at $u=u_0$. We choose the higher order terms in (3.15) in such a way that the form of (3.9) is preserved as far as terms of weight $\leqslant 5$ is concerned. This is accomplished by the choice

$$z=z^*+s(w^*)+2i\langle z^*, s'(\overline{w^*})\rangle z^*$$

$$w=w^*+u_0+2i\langle z^*, s(\overline{w^*})\rangle.$$

Since the hermitian form $\langle , \rangle$ is antilinear in the second argument this transformation is holomorphic. One computes

$$v-\langle z, z\rangle = v^* - \langle z^*, z^*\rangle + 4\,\mathrm{Re}\,\langle z^*, s''(0)\rangle\langle z^*, z^*\rangle^2 + \dots$$

if z, w lies on the manifold (3.9). The dots indicate terms of weight $\geqslant 6$ in z^*, $\overline{z^*}$, u^*. Thus, for $u^*=0$ we get, setting $z^*=z$,

$$F_{32}|_{u=u_0} = F^*_{32}|_{u^*=0} + 2\langle z^*, s''(0)\rangle\langle z^*, z^*\rangle^2.$$

Hence F^*_{32} depends on s, s', s'' only, and using that

$$(C(u_0+u)s(u))'' = C(u_0)s''(0) \quad \text{for } u=0$$

we see that
$$F^*_{32} + 2\langle z^*, C^{-1}(u_0)p''(0)\rangle\langle z^*, z^*\rangle^2$$

is independent of s which proves (3.14) with $B(u) = -2C^{-1}(u_0)$. Thus $B(0) = -2I$, and $B(u)$ is nonsingular for small values of $|u|$.

Therefore the equation $(\mathrm{tr})^2 F_{32}=0$ can be written as a differential equation

$$p'' = Q(p, \bar{p}, p', \bar{p}', u)$$

with an analytic right-hand side. Thus for given $p(0)$, $p'(0)$ there exists a unique analytic solution $p(u)$ for sufficiently small $|u|$. Choosing the curve γ in this manner we have $(\mathrm{tr})^2 F_{32}=0$.

To show that this differential equation $(\mathrm{tr})^2 F_{32}=0$ is independent of the parametrization and the frame e_α we subject the hypersurface (3.9) to the most general self mapping

$$z \to \sqrt{g'(w)}\,U(w)z$$
$$w \to g(w)$$

where $\mathrm{Im}\,g(u)=0$, $g(0)=0$, $g'(0)>0$, $\langle Uz, Uz\rangle = \langle z, z\rangle$ for real w. One checks easily that under such mapping F_{32} is replaced by

$$g'^{-3/2} F_{32}(U^{-1}z, \bar{U}^{-1}\bar{z}, g^{-1}(u))$$

and the equation $(\mathrm{tr})^2 F_{32}=0$ remains satisfied for $z=0$. Thus $(\mathrm{tr})^2 F_{32}$ is a differential equation for γ irrespective of the parametrization and the frame.

Next we fix the frame e_α so that $\mathrm{tr}\,F_{22}=0$. For this purpose we subject (3.9) with $(\mathrm{tr})^2 F_{32}=0$ to a coordinate transformation

$$z^* = U(w)z, \quad w^* = w$$

with a nonsingular matrix $U(w)$ which for $\mathrm{Im}\,w=0$ preserves the form $\langle z, z\rangle = \langle Uz, Uz\rangle$. We will define U via a differential equation

$$\frac{d}{du} U = UA \quad \text{with } \langle Az, z\rangle + \langle z, Az\rangle = 0 \tag{3.16}$$

and find from $U(w) = U(u) + ivU' + \ldots$ that

$$\begin{aligned}\langle z^*, z^*\rangle &= \langle (U + iU'\langle z, z\rangle + \ldots)z, (U + iU'\langle z, z\rangle + \ldots)z\rangle \\ &= \langle (I + iA\langle z, z\rangle + \ldots)z, (I + iA\langle z, z\rangle + \ldots)z\rangle \\ &= \langle z, z\rangle (1 + 2i\langle Az, z\rangle + \ldots)\end{aligned}$$

where the arguments of U, A are u and the dots indicate terms of order $\geqslant 6$ in $z, \bar{z}$. Thus

$$F^*_{22} = F_{22} + 2i\langle Az, z\rangle\langle z, z\rangle, \quad F^*_{32} = F_{32},$$

where on the left side we set $z^* = U(u)z$. Thus, since $\operatorname{tr} F_{22}$ is a hermitian form the equation $\operatorname{tr} F^*_{22} = 0$ determines $\langle iAz, z\rangle$ uniquely as a hermitian form, hence A is uniquely determined as an antihermitian matrix with respect to $\langle , \rangle$. Thus the differential equation (3.16) defines a $U(u)$, analytic in u, and preserving the form $\langle , \rangle$ if $U(0)$ does. More geometrically, (3.16) can be viewed as a first order differential equation

$$\frac{de_\alpha}{du} = \sum a^\beta_\alpha(u)\, e_\beta, \quad \langle e_\alpha, e_\beta\rangle = h_{\alpha\bar{\beta}}$$

for the frame. Note that the term F_{32} is not affected by this choice of the frame.

Finally, we are left with choosing the parametrization on the curve in such a way that $(\operatorname{tr})^3 F_{33} = 0$. For this purpose perform the transformation

$$z^* = (q'(w))^{1/2} z, \quad w^* = q(w)$$

with
$$q(0) = 0, \quad \overline{q(w)} = q(\bar{w}), \quad q'(0) > 0.$$

Thus
$$v^* = q'(u)\, v - \tfrac{1}{6} q''' v^3 + \ldots$$

$$\langle z^*, z^*\rangle = q'(u)\langle z, z\rangle - \tfrac{1}{2}\left(q''' - \frac{q''^2}{q'}\right)\langle z, z\rangle^3$$

which gives for z, w on the hypersurface

$$v^* - \langle z^*, z^*\rangle = q'(v - \langle z, z\rangle) + \left(\tfrac{1}{3} q''' - \tfrac{1}{2}\frac{q''^2}{q'}\right)\langle z, z\rangle^3 + \ldots$$

or
$$F^*_{33} = q' F_{33} + \left(\tfrac{1}{3} q''' - \tfrac{1}{2}\frac{q''^2}{q'}\right)\langle z, z\rangle^3.$$

Thus, $(\operatorname{tr})^3 F^*_{33} = 0$ gives rise to an analytic third order differential equation for the real function $q(u)$, uniquely determined by $q(0)=0$, $q'(0)>0$, $q''(0)$, which are assumed real. Thus we have a distinguished parameter ξ in the above curve which is determined up to real projective transformations $\xi \to \xi/(\alpha\xi+\beta)$, $\beta>0$.

Thus we have constructed a holomorphic transformation taking M into the normal form, and the existence proof has been reduced to that for ordinary differential equations. The choice of the initial values for $p'(0) \in C^n$, $U(0)$ and $\operatorname{Re} q'(0)$, $\operatorname{Re} q''(0)$ allows us to satisfy the normalization condition (2.9) of § 2. In fact, these $2n+n^2+1+1=(n+1)^2+1$ real parameters characterize precisely an element of the isotropic group H. Thus we have shown

THEOREM 3.5. *If M is a real analytic manifold the unique formal transformation of Theorem 2.2 taking M into a normal form and satisfying the normalization condition is given by convergent series, i.e. defines a holomorphic mapping.*

Two real analytic manifolds M_1, M_2 with distinguished points $p_1 \in M_1$, $p_2 \in M_2$ are holomorphically equivalent by a holomorphic mapping ϕ taking p_1 into p_2 if and only if $\langle M_k, p_k \rangle$ for $k=1, 2$ have the same normal forms for some choice of the normalization conditions. Thus the problem of equivalence is reduced to a finite dimensional one.

The arbitrary initial values for the differential equations $\operatorname{tr} F_{22}=0$, $(\operatorname{tr})^2 F_{32}=0$, $(\operatorname{tr})^3 F_{33}=0$ have a geometrical interpretation: At a fixed point $p \in M$ they correspond to

(i) a normalized frame $e_\alpha \in T_{\mathbf{C}}$, $\langle e_\alpha, e_\beta \rangle = h_{\alpha\bar{\beta}}$
(ii) a vector $e_{n+1} \in T_{\mathbf{R}} - T_{\mathbf{C}}$ corresponding to the tangent vector of the curve γ, and
(iii) a real number fixing the parametrization, corresponding to $\operatorname{Re} q''(0)$.

With the concepts of the following section this will be viewed as a frame in a line bundle over M.

As a consequence of these results above we see that the holomorphic mappings taking a nondegenerate hypersurface into themselves form a finite dimensional group. In fact, fixing a point the dimension of this group is at most equal to that of the isotropy group H, i.e. $(n+1)^2+1$. Adding the freedom of choice of a point gives $2n+1+(n+1)^2+1=(n+2)^2-1$ as an upper bound for the dimension of the group of holomorphic self mappings of M. This upper bound is realized for the hyperquadrics.

The above differential equations define a holomorphically invariant family of a parametrized curve γ transversal to the complex tangent bundle, with a frame e_α propagating along γ. The parameter ξ is fixed up to a projective transformation $\xi/(\alpha\xi+\beta)$ $(\beta \neq 0)$ keeping $\xi=0$ fixed. Thus cross ratios of 4 points on these curves are invariantly defined. We summarize: (i) $\operatorname{tr} F_{22}=0$ represents a first order differential equation for the frame e_α, (ii) $(\mathrm{tr})^2 F_{32}=0$ defines a second order differential equation for the distinguished curves γ, irrespective of parametrization and (iii) $(\mathrm{tr})^3 F_{33}=0$ defines a third order differential equation for the parametrization.

(c) The differential equations $\mathrm{tr}\, F_{22}=0$, $(\mathrm{tr})^2 F_{32}=0$, $(\mathrm{tr})^3 F_{33}=0$ remain meaningful for merely smooth manifolds. Indeed, if M is six times continuously differentiable one can achieve the above normal forms up to terms of order 6 inclusive, simply truncating the above series expansions. Clearly the resulting families of curves and frames are invariantly associated with the manifold under mappings holomorphic near M. Indeed since the differential equations are obtained by the expansions of § 2 up to terms of weight $\leqslant 6$ at any point one may approximate M at this point by a real analytic one and read off the holomorphic invariance of this system of differential equations. In this case the distinguished curves γ are, in general, only 3 times continuously differentiable but the normal form (see (2.11) via a holomorphic map, cannot be achieved, not even to sixth order in $z, \bar z$. This would require that the function $f(z,u)$, $g(z,u)$ defining the transformation and which can be taken as polynomials in z admit an analytic continuation to complex values of u. If the Levi form is indefinite one has to require an analytic continuation to both sides which can happen only in the exceptional case of analytic curves γ. If, however, the Levi-form is definite, i.e. in the pseudoconvex case one has to require only that $f(z,u)$, $g(z,u)$ admit one sided analytic continuations. However, we do not pursue this artificial question but record that the structure of differential equations for the curves γ and their associated frame is meaningful in the case of six times differentiable manifolds.

(d) In the case $n=1$ the normal form has a simpler form since the contraction (tr) becomes redundant. For this reason F_{22}, F_{23}, F_{32}, F_{33} all vanish and the normal form can be written

$$v = z\bar z + c_{42} z^4 \bar z^2 + c_{24} z^2 \bar z^4 + \sum_{k+l\geqslant 7} c_{kl} z^k \bar z^l \tag{3.18}$$

where again $\min(k,l)\geqslant 2$. This normal form is unique only up to the 5 dimensional group H given by

$$\left.\begin{aligned} z &\to \lambda(z+aw)\,\delta^{-1}, \quad w = 1 - 2i\bar a z - (r + i|a|^2)\,w \\ w &\to |\lambda|^2 w\delta^{-1} \end{aligned}\right\} \tag{3.19}$$

with $0 \neq \lambda \in \mathbf{C}$, $a \in \mathbf{C}$, $r \in \mathbf{R}$. It is easily seen that the property $c_{42}(0) \neq 0$ is invariant under these transformations. If $c_{42}(0)=0$ we call the origin an umbilical point. For a non-umbilical we can always achieve $c_{42}(0)=1$ since $z \to \lambda z$ leads to $c_{42}(0) \to \lambda^3 \bar\lambda c_{42}(0)$. By this normalization λ is fixed up to sign.

For a nonumbilical point we can use the parameters a, r to achieve

$$c_{43}(0)=0, \quad \mathrm{Re}\, c'_{42}(0)=0$$

so that the so normalized hypersurface can be approximated to order 7 in z, $\bar{z}$, u by the algebraic surface

$$v = z\bar{z} + 2\,\mathrm{Re}\,\{z^4\bar{z}^2(1+jz+iku\} \tag{3.20}$$

where $j \in \mathbf{C}$, $k \in \mathbf{R}$, and j^2, k are invariants at the origin.

The above statements follow from the fact that (3.19) with $\lambda=1$, $r=0$ leads to

$$c_{43}(0) \to c_{43}(0)+2ia, \quad c_{52}(0) \to c_{52}(0)+4i\bar{a}$$

so that $j = c_{52}(0)+\overline{2c_{43}(0)}$ is unchanged. We fix a so that $c_{43}(0)=0$ and consider (3.19) with $\lambda=1$, $a=0$ which gives rise to

$$\mathrm{Re}\, c'_{42}(0) \to \mathrm{Re}\, c'_{42}(0)+4r$$

Choosing $\mathrm{Re}\, c'_{42}(0)=0$ we obtain (3.20), where we still have the freedom to replace z by $-z$. Thus j^2 and k are indeed invariants.

The above choice (3.20) distinguishes a special frame at the origin, by prescribing a tangent vector $\partial/\partial u$ transversal to the complex tangent plane and a complex tangent vector pair $\pm\partial/\partial z$ in the complex tangent plane. These pairs of vectors can be assigned to any point of M which is non-umbilical. These considerations clearly are meaningful for seven times differentiable M.

The above vector fields, singular at umbilical points, can be viewed as analogous to the directions of principal curvature in classical differential geometry. This analogy suggests the question: Are there compact manifolds without umbilical points? Are there such manifolds diffeomorphic to the sphere S^3?

Clearly the sphere $|z|^2+|w|^2=1$ consists of umbilical points only as, except for one point, this manifold can be transformed into $v=z\bar{z}$ (cf. (1.4)). Therefore we can say by (3.18): *Any 3-dimensional manifold M in C^2 can at a point be osculated by the holomorphic image of the sphere $|z|^2+|w|^2=1$ up to order 5 but generally not to sixth order.* In the latter case we have an umbilical point.

For $n \geq 2$ the analogous definition of an umbilical point is different: A point p on M is called umbilical if the term F_{22} in the normal form vanishes. Again, it is easily seen that this condition is independent of the transformation (1.23) and we can say: *Any non-degenerate manifold M of real dimension $2n+1$ in $\mathbf{C}_{n+1}$ $(n \geq 2)$ can at a point be osculated by the holomorphic image of a hyperquadric $v=\langle z, z\rangle$ up to order 3, but generally not to order 4.* In case one has fourth order osculation one speaks of an umbilical point.

(e) The algebraic problems connected with the action of the isotropy group on the normal form are prohibitively complicated for large n. But for a strictly pseudoconvex 5-dimensional manifold in C_3 we obtain an interesting invariant connected with the 4th order terms F_{22}.

We assume $n=2$ and

$$\langle z, z\rangle = \sum_{\alpha=1}^{2} z^\alpha \overline{z^\alpha}$$

and consider a quartic $F_{22}(z, \bar{z})$ of type (2, 2) with tr $F_{22}=0$. If we subject the manifold

$$v^* = \langle z^*, z^*\rangle + F_{22}(z^*, \overline{z^*}) + \ldots$$

to the transformation (1.23) of the isotropy group of Q the fourth order term is replaced by

$$F_{22}(z^*, \overline{z^*}) = N_{22}(z, \bar{z}) \tag{3.21}$$

where
$$z^{*\beta} = C_\alpha{}^\beta z^\alpha, \quad C_\alpha{}^\sigma C_{\bar{\beta}}{}^{\bar{\sigma}} = \varrho \delta_{\alpha\bar{\beta}}, \quad \varrho > 0. \tag{3.22}$$

The question arises to find invariants of N_{22} under these transformations, which are evidently multiples of unitary transformations.

It turns out, and we will show, that one can find (3.22) such that N_{22} takes the form

$$N_{22} = \lambda_1\phi_1 + \lambda_2\phi_2 + \lambda_3\phi_3$$

where ϕ_1, ϕ_2, ϕ_3 are fixed quartics and $\lambda_1, \lambda_2, \lambda_3$ are three real numbers which we may order $\lambda_1 \leqslant \lambda_2 \leqslant \lambda_3$ and which satisfy

$$\lambda_1 + \lambda_2 + \lambda_3 = 0.$$

The λ_j may still be replaced by $\varrho\lambda_j$, so that

$$\frac{\lambda_3 - \lambda_2}{\lambda_2 - \lambda_1} = \mu$$

is a numerical invariant, provided we assume that the λ_j are distinct. In this case the matrix $C_\alpha{}^\beta$ is fixed up to a complex factor by these requirements. Geometrically speaking to every λ_j corresponds a pair of complex lines—if the λ_j are distinct—so that we have altogether three pairs of complex lines in the complex tangent space holomorphically invariantly associated with the manifold. We remark that $\lambda_3 = \max_j \lambda_j = 0$ characterizes an umbilical point, i.e. $F_{22}=0$.

The λ_j are reminiscent of eigenvalues of a quadratic form and, in fact, the above problem can be reduced to the equivalence problem of a quadratic form. One verifies by computation that any quartic F_{22} with tr $F_{22}=0$ is invariant under the involution

$$(z^1, z^2) \rightarrow (\overline{z^2}, -\overline{z^1}) \tag{3.23}$$

and conversely any such quartic differs from one with tr $F_{22}=0$ by a multiple of $\langle z, z\rangle^2$

The function $\langle z, z\rangle^{-2} F_{22}$ can be viewed as a function on the complex projective space $\mathbf{CP}^1$, that is on $\mathbf{S}^2$. We use the familiar mapping [3], derived from the stereographic projection:

$$\xi_1 = z^1\overline{z^2} + z^2\overline{z^1}$$

$$i\xi_2 = z^1\overline{z^2} - z^2\overline{z^1}$$

$$\xi_3 = z^1\overline{z^1} - z^2\overline{z^2}$$

so that

$$\sum_{\nu=1}^{3} \xi_\nu^2 = \langle z, z\rangle^2$$

to map $\langle z, z\rangle = 1$ onto $\mathbf{S}^2$. Then the above involution (3.23) goes into the antipodal maps and one verifies that F_{22} becomes a real quadratic form

$$F_{22}(z, \bar{z}) = \Phi(\xi) = \sum_{\nu,\mu=1}^{3} b_{\nu\mu}\xi_\nu\xi_\mu.$$

Moreover

$$\operatorname{tr} F_{22} = \tfrac{1}{2}\left(\sum_{\nu=1}^{3} b_{\nu\nu}\right)\langle z, z\rangle,$$

so that tr $F_{22}=0$ if and only if the trace of the quadratic form vanishes.

We subject F_{22} to the transformation (3.22). At first we take $\varrho=1$, so that $(C_\alpha{}^\beta)=C$ is unitary. We assume furthermore that det $C=1$ because of the homogeneous character of F_{22}. Then, as is well known every such C corresponds to a proper orthogonal transformation of the ξ-space, and every such orthogonal transformation belongs to two such unitary transformations, namely $\pm C$. Thus the equivalence problem is reduced to that of the quadratic form Φ under proper orthogonal transformations. Choosing this transformation so that Φ is mapped into diagonal form

$$\sum_{\nu=1}^{3} \lambda_\nu \xi_\nu^2$$

we have $\sum_{\nu=1}^{3} \lambda_\nu = 0$. Moreover, if the eigenvalues λ_ν are distinct and ordered the orthogonal transformation is up to $\xi_\nu \rightarrow \pm\xi_\nu$ uniquely determined by this requirement.

To complete the discussion we have to free ourselves from the restriction det $C=1$ and take the stretching $z\rightarrow\varrho z$ into account. Both factors are taken into account by a transformation $z\rightarrow\gamma z$, $w\rightarrow|\gamma|^2 w$ with γ a complex number which leads to $\lambda_j\rightarrow|\gamma|^2\lambda_j$.

Thus if we set

$$\phi_1 = \xi_1^2 = 2\{z^1 z^2 \overline{z^1 z^2} + \operatorname{Re}(z^1 \overline{z^2})^2\}$$
$$\phi_2 = \xi_2^2 = -2\{z^1 z^2 \overline{z^1 z^2} - \operatorname{Re}(z^1 \overline{z^2})^2\}$$
$$\phi_3 = \xi_3^2 = (z^1 \overline{z^1})^2 + (z^2 \overline{z^2})^2 - 2 z^1 z^2 \overline{z^1}\, \overline{z^2}$$

Then the above assertions follow. The pairs of complex lines which correspond to an eigendirection have the form

$$a_1 z^1 + a_2 z^2 = 0$$
$$\overline{a_2} z^1 - \overline{a_1} z^2 = 0$$

where a_1, a_2 are not both zero, i.e. the second line is obtained from the first by the involution (3.23).

4. Solution of an equivalence problem

Let G be the group of all nonsingular matrices of the form

$$\begin{pmatrix} u & 0 & 0 \\ v^\alpha & u_\beta{}^\alpha & 0 \\ v^{\bar\alpha} & 0 & u_{\bar\beta}{}^{\bar\alpha} \end{pmatrix}, \quad v^{\bar\alpha} = \overline{v^\alpha}, \quad u_{\bar\beta}{}^{\bar\alpha} = \overline{u_\beta{}^\alpha}, \tag{4.1}$$

where, as throughout this section, the small Greek indices run from 1 to n, u is real, and v^α, $u_\beta{}^\alpha$ are complex. G can be considered as a subgroup of $GL(2n+1, R)$. A G-structure in a manifold M of dimension $2n+1$ is a reduction of the group of its tangent bundle to G. Locally it is given by linear differential forms θ, θ^α, $\theta^{\bar\alpha}$, where θ is real and θ^α are complex, which are defined up to a transformation of G and satisfy the condition

$$\theta \wedge \theta^1 \wedge \dots \wedge \theta^n \wedge \theta^{\bar 1} \wedge \dots \wedge \theta^{\bar n} \neq 0. \tag{4.2}$$

Let T_x and T_x^*, $x \in M$, be respectively the tangent and cotangent spaces of M at x. The multiples of θ define a line E_x in T_x^* and their totality is a real line bundle over M, to be denoted by E. The annihilator $E_x^\perp = T_{x,c}$ in T_x, called the complex tangent space, has a complex structure.

The G-structure is called integrable if the Frobenius condition is satisfied: $d\theta$, $d\theta^\alpha$ belong to the differential ideal generated by θ, θ^β. Since θ is real, this condition implies

$$d\theta \equiv i h_{\alpha\bar\beta} \theta^\alpha \wedge \theta^{\bar\beta}, \quad \text{mod } \theta, \tag{4.3}$$

where
$$h_{\alpha\bar\beta} = \overline{h_{\beta\bar\alpha}} = h_{\bar\beta\alpha}. \tag{4.4}$$

An integrable G-structure is called nondegenerate if

$$\det (h_{\alpha\bar{\beta}}) \neq 0. \tag{4.5}$$

Integrable G-structures include the special cases:

(1) Real hypersurfaces in $\mathbf{C}_{n+1}$. Let z^α, w be the coordinates of $\mathbf{C}_{n+1}$. A real hypersurface M can be locally defined by

$$r(z^\alpha, \bar{z}^{\alpha}, w, \bar{w}) = 0, \quad r_w \neq 0, \tag{4.6}$$

where r is a smooth real-valued function. On M a G-structure is defined by putting

$$\theta = i\partial r, \quad \theta^\alpha = dz^\alpha. \tag{4.7}$$

(2) Complex-valued linear differential operators of the first order in $\mathbf{R}_{2n+1}$. Denote the operators by P_α and suppose the following conditions be satisfied: (a) P_α, $P_{\bar{\beta}}$ are linearly independent; (b) $[P_\alpha, P_\beta]$ is a linear combination of P_γ. We interpret the operators as complex vector fields and let L be the n-dimensional linear space spanned by P_α. Its annihilator $L^\perp$ is of dimension $n+1$. Condition (a) implies that $L^\perp \cap \bar{L}^\perp$ is one-dimensional. We can choose a real one-form $\theta \in L^\perp \cap \bar{L}^\perp$ and the forms θ, θ^α to span $L^\perp$. The G-structure so defined is integrable because of condition (b).

We shall define a complete system of local invariants of nondegenerate integrable G-structures.

We consider the real line bundle E, which consists of the multiples $u\theta$, u (>0) being a fiber coordinate. In E the form

$$\omega = u\theta \tag{4.8}$$

is intrinsically defined. By (4.3) its exterior derivative has the local expression

$$d\omega = iuh_{\alpha\bar{\beta}}\theta^\alpha \wedge \theta^{\bar{\beta}} + \omega \wedge \left(-\frac{du}{u} + \phi_0\right), \tag{4.9}$$

where θ^α, ϕ_0 are one-forms in M and ϕ_0 is real. This equation can be written

$$d\omega = ig_{\alpha\bar{\beta}}\,\omega^\alpha \wedge \omega^{\bar{\beta}} + \omega \wedge \phi, \tag{4.10}$$

where ω^α are linear combinations of θ^β, θ and $g_{\alpha\bar{\beta}} = g_{\bar{\beta}\alpha}$ are constants. The nondegeneracy of the G-structure is expressed by

$$\det (g_{\alpha\bar{\beta}}) \neq 0. \tag{4.11}$$

The forms ω, $\operatorname{Re} \omega^\alpha$, $\operatorname{Im} \omega^{\alpha}$ and ϕ constitute a basis of the cotangent space of E. The most general transformation on ω, ω^α, $\omega^{\bar{\alpha}}$, ϕ leaving the equation (4.10) (and the form ω) invariant has the matrix of coefficients

17 – 742902 *Acta mathematica* 133. Imprimé le 20 Février 1975

$$\begin{pmatrix} 1 & 0 & 0 & 0 \\ v^{\alpha} & u_{\beta}{}^{\alpha} & 0 & 0 \\ v^{\bar{\alpha}} & 0 & u_{\bar{\beta}}{}^{\bar{\alpha}} & 0 \\ s & ig_{\varrho\bar{\sigma}}\, u_{\beta}{}^{\varrho} v^{\bar{\sigma}} & -ig_{\varrho\bar{\sigma}}\, u_{\bar{\beta}}{}^{\bar{\sigma}} v^{\varrho}, & 1 \end{pmatrix} \tag{4.12}$$

where s is real and $u_{\beta}{}^{\alpha}$, v^{α} are complex satisfying the equations

$$g_{\alpha\bar{\beta}}\, u_{\varrho}{}^{\alpha} u_{\bar{\sigma}}{}^{\bar{\beta}} = g_{\varrho\bar{\sigma}}. \tag{4.13}$$

Let G_1 be the group of all the nonsingular matrices (4.12). It follows that E has a G_1-structure. Denote by Y its principal G_1-bundle. Then we have

$$G_1 \xrightarrow{j} Y \xrightarrow{\pi} E, \tag{4.14}$$

where j is inclusion of a fiber and π is projection. The quantities s, $u_{\alpha}{}^{\beta}$, v^{α} in (4.12), considered as new variables, are local fiber coordinates of Y. Observe that we have the dimensions

$$\dim G_1 = (n+1)^2, \quad \dim E = 2(n+1), \quad \dim Y = (n+2)^2 - 1. \tag{4.15}$$

In Y there are intrinsically (and hence globally) defined forms ω, ω^{α}, $\omega^{\bar{\alpha}}$, ϕ, and we will introduce new ones by intrinsic conditions, so that the total number equals the dimension of Y and they are everywhere linearly independent.

The condition that our G-structure is integrable implies

$$d\omega^{\alpha} = \omega^{\beta} \wedge \phi_{\beta\cdot}{}^{\alpha} + \omega \wedge \phi^{\alpha} \tag{4.16}$$

where $\phi_{\beta\cdot}{}^{\alpha}$, ϕ^{α} are not completely determined. We shall study the consequences of the equations (4.10), (4.16) by exterior differentiation. To be in a slightly more general situation the $g_{\alpha\bar{\beta}}$'s are allowed to be variable. It will be convenient to follow the practice of tensor analysis to introduce $g^{\alpha\bar{\beta}}$ by the equations

$$g_{\alpha\bar{\beta}}\, g^{\gamma\bar{\beta}} = \delta_{\alpha}{}^{\gamma}, \quad g_{\alpha\bar{\beta}}\, g^{\alpha\bar{\gamma}} = \delta_{\bar{\beta}}{}^{\bar{\gamma}} \tag{4.17}$$

and to use them to raise and lower indices. It will then be important to know the location of an index and this will be indicated by a dot, thus

$$u_{\alpha\cdot}{}^{\beta}\, g_{\beta\bar{\gamma}} = u_{\alpha\bar{\gamma}}, \quad u^{\beta}{}_{\cdot\alpha}\, g^{\alpha\bar{\gamma}} = u^{\beta\bar{\gamma}}, \text{ etc.} \tag{4.18}$$

The exterior differentiations of (4.10), (4.16) give respectively

$$i(dg_{\alpha\bar{\beta}} - \phi_{\alpha\bar{\beta}} - \phi_{\bar{\beta}\alpha} + g_{\alpha\bar{\beta}}\phi) \wedge \omega^{\alpha} \wedge \omega^{\bar{\beta}} + (-d\phi + i\omega_{\bar{\beta}} \wedge \phi^{\bar{\beta}} + i\phi_{\bar{\beta}} \wedge \omega^{\bar{\beta}}) \wedge \omega = 0, \tag{4.19}$$

$$(d\phi_{\beta\cdot}{}^{\alpha} - \phi_{\beta\cdot}{}^{\gamma} \wedge \phi_{\gamma\cdot}{}^{\alpha} - i\omega_{\beta} \wedge \phi^{\alpha}) \wedge \omega^{\beta} + (d\phi^{\alpha} - \phi \wedge \phi^{\alpha} - \phi^{\beta} \wedge \phi_{\beta\cdot}{}^{\alpha}) \wedge \omega = 0. \tag{4.20}$$

LEMMA 4.1. *There exist $\phi_\beta{}^\alpha$ which satisfy* (4.16) *and*

$$dg_{\alpha\bar\beta}+g_{\alpha\bar\beta}\phi-\phi_{\alpha\bar\beta}-\phi_{\bar\beta\alpha}=0,\quad \phi_{\bar\beta\alpha}=\overline{\phi}_{\beta\bar\alpha},\tag{4.21}$$

or

$$dg^{\alpha\bar\beta}-g^{\alpha\bar\beta}\phi+\phi^{\alpha\bar\beta}+\phi^{\bar\beta\alpha}=0.\tag{4.21 a}$$

Such $\phi_\beta{}^\alpha$ are determined up to additive terms in ω.

In fact, it follows from (4.19) that the expression in its first parentheses is a linear combination of ω^α, $\omega^{\bar\beta}$, ω, i.e.,

$$dg_{\alpha\bar\beta}-\phi_{\alpha\bar\beta}-\phi_{\bar\beta\alpha}+g_{\alpha\bar\beta}\phi=A_{\alpha\bar\beta\gamma}\omega^\gamma+B_{\alpha\bar\beta\bar\gamma}\omega^{\bar\gamma}+C_{\alpha\bar\beta}\omega,\tag{4.22}$$

where
$$A_{\alpha\bar\beta\gamma}=A_{\gamma\bar\beta\alpha},\quad B_{\alpha\bar\beta\bar\gamma}=B_{\alpha\bar\gamma\bar\beta}.\tag{4.23}$$

From the hermitian property of $g_{\alpha\bar\beta}$ we have also

$$\bar A_{\alpha\bar\beta\gamma}=B_{\beta\bar\alpha\bar\gamma},\quad \bar C_{\alpha\bar\beta}=C_{\beta\bar\alpha}.\tag{4.24}$$

The forms
$$\phi'_{\alpha\bar\beta}=\phi_{\alpha\bar\beta}+A_{\alpha\bar\beta\gamma}\omega^\gamma+\tfrac12 C_{\alpha\bar\beta}\omega\tag{4.25}$$

satisfy on account of (4.23) the equations (4.16) and (4.21). The second statement in the lemma can be verified without difficulty.

From now on we will suppose (4.21) to be valid. Equation (4.19) then gives

$$d\phi=i\omega_{\bar\beta}\wedge\phi^{\bar\beta}+i\phi_{\bar\beta}\wedge\omega^{\bar\beta}+\omega\wedge\psi,\tag{4.26}$$

where ψ is a real one-form.

LEMMA 4.2. *Let $\Phi_\beta{}^\alpha$ be exterior quadratic differential forms, satisfying*

$$\Phi_\beta{}^\alpha\wedge\omega^\beta\equiv0,\quad \Phi_{\alpha\bar\beta}+\Phi_{\bar\beta\alpha}\equiv0,\quad \text{mod }\omega.\tag{4.27}$$

Then we have
$$\Phi_{\alpha\bar\varrho}\equiv S_{\alpha\beta\bar\varrho\bar\sigma}\omega^\beta\wedge\omega^{\bar\sigma},\quad \text{mod }\omega,\tag{4.28}$$

where $S_{\alpha\beta\bar\varrho\bar\sigma}$ has the symmetry properties:

$$S_{\alpha\beta\bar\varrho\bar\sigma}=S_{\beta\alpha\bar\varrho\bar\sigma}=S_{\alpha\beta\bar\sigma\bar\varrho},\tag{4.29}$$

$$S_{\alpha\beta\bar\varrho\bar\sigma}=\bar S_{\varrho\sigma\bar\alpha\bar\beta}=S_{\bar\varrho\bar\sigma\alpha\beta}.\tag{4.30}$$

Computing mod ω, we have, from the first equation of (4.27),

$$\Phi_{\beta\bar\alpha}\equiv\chi_{\beta\bar\alpha\gamma}\wedge\omega^\gamma,$$

where $\chi_{\beta\bar\alpha\gamma}$ are one-forms. Its complex conjugate is

$$\Phi_{\bar\beta\alpha}\equiv\chi_{\bar\beta\alpha\bar\gamma}\wedge\omega^{\bar\gamma}.$$

By the second equation of (4.27) we have

$$\chi_{\alpha\bar{\beta}\gamma} \wedge \omega^{\gamma} + \chi_{\bar{\beta}\alpha\bar{\gamma}} \wedge \omega^{\bar{\gamma}} \equiv 0, \quad \text{mod } \omega.$$

The first term, $\chi_{\alpha\bar{\beta}\gamma} \wedge \omega^{\gamma}$, is therefore congruent to zero mod ω, $\omega^{\bar{\sigma}}$. But it is obviously congruent to zero mod ω^{ϱ}. Hence we have the conclusion (4.28). The symmetry properties (4.29) and (4.30) follow immediately from (4.27). Thus Lemma 4.2 is proved.

Equation (4.20) indicates the necessity of studying the expression

$$\Pi_{\alpha\cdot}{}^{\gamma} = d\phi_{\alpha\cdot}{}^{\gamma} - \phi_{\alpha\cdot}{}^{\beta} \wedge \phi_{\beta\cdot}{}^{\gamma}. \tag{4.31}$$

Using (4.21) we have

$$\Pi_{\beta\bar{\alpha}} = g_{\gamma\bar{\alpha}} d\phi_{\beta\cdot}{}^{\gamma} - \phi_{\beta\cdot}{}^{\gamma} \wedge \phi_{\gamma\bar{\alpha}} = d\phi_{\beta\bar{\alpha}} - \phi_{\beta\bar{\alpha}} \wedge \phi - \phi_{\bar{\alpha}\gamma} \wedge \phi_{\beta\cdot}{}^{\gamma}. \tag{4.32}$$

It follows that

$$\Pi_{\beta\bar{\alpha}} + \Pi_{\bar{\alpha}\beta} = d(\phi_{\beta\bar{\alpha}} + \phi_{\bar{\alpha}\beta}) - (\phi_{\beta\bar{\alpha}} + \phi_{\bar{\alpha}\beta}) \wedge \phi,$$

since

$$\phi_{\beta\bar{\gamma}} \wedge \phi_{\bar{\alpha}\cdot}{}^{\bar{\gamma}} = \phi_{\beta\cdot}{}^{\gamma} \wedge \phi_{\bar{\alpha}\gamma}.$$

Using the differentiation of (4.21), we get

$$\Pi_{\beta\bar{\alpha}} + \Pi_{\bar{\alpha}\beta} = g_{\beta\bar{\alpha}} d\phi. \tag{4.33}$$

By (4.20), (4.26), (4.33), it is found that

$$\Phi_{\beta\cdot}{}^{\gamma} \equiv \Pi_{\beta\cdot}{}^{\gamma} - i\omega_{\beta} \wedge \phi^{\gamma} + i\phi_{\beta} \wedge \omega^{\gamma} + i\delta_{\beta}{}^{\gamma}(\phi_{\sigma} \wedge \omega^{\sigma}), \quad \text{mod } \omega \tag{4.34}$$

or

$$\Phi_{\beta\bar{\alpha}} \equiv \Pi_{\beta\bar{\alpha}} - i\omega_{\beta} \wedge \phi_{\bar{\alpha}} + i\phi_{\beta} \wedge \omega_{\bar{\alpha}} + ig_{\beta\bar{\alpha}}(\phi_{\sigma} \wedge \omega^{\sigma}), \quad \text{mod } \omega \tag{4.34 a}$$

fulfill the conditions of Lemma 4.2. For such Φ the conclusions (4.28)–(4.30) of the Lemma are valid.

The forms $\phi_{\beta\cdot}{}^{\alpha}$, ϕ^{α}, ψ fulfilling equations (4.16), (4.21), and (4.26) are defined up to the transformation

$$\left.\begin{aligned} \phi_{\beta\cdot}{}^{\alpha} &= \phi'_{\beta\cdot}{}^{\alpha} + D_{\beta\cdot}{}^{\alpha}\omega, \\ \phi^{\alpha} &= \phi'^{\alpha} + D_{\beta\cdot}{}^{\alpha}\omega^{\beta} + E^{\alpha}\omega \\ \psi &= \psi' + G\omega + i(E_{\alpha}\omega^{\alpha} - E_{\bar{\alpha}}\omega^{\bar{\alpha}}), \end{aligned}\right\} \tag{4.35}$$

where G is real and

$$D_{\alpha\bar{\beta}} + D_{\bar{\beta}\alpha} = 0. \tag{4.36}$$

LEMMA 4.3. *The $D_{\beta\cdot}{}^{\alpha}$ can be uniquely determined by the conditions*

$$S_{\varrho\bar{\sigma}} \underset{\text{def}}{=} g^{\alpha\bar{\beta}} S_{\alpha\varrho\bar{\beta}\bar{\sigma}} = 0. \tag{4.37}$$

To prove Lemma 4.3 it suffices to study the effect on $S_{\alpha\bar{\beta}\varrho\bar{\sigma}}$ when the transformation (4.35) is performed. We put

$$S = g^{\alpha\bar{\beta}} S_{\alpha\bar{\beta}}, \quad D = D_{\alpha.}^{\ \alpha}. \tag{4.38}$$

Since $g^{\alpha\bar{\beta}}$ and $S_{\alpha\bar{\beta}}$ are hermitian and $D_{\alpha\bar{\beta}}$ is skew-hermitian, S is real and D is purely imaginary. Denoting the new coefficients by dashes, we find

$$S_{\alpha\beta.\bar{\sigma}}^{\ \ \gamma} = S'^{\ \ \gamma}_{\alpha\beta.\bar{\sigma}} + i(D_{\alpha.}^{\ \gamma}\, g_{\beta\bar{\sigma}} + D_{\beta.}^{\ \gamma}\, g_{\alpha\bar{\sigma}} - \delta_\beta^{\ \gamma} D_{\bar{\sigma}\alpha} - \delta_\alpha^{\ \gamma} D_{\bar{\sigma}\beta}). \tag{4.39}$$

It follows that

$$S_{\varrho\bar{\sigma}} = S'_{\varrho\bar{\sigma}} + i\{g_{\varrho\bar{\sigma}} D + D_{\varrho\bar{\sigma}} - (n+1) D_{\bar{\sigma}\varrho}\}. \tag{4.40}$$

Since we wish to make $S'_{\varrho\bar{\sigma}} = 0$, the lemma is proved if we show that there is one and only one set of $D_{\beta.}^{\ \gamma}$ satisfying (4.36) and

$$-iS_{\varrho\bar{\sigma}} = g_{\varrho\bar{\sigma}} D + (n+2) D_{\varrho\bar{\sigma}}. \tag{4.41}$$

In fact, contracting (4.41), we get

$$2(n+1) D = -iS. \tag{4.42}$$

Substitution of this into (4.41) gives

$$(n+2) D_{\varrho\bar{\sigma}} = -iS_{\varrho\bar{\sigma}} + \frac{i}{2(n+1)} S g_{\varrho\bar{\sigma}}. \tag{4.43}$$

It is immediately verified that the $D_{\varrho\bar{\sigma}}$ given by (4.43) satisfy (4.36) and (4.41). This proves Lemma 4.3.

By the condition (4.37) the $\phi_{\beta.}^{\ \gamma}$ are completely determined and we wish to compute their exterior derivatives. By (4.34) we can put

$$\Pi_{\beta.}^{\ \gamma} - i\omega_\beta \wedge \phi^\gamma + i\phi_\beta \wedge \omega^\gamma + i\delta_\beta^{\ \gamma}(\phi_\sigma \wedge \omega^\sigma) = S_{\beta\varrho.\bar{\sigma}}^{\ \ \gamma}\, \omega^\varrho \wedge \bar{\omega}^\sigma + \lambda_{\beta.}^{\ \gamma} \wedge \omega, \tag{4.44}$$

where $\lambda_{\beta.}^{\ \gamma}$ are one-forms. Substituting this into (4.20), we get

$$d\phi^\alpha - \phi \wedge \phi^\alpha - \phi^\beta \wedge \phi_{\beta.}^{\ \alpha} - \lambda_{\beta.}^{\ \alpha} \wedge \omega^\beta = \mu^\alpha \wedge \omega, \tag{4.45}$$

μ^α being also one-forms. From (4.44), (4.33), and (4.26), we get

$$(\lambda_{\beta\bar{\alpha}} + \lambda_{\bar{\alpha}\beta}) \wedge \omega = g_{\beta\bar{\alpha}}\, \omega \wedge \psi,$$

or

$$\lambda_{\beta\bar{\alpha}} + \lambda_{\bar{\alpha}\beta} + g_{\beta\bar{\alpha}} \psi \equiv 0, \quad \text{mod } \omega. \tag{4.46}$$

To utilize the condition (4.37) we shall take the exterior derivative of (4.44). We will need the following formulas, which follow immediately from (4.16), (4.45), (4.21):

$$d\omega_\alpha = d(g_{\alpha\bar{\beta}}\omega^{\bar{\beta}}) = -\omega^{\bar{\beta}} \wedge \phi_{\alpha\bar{\beta}} + \omega_\alpha \wedge \phi + \omega \wedge \phi_\alpha, \tag{4.47}$$

$$d\phi_\alpha = d(g_{\alpha\bar{\beta}}\phi^{\bar{\beta}}) = \phi_{\alpha\bar{\beta}} \wedge \phi^{\bar{\beta}} + \lambda_{\bar{\gamma}\alpha} \wedge \omega^{\bar{\gamma}} + \mu_\alpha \wedge \omega. \tag{4.48}$$

We take the exterior derivative of (4.44) and consider only terms involving $\omega^\varrho \wedge \omega^{\bar{\sigma}}$, ignoring those in ω. It gives

$$dS_{\beta\varrho.\bar{\sigma}}^{\gamma} - S_{\tau\varrho.\bar{\sigma}}^{\gamma}\phi_{\beta.}^{\tau} - S_{\beta\tau.\bar{\sigma}}^{\gamma}\phi_{\varrho.}^{\tau} + S_{\beta\varrho.\bar{\sigma}}^{\tau}\phi_{\tau.}^{\gamma} - S_{\beta\varrho.\bar{\tau}}^{\gamma}\phi_{\bar{\sigma}.}^{\bar{\tau}}$$
$$\equiv i(\lambda_{\beta.}^{\gamma}\, g_{\varrho\bar{\sigma}} + \lambda_{\varrho.}^{\gamma}\, g_{\beta\bar{\sigma}} - \delta_{\beta}^{\gamma}\,\lambda_{\bar{\sigma}\varrho} - \delta_{\varrho}^{\gamma}\,\lambda_{\bar{\sigma}\beta}) \quad \text{mod } \omega, \omega^{\alpha}, \omega^{\bar{\beta}} \tag{4.49}$$

and by contraction

$$dS_{\varrho\bar{\sigma}} - S_{\tau\bar{\sigma}}\phi_{\varrho.}^{\tau} - S_{\varrho\bar{\tau}}\phi_{\bar{\sigma}.}^{\bar{\tau}} \equiv i\{g_{\varrho\bar{\sigma}}\lambda_{\beta.}^{\beta} + \lambda_{\varrho\bar{\sigma}} - (n+1)\,\lambda_{\bar{\sigma}\varrho}\}, \quad \text{mod } \omega, \omega^{\alpha}, \omega^{\bar{\beta}}. \tag{4.50}$$

When (4.37) is satisfied, the left-hand side, and hence also the right-hand side, of (4.50) are congruent to zero. The congruence so obtained, combined with (4.46), gives

$$\lambda_{\varrho\bar{\sigma}} \equiv -\tfrac{1}{2} g_{\varrho\bar{\sigma}}\psi \quad \text{or} \quad \lambda_{\varrho.}^{\sigma} \equiv -\tfrac{1}{2}\delta_{\varrho}^{\ \sigma}\psi, \quad \text{mod } \omega, \omega^{\alpha}, \omega^{\bar{\beta}}.$$

Hence we can put

$$\lambda_{\varrho.}^{\sigma} = -\tfrac{1}{2}\delta_{\varrho}^{\ \sigma}\psi + V_{\varrho.\beta}^{\sigma}\,\omega^{\beta} + W_{\varrho.\bar{\beta}}^{\sigma}\,\omega^{\bar{\beta}} + a_{\varrho.}^{\sigma}\,\omega \tag{4.51}$$

or

$$\lambda_{\varrho\bar{\sigma}} = -\tfrac{1}{2} g_{\varrho\bar{\sigma}}\psi + V_{\varrho\bar{\sigma}\beta}\omega^{\beta} + W_{\varrho\bar{\sigma}\bar{\beta}}\omega^{\bar{\beta}} + a_{\varrho\bar{\sigma}}\omega. \tag{4.51a}$$

Substituting into (4.46), we get

$$V_{\varrho\bar{\sigma}\beta} + W_{\bar{\sigma}\varrho\beta} = 0. \tag{4.52}$$

We can therefore write (4.44) in the form

$$\Phi_{\beta.}^{\gamma} \underset{\text{def}}{=} d\phi_{\beta.}^{\gamma} - \phi_{\beta.}^{\sigma}\wedge\phi_{\sigma.}^{\gamma} - i\,\omega_{\beta}\wedge\phi^{\gamma} + i\phi_{\beta}\wedge\omega^{\gamma} + i\delta_{\beta}^{\ \gamma}\{\phi_{\sigma}\wedge\omega^{\sigma}\} + \tfrac{1}{2}\delta_{\beta}^{\ \gamma}\psi\wedge\omega$$
$$= S_{\beta\varrho.\bar{\sigma}}^{\gamma}\omega^{\varrho}\wedge\omega^{\bar{\sigma}} + V_{\beta.\varrho}^{\gamma}\omega^{\varrho}\wedge\omega - V_{.\beta\bar{\sigma}}^{\gamma}\omega^{\bar{\sigma}}\wedge\omega, \tag{4.53}$$

which is the formula for $d\phi_{\beta.}^{\gamma}$. Formula (4.53) defines $\Phi_{\beta.}^{\gamma}$ completely; it is consistent with earlier notations in Lemma 4.2 and in the subsequent discussions where $\Phi_{\beta.}^{\gamma}$ are defined only mod ω. Substituting into (4.20), we get

$$\Phi^{\alpha} \underset{\text{def}}{=} d\phi^{\alpha} - \phi\wedge\phi^{\alpha} - \phi^{\beta}\wedge\phi_{\beta.}^{\alpha} + \tfrac{1}{2}\psi\wedge\omega^{\alpha} = -V_{\beta.\gamma}^{\alpha}\omega^{\beta}\wedge\omega^{\gamma} + V_{.\beta\bar{\sigma}}^{\alpha}\omega^{\beta}\wedge\omega^{\bar{\sigma}} + \nu^{\alpha}\wedge\omega, \tag{4.54}$$

where ν^{α} are one-forms. Notice also that (4.49) simplifies to

$$dS_{\beta\varrho.\bar{\sigma}}^{\gamma} - S_{\tau\varrho.\bar{\sigma}}^{\gamma}\phi_{\beta.}^{\tau} - S_{\beta\tau.\bar{\sigma}}^{\gamma}\phi_{\varrho.}^{\tau} + S_{\beta\varrho.\bar{\sigma}}^{\tau}\phi_{\tau.}^{\gamma} - S_{\beta\varrho.\bar{\tau}}^{\gamma}\phi_{\bar{\sigma}.}^{\bar{\tau}} \equiv 0, \quad \text{mod } \omega, \omega^{\alpha}, \omega^{\bar{\beta}} \tag{4.55}$$

on account of (4.51) or (4.51a).

Consider again the transformation (4.35) with $D_{\beta.}^{\alpha} = 0$. The $\phi_{\beta.}^{\alpha}$ are now completely determined. From (4.53) its effect on $V_{\beta.\varrho}^{\gamma}$ is given by

$$V_{\beta.\varrho}^{\gamma} = V_{\beta.\varrho}^{\prime\gamma} - i\{\delta_{\varrho}^{\ \gamma}E_{\beta} + \tfrac{1}{2}\delta_{\beta}^{\ \gamma}E_{\varrho}\}. \tag{4.56}$$

Contracting, we have

$$V_{\beta.\varrho}^{\varrho} = V_{\beta.\varrho}^{\prime\varrho} - i\{n + \tfrac{1}{2}\}E_{\beta}. \tag{4.57}$$

This leads to the lemma:

LEMMA 4.4. *With (4.21) and (4.37) fulfilled as in Lemmas 4.1 and 4.2 there is a unique set of ϕ^α satisfying*

$$V^{\varrho}_{\beta.\varrho} = 0. \tag{4.58}$$

To find an expression for $d\psi$ we differentiate the equation (4.26). Using (4.16), (4.47), and (4.54), we get

$$\omega \wedge (-d\psi + \phi \wedge \psi + 2i\phi^\beta \wedge \phi_\beta - i\omega^\beta \wedge \nu_\beta - i\nu^\beta \wedge \omega_\beta) = 0.$$

Hence we can write

$$\Psi \underset{\text{def}}{=} d\psi - \phi \wedge \psi - 2i\phi^\beta \wedge \phi_\beta = -i\omega^\beta \wedge \nu_\beta - i\nu^\beta \wedge \omega_\beta + \varrho \wedge \omega, \tag{4.59}$$

where ϱ is a one-form.

With this expression for $d\psi$ (and expressions for other exterior derivatives found above) we differentiate (4.54) mod ω and retain only terms involving $\omega^\varrho \wedge \omega^{\bar\sigma}$. By the same argument used above, we derive the formula

$$\begin{aligned} dV^{\alpha}_{.\varrho\bar\sigma} - V^{\alpha}_{.\beta\bar\sigma}\phi^{\beta}_{\varrho.} + V^{\beta}_{.\varrho\bar\sigma}\phi^{\alpha}_{\beta.} - V^{\alpha}_{.\varrho\bar\tau}\phi^{\bar\tau}_{\bar\sigma.} - V^{\alpha}_{.\varrho\bar\sigma}\phi \\ = S_{\beta\varrho.\bar\sigma}^{\alpha}\phi^\beta + ig_{\varrho\bar\sigma}\nu^\alpha + \frac{i}{2}\delta_\varrho{}^\alpha \nu_{\bar\sigma} \quad \text{mod } \omega, \omega^\alpha, \omega^{\bar\beta}. \end{aligned} \tag{4.60}$$

Condition (4.58) is equivalent to

$$V^{\alpha}_{.\varrho\bar\sigma} g^{\varrho\bar\sigma} = 0. \tag{4.58 a}$$

Its differentiation gives, by using (4.21 a) and (4.60),

$$\nu^\gamma \equiv 0, \quad \text{mod } \omega, \omega^\alpha, \omega^{\bar\beta}.$$

Hence we can put

$$\nu^\gamma \equiv P_{\alpha.}^{\gamma}\omega^\alpha + Q_{\bar\beta.}^{\gamma}\omega^{\bar\beta} \quad \text{mod } \omega. \tag{4.61}$$

Substitution into (4.54) gives

$$\Phi^\alpha = -V^{\alpha}_{\beta.\gamma}\omega^\beta \wedge \omega^\gamma + V^{\alpha}_{.\beta\bar\sigma}\omega^\beta \wedge \omega^{\bar\sigma} + P^{\alpha}_{\beta.}\omega^\beta \wedge \omega + Q^{\alpha}_{\bar\beta.}\omega^{\bar\beta} \wedge \omega. \tag{4.62}$$

For future use we also write down the formula

$$\begin{aligned} \Phi_\alpha &= d\phi_\alpha - \phi_{\alpha\bar\beta} \wedge \phi^{\bar\beta} + \tfrac{1}{2}\psi \wedge \omega_\alpha \\ &= -V_{\bar\beta\alpha\bar\gamma}\omega^{\bar\beta} \wedge \omega^{\bar\gamma} - V_{\alpha\bar\gamma\beta}\omega^\beta \wedge \omega^{\bar\gamma} + Q_{\beta\alpha}\omega^\beta \wedge \omega + P_{\bar\beta\alpha}\omega^{\bar\beta} \wedge \omega. \end{aligned} \tag{4.63}$$

Since the indeterminacy in ω can be absorbed in ϱ, substitution of (4.61) into (4.59) gives

$$\Psi = i\{Q_{\alpha\beta}\omega^\alpha \wedge \omega^\beta - Q_{\bar\alpha\bar\beta}\omega^{\bar\alpha} \wedge \omega^{\bar\beta}\} - i\tilde{P}_{\varrho\bar\sigma}\omega^\varrho \wedge \omega^{\bar\sigma} + \varrho \wedge \omega, \tag{4.64}$$

where

$$\tilde{P}_{\alpha\bar\beta} = P_{\alpha\bar\beta} + P_{\bar\beta\alpha} = \tilde{P}_{\bar\beta\alpha}. \tag{4.65}$$

It remains to determine ψ, which can still undergo the transformation

$$\psi = \psi' + G\omega, \tag{4.66}$$

where G is real. Denoting the new coefficients by dashes, we get, from (4.54) and (4.62),

$$P'^{\alpha}_{\beta .} = P^{\alpha}_{\beta .} + \tfrac{1}{2}\delta_{\beta}{}^{\alpha} G, \tag{4.67}$$

which gives

$$P'^{\alpha}_{\alpha .} = P^{\alpha}_{\alpha .} + \frac{n}{2} G. \tag{4.68}$$

On the other hand, from (4.65) we have

$$\tilde{P}^{\alpha}_{\alpha .} = 2\,\mathrm{Re}\,(P^{\alpha}_{\alpha .}). \tag{4.69}$$

The equation

$$\tilde{P}'^{\alpha}_{\alpha} = \tilde{P}^{\alpha}_{\alpha .} + nG$$

involves only real quantities and we have the lemma:

LEMMA 4.5. *The real form ψ is completely determined by the condition*

$$\tilde{P}^{\alpha}_{\alpha .} = 0. \tag{4.70}$$

We differentiate the equation (4.64), using th fact that Ψ is defined by (4.59). Computing mod ω and considering only the terms involving $\omega^{\varrho} \wedge \omega^{\bar{\sigma}}$, we get

$$\begin{aligned} d\tilde{P}_{\varrho\bar{\sigma}} - \tilde{P}_{\tau\bar{\sigma}}\phi^{\tau}_{\varrho .} - \tilde{P}_{\varrho\bar{\tau}}\phi^{\bar{\tau}}_{\bar{\sigma} .} - \tilde{P}_{\varrho\bar{\sigma}}\phi \\ \equiv 2V^{\beta}_{\cdot\varrho\bar{\sigma}}\phi_{\beta} + 2V_{\beta\bar{\sigma}\varrho}\phi^{\beta} - g_{\varrho\bar{\sigma}}\eta, \quad \mathrm{mod}\ \omega, \omega^{\alpha}, \omega^{\bar{\beta}}. \end{aligned} \tag{4.71}$$

From (4.70) and using (4.21 a) and (4.58), we get

$$\eta \equiv 0, \quad \mathrm{mod}\ \omega, \omega^{\alpha}, \omega^{\bar{\beta}}.$$

Since Ψ is real, we can write (4.64) in the form

$$\Psi = i\{Q_{\alpha\beta}\omega^{\alpha} \wedge \omega^{\beta} - Q_{\bar{\alpha}\bar{\beta}}\omega^{\bar{\alpha}} \wedge \omega^{\bar{\beta}}\} - i\tilde{P}_{\varrho\bar{\sigma}}\omega^{\varrho} \wedge \omega^{\bar{\sigma}} + \{R_{\alpha}\omega^{\alpha} + R_{\bar{\alpha}}\omega^{\bar{\alpha}}\} \wedge \omega. \tag{4.72}$$

We summarize the discussions of this section in the theorem:

THEOREM 4.6. *Let the manifold M of dimension $2n+1$ be provided with an integrable nondegenerate G-structure. Then the real line bundle E over M has a G_1-structure, in whose associated principal G_1-bundle Y there is a completely determined set of one-forms ω, ω^{α}, ϕ, $\phi^{\beta}_{\alpha .}$, ϕ^{α}, ψ, of which ω, ϕ, ψ are real, which satisfy the equations* (4.10), (4.16), (4.21), (4.26), (4.37), (4.53), (4.54), (4.58), (4.59), (4.62), (4.70), (4.72). *The forms*

$$\omega, \omega^{\alpha}, \omega^{\bar{\alpha}}, \phi, \phi_{\alpha\bar{\beta}}, \phi^{\alpha}, \phi^{\bar{\alpha}}, \psi \tag{4.73}$$

are linearly independent.

In particular, suppose that the G-structure arises from a real analytic real hypersurface M in $\mathbf{C}_{n+1}$. Suppose there is a second real analytic hypersurface M' in $\mathbf{C}_{n+1}$ whose corresponding concepts are denoted by dashes. In order that there is locally a biholomorphic transformation of $\mathbf{C}_{n+1}$ to $\mathbf{C}'_{n+1}$ which maps M to M' it is necessary and sufficient that there is a real analytic diffeomorphism of Y to Y' under which the forms in (4.73) *are respectively equal to the forms with dashes.*

The necessity follows from our derivation of the forms in (4.73). To prove the sufficiency condition take the $2n+1$ local variables on M as complex variables. The ω, ω^α are linear combinations of dz^α, dw and are linearly independent over the complex numbers. From

$$\omega' = \omega, \quad \omega'^\alpha = \omega^\alpha$$

we see that the diffeomorphism has the property that dz'^α, dw' are linear combinations of dz^β, dw which implies that z'^α, w' are holomorphic functions of z^β, w.

The problem for $n=1$ was solved by E. Cartan [1]. In this case conditions (4.37), (4.58), (4.70) reduce to

$$S_{1\bar{1}1\bar{1}} = V_{1\bar{1}1} = \tilde{P}_{1\bar{1}} = 0.$$

Exterior differentiation of (4.53) then gives

$$P_{1\bar{1}} = 0.$$

Our formulas reduce to those given by Cartan.

As a final remark we wish to emphasize the algebraic nature of our derivation. Most likely the theorem is a special case of a more general theorem on filtered Lie algebras.

5. The connection

(a) The flat case. We apply the results of § 4 to the special case of the nondegenerate real hyperquadrics Q discussed in § 1. The notations introduced in both sections will be used. In particular, we suppose

$$g_{\alpha\bar{\beta}} = h_{\alpha\bar{\beta}} \tag{5.1}$$

and write the equation (1.1) of Q as

$$-\frac{i}{2}(w-\bar{w}) - g_{\alpha\bar{\beta}} z^\alpha z^{\bar{\beta}} = 0. \tag{5.2}$$

By (4,7) and (4.8) we have

$$\omega = u\{\tfrac{1}{2}\,dw - i g_{\alpha\bar{\beta}} z^{\bar{\beta}} dz^\alpha\}. \tag{5.3}$$

On the other hand, given Q consider Q-frames Z_A such that the point Z_0 lies on Q. We write

$$Z_0 = tY,\ Y = (1, z^1, \ldots, z^n, w) \tag{5.4}$$

Then
$$(dZ_0, Z_0) = \frac{i}{2}\pi_0^{n+1} = |t|^2 (dY, Y) = |t|^2\left(\frac{i}{2}dw + g_{\alpha\bar{\beta}} z^{\bar{\beta}} dz^\alpha\right). \tag{5.5}$$

By setting
$$u = |t|^2, \tag{5.6}$$

we have
$$\omega = \tfrac{1}{2}\pi_0{}^{n+1} = -i(dZ_0, Z_0). \tag{5.7}$$

The structure equation (1.32) for $d\pi_0{}^{n+1}$ shows that we can put

$$\omega^\alpha = \pi_0{}^\alpha,\ \phi = -\pi_0{}^0 + \pi_{n+1}{}^{n+1} = -\pi_0{}^0 - \bar{\pi}_0{}^0. \tag{5.8 a}$$

In fact, by setting

$$\phi^\alpha = 2\pi_{n+1}{}^\alpha,\ \phi_\alpha{}^\beta = \pi_\alpha{}^\beta - \delta_\alpha{}^\beta \pi_0{}^0,\ \psi = -4\pi_{n+1}{}^0, \tag{5.8 b}$$

we find with the aid of (1.30a) that the equations (1.32) are identical to the equations given in Theorem 4.6 of § 4 with

$$S_{\alpha\beta\bar{\varrho}\bar{\sigma}} = V_{\alpha\bar{\beta}\gamma} = P_{\alpha\bar{\beta}} = Q_{\alpha\beta} = R_\alpha = 0. \tag{5.9}$$

A (nondegenerate integrable) G-structure satisfying the conditions (5.9) is called *flat*. Conversely, it follows from the Theorem of Frobenius that every real analytic flat G-structure is locally equivalent to one arisen from a nondegenerate real hyperquadric in $\mathbf{C}_{n+1}$.

Under the change of Q-frame (1.15) we have, by (5.7),

$$\omega^* = |t|^2 \omega. \tag{5.10}$$

We therefore restrict ourselves to the subgroup H_1 of H characterized by the condition $|t| = 1$. The form ω is then invariant under H_1. From (1.29) we have

$$\pi_0{}^0 = 2i(dZ_0, Z_{n+1}),\quad \pi_0{}^\alpha = g^{\alpha\bar{\beta}}(dZ_0, Z_\beta). \tag{5.11}$$

By (5.8a) it follows that under a change of Q-frames by H_1, we have

$$\left.\begin{aligned} \omega^* &= \omega, \\ \omega^{*\alpha} &= t(it^\alpha \omega + t^\alpha_{.\beta}\omega^\beta), \\ \omega^{*\bar{\alpha}} &= t^{-1}(-it^{\bar{\alpha}}\omega + t^{\bar{\alpha}}_{.\bar{\beta}}\omega^{\bar{\beta}}), \\ \phi^* &= \mathrm{Re}\,(\tau t^{-1})\,\omega - 2it\tau_\alpha\omega^\alpha + 2it^{-1}\tau_{\bar{\alpha}}\omega^{\bar{\alpha}} + \phi. \end{aligned}\right\} \tag{5.12}$$

The matrix of the coefficients in (5.12) belongs to the group G_1 introduced in § 4. The mapping

$$H_1 \to G_1 \tag{5.13}$$

so defined is clearly a homomorphism. In fact, if K denotes the group defined in (1.9), we have the isomorphism

$$H_1/K \cong G_1. \tag{5.13 a}$$

Since $SU(p+1, q+1)/K \supset H_1/K$, we will consider G_1 as a subgroup of the former via the isomorphism (5.13a). This identification is essential in the treatment of the general case; the group $SU(p+1, q+1)$ is paramount in the whole theory.

We introduce the matrix notation

$$(h) = (h_{A\bar{B}}), \tag{5.14}$$

where $h_{A\bar{B}}$ are defined in (1.10a). The Lie algebra $\mathfrak{su}$ of $SU(p+1, q+1)$ is the algebra of all matrices

$$(l) = (l_{A\cdot}^{\ B}), \quad 0 \leqslant A, B \leqslant n+1, \tag{5.15}$$

satisfying

$$(l)(h) + (h)\,{}^t(\bar{l}) = 0, \quad \mathrm{Tr}\,(l) = 0. \tag{5.16}$$

The Lie algebra of H_1 is the subalgebra of $\mathfrak{su}$ satisfying the conditions

$$l_0^\alpha = l_0^{n+1} = \mathrm{Re}\,(l_0^0) = 0. \tag{5.17}$$

With this notation it follows from (1.30) and (5.16) that the matrix

$$(\pi) = (\pi_{A\cdot}^{\ B}) \tag{5.18}$$

is an $\mathfrak{su}$-valued one-form on the group $SU(p+1, q+1)$. The Maurer-Cartan equations (1.32) of the latter can be written

$$d(\pi) = (\pi) \wedge (\pi). \tag{5.19}$$

Let

$${}^t(Z) = (Z_0, \ldots, Z_{n+1}) \tag{5.20}$$

be a matrix of vectors of $\mathbf{C}_{n+2}$. Then equation (1.29) can be written

$$d(Z) = (\pi)(Z), \tag{5.21}$$

and equation (1.15) for the change of Q-frames becomes

$$(Z^*) = (t)(Z), \tag{5.22}$$

where the entries in (t) are supposed to be constants. If (π^*) is defined by

$$d(Z^*) = (\pi^*)(Z^*), \tag{5.23}$$

we have

$$(\pi^*) = (t)\,(\pi)\,(t)^{-1} \underset{\text{def}}{=} \mathrm{ad}\,(t)\,(\pi). \tag{5.24}$$

This equation will have an important generalization.

(b) General remarks on connections. Let Y be a principal G_1-bundle over a manifold E. Let Γ be a linear group which contains G_1 as a subgroup; in our case we will have

$$\Gamma = SU(p+1, q+1)/K \supset H_1/K \cong G_1. \tag{5.25}$$

In applications of connections it frequently occurs that one should consider in the bundle Y a connection relative to the larger group Γ. For instance, this is the case of classical Riemannian geometry, where we consider in the bundle of orthonormal frames a connection relative to the group of motions of euclidean space.

Let γ be the Lie algebra of Γ realized as a Lie algebra of matrices. Then G_1 acts on γ by the adjoint transformation

$$\mathrm{ad}\,(t)(l) = (t)(l)(t)^{-1}, \quad (t) \in G_1,\ (l) \in \gamma. \tag{5.26}$$

A Γ-connection in the bundle Y is a γ-valued one-form (π), the connection form, such that under a change of frame by the group G_1, (π) transforms according to the formula

$$(\pi^*) = \mathrm{ad}\,(t)(\pi), \quad (t) \in G_1. \tag{5.27}$$

Its curvature form is defined by

$$(\Pi) = d(\pi) - (\pi) \wedge (\pi) \tag{5.28}$$

and is therefore a γ-valued two-form following the same transformation law:

$$(\Pi^*) = \mathrm{ad}\,(t)(\Pi), \quad (t) \in G_1. \tag{5.29}$$

The adjoint transformation of G_1 on γ leaves the Lie algebra $\mathfrak{g}_1$ of G_1 invariant and induces an action on the quotient space $\gamma/\mathfrak{g}_1$. The projection of the curvature form on $\gamma/\mathfrak{g}_1$ is called the torsion form.

(c) Definition of the connection. This will be a geometrical interpretation of the results of § 4. Our first problem is to write the equations listed in the Theorem 4.1 of § 4, i.e., the equations (4.10), etc. in a convenient form, making use of the group $SU(p+1, q+1)$ and its Lie algebra $\mathfrak{su}$. The $g_{\alpha\bar\beta}$ are from now on supposed to be constants and we call attention to the convention (5.1). Following the flat case we solve the equations (5.7), (5.8a), (5.8b) and put

$$\left.\begin{aligned}
&\pi_0{}^{n+1} = 2\omega, && -(n+2)\,\pi_0{}^0 = \phi_\alpha{}^\alpha + \phi,\\
&\pi_0{}^\alpha = \omega^\alpha, && \pi_\alpha{}^{n+1} = 2i\omega_\alpha,\\
&\pi_{n+1}^{\alpha} = \tfrac{1}{2}\phi^\alpha, && \pi_\alpha{}^0 = -i\phi_\alpha,\\
&\pi_{\alpha.}^{\beta} = \phi_{\alpha.}^{\beta} + \delta_\alpha{}^\beta\pi_0{}^0, && \\
&\pi_{n+1}^{0} = -\tfrac{1}{4}\psi, && \pi_{n+1}^{n+1} = -\bar\pi_0{}^0.
\end{aligned}\right\} \tag{5.30}$$

The $\pi_A{}^B$ are one-forms in Y, and the matrix

$$(\pi) = (\pi_A{}^B), \quad 0 \leqslant A, B \leqslant n+1, \tag{5.31}$$

is $\mathfrak{su}$-valued, i.e.,

$$(\pi)(h) + (h)\,{}^t(\bar{\pi}) = 0, \quad \operatorname{Tr}(\pi) = 0. \tag{5.32}$$

Moreover, restricted to a fiber of Y, the non-zero π's give the Maurer-Cartan forms of H_1, as is already in the flat case.

As in the flat case it is immediately verified that using the form (π) the equations in the theorem of § 4 can be written

$$d(\pi) = (\pi) \wedge (\pi) + (\Pi), \tag{5.33}$$

where

$$(\Pi) = \begin{pmatrix} \Pi_0{}^0 & 0 & 0 \\ \Pi_\alpha{}^0 & \Pi_\alpha{}^\beta & 0 \\ \Pi_{n+1}^{\;0} & \Pi_{n+1}^{\;\beta} & -\bar{\Pi}_0{}^0 \end{pmatrix} \tag{5.34}$$

and

$$\left.\begin{aligned} (n+2)\,\Pi_0{}^0 &= -\Phi_{\alpha.}^{\alpha}, \quad \Pi_{n+1}^{\;0} = -\tfrac{1}{4}\Psi, \\ \Pi_\alpha{}^0 &= -i\Phi_\alpha, \quad \Pi_{n+1}^{\;\beta} = \tfrac{1}{2}\Phi^\beta, \\ \Pi_{\alpha.}^{\beta} &= \Phi_{\alpha.}^{\beta} - \frac{1}{n+2}\,\delta_\alpha{}^\beta\,\Phi_{\gamma.}^{\gamma}. \end{aligned}\right\} \tag{5.35}$$

where the right-hand side members are exterior two-forms in ω, ω^α, $\omega^{\bar{\beta}}$, defined in § 4. For any such form

$$\Theta \equiv a_{\alpha\bar{\beta}}\,\omega^\alpha \wedge \omega^{\bar{\beta}} + \text{terms quadratic in } \omega^\varrho \text{ or } \omega^{\bar{\sigma}}, \bmod \omega, \tag{5.36}$$

we set

$$\operatorname{Tr}\Theta = g^{\alpha\bar{\beta}}\,a_{\alpha\bar{\beta}}. \tag{5.37}$$

Then equations (4.37), (4.58), (4.70) can be expressed respectively by

$$\left.\begin{aligned} &\operatorname{Tr}\Pi_\alpha{}^\beta = 0, \ \operatorname{Tr}\Pi_0{}^0 = 0, \\ &\operatorname{Tr}\Pi_\beta{}^0 = \operatorname{Tr}\Pi_{n+1}^{\;\alpha} = 0, \\ &\operatorname{Tr}\Pi_{n+1}^{\;0} = 0, \end{aligned}\right\} \tag{5.38}$$

and their totality can be summarized in the matrix equation

$$\operatorname{Tr}(\Pi) = 0. \tag{5.39}$$

Under the adjoint transformation of H_1,

$$(\pi) \to \operatorname{ad}(t)(\pi),$$

$$(\Pi) \to \operatorname{ad}(t)(\Pi),$$

the condition (5.39) remains invariant. We submit ω, ω^α, ω^β, ϕ to the linear transformation with the matrix (4.12) and denote the new quantities by the same symbols with asterisks. Since (π) is uniquely determined by (5.39) according to theorem 4.1 in § 4 and since these conditions are invariant under the adjoint transformation by H_1, we have

$$(\pi^*) = \mathrm{ad}\,(t)(\pi), \quad t \in G_1. \tag{5.40}$$

Therefore (π) satisfies the conditions of a connection form and we have the theorem:

THEOREM 5.1. *Given a non-degenerate integrable G-structure on a manifold M of dimension $2n+1$. Consider the principal bundle Y over E with the group $G_1 \subset SU(p+1, q+1)/K$. There is in Y a uniquely defined connection with the group $SU(p+1, q+1)$, which is characterized by the vanishing of the torsion form and the condition* (5.39).

In terms of Q-frames Z_A which are meaningful under the group $SU(p+1, q+1)$, the connection can be written

$$DZ_A = \pi_{A.}^{\ B} Z_B. \tag{5.41}$$

These equations are to be compared with (5.21) where the differential is taken in the ordinary sense.

(d) Chains. Consider a curve λ which is everywhere transversal to the complex tangent hyperplane. Its tangent line can be defined by

$$\omega^\alpha = 0. \tag{5.42}$$

By (4.16) restricted to λ, we get

$$\phi^\alpha = b^\alpha \omega. \tag{5.43}$$

The curve λ is called a chain if $b^\alpha = 0$. The chains are therefore defined by the differential system

$$\omega^\alpha = \phi^\alpha = 0. \tag{5.44}$$

They generalize the chains on the real hyperquadrics in $\mathbf{C}_{n+1}$ (cf. (1.33)) and are here defined intrinsically. It is easily seen that through a point of M and tangent to a vector transversal to the complex tangent hyperplane there passes exactly one chain.

When restricted to a chain, equations (4.10), (4.26), (4.59), (4.72) give

$$d\omega = \omega \wedge \phi, \quad d\phi = \omega \wedge \psi, \quad d\psi = \phi \wedge \psi. \tag{5.45}$$

The forms ω, ϕ, ψ being real, these are the equations of structure of the group of real linear fractional transformations in one real variable. It follows that on a chain there is a preferred parameter defined up to a linear fractional transformation. In other words, on a chain the cross ratio of four points, a real value, is well defined.

6. Actual computation for real hypersurfaces

Consider the real hypersurface M in $\mathbf{C}_{n+1}$ defined by the equation (4.6). We wish to relate the invariants of the G-structure with the function $r(z^\alpha, z^{\bar\alpha}, w, \bar w)$, and thus also with the normal form of the equation of M established in § 2, 3. This amounts to solving the structure equations listed in the theorem of § 4, with the G-structure given by (4.7); the unique existence of the solution was the assertion of the theorem. We observe that it suffices to find a particular set of forms satisfying the structure equations, because the most general ones are then completely determined by applying the linear transformation with the matrix (4.12). In actual application it will be advantageous to allow $g_{\alpha\bar\beta}$ to be variable, which was the freedom permitted in § 4. Our method consists of first finding a set of solutions of the structure equations, without necessarily satisfying the trace conditions (4.37), (4.58), (4.70). By successive steps we will then modify the forms to fulfill these conditions.

We set

$$\omega = \theta = i\partial r, \quad \omega^\alpha = dz^\alpha. \tag{6.1}$$

Then (4.10) becomes

$$d\theta = i g_{\alpha\bar\beta} dz^\alpha \wedge dz^{\bar\beta} + \theta \wedge \phi. \tag{6.2}$$

It is fulfilled if

$$\left.\begin{aligned} g_{\alpha\bar\beta} &= -r_{\alpha\bar\beta} + r_w^{-1} r_\alpha r_{w\bar\beta} + r_{\bar w}^{-1} r_{\bar\beta} r_{\bar w\alpha} - (r_w r_{\bar w})^{-1} r_{w\bar w} r_\alpha r_{\bar\beta} \\ \phi &= -r_{\bar w}^{-1} r_{\bar w\alpha} dz^\alpha - r_w^{-1} r_{w\bar\beta} dz^{\bar\beta} + (r_w r_{\bar w})^{-1} r_{w\bar w}(r_\alpha dz^\alpha + r_{\bar\beta} dz^{\bar\beta}), \end{aligned}\right\} \tag{6.3}$$

where we use the convention

$$r_\alpha = \frac{\partial r}{\partial z^\alpha}, \quad r_{\bar\beta} = \frac{\partial r}{\partial z_{\bar\beta}}, \quad r_{\alpha\bar\beta} = \frac{\partial^2 r}{\partial z^\alpha \partial z^{\bar\beta}}, \quad \text{etc.} \tag{6.4}$$

Exterior differentiation of (6.2) gives

$$i(dg_{\alpha\bar\beta} + g_{\alpha\bar\beta}\phi) \wedge dz^\alpha \wedge dz^{\bar\beta} - \theta \wedge d\phi = 0. \tag{6.5}$$

This allows us to put

$$dg_{\alpha\bar\beta} + g_{\alpha\bar\beta}\phi = a_{\alpha\bar\beta\gamma} dz^\gamma + a_{\bar\beta\alpha\bar\gamma} dz^{\bar\gamma} + c_{\alpha\bar\beta}\theta, \tag{6.6}$$

$$d\phi = i c_{\alpha\bar\beta} dz^\alpha \wedge dz^{\bar\beta} + \theta \wedge \mu, \tag{6.7}$$

where

$$a_{\alpha\bar\beta\gamma} = a_{\gamma\bar\beta\alpha}, \quad c_{\alpha\bar\beta} = c_{\bar\beta\alpha}. \tag{6.8}$$

With θ, $g_{\alpha\bar\beta}$, ϕ given by (6.1), (6.3), equations (6.6), (6.7) determine completely $a_{\alpha\bar\beta\gamma}$, $c_{\alpha\bar\beta}$, and also μ, when we assume that μ is a linear combination of dz^α, $dz^{\bar\beta}$ only. The $a_{\alpha\bar\beta\gamma}$, $c_{\alpha\bar\beta}$, μ so defined involve partial derivatives of r up to order 3 inclusive.

With ω, ω^α, ϕ given by (6.1), (6.3), we see that the following forms satisfy (4.16), (4.21), and (4.26):

$$\left.\begin{aligned}\phi_{\beta.}^{\alpha(1)} &= a_{\beta.\gamma}^{\alpha}dz^{\gamma} + \tfrac{1}{2}c_{\beta.}^{\alpha}\theta,\\ \phi^{\alpha(1)} &= \tfrac{1}{2}c_{\beta.}^{\alpha}dz^{\beta},\\ \psi^{(1)} &= \mu.\end{aligned}\right\} \tag{6.9}$$

Its most general solution, to be denoted by $\phi_{\beta.}^{\alpha}$, ϕ^{α}, ψ, is related to the particular solution (6.9), the "first approximation", by

$$\left.\begin{aligned}\phi_{\beta.}^{\alpha(1)} &= \phi_{\beta.}^{\alpha} = d_{\beta.}^{\alpha}\theta,\\ \phi^{\alpha(1)} &= \phi^{\alpha} + d_{\beta.}^{\alpha}dz^{\beta} + e^{\alpha}\theta,\\ \psi^{(1)} &= \psi + g\theta + i(e_{\alpha}dz^{\alpha} - e_{\bar\beta}dz^{\bar\beta}),\end{aligned}\right\} \tag{6.10}$$

where $d_{\beta.}^{\alpha}$ satisfy
$$d_{\alpha\bar\beta} + d_{\bar\beta\alpha} = 0 \tag{6.11}$$

and g is real; cf. (4.35), (4.36). We will determine the coefficients in (6.10) by the conditions (4.37), (4.58), (4.70).

In view of (4.53) we set

$$\begin{aligned}d\phi_{\beta.}^{\gamma(1)} - \phi_{\beta.}^{\sigma(1)} \wedge \phi_{\sigma.}^{\gamma(1)} - ig_{\beta\bar\sigma}dz^{\bar\sigma} \wedge \phi^{\gamma(1)} + i\phi_{\beta}^{(1)} \wedge dz^{\gamma} + i\delta_{\beta}^{\gamma}(\phi_{\sigma}^{(1)} \wedge dz^{\sigma})\\ \equiv s_{\beta\varrho.\bar\sigma}^{\;\;\gamma(1)}\, dz^{\varrho} \wedge dz^{\bar\sigma}, \quad \operatorname{mod}\theta,\end{aligned} \tag{6.12}$$

by which $s_{\beta\varrho.\bar\sigma}^{\;\;\gamma(1)}$ are completely determined. Let

$$s_{\varrho\bar\sigma}^{(1)} = g^{\alpha\bar\beta}s_{\alpha\varrho\bar\beta\bar\sigma}^{(1)}, \qquad s^{(1)} = g^{\alpha\bar\beta}s_{\alpha\bar\beta}^{(1)}. \tag{6.13}$$

By (4.43) the condition (4.37) is fulfilled if we put

$$(n+2)\,d_{\varrho\bar\sigma} = -is_{\varrho\bar\sigma}^{(1)} + \frac{i}{2(n+1)}g_{\varrho\bar\sigma}s^{(1)}. \tag{6.14}$$

This equation determines $d_{\varrho\bar\sigma}$ and we have completely determined

$$\phi_{\beta.}^{\alpha} = \phi_{\beta.}^{\alpha(1)} - d_{\beta.}^{\alpha}\theta = a_{\beta.\gamma}^{\alpha}dz^{\gamma} + (\tfrac{1}{2}c_{\beta.}^{\alpha} - d_{\beta.}^{\alpha})\,\theta. \tag{6.15}$$

For the determination of ϕ^{α} we introduce the "second approximation":

$$\phi^{\alpha(2)} = \phi^{\alpha(1)} - d_{\beta.}^{\alpha}dz^{\beta}. \tag{6.16}$$

Again in view of (4.53), we set

$$\begin{aligned}d\phi_{\beta.}^{\gamma} - \phi_{\beta.}^{\sigma} \wedge \phi_{\sigma.}^{\gamma} - ig_{\beta\bar\sigma}dz^{\bar\sigma} \wedge \phi^{\gamma(2)} + i\phi_{\beta}^{(2)} \wedge dz^{\gamma} + i\delta_{\beta}^{\gamma}(\phi_{\sigma}^{(2)} \wedge dz^{\sigma}) + \tfrac{1}{2}\delta_{\beta}^{\gamma}\psi^{(1)} \wedge \theta\\ = s_{\beta\varrho.\bar\sigma}^{\;\;\gamma}dz^{\varrho} \wedge dz^{\bar\sigma} + v_{\beta.\varrho}^{\gamma(1)}dz^{\varrho} \wedge \theta - v_{\;\cdot\beta\bar\sigma}^{\gamma(1)}dz^{\bar\sigma} \wedge \theta,\end{aligned} \tag{6.17}$$

which defines the coefficients $s_{\beta\varrho.\bar\sigma}^{\;\;\gamma}$, $v_{\beta.\varrho}^{\gamma(1)}$. The former satisfy

$$s_{\varrho\bar{\sigma}} = g^{\alpha\bar{\beta}} s_{\alpha\varrho\bar{\beta}\bar{\sigma}} = 0. \tag{6.18}$$

By (4.57) we determine e_β by

$$-i(n+\tfrac{1}{2})\, e_\beta = v_{\beta.\varrho}^{\varrho(1)}, \tag{6.19}$$

so that (4.58) will be satisfied. We have then completely determined

$$\phi^\alpha = \phi^{\alpha(2)} - e^\alpha \theta \tag{6.20}$$

and we introduce the "second approximation"

$$\psi^{(2)} = \psi^{(1)} - i(e_\alpha dz^\alpha - e_{\bar{\beta}}\, dz^{\bar{\beta}}). \tag{6.21}$$

By (4.54) and (4.62) we set

$$\begin{aligned} &d\phi^\alpha - \phi \wedge \phi^\alpha - \phi^\beta \wedge \phi_{\beta.}^{\alpha} + \tfrac{1}{2}\psi^{(2)} \wedge dz^\alpha \\ &\quad = -v_{\beta.\gamma}^{\alpha}\, dz^\beta \wedge dz^\gamma + v_{.\beta\bar{\sigma}}^{\alpha}\, dz^\beta \wedge dz^{\bar{\sigma}} + p_{\beta.}^{\alpha(1)} dz^\beta \wedge \theta + q_{\bar{\beta}.}^{\alpha} dz^{\bar{\beta}} \wedge \theta. \end{aligned} \tag{6.22}$$

The condition (4.70) is fulfilled by setting

$$g = -\frac{2}{n}\,\mathrm{Re}\,(p_{\alpha.}^{\alpha(1)}) \tag{6.23}$$

and

$$\psi = \psi^{(2)} - g\theta. \tag{6.24}$$

The forms $\phi_{\beta.}^{\alpha}$, ϕ^α, ψ so determined in successive steps satisfy now all the structure equations, together with the trace conditions (4.37), (4.58), (4.70). Notice that our formulas allow the computation of the invariants from the function r. The determinations $d_{\beta.}^{\alpha}$, e^α, g involve respectively partial derivatives of r up to the fourth, fifth, and sixth orders inclusive.

The procedure described above can be applied when the equation of M is in the normal form of § 2, 3. Then we have

$$r = \frac{1}{2i}(w - \bar{w}) - \langle z, z\rangle - N_{22} - N_{32} - N_{23} - N_{42} - N_{24} - N_{33} - \dots, \tag{6.25}$$

where

$$\left.\begin{aligned} N_{22} &= b_{\alpha_1\alpha_2\bar{\beta}_1\bar{\beta}_2} z^{\alpha_1} z^{\alpha_2} z^{\bar{\beta}_1} z^{\bar{\beta}_2} \\ N_{32} &= \bar{N}_{23} = k_{\alpha_1\alpha_2\alpha_3\bar{\beta}_1\bar{\beta}_2} z^{\alpha_1} z^{\alpha_2} z^{\alpha_3} z^{\bar{\beta}_1} z^{\bar{\beta}_2} \\ N_{42} &= \bar{N}_{24} = l_{\alpha_1\dots\alpha_4\bar{\beta}_1\bar{\beta}_2} z^{\alpha_1} z^{\alpha_2} z^{\alpha_3} z^{\alpha_4} z^{\bar{\beta}_1} z^{\bar{\beta}_2} \\ N_{33} &= m_{\alpha_1\alpha_2\alpha_3\bar{\beta}_1\bar{\beta}_2\bar{\beta}_3} z^{\alpha_1} z^{\alpha_2} z^{\alpha_3} z^{\bar{\beta}_1} z^{\bar{\beta}_2} z^{\bar{\beta}_3} \end{aligned}\right\} \tag{6.26}$$

and N_{22} and N_{33} are real; the coefficients, which are functions of u, satisfy the usual symmetry relations and are completely determined by the polynomials. Moreover, we have the trace conditions

18 – 742902 *Acta mathematica* 133. Imprimé le 20 Février 1975

$$\operatorname{Tr} N_{22} = \operatorname{Tr}^2 N_{32} = 0, \tag{6.27}$$

$$\operatorname{Tr}^3 N_{33} = 0, \tag{6.28}$$

where the traces are formed with respect to $\langle , \rangle$.

The computation is lengthy and we will only state the following results:

(1) Along the u-curve Γ, i.e., the curve defined by

$$z^\alpha = v = 0, \tag{6.29}$$

we have $\phi^\alpha = 0$. This means that Γ is a chain. In fact, this is true whenever the conditions (6.27) are satisfied.

(2) Along Γ we find

$$s_{\alpha\beta\bar{\varrho}\bar{\sigma}} = -4 b_{\alpha\beta\bar{\varrho}\bar{\sigma}}, \tag{6.30}$$

$$v^{\beta}_{\alpha.\gamma} = -\frac{12 i}{n+2} h^{\beta\bar{\sigma}} k_{\alpha\gamma\bar{\sigma}}, \tag{6.31}$$

$$q^{\alpha}_{\bar{\beta}.} = -\frac{48}{(n+1)(n+2)} h^{\alpha\bar{\gamma}} l_{\bar{\gamma}\bar{\beta}}, \tag{6.32}$$

where the quantities are defined by

$$\langle z, z\rangle = h_{\alpha\bar{\beta}} z^\alpha z^{\bar{\beta}}. \tag{6.33}$$

$$\operatorname{Tr} N_{32} = k_{\alpha_1\alpha_2\bar{\beta}} z^{\alpha_1} z^{\alpha_2} z^{\bar{\beta}}. \tag{6.34}$$

$$\operatorname{Tr}^2 N_{24} = l_{\bar{\beta}_1\bar{\beta}_2} z^{\bar{\beta}_1} z^{\bar{\beta}_2}. \tag{6.35}$$

The situation is particularly simple for $n = 1$. Then conditions (6.27) and (6.28) imply

$$N_{22} = N_{32} = N_{33} = 0. \tag{6.36}$$

On the other hand, we have the remarks at the end of § 4; the invariant of lowest order is q_{11}. Equation (6.32) identifies it with the coefficient in N_{42}.

Appendix. Bianchi Identities

BY

S. M. WEBSTER

University of California, Berkeley, California, USA

In this appendix we will show that there are further symmetry relations on the curvature of the connection, which follow from the Bianchi identities and which simplify the structure equation.

The Bianchi identities for the connection defined in section 5c are obtained by taking the exterior derivative of the structure equation (5.33). This yields

$$0 = (\Pi) \wedge (\pi) - (\pi) \wedge (\Pi) + d(\Pi).$$

To write this more explicitly it is convenient to use the formulation given in the theorem of section 4. In the G_1 bundle Y over E we have the independent linear differential forms

$$\omega, \omega^{\alpha}, \omega^{\bar\beta}, \phi, \phi_{\alpha}{}^{\beta}, \phi^{\alpha}, \phi^{\bar\beta}, \psi,$$

the relations

$$\phi_{\alpha\bar\beta} + \phi_{\bar\beta\alpha} - g_{\alpha\bar\beta}\phi = 0$$

with the $g_{\alpha\bar\beta}$ constant, and the structure equations

$$d\omega = i g_{\alpha\bar\beta}\omega^{\alpha} \wedge \omega^{\bar\beta} + \omega \wedge \phi \tag{A.1}$$

$$d\omega^{\alpha} = \omega^{\beta} \wedge \phi_{\beta}{}^{\alpha} + \omega \wedge \phi^{\alpha} \tag{A.2}$$

$$d\phi = i\omega_{\bar\beta} \wedge \phi^{\bar\beta} + i\phi_{\bar\beta} \wedge \omega^{\bar\beta} + \omega \wedge \psi \tag{A.3}$$

$$d\phi_{\beta}{}^{\alpha} = \phi_{\beta}{}^{\sigma} \wedge \phi_{\sigma}{}^{\alpha} + i\omega_{\beta} \wedge \phi^{\alpha} - i\phi_{\beta} \wedge \omega^{\alpha} - i\delta_{\beta}{}^{\alpha}(\phi_{\sigma} \wedge \omega^{\sigma}) - \tfrac{1}{2}\delta_{\beta}{}^{\alpha}\psi \wedge \omega + \Phi_{\beta}{}^{\alpha} \tag{A.4}$$

$$d\phi^{\alpha} = \phi \wedge \phi^{\alpha} + \phi^{\beta} \wedge \phi_{\beta}{}^{\alpha} - \tfrac{1}{2}\psi \wedge \omega^{\alpha} + \Phi^{\alpha} \tag{A.5}$$

$$d\psi = \phi \wedge \psi + 2i\phi^{\beta} \wedge \phi_{\beta} + \Psi. \tag{A.6}$$

The curvature forms are given by

$$\Phi_{\beta}{}^{\alpha} = S_{\beta\varrho\cdot\bar\sigma}^{\alpha}\omega^{\varrho} \wedge \omega^{\bar\sigma} + V_{\beta\cdot\varrho}^{\alpha}\omega^{\varrho} \wedge \omega - V_{\cdot\beta\bar\sigma}^{\alpha}\omega^{\bar\sigma} \wedge \omega \tag{A.7}$$

$$\Phi^{\alpha} = -V_{\varrho\cdot\sigma}^{\alpha}\omega^{\varrho} \wedge \omega^{\sigma} + V_{\cdot\varrho\bar\sigma}^{\alpha}\omega^{\varrho} \wedge \omega^{\bar\sigma} + P_{\varrho}{}^{\alpha}\omega^{\varrho} \wedge \omega + Q_{\bar\sigma}{}^{\alpha}\omega^{\bar\sigma} \wedge \omega, \tag{A.8}$$

$$\Psi = iQ_{\varrho\sigma}\omega^{\varrho} \wedge \omega^{\sigma} - iQ_{\bar\varrho\bar\sigma}\omega^{\bar\varrho} \wedge \omega^{\bar\sigma} - i\tilde{P}_{\varrho\bar\sigma}\omega^{\varrho} \wedge \omega^{\bar\sigma} + (R_{\varrho}\omega^{\varrho} + R_{\bar\sigma}\omega^{\bar\sigma}) \wedge \omega, \tag{A.9}$$

where the coefficients satisfy the relations

$$S_{\beta\varrho\bar\alpha\bar\sigma} = S_{\varrho\beta\bar\alpha\bar\sigma} = S_{\beta\varrho\bar\sigma\bar\alpha},$$

$$S_{\beta\varrho\bar\alpha\bar\sigma} = \overline{S_{\alpha\sigma\bar\beta\bar\varrho}} = S_{\bar\alpha\bar\sigma\beta\varrho},$$

$$\tilde{P}_{\alpha\bar\beta} = P_{\alpha\bar\beta} + P_{\bar\beta\alpha},$$

and

$$V_{\beta\cdot\varrho}^{\varrho} = g^{\beta\bar\alpha}S_{\beta\varrho\bar\alpha\bar\sigma} = g^{\alpha\bar\beta}\tilde{P}_{\alpha\bar\beta} = 0.$$

Differentiating equations (A.1) through (A. 6) yields, respectively,

$$0 = (\phi_{\alpha\bar\beta} + \phi_{\bar\beta\alpha} - g_{\alpha\bar\beta}\phi) \wedge \omega^{\alpha} \wedge \omega^{\bar\beta} \tag{A. 1'}$$

$$0 = \omega^{\beta} \wedge \phi_{\beta\cdot}^{\alpha} + \omega \wedge \Phi^{\alpha} \tag{A. 2'}$$

$$0 = \omega \wedge \Psi - i(\Phi^{\alpha} \wedge \omega_{\alpha} - \Phi^{\bar{\alpha}} \wedge \omega_{\bar{\alpha}}) \tag{A. 3'}$$

$$0 = d\Phi_{\beta}{}^{\alpha} + \Phi_{\beta}{}^{\gamma} \wedge \phi_{\gamma}{}^{\alpha} - \phi_{\beta}{}^{\gamma} \wedge \Phi_{\gamma}{}^{\alpha} - i\omega_{\beta} \wedge \Phi^{\alpha} - i\Phi_{\beta} \wedge \omega^{\alpha} - \delta_{\beta}{}^{\alpha}\{i\Phi^{\bar{\sigma}} \wedge \omega_{\bar{\sigma}} + \tfrac{1}{2}\Psi \wedge \omega\} \tag{A. 4'}$$

$$0 = d\Phi^{\alpha} + \Phi^{\beta} \wedge \phi_{\beta}{}^{\alpha} - \phi^{\beta} \wedge \Phi_{\beta}{}^{\alpha} - \phi \wedge \Phi^{\alpha} - \tfrac{1}{2}\Psi \wedge \omega^{\alpha} \tag{A. 5'}$$

$$0 = d\Psi + 2i\Phi^{\beta} \wedge \phi_{\beta} - 2i\phi^{\beta} \wedge \Phi_{\beta} - \phi \wedge \Psi. \tag{A. 6'}$$

These are the Bianchi identities. The actual verification of these equations is rather long, but they result from differentiating and simply dropping all terms which do not contain one of the curvature forms $\Phi_{\beta}{}^{\alpha}$, Φ^{α}, Ψ or one of their differentials. Equations (A.1′), (A.2′), and (A.3′) are trivial because of the relations $\phi_{\alpha\bar{\beta}} + \phi_{\bar{\beta}\alpha} = g_{\alpha\bar{\beta}}\phi$ and $S_{\beta\varrho.\bar{\sigma}}^{\alpha} = S_{\varrho\beta.\bar{\sigma}}^{\alpha}$. Substituting (A.3′) into (A.4′) gives

$$0 = d\Phi_{\beta}{}^{\alpha} + \Phi_{\beta}{}^{\gamma} \wedge \phi_{\gamma}{}^{\alpha} - \phi_{\beta}{}^{\gamma} \wedge \Phi_{\gamma}{}^{\alpha} - i\omega_{\beta} \wedge \Phi^{\alpha} - i\Phi_{\beta} \wedge \omega^{\alpha} - \frac{i}{2}\delta_{\beta}{}^{\alpha}\{\Phi^{\sigma} \wedge \omega_{\sigma} + \Phi^{\bar{\sigma}} \wedge \omega_{\bar{\sigma}}\}. \tag{A. 4''}$$

Substituting the expression (A.7) for $\Phi_{\beta}{}^{\alpha}$ into (A.4″) gives, after differentiating and lowering the index α,

$$\begin{aligned} 0 = {} & DS_{\beta\varrho\bar{\alpha}\bar{\sigma}} \wedge \omega^{\varrho} \wedge \omega^{\bar{\sigma}} + B_{\beta\bar{\alpha}\varrho} \wedge \omega^{\varrho} \wedge \omega - \overline{B_{\alpha\bar{\beta}\varrho}} \wedge \omega^{\bar{\varrho}} \wedge \omega \\ & + iC_{\beta\bar{\alpha}\mu\varrho\bar{\sigma}}\omega^{\mu} \wedge \omega^{\varrho} \wedge \omega^{\bar{\sigma}} + i\overline{C_{\alpha\bar{\beta}\mu\varrho\bar{\sigma}}}\omega^{\bar{\mu}} \wedge \omega^{\bar{\varrho}} \wedge \omega^{\sigma} \\ & + iD_{\beta\bar{\alpha}\varrho\sigma}\omega^{\varrho} \wedge \omega^{\sigma} \wedge \omega + i\overline{D_{\alpha\bar{\beta}\varrho\sigma}}\omega^{\bar{\varrho}} \wedge \omega^{\bar{\sigma}} \wedge \omega \\ & + iE_{\beta\bar{\alpha}\varrho\bar{\sigma}}\omega^{\varrho} \wedge \omega^{\bar{\sigma}} \wedge \omega, \end{aligned} \tag{A. 10}$$

where we define

$$\begin{aligned} DS_{\beta\varrho\bar{\alpha}\bar{\sigma}} &= dS_{\beta\varrho\bar{\alpha}\bar{\sigma}} - S_{\mu\varrho\bar{\alpha}\bar{\sigma}}\phi_{\beta}^{\mu} - S_{\beta\mu\bar{\alpha}\bar{\sigma}}\phi_{\varrho}^{\mu} - S_{\beta\varrho\bar{\mu}\bar{\sigma}}\phi_{\bar{\alpha}}^{\bar{\mu}} - S_{\beta\varrho\bar{\alpha}\bar{\mu}}\phi_{\bar{\sigma}}^{\bar{\mu}} + S_{\beta\varrho\bar{\alpha}\bar{\sigma}}\phi \\ &= S_{\beta\varrho\bar{\alpha}\bar{\sigma}\mu}\omega^{\mu} + S_{\beta\varrho\bar{\alpha}\bar{\sigma}\bar{\mu}}\omega^{\bar{\mu}} + S_{\beta\varrho\bar{\alpha}\bar{\sigma}}\omega, \\ DV_{\beta\bar{\alpha}\varrho} &= dV_{\beta\bar{\alpha}\varrho} - V_{\mu\bar{\alpha}\varrho}\phi_{\beta}^{\mu} - V_{\beta\bar{\mu}\varrho}\phi_{\bar{\alpha}}^{\bar{\mu}} - V_{\beta\bar{\alpha}\mu}\phi_{\varrho}^{\mu} - S_{\beta\varrho\bar{\alpha}\bar{\sigma}}\phi^{\bar{\sigma}} \\ &= V_{\beta\bar{\alpha}\varrho\mu}\omega^{\mu} + V_{\beta\bar{\alpha}\varrho\bar{\mu}}\omega^{\bar{\mu}} + V_{\beta\bar{\alpha}\varrho *}\omega, \\ C_{\beta\bar{\alpha}\mu\varrho\bar{\sigma}} &= V_{\beta\bar{\alpha}\varrho}g_{\mu\bar{\sigma}} + V_{\mu\bar{\alpha}\varrho}g_{\beta\bar{\sigma}} + g_{\bar{\alpha}\mu}V_{\beta\bar{\sigma}\varrho} + g_{\beta\bar{\alpha}}V_{\mu\bar{\sigma}\varrho}, \\ D_{\beta\bar{\alpha}\varrho\sigma} &= Q_{\varrho\beta}g_{\sigma\bar{\alpha}} + \tfrac{1}{2}g_{\beta\bar{\alpha}}Q_{\varrho\sigma}, \\ E_{\beta\bar{\alpha}\varrho\bar{\sigma}} &= g_{\beta\bar{\sigma}}P_{\varrho\bar{\alpha}} - P_{\bar{\sigma}\beta}g_{\varrho\bar{\alpha}} + \tfrac{1}{2}g_{\beta\bar{\alpha}}(P_{\varrho\bar{\sigma}} - P_{\bar{\sigma}\varrho}). \end{aligned}$$

Comparing terms of the same type in (A.10), we get the following three relations:

$$\begin{aligned} S_{\beta\varrho\bar{\alpha}\bar{\sigma}\mu} - S_{\beta\mu\bar{\alpha}\bar{\sigma}\varrho} = -i\{ & V_{\beta\bar{\alpha}\varrho}g_{\mu\bar{\sigma}} - V_{\beta\bar{\alpha}\mu}g_{\varrho\bar{\sigma}} \\ & + (V_{\mu\bar{\alpha}\varrho} - V_{\varrho\bar{\alpha}\mu})g_{\beta\bar{\sigma}} + g_{\bar{\alpha}\mu}V_{\beta\bar{\sigma}\varrho} - g_{\bar{\alpha}\varrho}V_{\beta\bar{\sigma}\mu} + g_{\beta\bar{\alpha}}(V_{\mu\bar{\sigma}\varrho} - V_{\varrho\bar{\sigma}\mu})\}, \end{aligned} \tag{A. 11}$$

$$V_{\beta\bar{\alpha}\sigma\varrho} - V_{\beta\bar{\alpha}\varrho\sigma} = -i\{Q_{\varrho\beta}g_{\sigma\bar{\alpha}} - Q_{\sigma\beta}g_{\varrho\bar{\alpha}} + \tfrac{1}{2}g_{\beta\bar{\alpha}}(Q_{\varrho\sigma} - Q_{\sigma\varrho})\}, \tag{A. 12}$$

$$\overline{V_{\alpha\bar{\beta}\sigma\bar{\varrho}}} + V_{\beta\bar{\alpha}\varrho\bar{\sigma}} = i\{g_{\beta\bar{\sigma}}P_{\varrho\bar{\alpha}} - P_{\bar{\sigma}\beta}g_{\varrho\bar{\alpha}} + \tfrac{1}{2}g_{\beta\bar{\alpha}}(P_{\varrho\bar{\sigma}} - P_{\bar{\sigma}\varrho})\} + S_{\beta\varrho\bar{\alpha}\bar{\sigma}}. \tag{A.13}$$

Multiplying (A.11) by $g^{\beta\bar{\alpha}}g^{\mu\bar{\sigma}}$, summing over α, β, μ, and σ, and using the relations $g^{\beta\bar{\alpha}}S_{\beta\varrho\bar{\alpha}\bar{\sigma}\mu} = V^{\varrho}_{\beta.\varrho} = 0$ gives

$$V^{\beta}_{\beta.\sigma} = 0.$$

so that contracting β and $\bar{\alpha}$ only in (A.11) gives

$$V_{\mu\bar{\sigma}\varrho} = V_{\varrho\bar{\sigma}\mu}. \tag{A.14}$$

It then follows that $g^{\beta\bar{\alpha}}V_{\beta\bar{\alpha}\sigma\varrho} = 0$, so that contracting the indices β and $\bar{\alpha}$ in (A.12) and in (A.13) gives

$$Q_{\varrho\sigma} = Q_{\sigma\varrho} \quad \text{and} \quad P_{\varrho\bar{\sigma}} = P_{\bar{\sigma}\varrho}. \tag{A.15}$$

Equations (A.5′) and (A.6′) give further relations but no further symmetries of the curvature functions $S_{\beta\varrho\bar{\alpha}\bar{\sigma}}$, $V_{\beta\bar{\sigma}\varrho}$, $P_{\varrho\bar{\sigma}}$, $Q_{\varrho\sigma}$ or R_{α}.

We can now write the curvature forms Φ^{α} and Ψ as follows:

$$\Phi^{\alpha} = V^{\alpha}_{\varrho\bar{\sigma}}\,\omega^{\varrho} \wedge \omega^{\bar{\sigma}} + P_{\varrho}{}^{\alpha}\,\omega^{\varrho} \wedge \omega + Q_{\bar{\sigma}}{}^{\alpha}\,\omega^{\bar{\sigma}} \wedge \omega, \tag{A.8′}$$

$$\Psi = -2iP_{\varrho\bar{\sigma}}\,\omega^{\varrho} \wedge \omega^{\bar{\sigma}} + R_{\varrho}\,\omega^{\varrho} \wedge \omega + R_{\bar{\sigma}}\,\omega^{\bar{\sigma}} \wedge \omega. \tag{A.9′}$$

Since $V^{\beta}_{\beta.\varrho} = 0$ we now have $\Phi_{\alpha}{}^{\alpha} = 0$, so that in the equation (5.35) $\Pi_0{}^0 = 0$ and $\Pi_{\alpha}{}^{\beta} = \Phi_{\alpha}{}^{\beta}$.

References

[1] Cartan, E., Sur la géométrie pseudo-conforme des hypersurfaces de deux variables complexes, I. *Ann. Math. Pura Appl.*, (4) 11 (1932) 17–90 (or *Oeuvres* II, 2, 1231–1304); II, *Ann. Scuola Norm. Sup. Pisa,* (2) 1 (1932) 333–354 (or *Oeuvres* III, 2, 1217–1238).

[2]. Fefferman, C., The Bergman Kernel and Biholomorphic Mappings of Pseudoconvex Domains. *Invent. Math.*, 26 (1974), 1–65.

[3]. Hopf, H., Über die Abbildungen der dreidimensionalen Sphäre auf die Kugelfläche. *Math. Ann.*, 104, 1931, 637–665, § 5.

[4]. Moser, J., Holomorphic equivalence and normal forms of hypersurfaces. To appear in *Proc. Symp. in Pure Math.*, Amer. Math. Soc.

[5]. Nirenberg, L., *Lectures on linear partial differential equations.* Regional Conf. Series in Math.. No. 17 Amer. Math. Soc. 1973.

[6]. Poincaré, H., Les fonctions analytiques de deux variables et la représentation conforme. *Rend. Circ. Mat. Palermo* (1907), 185–220.

[7]. Tanaka, N., I. On the pseudo-conformal geometry of hypersurfaces of the space of n complex variables. *J. Math. Soc. Japan,* 14 (1962), 397–429; II. Graded Lie algebras and geometric structures, *Proc. US-Japan Seminar in Differential Geometry,* 1965, 147–150.

[8]. Wells, R. O., Function theory on differentiable submanifolds. *Contributions to Analysis,* Academic Press, 1974, 407–441.

Received May 15, 1974

Erratum

Real hypersurfaces in complex manifolds

by

S. S. CHERN and J. K. MOSER

University of California
Berkeley, CA, U.S.A.

New York University
New York, N.Y., U.S.A.

(Acta Math. 133 (1974), 219–271)

Corrections on p. 246–247.

(1) Formula (3.19) should read:

$$\left.\begin{array}{l} z \to \lambda(z+aw)\,\delta^{-1}, \quad \delta = 1-2i\bar{a}-(r+i|a|^2)\,w \\ w \to |\lambda|^2 w\delta^{-1} \end{array}\right\} \qquad (3.19)$$

(2) The displayed formula below (3.20) should read:
"... leads to

$$c_{43}(0) \to c_{43}(0) - \frac{2ia}{3}, \quad c_{52}(0) \to c_{52}(0) - 2i\bar{a}$$

so that $j = c_{52}(0) + 3\,\overline{c_{43}(0)}$ is unchanged."

(3) The next formula should read:

$$\operatorname{Re} c'_{42}(0) \to \operatorname{Re} c'_{42}(0) - 4r.$$

These errors (of the second author) were observed and rectified by James Faran. These changes do not affect any other statements of the paper.

19† – 838283

ON THE VOLUME DECREASING PROPERTY OF A CLASS OF REAL HARMONIC MAPPINGS.

By Shiing-shen Chern* and Samuel I. Goldberg.[†]

1. Introduction. Let M and N be smooth oriented riemannian manifolds of dimensions m and n, respectively, with the metrics ds_M^2 and ds_N^2. Let $f: M \to N$ be a smooth mapping. Then the inverse image $f^* ds_N^2$, considered as a form on M, is positive semidefinite and $\mathrm{Tr}(f^* ds_N^2)$, its trace relative to ds_M^2, is a function on M taking non-negative values. If D is a compact domain in M, the integral

$$E(f,D) = \tfrac{1}{2} \int_D \mathrm{Tr}(f^* ds_N^2) * 1, \tag{1}$$

where $*1$ is the volume element on M, is called the *energy of f over D*. The mapping f is called *harmonic over D* if the energy $E(f,D)$ has a critical value relative to all mappings which agree with f on the boundary of D. f is called a *harmonic mapping* if it is harmonic over any compact domain in M.

The conditions for a harmonic mapping can be expressed as follows: By the method of moving frames we write, *locally*,

$$\begin{aligned} ds_M^2 &= \omega_1^2 + \cdots + \omega_m^2, \\ ds_N^2 &= \omega_1^{*2} + \cdots + \omega_n^{*2}, \end{aligned} \tag{2}$$

where ω_i and ω_α^* are linear differential forms in M and N, respectively. (Throughout the paper we will use the convention that small Latin (resp. Greek) indices have the range $1,\ldots,m$ (resp. $1,\ldots,n$)). Under the mapping f we have

$$f^*\omega_\alpha^* = \sum_i a_{\alpha i}\omega_i. \tag{3}$$

Later on we will drop f^* in such formulas when its presence is clear from context. The covariant differential of $a_{\alpha i}$, which we will define in the next

Received November 15, 1972.
*Work done under partial support of National Science Foundation grant GP-34785X.
[†]Work done under partial support of National Science Foundation grant GP-22929.

American Journal of Mathematics, Vol. 97, No. 1, pp. 133–147 section, can be written as

$$Da_{\alpha i} = \sum_i a_{\alpha ij}\omega_j, \qquad a_{\alpha ij} = a_{\alpha ji}. \tag{4}$$

The tensor field with the components

$$\tau_\alpha = \sum_i a_{\alpha ii} \tag{5}$$

is the tension vector field of J. Eells and J. H. Sampson [2]. It was proved by them that the mapping f is harmonic if and only if

$$\tau_\alpha = 0. \tag{6}$$

Throughout this paper we will only make use of the condition (6) for harmonic mappings, and not its origin from a variational problem.

Our main results are concerned with the case that M and N are equidimensional, i.e., $m = n$. Let dv_M and dv_N be the volume elements of M and N, respectively, and let

$$u = (f^* dv_N / dv_M)^2. \tag{7}$$

We will study the restrictions on u under various assumptions on the curvatures of M and N.

The second named author would like to express his gratitude to his student Mr. Zvi Har'El whose help led to some improvements.

2. Fundamental Local Formulas. Relative to the decomposition (2) of ds_M^2 the structure equations in M are

$$\begin{aligned} d\omega_i &= \sum_j \omega_j \wedge \omega_{ji}, \\ d\omega_{ij} &= \sum_k \omega_{ik} \wedge \omega_{kj} + \Omega_{ij}. \end{aligned} \tag{8}$$

The forms ω_{ij}, which are anti-symmetric in i,j, are the connection forms. In fact, if e_i is the orthonormal frame dual to the coframe ω_i, the connection D in the tangent bundle is given by

$$De_i = \sum_j \omega_{ij} e_j, \qquad \omega_{ij} + \omega_{ji} = 0. \tag{9}$$

The Ω_{ij}, also anti-symmetric in i,j, are the curvature forms. They have the expression

$$\Omega_{ij} = -\frac{1}{2}\sum_{k,l} R_{ijkl}\omega_k\wedge\omega_l, \tag{10}$$

where the coefficients R_{ijkl} satisfy the well-known symmetry relations

$$\begin{aligned} &R_{ijkl} = -R_{jikl} = -R_{ijlk},\\ &R_{ijkl} = R_{klij},\\ &R_{ijkl} + R_{iklj} + R_{iljk} = 0. \end{aligned} \tag{11}$$

The Ricci tensor R_{ij} is defined by

$$R_{ij} = \sum_k R_{ikjk} \tag{12}$$

and the scalar curvature by

$$R = \sum_i R_{ii}. \tag{13}$$

Similar equations are valid in N and we will denote the corresponding quantities by the same notations with asterisks.

Suppose there is a smooth mapping $f: M \to N$. By taking the exterior derivative of (3) and making use of the structure equations in M and N, we get

$$\sum_i Da_{\alpha i}\wedge\omega_i = 0, \tag{14}$$

where

$$Da_{\alpha i} \underset{\text{def}}{=} da_{\alpha i} + \sum_j a_{\alpha j}\omega_{ji} + \sum_\beta a_{\beta i}\omega^*_{\beta\alpha} = \sum_j a_{\alpha ij}\omega_j \text{ (say)}. \tag{15}$$

From (14) it follows that

$$a_{\alpha ij} = a_{\alpha ji}. \tag{16}$$

The mapping f is called *harmonic* (resp. *totally geodesic*) if

$$\sum_i a_{\alpha ii} = 0 \text{ (resp. } a_{\alpha ij} = 0). \tag{17}$$

The tensor field $a_{\alpha ij}$ arises naturally from mappings of riemannian manifolds, not necessarily harmonic ones. It can be given a simple geometrical interpretation which generalizes the linear mapping on tangent spaces induced by f. In fact, let e_i (resp. e^*_α) be the frame dual to the coframe ω_j (resp. ω^*_β).

Then we have

$$f_* e_i = \sum_\alpha a_{\alpha i} e^*_\alpha. \tag{18}$$

Let γ be a geodesic in M with the arc length s. Along γ we have

$$\omega_i = \lambda_i \, ds \tag{19}$$

and $\Sigma\lambda_i e_i$ is the unit tangent vector to γ. The latter is thus an auto-parallel vector along γ and, by (9), we have

$$d\lambda_i + \Sigma\lambda_j \omega_{ji} = 0. \tag{20}$$

The image curve $f(\gamma)$ is in general not a geodesic on N; nor is the parameter s its arc length. However, the tangent vector of $f(\gamma)$ is

$$f_*(\Sigma\lambda_i e_i) = \sum a_{\alpha i}\lambda_i e^*_\alpha$$

and, by (15), it has an acceleration vector given by

$$D(\Sigma a_{\alpha i}\lambda_i e^*_\alpha) = \sum_\alpha (\Sigma da_{\alpha i}\lambda_i + \Sigma a_{\alpha i} d\lambda_i + \Sigma a_{\beta i}\lambda_i \omega^*_{\beta\alpha}) e^*_\alpha$$
$$= \Sigma a_{\alpha ij}\lambda_i\lambda_j e^*_\alpha \, ds.$$

Hence we have the following *geometrical interpretation of* $a_{\alpha ij}$: Let $x \in M$ and let T_x and $T_{f(x)}$ be the tangent spaces at x and $f(x)$, respectively. The mapping

$$f_{**}: T_x \rightarrow T_{f(x)} \tag{21}$$

defined by

$$f_{**}(v) = \Sigma a_{\alpha ij}\lambda_i\lambda_j e^*_\alpha, \qquad v = \sum_i \lambda_i e_i \tag{22}$$

is quadratic and has the property that if v is a unit vector, $f_{**}(v)$ is the acceleration vector of $f(\gamma)$ at $f(x)$, where γ is the geodesic tangent to v at x.

Differentiating the equation (15) and using the structure equations in M and N, we get

$$\Sigma Da_{\alpha ij} \wedge \omega_j = \Sigma a_{\alpha j}\Omega_{ji} + \Sigma a_{\beta i}\Omega^*_{\beta\alpha}, \tag{23}$$

where

$$Da_{\alpha ij} = da_{\alpha ij} + \Sigma a_{\beta ij}\omega^*_{\beta\alpha} + \Sigma a_{\alpha kj}\omega_{ki} + \Sigma a_{\alpha ik}\omega_{kj}. \tag{24}$$

By putting

$$Da_{\alpha ij} = \Sigma a_{\alpha ijk}\omega_k, \tag{25}$$

we get from (23) the commutativity relation

$$a_{\alpha ijk} - a_{\alpha ikj} = -\Sigma a_{\alpha l} R_{likj} - \Sigma a_{\beta i} a_{\gamma k} a_{\delta j} R^*_{\beta\alpha\gamma\delta}. \tag{26}$$

By (16) and (26) we easily calculate the Laplacian

$$\Delta a_{\alpha i} \underset{\text{def}}{=} \Sigma a_{\alpha ikk} = \Sigma a_{\alpha kik} = \Sigma a_{\alpha kki} + \Sigma a_{\alpha l} R_{li} - \Sigma a_{\beta k} a_{\gamma k} a_{\delta i} R^*_{\beta\alpha\gamma\delta}. \tag{27}$$

For a harmonic mapping the first term in the last expression is zero and we have

$$\Delta a_{\alpha i} = \Sigma a_{\alpha l} R_{li} - \Sigma a_{\beta k} a_{\gamma k} a_{\delta i} R^*_{\beta\alpha\gamma\delta}. \tag{28}$$

Example. The simplest case is a smooth mapping $f: E^m \to E^n$ of Euclidean spaces. Let $y_1, \ldots, y_m$ and $x_1, \ldots, x_n$ be the coordinates in the respective spaces and let f be given by the functions $x_\alpha(y_i)$. Then f is harmonic if and only if

$$\Delta x_\alpha = \sum_i \frac{\partial^2 x_\alpha}{\partial y_i^2} = 0, \tag{29}$$

i.e., the x_α's are harmonic functions in the ordinary sense.

PROPOSITION 2.1. *Let $f: M \to N$ be an isometry*, i.e., $f^*(ds_N^2) = ds_M^2$. *Then f is harmonic if and only if it is a minimal immersion.*

Proof. Since the mapping is an isometry we have $m \leqslant n$ and f is an immersion. Locally the frame fields can be so chosen that

$$\omega_i^* = \omega_i, \qquad \omega_r = 0, \qquad m+1 \leqslant r \leqslant n, \tag{30}$$

i.e.,

$$a_{ij} = \delta_{ij}, \qquad a_{rj} = 0. \tag{31}$$

Taking the exterior derivative of the first equation of (30) and using the structure equations, we get, after a well-known derivation,

$$\omega_{ik}^* = \omega_{ik}. \tag{32}$$

Substitution of the values (31) into (15) gives

$$a_{kij} = 0, \qquad \omega_{ir}^* = \Sigma a_{rij} \omega_j.$$

The forms

$$\Sigma \omega_i \omega_{ir}^* = \sum_4 a_{rij} \omega_i \omega_j \tag{33}$$

are the second fundamental forms of M, the vanishing of whose traces is the condition for a minimal immersion. Hence the proposition follows immediately.

Another example of a harmonic mapping arising naturally from a geometrical problem concerns the Gauss map of an immersion $f: M \to E^n$. To state the theorem we first give a brief review of the geometry of a submanifold in E^n.

Let 0 be a fixed point of E^n. We identify $SO(n)$ with the manifold of all orthonormal frames with origin at 0, the quotient space $SO(n)/SO(n-m)$ with the Stiefel manifold St_m of orthonormal m-frames, and the quotient space $SO(n)/SO(m)\times SO(n-m)$ with the Grassman manifold G_m of oriented m-dimensional linear spaces through 0. Let P be the principal bundle of orthonormal frames of M and Q the principal bundle associated with the normal bundle of M. Then we have the basic diagram

$$\begin{array}{ccc} P\oplus Q & \xrightarrow{g_2} & SO(n) \\ \downarrow & & \downarrow \pi_1 \\ P & \xrightarrow{g_1} & St_m \\ \pi \downarrow & & \downarrow \pi_2 \\ M & \xrightarrow{g} & G_m \end{array} \tag{34}$$

At the right-hand side the projections are defined as follows: If $\{e_1,\ldots,e_n\} \in SO(n)$, then $\pi_1(\{e_1,\ldots,e_n\}) = \{e_1,\ldots,e_m\}$ and $\pi_2\{e_1,\ldots,e_m\}$ is the oriented space spanned by $e_1,\ldots,e_m$. The projections at the left-hand column are also obviously defined. The maps g_1, g_2, g are the Gauss maps and are defined by parallelism in the ambient space E^n. The diagram is clearly commutative.

For $\{e_1,\ldots,e_n\} \in SO(n)$ let

$$de_\alpha = \sum \theta_{\alpha\beta} e_\beta. \tag{35}$$

Then the $\theta_{\alpha\beta}$ are anti-symmetric:

$$\theta_{\alpha\beta} + \theta_{\beta\alpha} = 0 \tag{36}$$

and satisfy the Maurer-Cartan equations

$$d\theta_{\alpha\beta} = \sum \theta_{\alpha\gamma} \wedge \theta_{\gamma\beta}. \tag{37}$$

The quadratic differential form

$$d\sigma^2 = \sum_{i,r} \theta_{ir}^2, \qquad m+1 \leqslant r,s \leqslant n \tag{38}$$

defines a canonical riemannian metric on G_m, which is invariant under the action of $SO(n)$. Relative to the decomposition (38) the connection forms $\theta_{js,ir}$ of the metric $d\sigma^2$ can be read off from the equation (37), i.e.,

$$\begin{aligned} d\theta_{ir} &= \Sigma\theta_{ij}\wedge\theta_{jr} + \Sigma\theta_{is}\wedge\theta_{sr} \\ &= \Sigma\theta_{js}\wedge(-\delta_{rs}\theta_{ij} - \delta_{ij}\theta_{rs}) \end{aligned}$$

and we obtain

$$\theta_{js,ir} = -\delta_{rs}\theta_{ij} - \delta_{ij}\theta_{rs}. \tag{39}$$

Let

$$\omega_{\alpha\beta} = g_2^*\theta_{\alpha\beta} \tag{40}$$

and let ω_i be the coframe dual to the frame e_i. Then we have

$$\omega_{ir} = \Sigma h_{irk}\omega_k, \qquad h_{irk} = h_{kri} \tag{41}$$

and

$$II_r = \Sigma\omega_i\omega_{ir} = \Sigma h_{irk}\omega_i\omega_k \tag{42}$$

as the second fundamental forms of M. The vector

$$H = \frac{1}{n}\sum_{i,r} h_{iri}e_r \tag{43}$$

is the mean curvature vector of M. We have

$$\begin{aligned} Dh_{irk} &= dh_{irk} + \Sigma h_{jrk}\omega_{ji} + \Sigma h_{irj}\omega_{jk} + \Sigma h_{isk}\omega_{sr} \\ &= \Sigma h_{irkj}\omega_j, \end{aligned} \tag{44}$$

where

$$h_{irkj} = h_{irjk}. \tag{45}$$

This, combined with (41), gives the important fact that h_{irjk} is symmetric in any two of the indices i, j, k. M is said to be of constant mean curvature if

$$DH = \frac{1}{n}\Sigma h_{irik}\omega_k e_r = 0, \tag{46}$$

D being the canonical connection in the normal bundle. Condition (46) is equivalent to

$$\sum_i h_{irik} = 0. \tag{46a}$$

On the other hand, equation (41) is exactly our equation (3) in the case of the Gauss map g. By comparison of (15), (39), and (44), the condition for g to be a harmonic mapping is easily seen to be

$$\sum_k h_{irkk}=0. \tag{47}$$

Since this is equivalent to (46a), we have the theorem:

PROPOSITION 2.2. *The Gauss map of an immersed manifold M in E^n is harmonic if and only if M is of constant mean curvature.*

This theorem has been proved by E. A. Ruh and J. Vilms [3].

3. A Formula in the Equidimensional Case. We consider the case that M and N have the same dimension, i.e., $m=n$. The ratio A of the volume elements has the expression

$$A=\det(a_{\alpha i}) \tag{48}$$

and we have $u=A^2$, as defined in (7). We wish to calculate the Laplacian of u. Let $(B_{i\alpha})$ be the adjoint of the matrix $(a_{\alpha j})$, so that

$$\Sigma B_{i\alpha}a_{\alpha j}=\Sigma a_{i\alpha}B_{\alpha j}=\delta_{ij}A. \tag{49}$$

Then

$$dA=\Sigma B_{i\alpha}da_{\alpha i}$$

and we have, by (15),

$$dA=\Sigma B_{i\alpha}a_{\alpha ij}\omega_j=\Sigma A_j\omega_j \text{ (say)}. \tag{50}$$

Let

$$du=\Sigma u_k\omega_k, \tag{51}$$

where

$$u_k=2A\Sigma B_{i\alpha}a_{\alpha ik}. \tag{52}$$

Exterior differentiation of (51) gives

$$\sum_k (du_k-\Sigma u_i\omega_{ki})\wedge\omega_k=0.$$

Hence we can write

$$du_k-\Sigma u_i\omega_{ki}=\Sigma u_{kj}\omega_j, \tag{53}$$

where

$$u_{jk}=u_{kj}. \tag{54}$$

The Laplacian of u is by definition equal to

$$\Delta u = \Sigma u_{kk}. \tag{55}$$

By differentiating (52), using (53), and simplifying, we get

$$\tfrac{1}{2} u_{kj} = 2A_j A_k - \Sigma B_{i\alpha} B_{l\beta} a_{\alpha lj} a_{\beta ik} + A \Sigma B_{i\alpha} a_{\alpha ikj}. \tag{56}$$

By the use of (27) it follows that, for a harmonic mapping,

$$\tfrac{1}{4}\Delta u = \Sigma A_k^2 + \frac{u}{2}(R - \Sigma a_{\beta k} a_{\gamma k} R^*_{\beta\gamma}) - \tfrac{1}{2}C, \tag{57}$$

where C is a scalar invariant of the mapping defined by

$$C = \Sigma B_{i\alpha} B_{l\beta} a_{\alpha lj} a_{\beta ij}. \tag{58}$$

If $u > 0$, we have

$$\Delta \log u = -\frac{1}{u^2}\langle du, du\rangle + \frac{1}{u}\Delta u. \tag{59}$$

From (57) and (59) it follows that

$$\tfrac{1}{2}\Delta \log u = R - \Sigma a_{\beta k} a_{\gamma k} R^*_{\beta\gamma} - \frac{C}{u}, \qquad u > 0. \tag{60}$$

Consider the case of a totally geodesic mapping. From (58) we have $C = 0$. Assuming M and N are connected, it follows from (52) that $u = \text{const}$. If $u \neq 0$, formula (60) gives

$$R = \Sigma a_{\beta k} a_{\gamma k} R^*_{\beta\gamma}. \tag{61}$$

If N is Einsteinian, then $R = (R^*/n)\Sigma \alpha^2_{\beta k}$. The matrix $(\Sigma a_{\alpha i} a_{\alpha j})$ being positive semi-definite, if $R \neq 0$ and R^* and R have the same sign $|R| \geq |R^*| u^{1/n}$. Let $\lambda_1, \ldots, \lambda_n$ be the eigenvalues of $(\Sigma a_{\alpha i} a_{\alpha j})$. Then, $\lambda_i \geq 0$, $i = 1, \ldots, n$ and $f^* ds_N^2 = \Sigma \lambda_i \omega_i^2 \leq (\lambda_1 + \cdots + \lambda_n)\Sigma \omega_i^2$. Putting $\delta = \Sigma a_{\alpha i}^2$, we see that δ is a well-defined non-negative function on M giving an upper bound of the ratio of distances on M and N, i.e., $f^* ds_N^2 \leq \delta ds_M^2$. We have therefore the following theorem:

PROPOSITION 3.1. *Let $f: M \to N$ be a totally geodesic mapping, with $\dim M = \dim N$. Then u is a constant. The mapping f is either:* 1) *totally degenerate* $(u = 0)$ *or:* 2) *nowhere degenerate* $(u \neq 0)$ *in which case the relation* (61) *holds. The latter case cannot occur if $R > 0$ (resp. $R < 0$) and the Ricci curvature of N is non-positive (resp. non-negative). In the case where N is Einsteinian, $R \neq 0$ and R^* and R have the same sign, then if $|R^*| \geq |R|$, f is* *volume decreasing. If f is not totally degenerate and $|R^*| \geqslant n|R|$, then f is distance decreasing.*

Remark. The scalar C may be interpreted geometrically as a weighted measure of the deviation of the square length of the tensor C_{ijk} from the square length of its symmetric part, where C_{ijk} is the pullback of $a_{\alpha ij}$ under f.

From (57) and (60) it is clear that the behavior of u depends much on the sign of C. Some information is furnished by the following results:

PROPOSITION 3.2. *Let $f: E^2 \to E^2$ be a harmonic mapping given by a pair of conjugate harmonic functions $x_\alpha(p)$, $p \in E^2$, $\alpha = 1,2$. Then $C = 0$.*

PROPOSITION 3.3. *Let $x: M \to E^3$ be an oriented surface immersed in the euclidean space E^3 with constant mean curvature, and let $g: M \to G_2$ be its Gauss map* (Cf. (34)). *Then g is a harmonic mapping with the property $C \geqslant 0$. Moreover, on a surface of constant mean curvature in E^3 the function $\log K^2$, where K is the gaussian curvature, is superharmonic wherever $K \neq 0$.*

Proposition 3.2 is based on an elementary computation and we will omit it here. To prove Proposition 3.3 we write equations (41) and (44) as

$$\begin{aligned} \omega_{i3} &= \Sigma h_{ik}\omega_k, \\ Dh_{ik} &= \Sigma h_{ikj}\omega_j, \qquad 1 \leqslant i,j,k \leqslant 2, \end{aligned} \tag{62}$$

where h_{ikj} is symmetric in any two of its indices. Let (H_{ij}) be the adjoint matrix of (h_{jk}). The condition for M to be of constant mean curvature is

$$h_{11k} + h_{22k} = 0. \tag{63}$$

Using these properties of h_{ijk} we find

$$C = \Sigma H_{ij} H_{kl} h_{ilm} h_{jkm} = (h_{111}^2 + h_{112}^2)(H_{11} + H_{22})^2 \geqslant 0. \tag{64}$$

Moreover, we have in this case

$$u = K^2, \qquad R = 2K = 2(h_{11}h_{22} - h_{12}^2),$$

$$R^*_{\alpha\beta} = \delta_{\alpha\beta}, \qquad a_{\alpha k} = h_{\alpha k},$$

so that

$$\tfrac{1}{2}\Delta \log K^2 = -(h_{11} - h_{22})^2 - 4h_{12}^2 - \frac{C}{K^2} \leqslant 0. \tag{65}$$

This proves the last statement in Proposition 3.3.

Formula (65) may also be written as

$$\tfrac{1}{2}\Delta \log K^2 = -4(\mu^2 - K) - \frac{C}{K^2}, \tag{65a}$$

where $\mu = \frac{1}{2}(h_{11} + h_{22})$ is the mean curvature.

Remark. The superharmonicity of $\log K^2$ for a surface of constant mean curvature seems to be a non-trivial result, but we are unable to draw a geometrical conclusion from it.

PROPOSITION 3.4. *A closed orientable surface in E^3 of constant mean curvature with $K^2 \geqslant C/2\mu^2$ is a sphere.*

Proof. With the data as above, (57) becomes

$$\tfrac{1}{2}\Delta K^2 = 4(\mu^2 - K)\left(\frac{C}{2\mu^2} - K^2\right) - C \leqslant 0.$$

Since the surface is compact K must be a positive constant.

4. Harmonic Mappings from a Space of Constant Curvature. We consider the special case that M is of constant curvature K. For its metric we use the Riemann normal form

$$ds_M^2 = \lambda(r)^2(dx_1^2 + \cdots + dx_n^2), \tag{66}$$

where

$$\lambda(r) = \left(1 + \frac{K}{4}r^2\right)^{-1},$$
$$r^2 = x_1^2 + \cdots + x_n^2, \qquad r \geqslant 0. \tag{67}$$

For a real-valued function $w(x_1, \ldots, x_n)$ on M we find

$$\lambda^2 \Delta w = \Sigma \frac{\partial^2 w}{\partial x_i^2} + \frac{n-2}{r}(\log\lambda)' \, \Sigma x_i \frac{\partial w}{\partial x_i}, \tag{68}$$

where the dash denotes differentiation with respect to r. If w is itself a function of r, $w = w(r)$, then

$$\lambda^2 \Delta w = w'' + (n-1)\frac{w'}{r} + (n-2)(\log\lambda)' w'. \tag{69}$$

We will prove the following theorem:

PROPOSITION 4.1. *Let M be the open ball $r<1$ with the hyperbolic metric of constant curvature $K=-4$. Let N be an Einstein manifold of the same dimension with scalar curvature $R^*\leqslant -4n(n-1)$. Let $f:M\to N$ be a harmonic mapping satisfying the condition $C\leqslant 0$. Then f is volume decreasing* (i.e., $u\leqslant 1$).

To prove the theorem we exhaust the open ball $r<1$ by a family of concentric open balls U_α defined by $r<\alpha<1$ and in each U_α choose a function v_α satisfying a differential inequality. By (69) we find, under the hypothesis $K=-4$,

$$(1-r^2)^{-2}\Delta\log(\alpha^2-r^2)=-\frac{2n\alpha^2}{(\alpha^2-r^2)^2}+\frac{2(n-2)r^2(1-2\alpha^2+r^2)}{(1-r^2)(\alpha^2-r^2)^2}. \tag{70}$$

It follows that

$$\tfrac{1}{2}\Delta\log\frac{1-r^2}{\alpha^2-r^2}\leqslant -n-(n-2)r^2+[n\alpha^2-(n-2)r^2]\left(\frac{1-r^2}{\alpha^2-r^2}\right)^2$$

$$+(n-2)r^2\left[1+\left(\frac{1-r^2}{\alpha^2-r^2}\right)^2\right]\leqslant -n+n\left(\frac{1-r^2}{\alpha^2-r^2}\right)^2. \tag{71}$$

We set

$$v_\alpha=4(n-1)\log\left(\frac{1-r^2}{\alpha^2-r^2}\right). \tag{72}$$

Then

$$\tfrac{1}{2}\Delta v_\alpha\leqslant -4n(n-1)+4n(n-1)\exp\left(\frac{v_\alpha}{n}\right),\qquad n\geqslant 2. \tag{73}$$

On the other hand, we put

$$v=\log u,\quad u>0. \tag{74}$$

Then by (60) we have

$$\tfrac{1}{2}\Delta v=-4n(n-1)-\frac{R^*}{n}\Sigma a_{\alpha k}^2-C\exp(-v). \tag{75}$$

Since the matrix $(\Sigma a_{\alpha k}a_{\beta k})$ is positive semidefinite, we have the inequality

$$\frac{1}{n}\Sigma a_{\alpha k}^2\geqslant(\det(a_{\alpha k}))^{\frac{2}{n}}=u^{\frac{1}{n}}. \tag{76}$$

This gives, using the hypotheses $C \leqslant 0$,

$$\tfrac{1}{2}\Delta v \geqslant -4n(n-1) - R^* \exp\left(\frac{v}{n}\right). \tag{77}$$

To complete the proof of the theorem consider the open subset E of U_α defined by $u > \exp v_\alpha$. In E we have $u > 0$ and $v > v_\alpha$. By (73) and (77) we have

$$\tfrac{1}{2}\Delta(v - v_\alpha) \geqslant -R^* \exp\left(\frac{v}{n}\right) - 4n(n-1)\exp\left(\frac{v_\alpha}{n}\right),$$

which is positive in E. This implies that the function $v - v_\alpha$ cannot have a maximum in E. Hence $v - v_\alpha$ must approach its least upper bound on a sequence tending to the boundary of E. This sequence cannot have a limit point x_0 in U_α, for at x_0, $v - v_\alpha > 0$ and x_0 would belong to E and be a maximum for $v - v_\alpha$. It also cannot be divergent, for then $v_\alpha \to \infty$ while v is bounded. It follows that the only possibility is that E is vacuous, so that $v \leqslant v_\alpha$ in U_α.

As $\alpha \to 1$, we conclude $v \leqslant 0$ or $u \leqslant 1$.

Remark. If $n = 2$, $N = E^2$, and f is given by a pair of conjugate harmonic functions, then by Proposition 3.2, $C = 0$. It follows from the above theorem that f is volume decreasing if $R^* \leqslant -8$. Noticing that the scalar curvature is twice the Gaussian curvature for $n = 2$, this gives the classical Schwarz-Ahlfors lemma.

To complete our treatment we will state a theorem analogous to Proposition 4.1 for the case that M is the euclidean space:

PROPOSITION 4.2. *Let E^n be the n-dimensional euclidean space and N an Einstein manifold of the same dimension with negative scalar curvature which is bounded away from zero. Let $f: E^n \to N$ be a harmonic mapping satisfying the condition $C \leqslant 0$. Then f is volume decreasing.*

The proof is similar and the analytical details are much simpler. It suffices to mention that one exhausts E^n by the open balls U_α defined by $r < \alpha$, $\alpha \geqslant n\sqrt{2n}/\epsilon$, ϵ: an arbitrary positive number, $\alpha \to \infty$, and takes in U_α the exhaustion function

$$v_\alpha = 2n \log \frac{\alpha^2}{\alpha^2 - r^2}. \tag{78}$$

Then v_α fulfills the inequality

$$\tfrac{1}{2}\Delta v_\alpha \leqslant \frac{\epsilon^2}{n} \exp\left(\frac{v_\alpha}{n}\right). \tag{79}$$

Let N be an Einstein space with scalar curvature $R^* \leqslant -\epsilon^2$. Then,

$$\tfrac{1}{2}\Delta \geqslant \frac{\epsilon^2}{n}\exp\left(\frac{v}{n}\right) - C\exp(-v)$$

$$\geqslant \frac{\epsilon^2}{n}\exp\left(\frac{v}{n}\right).$$

Let x_0 be a point in E^n for which $u(x_0) > 0$. Let λ be a sufficiently large constant such that $\lambda^{2n}u(x_0) > 1$ and consider the homothetic change of metric in N given by $\hat{g} = \lambda^2 g$. Then, $(N, \hat{g})$ is Einsteinian as before with negative scalar curvature. By Proposition 4.2, $\hat{u}(x_0) \leqslant 1$. But $\hat{u}(x_0) = \lambda^{2n}\, u(x_0)$, so we have a contradiction.

Proposition 4.3. *Under the conditions in Proposition 4.2, f is totally degenerate.*

5. Further Results and Remarks.

Proposition 5.1. *Let M be topologically the two-sphere S^2 and let $f: M \to N$ be a harmonic immersion. Then f is a minimal immersion.*

Proof. We choose orthonormal frame fields in M and N and write (3) explicitly as

$$\omega_\alpha^* = a_{\alpha 1}\omega_1 + a_{\alpha 2}\omega_2, \qquad 1 \leqslant \alpha \leqslant n. \tag{80}$$

The ω_1, ω_2 are defined up to a rotation through an angle θ, under which $\omega_1 + i\omega_2$ goes into $(\exp i\theta)\,(\omega_1 + i\omega_2)$ and $a_{\alpha 1} - ia_{\alpha 2}$ into $(\exp(-i\theta))\,(a_{\alpha 1} - ia_{\alpha 2})$. Moreover, the ω_α^* are determined up to an orthogonal transformation in n variables. It follows that the complex-valued quadratic differential form

$$\Phi = \Sigma(a_{\alpha 1} - ia_{\alpha 2})^2(\omega_1 + i\omega_2)^2 \tag{81}$$

is independent of the choices of the frame fields and is globally defined in M.

Let z be a complex local coordinate on M, so that

$$\omega_1 + i\omega_2 = \lambda\, dz. \tag{82}$$

Then

$$\Phi = \Sigma(a_{\alpha 1} - ia_{\alpha 2})^2\lambda^2 dz^2 = h(z)dz^2 \text{ (say)}. \tag{83}$$

We wish to show that h is a holomorphic function in z. In fact, exterior differentiation of (82) gives

$$d\lambda + \lambda i\omega_{12} \equiv 0, \bmod dz. \tag{84}$$

From (15) we find, by using the condition for a harmonic mapping,

$$d(a_{\alpha 1} - ia_{\alpha 2}) \equiv i(a_{\alpha 1} - ia_{\alpha 2})\omega_{12} - \Sigma(a_{\beta 1} - ia_{\beta 2})\omega^*_{\beta\alpha}, \bmod dz,$$

from which it follows that

$$d\left(\Sigma(a_{\alpha 1} - ia_{\alpha 2})^2\right) \equiv 2i\Sigma(a_{\alpha 1} - ia_{\alpha 2})^2 \omega_{12}, \bmod dz. \tag{85}$$

Combining (84) and (85), we get $dh \equiv 0, \bmod dz$, i.e., h is holomorphic in z. Since M is a two-sphere, we have $h = 0$.

This implies

$$\Sigma a^2_{\alpha 1} = \Sigma a^2_{\alpha 2}, \qquad \Sigma a_{\alpha 1} a_{\alpha 2} = 0. \tag{86}$$

Hence M is conformally immersed in N. By a conformal change of the metric of M, f remains a harmonic mapping. It follows from Proposition 2.1 that f is a minimal immersion.

We wish to take this opportunity to correct an error in the paper [1], from which this paper originated. The inequality in formula (70) is incorrect. It should be replaced by

$$\tfrac{1}{2}\Delta v_\rho < -2(n+1) + 2n(n+1)\exp v_\rho.$$

As a consequence, Theorem 3 of that paper should be modified to read as follows:

Let $f: D_1 \to N$ be a holomorphic mapping where D_1 is the unit n-ball with the standard Kahler metric and where N is an n-dimensional hermitian Einstein manifold with scalar curvature $\leqslant -2n^2(n+1)$. Then f is volume decreasing.

UNIVERSITY OF CALIFORNIA, BERKELEY
UNIVERSITY OF ILLINOIS, URBANA

REFERENCES.

1. S. S. Chern, "On holomorphic mappings of hermitian manifolds of the same dimension," *Proceedings of the Symposium Pure Mathematics*, American Mathematical Society (1968), pp. 157–170. MR 40 # 3482.
2. James Eells Jr. and J. H. Sampson, "Harmonic mappings of Riemannian manifolds," *American Journal of Mathematics*, 86 (1964), pp. 109-160. MR 29 #1603.
3. E. A. Ruh and J. Vilms, "The tension field of the Gauss map," *Transactions of the American Mathematical Society*, 149 (1970), pp. 569–573. MR 41 #4400.

MATH. SCAND. 36 (1975), 74–82

ON THE PROJECTIVE STRUCTURE OF A REAL HYPERSURFACE IN C_{n+1}

SHIING-SHEN CHERN

To Werner Fenchel on his 70th birthday.

Let C_{n+1} be the complex number space of dimension $n+1$ with the coordinates $z^1,\ldots,z^{n+1}$. A real hypersurface M is defined analytically by the equation

$$(1) \qquad r(z^j,\bar{z}^j) = 0, \quad 1 \leqq j \leqq n+1 ,$$

where r is a real-valued function. We suppose r to be smooth with $\operatorname{grad} r \neq 0$. For $n=1$, B. Segre observed that the corresponding equation

$$(2) \qquad r(z^j,a^j) = 0 ,$$

where $\bar{z}^j$ is replaced by arbitrary parameters a^j, defines a two-parameter family of curves in the complex plane whose invariants were first studied by A. Tresse in 1896 [5]. These invariants clearly provide local invariants of the hypersurface M itself. But they do not give a complete system of invariants of M under biholomorphic transformations in C_{n+1}, as remarked by Elie Cartan [1]. The latter problem has been the object of a recent study by J. Moser and the author [3].

The purpose of this note is to carry out Segre's idea for general n and relate it to the invariants given in [3]. Equation (2) defines an $(n+1)$-parameter family of hypersurfaces when M is non-degenerate. Generalizing the work of Tresse, M. Hachtroudi showed that a projective connection can be defined intrinsically in the space of hyperplane elements of C_{n+1} [4]. The definition is a generalization, by no means obvious, of the construction of classical projective geometry from the data of its hyperplanes; cf. also Chern [2], Yen [6] for a further generalization. We will show that the definition of Hachtroudi's connection is closely related to that of the connection in [3]. This study has the advantage that it works only with the variables z^j and their holomorphic functions; the conjugate variables $\bar{z}^j$ are not involved. Could this fact be of significance for the results to play a rôle in abstract algebraic geometry?

---

* This work was partially supported by NSF grant GP-34785-X.

Received January 6, 1975.

1. The Equivalence Problem.

We put $w=z^{n+1}$. The hypersurfaces (2) can be considered as the integral hypersurfaces of the completely integrable differential system

$$dw-p_\alpha dz^\alpha = 0, \; dp_\alpha - r_{\alpha\beta}dz^\beta = 0\,, \tag{3}$$

where $r_{\alpha\beta}=r_{\beta\alpha}$ are holomorphic functions of z^α, w, p_β. (Throughout this paper small Greek indices will run from 1 to n and the summation convention will be adopted.) The latter variables, i.e., z^α, w, p_β, can be interpreted as the coordinates in the space of hyperplane elements in C_{n+1}.

We allow biholomorphic changes of coordinates defined by

$$\begin{aligned} z^{*\alpha} &= z^{*\alpha}(z^\beta,w)\,,\\ w^* &= w^*(z^\beta,w)\,, \end{aligned} \tag{4}$$

and the transformation on p_α is given by expressing that $dw^*-p_\alpha{}^*dz^{*\alpha}$ is a multiple of $dw-p_\alpha dz^\alpha$. It follows that the form $dw-p_\alpha dz^\alpha$ is defined up to a multiple and the sets of forms

$$dw-p_\alpha dz^\alpha,\; dz^\beta \tag{5a}$$

and

$$dw-p_\alpha dz^\alpha,\; dp_\alpha - r_{\alpha\beta}dz^\beta \tag{5b}$$

are each defined up to a linear transformation. Following the general procedure in studying equivalence problems, we set

$$\begin{aligned} \omega &= u(dw-p_\alpha dz^\alpha)\,,\\ \omega^\alpha &= u_\beta{}^\alpha dz^\beta + u^\alpha(dw-p_\beta dz^\beta)\,,\\ \omega_\alpha &= v_\alpha(dw-p_\beta dz^\beta)+v_\alpha{}^\beta(dp_\beta - r_{\beta\gamma}dz^\gamma)\,, \end{aligned} \tag{6}$$

where

$$u, u_\alpha{}^\beta, v_\alpha{}^\beta, u^\alpha, v_\alpha \tag{7}$$

are new variables satisfying

$$u \neq 0,\; \det(u_\alpha{}^\beta) \neq 0,\; \det(v_\alpha{}^\beta) \neq 0\,.$$

Then the forms in (6) are invariant in the space of all the variables: the ones in (7), together with z^α, w, p_β. Computing mod ω, we find

$$\begin{aligned} d\omega &\equiv u\,dz^\alpha \wedge dp_\alpha\,,\\ i\omega^\alpha \wedge \omega_\alpha &\equiv iu_\beta{}^\alpha v_\alpha{}^\varrho dz^\beta \wedge (dp_\varrho - r_{\varrho\gamma}dz^\gamma)\,. \end{aligned}$$

The condition

$$d\omega \equiv i\omega^\alpha \wedge \omega_\alpha$$

is therefore equivalent to

$$u\delta_\alpha{}^\beta = iu_\alpha{}^\gamma v_\gamma{}^\beta . \tag{8}$$

We suppose (8) fulfilled, and set

$$d\omega = i\omega^\alpha \wedge \omega_\alpha + \omega \wedge \varphi , \tag{9}$$

where φ is defined up to the change

$$\varphi \to \varphi + t\omega . \tag{10}$$

We will take t as another new variable. Our variables are now

$$u(\neq 0),\ u_\alpha{}^\beta,\ u^\alpha,\ v_\alpha,\ t,\ z^\alpha,\ w,\ p_\alpha , \tag{11}$$

which are $(n+2)^2-1$ in number, the $v_\alpha{}^\beta$ being determined by (8), and we have the invariant forms ω, ω^α, ω_α, φ. Our purpose is to show that it is possible to introduce

$$(n+2)^2-1-(2n+2) = n^2+2n+1$$

other invariant forms, characterized by intrinsic conditions, so that the totality forms an independent set.

Clearly we can write

$$d\omega^\alpha = \omega^\beta \wedge \varphi_\beta{}^\alpha + \omega \wedge \varphi^\alpha . \tag{12}$$

LEMMA 1. *The forms $\varphi_\alpha{}^\beta$ in (12) can be so chosen that*

$$d\omega_\alpha \equiv \varphi_\alpha{}^\beta \wedge \omega_\beta + \omega_\alpha \wedge \varphi, \quad \mathrm{mod}\,\omega . \tag{13}$$

They are then determined up to additive terms in ω.

In fact, exterior differentiation of (9) gives the equation

$$i\omega^\alpha \wedge (-d\omega_\alpha + \varphi_\alpha{}^\beta \wedge \omega_\beta + \omega_\alpha \wedge \varphi) + \omega \wedge (-d\varphi + i\varphi^\alpha \wedge \omega_\alpha) = 0 . \tag{14}$$

It follows that

$$-d\omega_\alpha + \varphi_\alpha{}^\beta \wedge \omega_\beta + \omega_\alpha \wedge \varphi \equiv 0, \quad \mathrm{mod}\,\omega, \omega^\beta . \tag{14a}$$

Since the system

$$\omega = 0, \ \omega_\alpha = 0$$

is completely integrable, the left-hand side of (14a) is also $\equiv 0$, $\mathrm{mod}\,\omega, \omega_\beta$. Hence we have

$$-d\omega_\alpha + \varphi_\alpha{}^\beta \wedge \omega_\beta + \omega_\alpha \wedge \varphi \equiv a_{\alpha\beta}{}^\gamma \omega^\beta \wedge \omega_\gamma, \quad \mathrm{mod}\,\omega .$$

Substituting into (14), we get

$$a_{\alpha\beta}{}^\gamma = a_{\beta\alpha}{}^\gamma .$$

Writing $\varphi_\alpha{}^\beta$ for $\varphi_\alpha{}^\beta - a_{\alpha\gamma}{}^\beta\omega^\gamma$ fulfills the equation (13) and leaves (12) unchanged. The second statement in the lemma is immediate.

We shall therefore put

$$d\omega_\alpha = \varphi_\alpha{}^\beta \wedge \omega_\beta + \omega_\alpha \wedge \varphi + \omega \wedge \varphi_\alpha . \tag{15}$$

Using (14) we let

$$d\varphi = i\omega^\alpha \wedge \varphi_\alpha + i\varphi^\alpha \wedge \omega_\alpha + \omega \wedge \psi , \tag{16}$$

where ψ is a new one-form. The forms $\varphi_\alpha{}^\beta$, φ^α, φ_α, ψ are determined up to the transformation

$$\begin{aligned} \varphi_\alpha{}^\beta &= \varphi_\alpha{}^{*\beta} + b_\alpha{}^\beta\omega , \\ \varphi^\alpha &= \varphi^{*\alpha} + b_\beta{}^\alpha\omega^\beta + c^\alpha\omega , \\ \varphi_\alpha &= \varphi_\alpha{}^* - b_\alpha{}^\beta\omega_\beta + d_\alpha\omega , \\ \psi &= \psi^* + i(d_\alpha\omega^\alpha - c^\alpha\omega_\alpha) + e\omega . \end{aligned} \tag{17}$$

We shall determine the coefficients $b_\alpha{}^\beta$, c^α, d_α, e by intrinsic conditions imposed on the exterior derivatives of the forms.

For this purpose we take the exterior derivatives of the equations (12) and (15). The resulting equations can be written

$$\begin{aligned} \omega^\beta \wedge \Phi_\beta{}^\alpha + \omega \wedge \Phi^\alpha &= 0 , \\ \Phi_\alpha{}^\beta \wedge \omega_\beta - \omega \wedge \Phi_\alpha &= 0 , \end{aligned} \tag{18}$$

where we set

$$\begin{aligned} \Phi_\alpha{}^\beta &= d\varphi_\alpha{}^\beta - \varphi_\alpha{}^\gamma \wedge \varphi_\gamma{}^\beta - i\omega_\alpha \wedge \varphi^\beta + i\varphi_\alpha \wedge \omega^\beta + i\delta_\alpha{}^\beta(\varphi_\sigma \wedge \omega^\sigma) + \tfrac{1}{2}\delta_\alpha{}^\beta\psi \wedge \omega , \\ \Phi^\alpha &= d\varphi^\alpha - \varphi \wedge \varphi^\alpha - \varphi^\beta \wedge \varphi_\beta{}^\alpha + \tfrac{1}{2}\psi \wedge \omega^\alpha , \\ \Phi_\alpha &= d\varphi_\alpha - \varphi_\alpha{}^\beta \wedge \varphi_\beta + \tfrac{1}{2}\psi \wedge \omega_\alpha . \end{aligned} \tag{19}$$

From (18) it follows that

$$\Phi_\alpha{}^\beta = S_{\alpha\varrho}{}^{\beta\sigma}\omega^\varrho \wedge \omega_\sigma + \omega \wedge \psi_\alpha{}^\beta , \tag{20}$$

where

$$S_{\alpha\varrho}{}^{\beta\sigma} = S_{\varrho\alpha}{}^{\beta\sigma} = S_{\alpha\varrho}{}^{\sigma\beta} \tag{21}$$

and $\psi_\alpha{}^\beta$ is a one-form. Equation (18) then gives

$$\begin{aligned} \Phi^\alpha &= \omega^\beta \wedge \psi_\beta{}^\alpha + \omega \wedge \lambda^\alpha , \\ \Phi_\alpha &= \psi_\alpha{}^\beta \wedge \omega_\beta + \omega \wedge \mu_\alpha , \end{aligned} \tag{22}$$

where λ^α, μ_α are one-forms.

Applying the transformation (17) and denoting the new coefficients by asterisks, we get

$$S^*{}_{\alpha\varrho}{}^{\beta\sigma} = S_{\alpha\varrho}{}^{\beta\sigma} - i(\delta_\varrho{}^\sigma b_\alpha{}^\beta + \delta_\alpha{}^\sigma b_\varrho{}^\beta + \delta_\varrho{}^\beta b_\alpha{}^\sigma + \delta_\alpha{}^\beta b_\varrho{}^\sigma) . \tag{23}$$

Putting

$$S_\alpha{}^\beta = S_{\alpha\rho}{}^{\beta\rho},\ S^{*\beta}_\alpha = S^{*\beta\rho}_{\alpha\rho}, \tag{24}$$

the contraction of (23) gives

$$S^{*\beta}_\alpha = S_\alpha{}^\beta - i(n+2)b_\alpha{}^\beta - i\delta_\alpha{}^\beta b_\rho{}^\rho. \tag{25}$$

LEMMA 2. *The forms $\varphi_\alpha{}^\beta$ are determined uniquely by the condition*

$$S_\alpha{}^\beta = 0. \tag{26}$$

In fact, setting $S^{*\beta}_\alpha = 0$ in (25), we find

$$i(n+2)b_\alpha{}^\beta = S_\alpha{}^\beta - \tfrac{1}{2}(n+1)^{-1}\delta_\alpha{}^\beta S_\rho{}^\rho. \tag{27}$$

From now on we suppose (26) to be fulfilled.

Exterior differentiation of (16) gives

$$-i\omega^\alpha \wedge \Phi_\alpha + i\Phi^\alpha \wedge \omega_\alpha - \omega \wedge \Psi = 0, \tag{28}$$

where we set

$$\Psi = d\psi - \varphi \wedge \psi - 2i\varphi^\alpha \wedge \varphi_\alpha. \tag{29}$$

Exterior differentiation of the first equation of (19) gives

$$\begin{aligned} d\Phi_\alpha{}^\beta + \Phi_\alpha{}^\sigma \wedge \varphi_\sigma{}^\beta - \varphi_\alpha{}^\sigma \wedge \Phi_\sigma{}^\beta - i\omega_\alpha \wedge \Phi^\beta - i\Phi_\alpha \wedge \omega^\beta \\ - i\delta_\alpha{}^\beta \Phi_\sigma \wedge \omega^\sigma - \tfrac{1}{2}\delta_\alpha{}^\beta \Psi \wedge \omega = 0. \end{aligned} \tag{30}$$

Contracting, we get

$$d\Phi_\alpha{}^\alpha - i(n+1)\Phi_\sigma \wedge \omega^\sigma - i\Phi^\alpha \wedge \omega_\alpha - \tfrac{1}{2}n\Psi \wedge \omega = 0. \tag{31}$$

On the other hand, by (20) we have, as a result of (26),

$$\Phi_\alpha{}^\alpha = \omega \wedge \psi_\alpha{}^\alpha.$$

Substitution of this and (22) into (31) gives

$$(n+2)\psi_\alpha{}^\beta + \delta_\alpha{}^\beta \psi_\rho{}^\rho \equiv 0, \quad \operatorname{mod} \omega, \omega^\rho, \omega_\sigma,$$

whence

$$\psi_\alpha{}^\beta \equiv R_\alpha{}^\beta{}_\gamma \omega^\gamma + T_\alpha{}^{\beta\gamma}\omega_\gamma, \quad \operatorname{mod} \omega. \tag{32}$$

We can therefore write (20) as

$$\Phi_\alpha{}^\beta = S_{\alpha\rho}{}^{\beta\sigma}\omega^\rho \wedge \omega_\sigma + R_\alpha{}^\beta{}_\gamma \omega \wedge \omega^\gamma + T_\alpha{}^{\beta\gamma}\omega \wedge \omega_\gamma. \tag{33}$$

LEMMA 3. *The forms φ^α and φ_α are determined uniquely by the conditions*

$$R_\alpha{}^\alpha{}_\beta = 0,\ T_\alpha{}^{\alpha\beta} = 0. \tag{34}$$

When they are fulfilled, we have

(35) $$R_{\alpha\ \gamma}^{\ \beta} = R_{\gamma\ \alpha}^{\ \beta}, \quad T_{\ \alpha}^{\beta\gamma} = T_{\ \alpha}^{\gamma\beta},$$

and (22) *can be written*

(36) $$\Phi^{\alpha} = T_{\beta}^{\ \alpha\gamma}\omega^{\beta} \wedge \omega_{\gamma} + \omega \wedge \lambda^{\alpha},$$

$$\Phi_{\alpha} = R_{\alpha\ \gamma}^{\ \beta}\omega^{\gamma} \wedge \omega_{\beta} + \omega \wedge \mu_{\alpha}.$$

In fact applying the transformation (17) to the first equation of (19), noticing that $b_{\alpha}^{\ \beta}=0$ and $\varphi_{\alpha}^{\ \beta}$ are completely determined, we find

$$R_{\beta\ \gamma}^{*\alpha} = R_{\beta\ \gamma}^{\ \alpha} - i\delta_{\gamma}^{\ \alpha}d_{\beta} - \tfrac{1}{2}i\delta_{\beta}^{\ \alpha}d_{\gamma},$$

$$T_{\beta}^{*\alpha\gamma} = T_{\beta}^{\ \alpha\gamma} - i\delta_{\beta}^{\ \gamma}c^{\alpha} - \tfrac{1}{2}i\delta_{\beta}^{\ \alpha}c^{\gamma},$$

so that

$$R_{\alpha\ \gamma}^{*\alpha} = R_{\alpha\ \gamma}^{\ \alpha} - i(\tfrac{1}{2}n+1)d_{\gamma},$$

$$T_{\alpha}^{*\alpha\gamma} = T_{\alpha}^{\ \alpha\gamma} - i(\tfrac{1}{2}n+1)c^{\gamma}.$$

Hence c^{γ} and d_{γ} can be determined to achieve (34).

With conditions (34) we have, from (33),

(37) $$\Phi_{\alpha}^{\ \alpha} = 0.$$

Equations (28) and (31) then give

(38) $$\omega^{\alpha} \wedge \Phi_{\alpha} : \omega_{\alpha} \wedge \Phi^{\alpha} : \omega \wedge \Psi = 1 : -1 : -2i.$$

This shows that Ψ is of the form

(39) $$\Psi \equiv Q_{\alpha}^{\ \beta}\omega^{\alpha} \wedge \omega_{\beta}, \quad \mathrm{mod}\,\omega,$$

and hence

(40) $$\omega^{\alpha} \wedge \Phi_{\alpha} = \tfrac{1}{2}i\omega \wedge Q_{\alpha}^{\ \beta}\omega^{\alpha} \wedge \omega_{\beta},$$

$$\omega_{\alpha} \wedge \Phi^{\alpha} = -\tfrac{1}{2}i\omega \wedge Q_{\alpha}^{\ \beta}\omega^{\alpha} \wedge \omega_{\beta}.$$

From (22), (32), and (40) it follows that, mod ω,

$$0 \equiv \omega^{\alpha} \wedge \Phi_{\alpha} \equiv \omega^{\alpha} \wedge \psi_{\alpha}^{\ \beta} \wedge \omega_{\beta} \equiv \omega^{\alpha} \wedge (R_{\alpha\ \gamma}^{\ \beta}\omega^{\gamma} + T_{\ \alpha}^{\beta\gamma}\omega_{\gamma}) \wedge \omega_{\beta}.$$

This relation implies (35), and hence also (36). This establishes Lemma 3.

From (36) and (40) we also get

$$\omega \wedge \omega^{\alpha} \wedge (\mu_{\alpha} + \tfrac{1}{2}iQ_{\alpha}^{\ \beta}\omega_{\beta}) = 0,$$

$$\omega \wedge \omega_{\beta} \wedge (\lambda^{\beta} + \tfrac{1}{2}iQ_{\alpha}^{\ \beta}\omega^{\alpha}) = 0.$$

We can therefore set

(41) $$\lambda^{\alpha} = -\tfrac{1}{2}iQ_{\beta}^{\ \alpha}\omega^{\beta} + L^{\alpha\beta}\omega_{\beta},$$

$$\mu_{\alpha} = P_{\alpha\beta}\omega^{\beta} - \tfrac{1}{2}iQ_{\alpha}^{\ \beta}\omega_{\beta},$$

with

(42) $$L^{\alpha\beta} = L^{\beta\alpha}, \; P_{\alpha\beta} = P_{\beta\alpha}\,.$$

Substituting into (36), we have the expressions

(43) $$\begin{aligned} \Phi^{\alpha} &= T_{\beta}{}^{\alpha\gamma}\omega^{\beta} \wedge \omega_{\gamma} - \tfrac{1}{2}iQ_{\beta}{}^{\alpha}\omega \wedge \omega^{\beta} + L^{\alpha\beta}\omega \wedge \omega_{\beta}\,, \\ \Phi_{\alpha} &= R_{\alpha}{}^{\beta}{}_{\gamma}\omega^{\gamma} \wedge \omega_{\beta} + P_{\alpha\beta}\omega \wedge \omega^{\beta} - \tfrac{1}{2}iQ_{\alpha}{}^{\beta}\omega \wedge \omega_{\beta}\,. \end{aligned}$$

By using the second and third equations of (19) we immediately get the lemma:

LEMMA 4. *The form ψ is completely determined by the condition*

(44) $$Q_{\alpha}{}^{\alpha} = 0\,.$$

To find the expression for Ψ, we write down the equation obtained by exterior differentiation of (29), which is

(45) $$d\Psi - \varphi \wedge \Psi + 2i\Phi^{\alpha} \wedge \varphi_{\alpha} - 2i\varphi^{\alpha} \wedge \Phi_{\alpha} = 0\,.$$

By (39) we set

(46) $$\Psi = Q_{\alpha}{}^{\beta}\omega^{\alpha} \wedge \omega_{\beta} + \omega \wedge \nu\,.$$

Substituting this into (45) and making use of (43), we get

$$\{dQ_{\alpha}{}^{\beta} - Q_{\varrho}{}^{\beta}\varphi_{\alpha}{}^{\varrho} + Q_{\alpha}{}^{\sigma}\varphi_{\sigma}{}^{\beta} - 2Q_{\alpha}{}^{\beta}\varphi + i\delta_{\alpha}{}^{\beta}\nu + 2iT_{\alpha}{}^{\sigma\beta}\varphi_{\sigma} - 2iR_{\sigma}{}^{\beta}{}_{\alpha}\varphi^{\sigma}\}\omega^{\alpha} \wedge \omega_{\beta} \equiv 0, \quad \operatorname{mod} \omega\,.$$

It follows that the expression between the braces is $\equiv 0 \operatorname{mod} \omega, \omega^{\varrho}, \omega_{\sigma}$. Contracting α, β and using (34), (35), (44), we conclude that ν is $\equiv 0$ $\operatorname{mod} \omega, \omega^{\alpha}, \omega_{\beta}$. We can therefore put

(47) $$\Psi = Q_{\alpha}{}^{\beta}\omega^{\alpha} \wedge \omega_{\beta} + H_{\alpha}\omega \wedge \omega^{\alpha} + K^{\alpha}\omega \wedge \omega_{\alpha}\,.$$

We summarize our results in the following theorem:

THEOREM. *Given in C_{n+1} an $(n+1)$-parameter family of hypersurfaces defined by the completely integrable differential system* (3), *there exist in the space of the variables* (11), *the same number of invariant differential forms*

(48) $$\omega, \omega^{\alpha}, \omega_{\alpha}, \varphi, \varphi_{\alpha}{}^{\beta}, \varphi^{\alpha}, \varphi_{\alpha}, \psi\,,$$

linearly independent, whose exterior derivatives are given by the "structure equations" (9), (12), (15), (16), (19), (29). *The "curvature forms" $\Phi_{\alpha}{}^{\beta}$, Φ^{α}, Φ_{α}, Ψ have expressions given by* (33), (43), (47), *whose coefficients satisfy* *the symmetry relations* (21), (34), (35), (42), (44). *The forms* (48) *are completely determined by these conditions.*

When $n=1$, $S_{\alpha\varrho}{}^{\beta\sigma}$, $R_{\alpha}{}^{\beta}{}_{\gamma}$, $T_{\alpha}{}^{\beta\gamma}$, $Q_{\alpha}{}^{\beta}$ all vanish and the lowest-order invariants are L^{11}, P_{11}.

2. Geometrical Construction.

The tangent space T_z at a point $z \in C_{n+1}$ is a complex vector space of dimension $n+1$. We consider it as a part of a projective space PT_z by adding to it a hyperplane at infinity. In turn PT_z is considered as the quotient space of $V_z^* = (V_{n+2} - \{0\})_z$ by the action on $V_{n+2} - \{0\}$ by the multiplication of a non-zero complex number, where V_{n+2} is the complex vector space of dimension $n+2$. To a point $\xi \in PT_z$ the components of the corresponding points of $V_{n+2} - \{0\}$, defined up to a non-zero factor, are called the homogeneous coordinates of ξ. In particular, we let a point $(y^1,\ldots,y^{n+1}) \in T_z$ to have the homogeneous coordinates $(y^1,\ldots,y^{n+1},1)$ and a vector $(v^1,\ldots,v^{n+1})$, which can be considered as the difference of two points, to have the homogeneous coordinates $(v^1,\ldots,v^{n+1},0)$.

A projective frame in PT_z consists of an ordered set $Z_0, Z_1, \ldots, Z_{n+1} \in V_z^*$, linearly independent and defined up to a common factor. On the other hand, a frame in T_z consists of the origin z and an ordered set of $n+1$ vectors. Using the above convention, to a frame in T_z corresponds a uniquely determined projective frame in PT_z.

In the discussion of the last section the forms ω, ω^α constitute a coframe at $z \in C^*_{n+1}$. They determine uniquely a dual frame, and hence a projective frame $Z_0, \ldots Z_{n+1}$ in PT_z. As in [3], p. 260, we put

$$(49)\quad \pi_0{}^{n+1} = 2\omega,\ \pi_0{}^{\alpha} = \omega^\alpha,\ \pi_{n+1}{}^{n+1} - \pi_0{}^0 = \varphi,\ \pi_\alpha{}^0 = -i\varphi_\alpha,\ \pi_\alpha{}^{n+1} = 2i\omega_\alpha,$$

$$\pi_{n+1}{}^{\alpha} = \tfrac{1}{2}\varphi^\alpha,\ \pi_\alpha{}^\beta - \delta_\alpha{}^\beta\pi_0{}^0 = \varphi_\alpha{}^\beta,\ \pi_{n+1}{}^0 = -\tfrac{1}{4}\psi .$$

Then it can be verified that the equations

$$(50)\qquad DZ_A = \pi_A{}^B Z_B,\quad 0 \leqq A,B \leqq n+1 ,$$

define a projective connection. Except in notation this is essentially the one defined by M. Hachtroudi [4].

Finally we wish to make a remark on the relation of this connection with the one defined in [3]. We have chosen the notations so that the structure equations are identically the same. This implies that the projective connection underlies the connection in [3].

BIBLIOGRAPHY

1. E. Cartan, *Sur la géométrie pseudo-conforme des hypersurfaces de deux variables complexes*, I. Ann. Math. Pura Appl. (4) 11 (1932), 17–90, (or Oeuvres II, 2, 1231–1304); II, Ann. Scuola Norm. Sup. Pisa (2) 1 (1932), 333–354 (or Oeuvres III, 2, 1217–1238).
2. S. S. Chern, *A generalization of the projective geometry of linear spaces*, Proc. Nat. Acad. Sci. USA, 29 (1943), 38–43.
3. S. S. Chern and J. K. Moser, *Real hypersurfaces in complex manifolds*, Acta Math. 133 (1974), 219–271.
4. M. Hachtroudi, *Les espaces d'éléments à connexion projective normale*, Hermann, Paris 1937.
5. B. Segre, I. *Intorno al problema di Poincaré della rappresentazione pseudo-conforme*, Rend. Acc. Lincei 13 (1931), 676–683; II. *Questioni geometriche legate colla teoria delle funzioni di due variabili complesse*, Rend. Semin. Mat. Roma 7 (1931).
6. Chih-ta Yen, *Sur la connexion projective normale associée à un feuilletage du deuxième ordre*, Ann. Math. Pura Appl. (4) 34 (1953). 55–94.

UNIVERSITY OF CALIFORNIA, BERKELEY, U.S.A.

Inventiones math. 35, 285–297 (1976)

Inventiones mathematicae

Duality Properties of Characteristic Forms

Shiing-shen Chern (Berkeley) and James White (Los Angeles)

To Jean-Pierre Serre

1. Introduction

This paper proves various statements and propositions concerning fiber bundles with a Lie group, connections, curvature forms, and characteristic classes, and in particular concerning the recent work of Chern-Simons on secondary characteristic classes [4] and of Cheeger-Simons on differential characters [2]. Our main interest in this work is not so much in the characteristic classes but in the differential forms which represent them. Indeed, the aim of this paper is to study the duality properties of the characteristic forms, both the primary ones and the secondary ones. We begin by reviewing some basic notions. For general references, the reader is referred to [3], [5], [6] and [7].

Let G be a Lie group with Lie algebra $\mathcal{G}$. The Maurer-Cartan form on G, denoted symbolically by $\alpha = ds \cdot s^{-1}$, $s \in G$, is a $\mathcal{G}$-valued right-invariant linear differential form satisfying the Maurer-Cartan equation

$$d\alpha = \tfrac{1}{2}[\alpha, \alpha].$$

The inner automorphism $s \to asa^{-1}$, s, $a \in G$, a fixed, leaves the unit element invariant and induces the adjoint endomorphism on $\mathcal{G}$, denoted by $\mathrm{ad}(a)$.

Let π: $P \to M$ be a principal G-bundle, where P and M are C^∞ manifolds, and π is a C^∞ mapping. A connection in P is a $\mathcal{G}$-valued one-form φ which restricts to the Maurer-Cartan form on every fiber and which satisfies the condition $\varphi(az) = \mathrm{ad}(a)\,\varphi(z)$, $z \in P$. The curvature form Φ of this connection is defined by the equation

$$\Phi = d\varphi - \tfrac{1}{2}[\varphi, \varphi],$$

so that Φ is a $\mathcal{G}$-valued two form on P [3].

From the curvature form Φ we can construct differential forms in the base M by considering complex-valued k-linear functions $F(X_1, \ldots, X_k)$, $X_i \in \mathcal{G}$, $1 \leq i \leq k$,

The first author supported in part by NSF Grant MPS 74-23180 and the second author by MPS 71-02597.

such that

$$F(\mathrm{ad}(a)X_1, \ldots, \mathrm{ad}(a)X_k) = F(X_1, \ldots, X_k), \qquad a \in G.$$

In contrast to [3] and [4] we will not always suppose F to be symmetric in its arguments. For brevity we will write

$$F(\Phi) = F(\overbrace{\Phi, \ldots, \Phi}^{k}).$$

Thus, $F(\Phi)$ is a form of degree $2k$ in M and is closed. In the sense of de Rham theory, it determines a cohomology class $\{F(\Phi)\} \in H^{2k}(M; R)$, a class which depends only on the bundle P and not on the choice of connection φ. However, in this paper we shall study directly the geometry of the forms $F(\Phi)$.

The second section of this paper is devoted to two theorems concerning $F(\Phi)$. Theorem 1 states that $F(\Phi)$ is an exact form in P. The proof differs slightly from the proof given in [4] in that $F(\Phi)$ is not assumed symmetric. Further, it is important for applications that the theorem actually provides the form whose exterior derivative is $F(\Phi)$. The theorem may be stated explicitly as follows. Define

$$\varphi_t = t\varphi$$

$$\Phi_t = d\varphi_t - \tfrac{1}{2}[\varphi_t, \varphi_t] = t\Phi + \tfrac{1}{2}(t - t^2)[\varphi, \varphi], \tag{1}$$

and

$$F_i = F(\Phi_t, \ldots, \Phi_t, \underset{\substack{\uparrow \\ i^{\text{th}}\ \text{place}}}{\varphi}, \Phi_t, \ldots, \Phi_t). \tag{2}$$

We introduce the "transgressed" or secondary characteristic form

$$TF(\varphi) = \int_0^1 \Big(\sum_{1 \le i \le k} F_i \Big)\, dt, \tag{3}$$

which is a form of degree $2k-1$ in P, depending on the connection φ.

Theorem 1. $dTF(\varphi) = F(\Phi)$.

Now the characteristic forms $F(\Phi)$ give rise to numerical invariants by integration. When M is a compact oriented manifold of dimension $2k$, we define the "characteristic number"

$$F[M] = F(\Phi)[M] = \int_M F(\Phi),$$

where $[M]$ is the fundamental class of M. Analogous to this, Simons proposed the idea of attaching a number to a compact oriented manifold M of dimension $2k-1$, somehow arising from the integration of $TF(\varphi)$. Since $TF(\varphi)$ is defined in P, some method must be used to pull it back to M. The particulars are as follows: By a theorem of Narashimhan-Ramanan [8], there is a commutative diagram

$$\begin{array}{ccc} P & \xrightarrow{\tilde{g}} & P_G \\ {\scriptstyle\pi}\downarrow & & \downarrow{\scriptstyle\pi_G} \\ M & \xrightarrow{g} & B_G, \end{array} \tag{4}$$

where the right-hand side is a universal bundle with a universal connection φ_G and curvature form Φ_G and where the connection of the left-hand side is induced by the mappings. For simplicity we assume the cohomology class $\{F(\Phi_G)\}$ is integral in B_G. Since B_G has no odd dimensional cohomology with rational coefficients, there is a $2k$-dimensional chain V, defined up to an integral cycle, such that

$$l g_*[M] = \partial V, \quad l \in Z.$$

The number

$$\frac{1}{l}\int_V F(\Phi_G)$$

is defined mod Z, and is called a *Simons character*; it will be denoted by

$$S_F[M] \in R/Z.$$

Theorem 2 of our paper states a relation of the transgressed form to the Simon character.

Theorem 2. *Suppose there exists* $Y \in H_{2k-1}(P; Z)$ *such that* $\pi_* Y = [M]$. *Then,* $S_F[M]$ *is the reduction* mod Z *of the integral* $\int_Y TF(\varphi)$.

The third and fourth sections of this paper are devoted to the duality properties of the primary and secondary characteristic forms associated with certain special bundles. We are interested in the case of a manifold M with two vector bundles E, $E^\perp$ whose Whitney sum $E \oplus E^\perp$ is trivial. On the latter there exists a flat connection given by the Maurer-Cartan form. This connection induces connections on E, $E^\perp$ respectively and on their associated principal bundles $\pi: P \to M$, $\pi^\perp: P^\perp \to M$. Thus, we are interested in the special principal bundles where $G = GL(n; C)$ or $GL(n; R)$ or where $G = SO(2m)$.

First, we consider the case where $G = GL(n; C)$ or $GL(n; R)$. Then $\mathscr{G}$ may be identified with the vector spaces of $(n \times n)$-matrices, respectively complex or real-valued, and we can write $\Phi = (\tilde{\Phi}^j_i)$. We define the characteristic forms:

$$c_k = \frac{1}{k!}\left(\frac{i}{2\pi}\right)^k \sum \delta \begin{pmatrix} i_1 & \dots & i_k \\ j_1 & \dots & j_k \end{pmatrix} \tilde{\Phi}^{j_1}_{i_1} \wedge \dots \wedge \tilde{\Phi}^{j_k}_{i_k} \tag{5}$$

$$b_k = \left(\frac{i}{2\pi}\right)^k \sum \tilde{\Phi}^{i_2}_{i_1} \wedge \tilde{\Phi}^{i_3}_{i_2} \wedge \dots \wedge \tilde{\Phi}^{i_1}_{i_k} = \left(\frac{i}{2\pi}\right)^k \mathrm{Tr}(\tilde{\Phi}^k), \tag{6}$$

where the symbol δ in the first equation is the Kronecker index which is equal to $+1$ or -1 according as whether $j_1 \dots j_k$ is an even or odd permutation of

$i_1 \ldots i_k$, and is otherwise zero, and where both sums are over $i_1, \ldots, i_k$ from 1 to n. Now, the b's and the c's are related by the identities:

$$b_k - b_{k-1} c_1 + b_{k-2} c_2 - \cdots + (-1)^{k-1} b_1 c_{k-1} + (-1)^k k c_k = 0, \quad 1 \leqq k \leqq n. \tag{7}$$

In fact, this can be recognized as the classical Newton identities when Φ is diagonal. The general case follows from supposing Φ to be in Jordan normal form. In both expressions (5) and (6) the numerical factors out front are chosen so that the cohomology classes $\{c_k\}$ and $\{b_k\}$ are both in $H^{2k}(M; Z)$.

In case $G = GL(n; R)$, we define the characteristic forms

$$p_k = (-1)^k c_{2k}, \quad q_k = (-1)^k b_{2k}, \quad 1 \leqq k \leqq \left[\frac{n}{4}\right]$$

which are real forms of degree $4k$.

In the theorems that follow we consider the following general situation. Let M be the Grassmann manifold $G(n, r, \mathbb{C})$ (or $G(n, r, R)$) of n-planes through the origin in Euclidean complex (or real) $(n+r)$-space. Let E, $E^\perp$ be the vector bundles of fiber dimensions n and r, such that at a point $x \in G(n, r, \mathbb{C})$ (or $G(n, r, R)$) E_x is the n-plane x and $E_x^\perp$ is the complementary r-plane.

More generally we may let M be a manifold of pairs of planes, n dimensional and r dimensional, passing through the origin and intersecting transversally. In fact, the manifold of all such pairs is a generalized Grassmann manifold which we denote by $\hat{G}(n, r, \mathbb{C})$ (or $\hat{G}(n, r, R)$). Then E and $E^\perp$ may be chosen analogously to the usual case.

Theorem 3. *Let E, $E^\perp \to M$ be two complex vector bundles of fiber dimensions n and r respectively, whose Whitney sum is trivial and has the connection given by the Maurer-Cartan form of $GL(n+r; C)$. Let b_k and $b_k^\perp$ be the characteristic forms of the associated principal bundles π: $P \to M$, $\pi^\perp$: $P^\perp \to M$ relative to the induced connections. Then $b_k + b_k^\perp = 0$, $1 \leqq k \leqq \max(n, r)$, where b_k or $b_k^\perp$ is set equal to zero when the subscript is larger than the fiber dimension.*

Remark. This is the Whitney duality theorem in terms of forms. It shows that under Whitney sum the b's behave more simply than the c's. Recall the usual duality theorem:

$$(1 + c_1 + c_2 + \cdots)(1 + c_1^\perp + c_2^\perp + \cdots) = 1.$$

This follows from the fact that $b_k + b_k^\perp = 0$ and Eq. (7).

Theorem 3 can be strengthened to a theorem on transgressed forms. There is a natural map

$$\psi\colon\ Q = \bigcup_{x \in M} \pi^{-1}(x) \times \pi^{\perp^{-1}}(x) \to GL(n+r; \mathbb{C}).$$

Define forms σ_k by

$$(-2\pi i)^k \binom{2k-1}{k} \sigma_k = \sum \tilde{\varphi}_{A_1}^{A_2} \wedge \tilde{\varphi}_{A_2}^{A_3} \wedge \cdots \wedge \tilde{\varphi}_{A_{2k-1}}^{A_1}$$
$$= \mathrm{Tr}(\tilde{\varphi}^{2k-1}), \quad 1 \leqq k \leqq n+r.$$

These forms are closed in $GL(n+r;\mathbb{C})$ and, when restricted to the subgroup $U(n+r)$, are the integral-valued primitive forms of degree $2k-1$.

Theorem 4a. *The form $Tb_k + Tb_k^{\perp}$ differs from $\psi^* \sigma_k$ by an exact form.*

Theorem 4a has a real counterpart. We assume that E, $E^{\perp}$ are real vector bundles and

$$\psi\colon Q \to GL(n+r;R).$$

Theorem 4b. *The form $\frac{1}{2}(Tq_k + Tq_k^{\perp})$ differs from $\psi^* \sigma_{2k}$ by an exact form.*

Remark. When restricted to the subgroup $O(n+r)$, σ_{2k} is the integral-valued primitive form of degree $4k-1$.

Clearly Theorem 3 follows from Theorem 4 by exterior differentiation.

There are two corollaries that follow immediately from Theorems 2 and 4. First, let M be a compact oriented manifold of dimension $2k-1$ over which there are two vector bundles E, $E^{\perp}$ of fiber dimensions n and r respectively, such that $E \oplus E^{\perp}$ is trivial and has the connection given by the Maurer-Cartan form of $GL(n+r;\mathbb{C})$. Then there is a natural map of M into $\hat{G}(n,r,\mathbb{C})$ and the information of Theorem 4 may be pulled-back to give

Corollary a. *Suppose there exist Y, $Y^{\perp}$ in $H_{2k-1}(P;Z)$, $H_{2k-1}(P^{\perp};Z)$ such that $\pi_* Y = [M]$, $\pi_*^{\perp} Y^{\perp} = [M]$. Then, $S_{b_k}[M] + S_{b_k^{\perp}}[M] = 0$, where $S_F[M]$ is the Simons character associated with F.*

The real counterpart of Corollary a) deals with a real manifold of dimension $4k-1$ over which are real bundles.

Corollary b. *Suppose there exists Y, $Y^{\perp}$ in $H_{4k-1}(P;Z)$, $H_{4k-1}(P^{\perp};Z)$ such that $\pi_* Y = [M]$, $\pi_*^{\perp} Y^{\perp} = [M]$. Then $\frac{1}{2}(S_{q_k}[M] + S_{q_k^{\perp}}[M]) = 0$.*

Section 4 discusses the case where $G = SO(2m)$. Then, $\mathscr{G}$ can be identified with the vector space of $(2m \times 2m)$ anti-symmetric matrices, and we can write $\Phi = (\Phi_{jk})$, with $\Phi_{jk} + \Phi_{kj} = 0$. We define the pfaffian or Euler form

$$\chi = \frac{(-1)^m}{2^{2m}\pi^m m!} \sum \delta_{i_1 \ldots i_{2m}} \Phi_{i_1 i_2} \wedge \cdots \wedge \Phi_{i_{2m-1} i_{2m}},$$

where $\delta_{i_1 \ldots i_{2m}} = \delta \begin{pmatrix} 1 \ldots 2m \\ i_1 \ldots i_{2m} \end{pmatrix}$. In this case, we let M be the Grassmann manifold $\tilde{G}(2n, 2r)$ of oriented $2n$-planes through the origin of an oriented Euclidean $(2n+2r)$-space R_{2n+2r}, endowed with the standard inner product, and let E, $E^{\perp}$ be respectively the complementary $2n$-plane and $2r$-plane bundles over M. Further, we define E_0 and $E_0^{\perp}$ to be the unit vector bundles arising naturally from E, $E^{\perp}$.

Theorem 5. *Let E, $E^{\perp} \to M$ be two real vector bundles of fiber dimensions $2n$ and $2r$ respectively, whose Whitney sum is trivial (cf. above). Let χ and $\chi^{\perp}$ be the Euler forms of the vector (unit vector) bundles $\pi\colon E(E_0) \to M$, $\pi^{\perp}\colon E^{\perp}(E_0^{\perp}) \to M$. Then $\chi \wedge \chi^{\perp} = 0$.*

Remark. One may think of $\chi \wedge \chi^{\perp}$ as a form on M and hence on $E \oplus E^{\perp}$.

As with Theorem 3, Theorem 5 may be strengthened to a theorem on transgressed forms. However, the flavor of the Euler form duality is substantially different from the Chern form duality in that one does not need the full frame bundle to find the form $T\chi$ whose exterior derivative is χ. (Of course, one can find $T\chi$ on the full frame bundle, using Theorem 1.) Indeed, using the generalized Gauss-Bonnet theorem one may find a form $T_1\chi$ in E_0 such that $dT_1\chi=\chi$. For a complete definition of $T_1\chi$, cf. Section 4. Thus, the Euler duality theorem is a statement on the vector-bundle level and proceeds as follows.

Now, $\chi^{\perp}$ is a form in M and, as we have stated $T_1\chi$ is a form in E_0, so that we may consider the $(2n+2r-1)$-form $T_1\chi\wedge\chi^{\perp}$ as a form in E_0. We define a map $e: E_0\to S^{2n+2r-1}$, the unit sphere in R_{2n+2r}, by assigning to each unit vector in E_0 its translate to the origin in R_{2n+2r}. Then, if $[S]$ is the fundamental form of $S^{2n+2r-1}$, $e^*[S]$ is a form in E_0. Our duality theorem for the "transgressed" Euler form is

Theorem 6. *The form $T_1\chi\wedge\chi^{\perp}$ differs from $c_0e^*[S]$ by an exact form where c_0 is a constant depending only on n and r.*

Remarks. Theorem 5 follows from Theorem 6 by exterior differentiation. c_0 is given explicitly in Section 4.

We conclude the introduction by remarking that we have stated our theorems in the most general manner, choosing the base manifold M to be Grassmann manifolds. For applications, however, one usually studies immersions into complex (or real) Euclidean space of complex (or real) dimension $n+r$. Then, M is an immersed manifold, E, $E^{\perp}$ are vector-bundles of dimension n and r whose Whitney sum is trivial in the sense that for each $x\in M$, E_x is an n-plane in the Euclidean space, $E_x^{\perp}$ is the complementary r-plane, and $(E\oplus E^{\perp})_x$ is the entire Euclidean space. One may apply the theorems in this case by using the Gauss map of M into the Grassmann manifold and pulling back the statements of the theorems. Perhaps the most important case to consider is that in which M is n-dimensional, E is the tangent bundle, and $E^{\perp}$ the normal bundle. A series of applications will appear in a forthcoming paper by the authors.

2. Proofs of Theorems 1 and 2

We proceed now to the proofs of Theorems 1 and 2. The proof of Theorem 1 is analogous to that in [3] where F is assumed symmetric. We first note that

$$\frac{d}{dt}F(\Phi_t)=\sum_i F(\Phi_t,\ldots,\Phi_t,\underbrace{d\varphi-[\varphi_t,\varphi]}_{i^{\text{th}}\text{ place}},\Phi_t,\ldots,\Phi_t).$$

Exterior differentiation of (1) yields

$$d\Phi_t=-[\Phi_t,\varphi_t]=[\varphi_t,\Phi_t]. \tag{8}$$

Since F is invariant under the action of the adjoint map, the infinitesimal form of Eq. (2) gives, for every i,

$$\sum_{j<i} F(\Phi_t, \ldots, \underset{j^{\text{th}}}{\Phi_t, [\varphi_t, \Phi_t]}, \Phi_t, \ldots, \underset{i^{\text{th}}}{\Phi_t, \varphi}, \Phi_t, \ldots, \Phi_t)$$

$$-\sum_{i<j} F(\Phi_t, \ldots, \underset{i^{\text{th}}}{\Phi_t, \varphi}, \Phi_t, \ldots, \underset{j^{\text{th}}}{\Phi_t, [\varphi_t, \Phi_t]}, \Phi_t, \ldots, \Phi_t)$$

$$+F(\Phi_t, \ldots, \underset{i^{\text{th}}}{\Phi_t, [\varphi_t, \varphi]}, \Phi_t, \ldots, \Phi_t)=0.$$

By (2) and (8) it follows that

$$dF_i = F(\Phi_t, \ldots, \Phi_t, d\varphi - [\varphi_t, \varphi], \Phi_t, \ldots, \Phi_t),$$

so that

$$\frac{d}{dt} F(\Phi_t) = d\Big(\sum_{1\leq i\leq k} F_i\Big).$$

Theorem 1 follows by integrating this equation with respect to t from 0 to 1.

To prove Theorem 2 we use the universal bundle and commutative diagram (4). First,

$$\int_Y TF(\varphi) = \int_Y \tilde{g}^* TF(\varphi_G) = \int_{\tilde{g}_* Y} TF(\varphi_G).$$

Since $H_{2k-1}(P_G; Z)=0$, there is an integral chain W such that $\partial W = \tilde{g}_* Y$. Then the above integral is equal to

$$\int_W dTF(\varphi_G) = \int_W F(\Phi_G) = \int_{(\pi_G)_* W} F(\Phi_G)$$

since $F(\Phi_G)$ is a form in B_G. If $\pi_* Y = [M]$, as in the hypothesis of Theorem 2, we have

$$\partial(\pi_G)_* W = (\pi_G)_* \partial W = (\pi_G)_* \tilde{g}_* Y = g_* \pi_* Y = g_* [M].$$

This shows that the integral $\int_Y TF(\varphi)$, when reduced mod Z, is equal to Simons character $S_F[M]$.

On the other hand, the above argument also shows that, if $\pi_* Y=0$, $(\pi_G)_* W$ is a cycle and, $F(\Phi_G)$ being an integral class, $\int_Y TF(\varphi)$ is an integer. The hypothesis of Theorem 2 is fulfilled if the bundle has a section $s\colon M \to P$, such that $\pi \circ s =$ identity. It suffices to take $Y = s_* [M]$.

3. Duality Properties of Chern and Pontryagin Forms

In this section we prove Theorems 3 and 4. Let C_{n+r} denote the complex number space of dimension $n+r$. A frame in C_{n+r} is an ordered set of $n+r$ linear independent vectors $e_1, \ldots, e_{n+r}$. The space of all frames can be identified with $GL(n+r; \mathbb{C})$.

If we write

$$de_A = \sum_B \tilde{\varphi}_A^B e_B, \qquad 1 \leq A, B, C \leq n+r,$$

the matrix $\tilde{\varphi} = (\tilde{\varphi}_A^B)$ is the Maurer-Cartan form of $GL(n+r; \mathbb{C})$. Its Maurer-Cartan equations are

$$d\tilde{\varphi}_A^B = \sum_C \tilde{\varphi}_A^C \wedge \tilde{\varphi}_C^B. \tag{9}$$

If C_{n+r} is endowed with a positive definite hermitian form, we can speak of unitary frames. The space of all unitary frames can be identified with $U(n+r)$. Let j: $U(n+r) \to GL(n+r; \mathbb{C})$ be the obvious inclusion. Then $j^* \tilde{\varphi}$ is skew-hermitian, i.e. $j^{*t} \tilde{\varphi} + j^* \bar{\tilde{\varphi}} = 0$.

In $GL(n+r; \mathbb{C})$, the forms

$$\sigma_k = \left(\frac{i}{2\pi}\right)^k \frac{k!(k-1)!}{(2k-1)!} \operatorname{Tr}(\tilde{\varphi}^{2k-1}), \qquad 1 \leq k \leq n+r$$

are well defined closed forms of degree $2k-1$. According to the work of Hopf-Samelson, $j^* \sigma_k$ is primitive in $U(n+r)$, and, furthermore, the coefficient is chosen so that the form is integral.

Now suppose M is the Grassmann manifold $\hat{G}(n, r, C)$. Let $E, E^\perp$ be as in Theorem 3 with associated principal bundles $P, P^\perp$. Thus, if $x \in M$, a frame of $P_x = \pi^{-1}(x)$ and a frame of $P_x^\perp = \pi^{\perp -1}(x)$ combine to form a frame in C_{n+r}. Hence, there is a mapping

$$\psi\colon Q = \bigcup_{x \in M} P_x \times P_x^\perp \to GL(n+r; \mathbb{C}),$$

and there are obvious projections p: $Q \to P$, $p^\perp$: $Q \to P^\perp$ by taking respectively the first n and last r vectors of a frame. We set

$$\varphi_A^B = \psi^* \tilde{\varphi}_A^B.$$

Then the structure equations (9) yield

$$d\varphi_A^B = \sum_C \varphi_A^C \wedge \varphi_C^B.$$

In particular, letting $1 \leq i, j, k \leq n$, $n+1 \leq \alpha, \beta, \gamma \leq n+r$, we have

$$d\varphi_i^k = \sum_j \varphi_i^j \wedge \varphi_j^k + \Phi_i^k$$

$$d\varphi_\alpha^\beta = \sum_\gamma \varphi_\alpha^\gamma \wedge \varphi_\gamma^\beta + \Phi_\alpha^\beta,$$

where

$$\Phi_i^k = \sum_\alpha \varphi_i^\alpha \wedge \varphi_\alpha^k, \qquad \Phi_\alpha^\beta = \sum_j \varphi_\alpha^j \wedge \varphi_j^\beta.$$

The forms φ_i^k, Φ_i^k can be considered as forms in P; in other words, there are uniquely determined forms in P which have them as their images under p^*. They are the connection and curvature forms of an induced connection in P. Similarly the forms φ_α^β define a connection in $E^\perp$ with Φ_α^β as the curvature forms.

By definition

$$b_k(P)=b_k=\left(\frac{i}{2\pi}\right)^k \sum \Phi_{i_1}^{i_2}\wedge\Phi_{i_2}^{i_3}\wedge\cdots\wedge\Phi_{i_k}^{i_1}$$

$$=\left(\frac{i}{2\pi}\right)^k \sum \varphi_{i_1}^{\alpha_1}\wedge\varphi_{\alpha_1}^{i_2}\wedge\varphi_{i_2}^{\alpha_2}\wedge\varphi_{\alpha_2}^{i_3}\wedge\cdots\wedge\varphi_{i_k}^{\alpha_k}\wedge\varphi_{\alpha_k}^{i_1},$$

$$b_k(P^\perp)=b_k^\perp=\left(\frac{i}{2\pi}\right)^k \sum \Phi_{\alpha_1}^{\alpha_2}\wedge\Phi_{\alpha_2}^{\alpha_3}\wedge\cdots\wedge\Phi_{\alpha_k}^{\alpha_1}$$

$$=\left(\frac{i}{2\pi}\right)^k \sum \varphi_{\alpha_1}^{i_1}\wedge\varphi_{i_1}^{\alpha_2}\wedge\varphi_{\alpha_2}^{i_2}\wedge\varphi_{i_2}^{\alpha_3}\wedge\cdots\wedge\varphi_{\alpha_k}^{i_k}\wedge\varphi_{i_k}^{\alpha_1}.$$

By moving the final one-form $\varphi_{i_k}^{\alpha_1}$ to the beginning in each term of the sum in $b_k^\perp$, we see immediately that $b_k+b_k^\perp=0$. This proves Theorem 3.

Remark. By interpreting c_k as the kth elementary symmetric function of the formal eigenvalues of $\frac{i}{2\pi}\Phi$ and b_k as the sum of the kth powers of these eigenvalues, it is immediate that the class of $b_k+b_k^\perp$ is zero. The formula of Theorem 3 is, however, non-trivial, because $E\oplus E^\perp$ is not endowed with the sum connection.

To prove Theorem 4a, we first observe that when $G=GL(n;\mathbb{C})$ or $GL(n;R)$, we have

$$\Phi=d\varphi-\varphi\wedge\varphi$$

$$\Phi_t=t\{(1-t)\,\varphi\wedge\varphi+\Phi\}$$

where φ and Φ are $(n\times n)$-matrices. Hence by (3) and (6)

$$Tb_k=\left(\frac{i}{2\pi}\right)^k k\int_0^1 t^{k-1}\,\mathrm{Tr}\,[\varphi\wedge\overbrace{\{(1-t)\,\varphi\wedge\varphi+\Phi\}\wedge\cdots\wedge\{(1-t)\,\varphi\wedge\varphi+\Phi\}}^{k-1}]\,dt.$$

In particular, we find that $\mathrm{Tr}\,[\overbrace{\varphi\wedge\cdots\wedge\varphi}^{2k-1}]$ has the coefficient

$$\left(\frac{i}{2\pi}\right)^k k\int_0^1 t^{k-1}(1-t)^{k-1}\,dt=\frac{k!(k-1)!}{(2k-1)!}\left(\frac{i}{2\pi}\right)^k.$$

To prove Theorem 4a, we observe that the relation to be proved is a relation in φ_A^B. The corresponding relation in $\tilde\varphi_A^B$ pertains to the group $GL(n+r;\mathbb{C})$. This will be proved and pulled-back by the map ψ^*. Thus, we must show that as forms in $GL(n+r;\mathbb{C})$

$$Tb_k+Tb_k^\perp\sim\sigma_k. \tag{10}$$

First, we will prove this restricting to the unitary group $U(n+r)$. The two sides of (10) are closed forms: the right-hand side is closed because it is a primitive form and the left-hand side is closed by Theorem 3. To prove (10), it suffices, therefore, to show that the two sides, when evaluated over a $(2k-1)$ dimensional homology basis of $U(n+r)$, are equal. The latter consists of the $(2k-1)$ dimensional cycles in the Pontryagin product [9]

$$(e\cup A_1)\cdot(e\cup A_2)\dots(e\cup A_k),$$

where A_i is a cycle of dimension $2i-1$ acting on an i-dimensional space. There are two possibilities. a) The cycle is decomposable, i.e. it can be written in the form $C=A_{i_1}\circ\dots\circ A_{i_m}$ (Pontryagin product $m\geqq 2$) $i_1\leqq i_2\leqq\dots\leqq i_m$. If this is the case, set $A_{i_1}\circ\dots\circ A_{i_{m-1}}=B$. We may suppose that A_{i_m} and B act in orthogonal subspaces of dimensions i_m and $n+r-i_m$ respectively. Then the Maurer-Cartan form on C appears as

$$\tilde{\varphi}=\underbrace{\begin{pmatrix}\tilde{\varphi}_1 & 0\\ 0 & \tilde{\varphi}_2\end{pmatrix}}_{i_m}\}i_m.$$

Since the two sides of (10) involve cyclic summations of the indices, we easily see that both are zero when restricted to the cycle C. b) The cycle is A_k, i.e. the cycle is not decomposable. Since the bundles E and $E^\perp$ are interchangeable, we suppose without loss of generality that $n\geqq r$. Let A_k act in the n-dimensional space spanned by the first n-vectors. Then, the last r vectors being fixed, we have

$$\tilde{\varphi}^A_\alpha=\tilde{\varphi}^\alpha_A=0,\quad n+1\leqq\alpha\leqq n+r$$
$$1\leqq A\leqq n+r.$$

Hence, $\tilde{\Phi}^j_i=0$ for all pairs i,j. This gives

$$Tb_k=\left(\frac{i}{2\pi}\right)^k\frac{k!(k-1)!}{(2k-1)!}\sum\tilde{\varphi}^{j_2}_{j_1}\wedge\dots\wedge\tilde{\varphi}^{j_1}_{j_{2k-1}}.$$

$$Tb^\perp_k=0.$$

Hence, the two sides of (10) are equal when restricted to A_k. Combining cases a) and b) we have proved that the difference of the two sides of (10) is an exact form $d\Theta$, where Θ is a form of degree $2k-2$ in $U(n+r)$.

Now the forms on the two sides of (10) are invariant under right translations by $U(n+r)$ and left translations by the subgroup

$$\begin{pmatrix}U(n) & 0\\ 0 & U(r)\end{pmatrix}.$$

Since both are compact groups, by applying integrations if necessary, we can suppose that Θ is also invariant under such actions. This means that Θ is an exterior polynomial in $j^*\tilde{\varphi}^B_A$ with constant coefficients and is invariant under left translations by the above subgroup. By the first main theorem on vector invariants it follows that the expressions Tb_k, $Tb^\perp_k$, Θ, when considered in $GL(n+r;\mathbb{C})$, are exterior polynomials in $\tilde{\varphi}^B_A$ with constant coefficients and are invariant under left translations of the group

$$\begin{pmatrix}GL(n;\mathbb{C}) & 0\\ 0 & GL(r;\mathbb{C})\end{pmatrix}.$$

Then equation

$$Tb_k+Tb^\perp_k-\tilde{\sigma}_k=d\Theta$$

is a formal consequence of the Maurer-Cartan equation and is hence valid in $GL(n+r;\mathbb{C})$. Theorem 4a follows immediately.

Now we consider the case of real vector bundles $E, E^{\perp}\to\hat{G}(n,r,R)$ such that their Whitney sum is trivial and is endowed with the connection from projection into $GL(n+r,R)$. Let $\tau(E)$ and $\tau(E^{\perp})$ denote the complexifications of E and $E^{\perp}$. It will be consistent to use τ also for the injections

$$\tau\colon GL(n+r;R)\to GL(n+r;\mathbb{C})$$

$$\tau\colon O(n+r)\to U(n+r).$$

By [4], one can show that $\tau^*\sigma_k$ is cohomologous to zero if k is odd and is twice the primitive class of $O(n+r)$ if k is even. The latter result gives

$$\tfrac{1}{2}\{Tq_k+Tq_k^{\perp}\}$$

is cohomologous to $\psi^*\sigma_{2k}$, where σ_{2k} is the $(4k-1)$-dimensional integral primitive class of $O(n+r)$. Theorem 4b follows immediately.

4. Duality Properties of the Euler Form

In this section we prove Theorems 5 and 6. In actuality Theorem 5 is well-known, for since $E\oplus E^{\perp}$ is trivial

$$0=\chi(E\oplus E^{\perp})=\chi(E)\cup\chi^{\perp}(E^{\perp}),$$

and hence on the class level, the theorem is true. It follows on the form level since the Grassmann manifold is a symmetric riemannian homogeneous manifold. Thus, we turn our attention to Theorem 6.

Let R_{2n+2r} be the real Euclidean space of dimension $2n+2r$, endowed with the fixed orientation and the natural inner product. An orthonormal frame in R_{2n+2r} is an ordered set of $2n+2r$ orthonormal vectors $e_1,\dots,e_{2n+2r}$. If we write $de_A=\sum_B\varphi_{AB}e_B$, $1\leqq A,B,C\leqq 2n+2r$, where $\varphi_{AB}+\varphi_{BA}=0$, then the Maurer-Cartan equations are

$$d\varphi_{AB}=\sum_C\varphi_{AC}\wedge\varphi_{CB}.$$

If we assume that the first $2n$-vectors of a frame span E_x and the last $2r$-vectors span $E_x^{\perp}$ for every $x\in M$, we may write the Euler forms of E and $E^{\perp}$ (or E_0 and $E_0^{\perp}$) respectively as

$$\chi=\frac{(-1)^n}{2^{2n}\pi^n n!}\sum\delta_{i_1\dots i_{2n}}\Phi_{i_1i_2}\wedge\cdots\wedge\Phi_{i_{2n-1}i_{2n}}$$

$$\chi^{\perp}=\frac{(-1)^r}{2^{2r}\pi^r r!}\sum\delta_{\alpha_1\dots\alpha_{2r}}\Phi_{\alpha_1\alpha_2}\wedge\cdots\wedge\Phi_{\alpha_{2r-1}\alpha_{2r}}$$

where $1\leqq i,j\leqq 2n$, $2n+1\leqq\alpha,\beta\leqq 2n+2r$, and where

$$\Phi_{ij}=\sum_\alpha\varphi_{i\alpha}\wedge\varphi_{\alpha j}$$

$$\Phi_{\alpha\beta}=\sum_i\varphi_{\alpha i}\wedge\varphi_{i\beta}.$$

To prove Theorem 6, we first write the two forms $T_1\chi\wedge\chi^\perp$ and $e^*[S]$ in terms of the forms φ and Φ. Recall that $e\colon E_0\to S^{2n+2r-1}$ assigns to each unit vector in E_0 its translate to the origin. If we choose e_1 along this vector, we may write

$$e^*[S]=\frac{1}{O_{2n+2r-1}}\varphi_{12}\wedge\cdots\wedge\varphi_{1\,2n+2r}$$

where $O_{2n+2r-1}$ is the volume of the $S^{2n+2r-1}$.

On the other hand $T_1\chi$ may be defined via the generalized Gauss-Bonnet Theorem as follows: let

$$\Phi_k=\delta_{i_2\ldots i_{2n}}\varphi_{1i_2}\wedge\cdots\wedge\varphi_{1i_{2n-2k}}\wedge\Phi_{i_{2n-2k+1}i_{2n-2k+2}}\wedge\cdots\wedge\Phi_{i_{2n-1}i_{2n}}$$

where $\delta_{i_2\ldots i_{2n}}$ is $+1$ or -1 according as whether $i_2\ldots i_{2n}$ is an even or odd permutation of $2,\ldots,2n$ and is 0 otherwise. Then $T_1\chi$ may be written

$$T_1\chi=\frac{1}{\pi^n}\sum_{k=0}^{n-1}(-1)^{k+1}\frac{1}{1\cdot 3\ldots(2n-2k-1)\,2^{n+k}k!}\Phi_k.$$

Hence, as one can see, $T_1\chi\wedge\chi^\perp$ is a rather complicated expression in the φ's. However, we will show that $T_1\chi\wedge\chi^\perp$ and $e^*[S]$ differ by an exact form, by showing that they are equal when evaluated on a homology basis of $H_{2n+2r-1}(E_0)$. In so doing we will be able to reduce $T_1\chi\wedge\chi^\perp$ to a very simple form.

In order to find a homology basis of $H_{2n+2r-1}(E_0)$, we use the Gysin sequence:

$$\cdots\to H_{2n+2r}(M)\to H_{2r}(M)\to H_{2n+2r-1}(E_0)\to H_{2n+2r-1}(M)\to\cdots$$

and its dual:

$$\cdots\to H^{2n+2r-1}(M)\to H^{2n+2r-1}(E_0)\to H^{2r}(M)\xrightarrow{\cup\chi}H^{2n+2r}(M)\to\cdots$$

where the last map in the cohomology sequence is cup product with χ the Euler class of E_0. By [1], $H_{2n+2r-1}(M)=0$; indeed, all odd dimensional cohomology of $\tilde{G}(2n,2r)$ vanishes. Further, the kernel of the map

$$H^{2r}(M)\xrightarrow{\cup\chi}H^{2n+2r}(M)$$

consists precisely of the multiples of $\chi^\perp$. Hence, it follows using the homology sequence, that $H_{2n+2r-1}(E_0)$ has a basis consisting of the image of the homology dual of $\chi^\perp$ under the map

$$H_{2r}(M)\to H_{2n+2r-1}(E_0),$$

where by the homology dual we mean that class on which $\chi^\perp$ evaluated yields 1. As a Schubert manifold this class may be represented by the oriented $2n$-planes of $\tilde{G}(2n,2r)$ which contain a fixed $(2n-1)$-plane.

If we choose e_1 in the $2n$-plane orthogonal to the fixed $(2n-1)$-plane, $\varphi_{iA}=0$ for $2\leqq i\leqq 2n$, $2\leqq A\leqq 2n+2r$. Hence, the form $T_1\chi\wedge\chi^\perp$ reduces to a constant multiple c_0 of $\varphi_{12}\wedge\cdots\wedge\varphi_{1\,2n+2r}$. Thus, on a basis of $H_{2n+2r-1}(E_0)$, $T_1\chi\wedge\chi^\perp$ and $c_0\,e^*[S]$ are equal. This proves Theorem 6.

Finally, to compute c_0 we need only find the coefficient of $\varphi_{12}\wedge\cdots\wedge\varphi_{1\,2n+2r}$ in the expression $T_1\chi\wedge\chi^\perp$. A simple computation yields this to be

$$\frac{-(2n-1)!(2r!)}{2^{n+2r}\pi^{n+r}r!}.$$

Hence

$$c_0 = \frac{-(2n-1)!(2r)!}{2^{n+2r}\pi^{n+r}r!} \cdot O_{2n+2r-1}.$$

References

1. Borel, A.: Sur la cohomologie des espaces fibrés principaux et des espaces homogènes de groupes de Lie compacts. Ann. Math. **57**, 115–207 (1953)
2. Cheeger, J., Simons, J.: Differential characters and characteristic invariants, mimeographed lecture notes. A.M.S. Summer Institute, Stanford 1973
3. Chern, S.S.: Geometry of characteristic classes. Proc. of 13th Biennial Seminar. Canadian Math. Congress 1–40 (1972)
4. Chern, S.S., Simons, J.: Characteristic forms and geometric invariants. Ann. Math. **99**, 48–69 (1974)
5. Kobayashi, S., Nomizu, K.: Foundations of differential geometry, Vol. I, II. New York: Interscience 1969
6. Koszul, J.L.: Travaux de S.S. Chern et J. Simons sur les classes caractéristiques. Seminaire Bourbaki. No. 440 1973
7. Milnor, J., Stasheff, J.D.: Characteristic classes. Ann. math. Studies, No. 76. Princeton 1974
8. Narasimham, M.S., Ramanan, S.: Existence of universal connections. Amer. J. Math. **83**, 563–572 (1961); **85**, 223–231 (1963)
9. Pontryagin, L.: Homologies in compact Lie groups. Mat. Sbornik. **6**(48), 389–422 (1939)

Shiing-shen Chern
University of California, Berkeley
Department of Mathematics
Berkeley, California 54720
USA

James White
University of California, Los Angeles
Department of Mathematics
Los Angeles, California 90024
USA

Received October 27, 1975

On Projective Connections and Projective Relativity†

SHIING-SHEN CHERN
Department of Mathematics, University of California, Berkeley, California 94720

The theory of projective connections was first introduced by Elie Cartan in 1924.[1,2] It is a generalization of classical projective geometry in that the lines are replaced by the integral curves of a system of ordinary differential equations of the second order, which are the paths of a symmetric affine connection. A fundamental theorem says that the paths, unparametrized, determine uniquely a normal projective connection (cf. below). At about the same time the projective geometry of paths was developed by the Princeton school under O. Veblen and T. Y. Thomas.[5,6] Its emphasis was to study the effect of projective changes of the affine connection and to make use of projective normal coordinates, whose definition and existence posed a serious problem. In his monograph[7] Veblen developed the projective theory of relativity, which incorporated the unified field theory of Einstein–Mayer and the five-dimensional theory of Kaluza–Klein. Projective concepts should enter naturally in a unified field thoery, because the primary interest in dynamics is the motion of a charged particle and the rôle of the trajectories is undoubtedly fundamental.

This paper aims at a review of projective connections in view of recent developments in differential geometry. Our attention will be on an affine tensor, which behaves in a simple manner under projective transformations. We will discuss, from the mathematical standpoint, its possible significance in a projective theory of relativity. No metric will be used.

Fifty years ago Ta-You and I were students in the same class on theoretical mechanics under Professor Yu-Tai Yao of Nankai University, Tientsin, China. May the following pages, devoted to a topic not unrelated to our early common interest, be a token expression of our long friendship and scientific activity.

† Work done under partial support of NSF grant MCS 74-23180.

1 AFFINE CONNECTIONS[3]

Let M be a C^∞-manifold and

$$\pi: E \to M \tag{1}$$

be a real vector bundle of fiber dimension $n + 1$. A frame is an ordered set of $n + 1$ linearly independent vectors of the same fiber. To define a connection in the bundle we take a neighborhood U of M and a frame field $A_0(x), \ldots, A_n(x)$, $x \in U$, in U. (In this paper all functions are smooth.) Then the connection is given by the "infinitesimal displacement"

$$DA_\alpha = \sum_\beta \omega_\alpha^\beta A_\beta, \tag{2}$$

where ω_α^β are linear differential forms in U and the indices run over the range

$$0 \leqq \alpha, \beta, \gamma, \ldots, \leqq n. \tag{3}$$

The curvature of the connection is given by the two-forms

$$\Omega_\alpha^\beta = \mathrm{d}\omega_\alpha^\beta - \sum_\gamma \omega_\alpha^\gamma \wedge \omega_\gamma^\beta. \tag{4}$$

It will be convenient to introduce the matrices

$$A = \begin{pmatrix} A_0 \\ \vdots \\ A_n \end{pmatrix}, \quad \omega = (\omega_\alpha^\beta), \quad \Omega = (\Omega_\alpha^\beta), \tag{5}$$

and write Eqs. (2) and (4) in the matrix form

$$DA = \omega A, \tag{2a}$$

$$\Omega = \mathrm{d}\omega - \omega \wedge \omega. \tag{4a}$$

These considerations hold in the neighborhood U of M and depend on the choice of the frame field A. It is of fundamental importance to study the effect of a change of the frame field. The latter is given by

$$A^* = gA, \tag{6}$$

where

$$g = (g_\alpha^\beta), \quad \det g \neq 0, \tag{7}$$

is a matrix of functions. In terms of the new frame field A^* let

$$DA^* = \omega^* A^* \tag{8}$$

be the infinitesimal displacement. Differentiating (6), we have

$$DA^* = \mathrm{d}gA + g\omega A = (\mathrm{d}gg^{-1} + g\omega g^{-1})A^*,$$

so that

$$\mathrm{d}g + g\omega = \omega^* g. \tag{9}$$

Taking the exterior derivative of this equation, we get

$$g\Omega = \Omega^* g. \tag{10}$$

Eqs. (9) and (10) are of fundamental importance in the study of affine connections. They give the behavior of the *connection matrix* ω and the *curvature matrix* Ω under a change of the frame field.

2 AFFINE CONNECTIONS AND PROJECTIVE CONNECTIONS

From the bundle (1) we omit the zero section and identify the points of each fiber which differ from each other by a factor. The result is a bundle PE of projective spaces:

$$E - \{0\} \to PE \to M, \tag{11}$$

whose fibers are projective spaces of dimension n. From the point of view of group theory the group of the bundle E is the general linear group $GL(n+1; R)$ in $n+1$ real variables and that of PE is $PL(n+1; R) = GL(n+1; R)/\text{center}$. A PL-frame, or projective frame, is a class of frames A_α defined up to the transformation

$$A_\alpha \to A_\alpha^* = \lambda A_\alpha, \quad \lambda \neq 0. \tag{12}$$

When the connection ω is given in E, the transformation (12) has the effect

$$\omega_\alpha^{*\beta} = \omega_\alpha^\beta + \frac{\mathrm{d}\lambda}{\lambda}\delta_\alpha^\beta. \tag{13}$$

It follows that the forms

$$\omega_\alpha^\beta - \delta_\alpha^\beta \omega_0^0, \quad (\alpha, \beta) \neq (0, 0) \tag{14}$$

are invariant under (12). They are said to define a *projective connection* in PE. This definition agrees with the general notion of a connection in Ref. 3.

By (4), the forms in (14) have their exterior derivatives given by

$$\mathrm{d}(\omega_\alpha^\beta - \delta_\alpha^\beta\omega_0^0) = \sum_\gamma (\omega_\alpha^\gamma - \delta_\alpha^\gamma\omega_0^0) \wedge (\omega_\gamma^\beta - \delta_\gamma^\beta\,\omega_0^0) - \delta_\alpha^\beta \sum_k \omega_0^k \wedge \omega_n^0 + \Pi_\alpha^\beta, \tag{15}$$

where, as in the future, small Latin indices have the range

$$1 \leqq i, j, k, \ldots, \leqq n, \tag{16}$$

and

$$\Pi_\alpha^\beta = \Omega_\alpha^\beta - \delta_\alpha^\beta \Omega_0^0 \tag{17}$$

define the *curvature of the projective connection.*

3 PROJECTIVE CONNECTIONS IN THE TANGENT BUNDLE

Let M be a C^∞-manifold of dimension n. All its tangent spaces T_x, $x \in M$, form the tangent bundle TM of M. If x^i are local coordinates at x, T_x is centered at x and spanned by the vectors $\partial/\partial x^i$. We extend T_x into a projective space ΠT_x, of dimension n, by considering all combinations

$$\lambda x + \sum_i \lambda^i \frac{\partial}{\partial x^i}, (\lambda, \lambda^1, \ldots, \lambda^n) \neq (0, \ldots, 0), \tag{18}$$

and identifying those differing from each other by a factor, T_x consisting of the points $x + \sum_i \lambda^i(\partial/\partial x^i)$. The result can be expressed by the diagram

$$\begin{array}{ccc} \Pi TM & \xleftarrow{\ j\ } & TM \\ & \searrow \quad \swarrow & \\ & M & \end{array} \tag{19}$$

where TM and ΠTM are bundles over M and j is the inclusion mapping.

We consider projective frames A_α such that $A_0 = x$ and projective connections such that

$$DA_0 = \mathrm{d}x = \sum_i \mathrm{d}x^i \frac{\partial}{\partial x^i}. \tag{20}$$

The change of frames (6) will then be restricted to those leaving the point A_0 invariant. This means that in (7) we have

$$g_0^k = 0. \tag{21}$$

Then (10) gives

$$g_0^0 \Omega_0^\beta = \sum_\alpha \Omega_0^{*\alpha} g_\alpha^\beta$$

and

$$g_0^0 \Omega_0^k = \sum_i \Omega_0^{*i} g_i^k.$$

(Notice the ranges of indices fixed in (3) and (16), which will be adopted throughout this paper.) It follows that

$$\Omega_0^k = 0 \tag{22}$$

is invariant under a change of the frame field. A projective connection satisfying (22) is said to be *without torsion.*

We consider *projective connections without torsion.* Then (10) gives

$$\begin{aligned} \Omega_0^0 &= \Omega_0^{*0}, \\ \sum_j g_i^j \Pi_j^k &= \sum_j \Pi_i^{*j} g_j^k, \\ \sum_j g_i^j \Omega_j^0 &= \sum_j \Pi_i^{*j} g_j^0 + \Omega_j^{*0} g_0^0. \end{aligned} \tag{23}$$

We put

$$\Pi_\alpha^\beta = \frac{1}{2} \sum_{i,k} W_{\alpha ik}^\beta \omega_0^i \wedge \omega_0^k, \tag{24}$$

where

$$W_{\alpha ik}^\beta + W_{\alpha ki}^\beta = 0. \tag{25}$$

We also put

$$W_{ik} = \sum_j W_{ikj}^j. \tag{26}$$

By (9) we have

$$g_0^0 \omega_0^k = \sum_j \omega_0^{*j} g_j^k. \tag{27}$$

It follows from the second equation of (23) and (27) that the condition

$$W_{ik} = 0 \tag{28}$$

is independent of the choice of the frame field. A torsionless projective connection satisfying (28) is called *normal.*

A fundamental theorem in projective connections is: *Given a symmetric affine connection, there is a uniquely determined normal projective connection with the same paths.*

We wish to determine explicitly this normal projective connection, given the affine connection $\Gamma_{ik}^j = \Gamma_{ki}^j$. Take the projective frames consisting of the points $x, \partial/\partial x^1, \ldots, \partial/\partial x^n$ in ΠT_x. (Such frames are called semi-natural in the terminology of E. Cartan.) We have (Ref. 2, p. 177)

$$\omega_0^0 = 0, \quad \omega_0^i = dx^i, \quad \omega_i^j = \sum_k \Gamma_{ik}^j \, dx^k, \tag{29}$$

and it remains to determine the forms

$$\omega_i^0 = \sum_k \Pi_{ik}^0 \, dx^k, \tag{30}$$

so that conditions (28) are fulfilled.

Let

$$d\omega_i^j - \sum_k \omega_i^k \wedge \omega_k^j = \Theta_i^j = \frac{1}{2}\sum_{k,l} R_{ikl}^j \, dx^k \wedge dx^l, \tag{31}$$

where

$$R_{ikl}^j + R_{ilk}^j = 0. \tag{32}$$

Thus R_{ikl}^j are the components of the affine curvature tensor. Substituting into (15), we get

$$\Theta_i^j - \omega_i^0 \wedge dx^j + \delta_i^j \sum_k dx^k \wedge \omega_k^0 = \frac{1}{2}\sum_{k,l} W_{ikl}^j \, dx^k \wedge dx^l, \tag{33}$$

which gives

$$R_{ikl}^j - \delta_l^j \Pi_{ik}^0 + \delta_k^j \Pi_{il}^0 + \delta_i^j(\Pi_{kl}^0 - \Pi_{lk}^0) = W_{ikl}^j. \tag{34}$$

Contracting j, l, we get, for a normal projective connection,

$$R_{ik} - n\Pi_{ik}^0 + \Pi_{ki}^0 = W_{ik} = 0, \tag{35}$$

where

$$\sum R_{ikj}^j = R_{ik} \tag{36}$$

is the Ricci tensor of the affine connection. Solution of (35) for Π_{ik}^0 gives

$$(n^2 - 1)\Pi_{ik}^0 = nR_{ik} + R_{ki} \tag{37}$$

and

$$(n^2 - 1)\omega_i^0 = \sum_k (nR_{ik} + R_{ki}) \, dx^k. \tag{37a}$$

When expressions (37) are substituted into (34), the left-hand side becomes the *Weyl projective curvature tensor*.

From (15) we find

$$(n^2 - 1)\Omega_i^0 = \frac{1}{2}\sum_{k,l} \{(-R_{ki|l} + R_{li|k}) + n(-R_{ik|l} + R_{il|k})\} \, dx^k \wedge dx^l, \tag{38}$$

where $R_{ik|l}$ is the covariant derivative of R_{ik} relative to the affine connection Γ_{lk}^j, i.e.,

$$R_{ik|l} = \frac{\partial R_{ik}}{\partial x^l} - \sum_j (\Gamma_{il}^j R_{jk} + \Gamma_{kl}^j R_{ij}). \tag{39}$$

The Ω_i^0, although not projectively invariant, behave in a simple way under a change of the frame field, as given by (23). Perhaps the main aim of this paper is to point out the possible importance of the tensor given by the coefficient of $dx^k \wedge dx^l$ at the right-hand side of (38).

4 PHYSICAL INTERPRETATION?

In a physical world M the dimension $n = 4$ and the equations

$$\Omega_i^0 = 0, \tag{40}$$

when explicitly written, are

$$4(-R_{ik|l} + R_{il|k}) - R_{ki|l} + R_{li|k} = 0. \tag{41}$$

If, in addition, the Ricci curvature of the affine connection is symmetric, i.e.,

$$R_{ik} = R_{ki}, \tag{42}$$

as in the case when the affine connection arises from a Lorentzian metric, (41) reduces to

$$R_{ik|l} = R_{il|k}. \tag{43}$$

It is my understanding that Eqs. (43) have received much attention in the general theory of relativity. Our discussion provides their projective background. Thus Eqs. (40) could be considered as a generalization of the field equations.

The lack of symmetry of the Ricci tensor could be attributed to the presence of electro-magnetic phenomenon. In fact, suppose the group $PL(n + 1; R)$ of the bundle ΠTM be extended to $GL(n + 1; R)$, i.e., suppose there be the line bundle

$$HTM \rightarrow \Pi TM, \tag{44}$$

where the fibers of HTM are the $(n + 1)$-dimensional vector spaces with the origins deleted. Suppose there be an affine connection in HTM which gives rise to the projective connection in ΠTM, as described in § 2. Then ω_0^0 is a connection form of the line bundle (44). This additional data might contain the mathematical basis of a unified field theory of gravitation and electro-magnetism. A natural proposal for the field equations consists of the following:

$$\Omega_0^0 = 0, \Omega_i^0 = 0. \tag{45}$$

The first of these equations gives

$$\begin{aligned} d\omega_0^0 &= \sum_k \omega_0^k \wedge \omega_k^0 = \frac{1}{n+1} \sum_{i,k} R_{ik}\, dx^i \wedge dx^k \\ &= \frac{1}{2(n+1)} \sum_{i,k} (R_{ik} - R_{ki})\, dx^i \wedge dx^k, \end{aligned} \tag{46}$$

which, up to a numerical factor, is the characteristic form of the connection ω_0^0. It could be interpreted as the *electro-magnetic strength*. On the other hand, the second Eq. in (45) is explicitly given by (41).

When the electro-magnetic field is not present, the Ricci tensor is symmetric, and the field equations become (43).

References

1. E. Cartan, Sur les variétés à connexion projective, *Bull. Soc. Math. de France*, **52**, 205–241 (1924); or *Oeuvres Complètes, Partie 3*, tome 1, 825–861.
2. E. Cartan, *Lecons sur la théorie des espaces à connexion projective*, Paris (1937).
3. S. Chern, Geometry of characteristic classes, *Proc. 13th Biennial Sem.*, Canadian Math. Congress, 1–40 (1972).
4. L. P. Eisenhart, *Non-Riemannian Geometry*, New York (1927).
5. O. Veblen and T. Y. Thomas, The geometry of paths, *Trans. Amer. Math. Soc.*, **25**, 551–608 (1923).
6. O. Veblen and J. M. Thomas, Projective normal coordinates for the geometry of paths, *Proc. Nat. Acad. Sci., USA*, **11**, 204–207 (1925).
7. O. Veblen, Projektive Relativitätstheorie, Ergeb. der Math., Bd. 2, Berlin (1933).
8. H. Weyl, Zur Infinitesimalgeometrie: Einordnung der projektiven und der konformen Auffassung, Gött. Nachrichten, 99–112 (1921), ges. Abh., Bd. 2, 195–207.

Reprinted from
Science of Matter, (1978) 225–232

Sonderdruck aus
Jber. d. Dt. Math.-Verein.
80 (1978) 13–110

Abel's Theorem and Webs

By S. S. CHERN*) and PHILLIP GRIFFITHS**)

Dedicated to Professors
G. Bol, E. Kaehler, and E. Sperner

Table of Contents

---

*) Research partially supported by NSF Grant MCS 74-23180, A01.
**) Miller Institute for Basic Research in Science at University of California, Berkeley and Harvard University. Research partially supported by NSF Grant MCS 72-05154.

Index of notations and list of structure equations

General

$\boldsymbol{P}^n$ = projective space
$\boldsymbol{G}(k,n)$ = Grassmann manifold of projective k-planes in $\boldsymbol{P}^n$
$\boldsymbol{P}^{n*}$ = dual projective space of hyperplanes in $\boldsymbol{P}^n$

§ I

p = point in $\boldsymbol{P}^n$ with homogeneous coordinates $[z_0, \dots, z_n]$;
$\{p_1, \dots, p_d\}$ = linear span of $p_1, \dots, p_d \in \boldsymbol{P}^n$;
$S \oplus S'$ = span in $\boldsymbol{P}^n$ of linear subspaces S and S';
ξ = point in $\boldsymbol{P}^{n*}$ defining a hyperplane $\sum_{a=0}^{n} \xi_a z_a = 0$;
$r(Q)$ = rank of a quadric Q in $\boldsymbol{P}^n$;
$\pi = \pi(C)$ = genus of algebraic curve C;
$h^i(\ )$, $\barwedge$, $|V|$, Ω^1, $[D]$, $\mathfrak{O}(D)$, K_M are standard notations from algebraic geometry explained in § I;
$\varkappa: C \to \boldsymbol{P}^{\pi-1}$ canonical mapping with image $C_\varkappa$.

§ II

$x \in U$ an open set in $\boldsymbol{R}^n$ or $\boldsymbol{C}^n$;
$\{u_i(x) = \text{constant}\}$ defines a d-web in U;
$\boldsymbol{P}(T_x^*)$ = projectivized cotangent space at $x \in U$;
$\omega^i(x) \in \boldsymbol{P}(T_x^*)$ denotes the ith web normal;
$d\omega^i = \pi^i \wedge \omega^i$ (no summation) defines π^i;
rank of a d-web is $r \leqq \pi(d,n)$ = Castelnuovo's number;
$Z_i(x) \in \boldsymbol{P}^{d-n-1}(x) \subset \boldsymbol{P}^{r-1}$;
$\{Z_1(x), \dots, Z_d(x)\} = \boldsymbol{P}^{d-n-1}(x)$

§ III

$D_x \subset \boldsymbol{P}(T_x^*)$ and $E_x \subset \boldsymbol{P}^{d-n-1}(x)$ are rational normal curves;
$D_x \barwedge E_x$ projectivity carrying $\omega^i(x)$ to $Z_i(x)$;
$\phi^1, \dots, \phi^n$ = moving coframe for $D_x \subset \boldsymbol{P}(T_x^*)$;
$d\phi^\alpha = \sum_\beta \phi^\beta \wedge \phi^\alpha_\beta$ is a symmetric connection;

$P^{n-2}(x,\sigma) = \{Z \in P^{d-n-1}(x) : dZ/d\sigma \in P^{d-n-1}(x)\}$
(*Note:* Suppose $F: U \to G(k,n)$ is a smooth mapping given by $x \to P^k(x) \subset P^n$. For a point $Z \in P^k(x)$ and tangent direction $\sigma \in P(T_x^*)$ we choose a curve $x(t)$ with $x(0) = x$ and with tangent $x'(0)$ in the direction σ and $Z(t) \in P^k(x(t))$ with $Z(0) = Z$. For a lifting $\tilde{Z}(t)$ of $Z(t)$ to R^{n+1} we let

$$dZ/d\sigma = \text{projection to } P^n \text{ of } \tilde{Z}'(0),$$

and note that $dZ/d\sigma$ is well-defined modulo $P^k(x)$. In particular the condition

$$dZ/d\sigma \in P^k(x)$$

is intrinsic. Geometrically, this means that as we move along the curve $x(t)$ the linear space $P^k(x(t)) \subset P^k(x)$ contains Z up to 2nd order.)
$dZ_i + \pi^i Z_i = Z_i'\omega^i$ (no summation);
$\omega^i(x) = \sum_\alpha t_i^\alpha(x)\phi^\alpha(x)$; and
$dt_i^\alpha = \pi^i t_i^\alpha + \sum_\beta t_i^\beta \phi_\alpha^\beta + \sum_\beta t_{i\alpha\beta}\phi^\beta$ where $t_{i\alpha\beta} = t_{i\beta\alpha}$

§ IV
$\mathscr{F}(P^n)$ = manifold of frames $F = \{Z_0,\ldots,Z_n\}$, $Z_a \in R^{n+1}$;
$\{\theta^\alpha, \phi_\beta^\alpha, \phi_\alpha^0\}$ = projective connection matrices;
$\Theta^\alpha = d\theta^\alpha - \theta^\beta \wedge \phi_\beta^\alpha$ = projective torsion;
$\Phi_\beta^\alpha = d\phi_\beta^\alpha - \phi_\beta^\gamma \wedge \phi_\gamma^\alpha - \theta_\beta^0 \wedge \theta^\alpha + \delta_\beta^\alpha \theta^\gamma \wedge \theta_\gamma^0$
$= 1/2\{R_{\beta\gamma\lambda}^\alpha \theta^\gamma \wedge \theta^\lambda\}$, $R_{\beta\gamma\lambda}^\alpha + R_{\beta\lambda\gamma}^\alpha = 0$, denotes the projective curvature;
$\Theta_\beta^0 = d\theta_\beta^0 - \phi_\beta^\alpha \wedge \phi_\alpha^0$;
$\Theta^\alpha = 0$ and $R_{\alpha\beta\gamma}^\alpha = 0$ define normal a projective connection.

Finally, we shall use the following

Ranges of indices

$1 \leqq i \leqq d$,
$1 \leqq \alpha, \beta, \gamma \leqq n$,
$1 \leqq \varrho, \sigma, \tau \leqq 2n$,
$n + 1 \leqq s \leqq d$,
$0 \leqq a, b, c \leqq n$.

Introduction

In recent years there have been important developments in the study of the global properties of foliations. A foliation is, briefly speaking, a local slicing of a manifold M (supposed to be C^∞). If $x^1,\ldots,x^n$ are local coordinates on M, a foliation of dimension $n - k$, or codimension k, has leaves defined by equations of the form

(1) $$F_1(x) = \text{const}, \ldots, F_k(x) = \text{const}, \quad x = (x^1,\ldots,x^n),$$

where $F_1,\ldots,F_k$ are smooth functions such that the Jacobian matrix

$$(\partial F_i/\partial x^\alpha),\quad 1 \leqq i \leqq k,\quad 1 \leqq \alpha \leqq n, \tag{2}$$

is of rank k everywhere. The functions $F_1,\ldots,F_k$ are defined up to an arbitrary C^∞ transformation. A *d-web* consists of d foliations. Throughout we make the additional assumption that the leaves are everywhere in general position.

It was Blaschke who began in the thirties a systematic study of webs. Web geometry has intimate contacts with the foundations of geometry, differential geometry, and algebraic geometry. From 1927 to 1938 Blaschke and his co-workers and students published 66 papers under the general heading "Topologische Fragen der Differentialgeometrie". These and other results were given a unified account in the book "Geometrie der Gewebe", by W. Blaschke and G. Bol [1].

From a projective variety a web can be constructed as follows: Let V^k be an algebraic variety of dimension k and degree d in a projective space $\boldsymbol{P}^n$ of dimension n. Then V^k meets a linear space $\boldsymbol{P}^{n-k}$ of dimension $n-k$ in d points. Consider the Grassmann manifold $\boldsymbol{G}(n-k,n)$ of all $\boldsymbol{P}^{n-k}$'s in $\boldsymbol{P}^n$; its dimension is $k(n-k+1)$. The $\boldsymbol{P}^{n-k}$'s through a point of $\boldsymbol{P}^n$ form a submanifold of dimension $k(n-k)$. Thus the d points of intersection of V^k with a given $\boldsymbol{P}^{n-k}$ associate to $\boldsymbol{P}^{n-k}$ d submanifolds of $\boldsymbol{G}(n-k,n)$, of dimension $k(n-k)$ or codimension k, which contain $\boldsymbol{P}^{n-k}$. This shows that V^k defines in $\boldsymbol{G}(n-k,n)$, or at least in a neighborhood of it, a d-web of codimension k. In particular, for $k=1$, an algebraic curve of degree d in $\boldsymbol{P}^n$ defines in the dual space $\boldsymbol{P}^{n*}$ $(=\boldsymbol{G}(n-1,n))$ a d-web of codimension 1. We will call *algebraic* the web defined from V^k by the above construction. In this sense web geometry generalizes the geometry of projective varieties.

The relationship between web geometry and algebraic geometry goes much deeper. In this paper we will restrict ourselves to the study of webs of codimension one, and mostly to their local properties. The d foliations defining the web will each be defined by an equation

$$u_i(x^1,\ldots,x^n) = \text{const},\quad 1 \leqq i \leqq d, \tag{3}$$

where we suppose u_i to be a smooth function, with gradient $\neq 0$. The function u_i can be replaced by a function $v_i(u_i)$ with $v_i' \neq 0$ without changing the ith foliation.

An equation of the form

$$\sum_i f_i(u_i)du_i = 0 \tag{4}$$

is called an *abelian* equation. An abelian equation remains an abelian equation under the above change $u_i \to v_i(u_i)$. The validity of an abelian equation is a strong property on the web. The number of linearly independent abelian equations (over the constants) is called the *rank* of the web. We have the following theorem [0]):

Let r be the rank of a d-web of codimension 1 *in* $\boldsymbol{R}^n$, $d \geqq n + 1$. *Then*

$$r \leqq \pi(d,n), \tag{5}$$

where

$$\pi(d,n) = 1/2(n-1)\{(d-1)(d-n) + s(n-s-1)\}, \tag{6}$$

s being defined by

$$s \equiv (-d+1) \bmod (n-1), \quad 0 \leqq s \leqq n-2. \tag{7}$$

$\pi(d,n)$ *is an integer.*

The integer $\pi(d,n)$ plays a role in the theory of algebraic curves. In 1889 Castelnuovo proved that a non-degenerate algebraic curve of degree d in a complex projective space $\boldsymbol{P}^n$ of dimension n has a genus $\leqq \pi(d,n)$. (A curve is non-degenerate if it does not lie in a lower dimensional projective space.) Those for which the maximum genus is attained are called *extremal curves*. They were investigated by Castelnuovo. In particular, they lie on special ruled surfaces, the Castelnuovo surfaces.

The algebraic web constructed from a Castelnuovo extremal curve C is of maximum rank $\pi(d,n)$. For let ω_λ, $1 \leqq \lambda \leqq \pi = \pi(d,n)$, be the linearly independent abelian differentials on C. By Abel's theorem, we have

$$\sum_i \int_{P_0}^{P_i} \omega_\lambda = \text{const}, \tag{8}$$

where P_0 is a fixed point on C and P_i are the points of intersection of C by a hyperplane. In differentials this relation can be written

$$\sum_i \omega_\lambda(P_i) = 0, \quad 1 \leqq \lambda \leqq \pi, \tag{9}$$

which are the π abelian equations of the web. Needless to say, this argument applies to the complex domain. A careful analysis gives a d-web of codimension 1 in a neighborhood of $\boldsymbol{R}^n$ with rank $\pi(d,n)$. (Cf. [3].)

A fundamental problem in web geometry is whether a d-web of codimension 1 in $\boldsymbol{R}^n$ of rank $\pi(d,n)$ is necessarily algebraic and is obtained from an extremal curve by the above construction. The problem can be separated into two parts:

---

[0]) The result appears in the paper [3] by the first author. It is number **T 60** in the series "Topologische Fragen der Differentialgeometrie" mentioned above, and constituted part of the first author's dissertation written in Hamburg under the supervision of Blaschke.

a) Linearization problem. Can all the leaves of the web become hyperplanes under a change of coordinates?

b) Algebraization problem. If all the leaves of a web are hyperplanes, under what conditions will they belong to an algebraic variety?

In the plane ($n = 2$) the answer to the linearization problem is negative. In fact, Bol gave an example of a plane 5-web of maximum rank 6 whose leaves cannot be mapped into straight lines by a change of coordinates. It is of interest to remark that one of the abelian equations in Bol's example involves Euler's dilogarithm, which plays a role in several problems of current interest: the volume of an odd-dimensional simplex in a non-euclidean space, the combinatorial formula for the first Pontrjagin number of a compact oriented 4-manifold, etc.

The book of Blaschke-Bol quoted above culminates with the following theorem of Bol:

A d-web of codimension 1 *in* $\boldsymbol{R}^3$, $d \neq 5$, *of maximum rank* $\pi(d,3)$ *is equivalent to a web with plane leaves. The leaves belong to an algebraic curve of degree d in the dual projective space* $\boldsymbol{P}^{3*}$.

By "equivalence" is meant "local equivalence", i.e., change of coordinates in a sufficiently small neighborhood. The same is understood in the following theorem, which is the main result of this paper:

Theorem. *Consider a d-web of codimension* 1 *in* $\boldsymbol{R}^n$ *of maximum rank* $\pi(d,n)$. *Suppose that* $n \geqq 3$, $d \geqq 2n$. *Then the web is linearizable, i.e., equivalent to a web whose leaves are hyperplanes.*

The idea of the proof is to compare the geometry in a neighborhood $U \subset \boldsymbol{R}^n$ where the web is given with that of an auxiliary projective space. In fact, let

$$\sum_i f_i^\lambda(u_i)\, du_i = 0, \quad 1 \leqq \lambda \leqq \pi = \pi(d,n) \tag{10}$$

be the linearly independent abelian equations. By interpreting

$$Z_i(x) = [f_i^1(u_i), \ldots, f_i^\pi(u_i)], \quad 1 \leqq i \leqq d \tag{11}$$

as the homogeneous coordinates in an auxiliary projective space $\boldsymbol{P}^{\pi-1}$, we obtain a mapping $U \to \boldsymbol{G}(d-n-1, \pi-1)$ with $x \to \{Z_1(x), \ldots, Z_d(x)\} = \boldsymbol{P}^{d-n-1}(x)$, to be called the *Poincaré mapping.* The d points $Z_i(x)$, $x \in U$, determine a normal curve $E(x)$ in the linear space $\boldsymbol{P}^{d-n-1}(x)$ of $\boldsymbol{P}^{\pi-1}$. To a curve in U corresponds ∞^1 normal curves $E(x)$. Our main lemma is to show that the curves in U whose corresponding normal curves pass through $n-1$ fixed points are the integral curves of a system of ordinary differential equations of the second order. This defines intrinsically a family of paths.

On the other hand, in the projectivized cotangent space $P(T^*)$ (of dimension $n-1$) at every point $x \in U$ the web of maximum rank defines a normal curve C_x, which contains the tangent hyperplanes of the leaves. (The points of $P(T^*)$ and the hyperplanes of the tangent space T_x can be identified by the pairing of T_x and T_x^*.) The curves C_x and $E(x)$ are in a projective correspondence. When $n = 3$, the normal curves C_x are conics and the field of C_x for $x \in U$ leads intrinsically to a Weyl geometry. The general case is considerably more complicated. We believe we are justified in saying that it is Elie Cartan's method of moving frames that effectively leads to the goal. The computations involve some mysterious and unexpected simplifications, to be expected of a good geometrical problem.

The path geometry so introduced has ∞^2 totally geodesic hypersurfaces which include the leaves of the web. The presence of these totally geodesic hypersurfaces implies that the normal projective connection associated to the path geometry is flat, and that the paths can be mapped into straight lines by a local change of coordinates. This mapping carries the totally geodesic hypersurfaces into hyperplanes.

Our proof contains also a simplification of Bol's proof in the case $n = 3$, by proving a sufficient condition for the flatness of a projective connection. (Cf. § IV. B.) Bol had to resort to a theorem of Enriques in algebraic geometry to complete his proof. Our proof of the main theorem is purely differential-geometric.

Once the linearization problem is solved, the affirmative answer to the algebraization problem is given by a converse of Abel's theorem. For details we refer the reader to a recent paper by the second author [4]. The relevant theorem goes back to Sophus Lie and can be stated as follows:

Suppose A_0 is a hyperplane in P^n and $C_1, \ldots, C_d$ are arcs each meeting A_0 in a single point. Suppose there are abelian forms $\omega_i \not\equiv 0$ on C_i such that

$$\sum_i \omega_i(A \cdot C_i) = 0, \quad 1 \leqq i \leqq d, \tag{12}$$

is valid for hyperplanes A varying in a neighborhood of A_0. Then there is an algebraic curve C in P^n and an abelian form ω on C such that $C_i \subset C$, $\omega | C_i = \omega_i$.

By duality in the projective space P^n this theorem can be translated to the following theorem on webs:

Consider a d-web in a neighborhood of R^n whose leaves are hyperplanes such that an abelian equation (4) holds, with $f_i(u_i) \not\equiv 0$. Then the leaves belong to an algebraic curve of degree d in the dual projective space.

Observe that in this theorem it is sufficient to have one abelian equation.

Added in proof: A summary of the proof of our main result appears in the paper "Linearization of Webs of Codimension One and Maximum Rank", to appear in Proc. of the Int. Symp. on Alg. Geom., Kyoto, Japan (1977).

I. Extremal algebraic curves

A. Rational normal curves

i. *Definition and basic properties.* An *algebraic curve* is given by a holomorphic mapping

$$f : C \to \boldsymbol{P}^n$$

from a compact Riemann surface C into the complex projective space of dimension n. The curve is *non-degenerate* in case the image does not lie in a lower dimensional linear space, and it is *normal* in case the image is not the projection of a curve in some $\boldsymbol{P}^{n'}$ for $n' > n$. In this paper it will usually be the case that f is a smooth embedding, and we shall identify C with its image $f(C)$ when there is no ambiguity.

In order to facilitate our discussion of algebraic curves it will be convenient to use the language of line bundles and Chern classes – cf. Chapters 0 and I of [5].

Over a compact complex manifold M we consider a holomorphic line bundle $L \to M$. We denote its *Chern class* by $c_1(L) \in H^2(M, \boldsymbol{Z})$. The two most important cases for us are when M is $\boldsymbol{P}^n$ or a compact Riemann surface; then $H^2(M, \boldsymbol{Z}) \cong \boldsymbol{Z}$ and the Chern class will be an integer called the *degree* and denoted by deg (L). The space $H^0(M, \mathfrak{O}(L))$ of global holomorphic sections is a finite dimensional vector space, frequently denoted by $H^0(L)$. Two sections s, s' in $H^0(L)$ have the same zero divisors if and only if $s = \lambda s'$ $(\lambda \in \boldsymbol{C})$ is a constant multiple of s'. A *linear system* is the projective space $\boldsymbol{P}(V)$ of lines through the origin in a linear subspace $V \subset H^0(L)$; thus $\boldsymbol{P}(V)$ may be identified with the divisors of sections $s \in V$. We shall sometimes use the classical notation $|V|$ instead of $\boldsymbol{P}(V)$. The linear system is *complete* in case $V = H^0(L)$. A *base point* is a point $p \in M$ where all sections $s \in V$ are zero. A linear system with no base points defines a holomorphic mapping

$$f : M \to \boldsymbol{P}(V^*)$$

by

$$f(p) = \{s \in \boldsymbol{P}(V) : s(p) = 0\} .$$

In other words, $p \in M$ is mapped to the hyperplane in $\boldsymbol{P}(V)$ consisting of all divisors in the linear system which pass through p.

The unique line bundle of degree $+1$ over $\boldsymbol{P}^n$ will be denoted by $H \to \boldsymbol{P}^n$ and called the *hyperplane line bundle.* Using homogeneous coordinates $z = [z_0, \ldots, z_n]$ the space $H^0(\boldsymbol{P}^n, \mathfrak{O}(H))$ of global holomorphic sections is naturally identified with the linear functions

$$\xi(z) = \sum_{\nu=0}^{n} \xi_\nu z_\nu$$

on P^n. The divisor ξ of this section is the linear hyperplane $\xi(z) = 0$, and consequently the complete linear system is the dual projective space P^{n*} of hyperplanes in P^n. This linear system is base point free, and using *projective duality*

$$(P^{n*})^* \cong P^n$$

to identify the hyperplanes passing through $p \in P^n$ with p itself makes the corresponding map $P^n \to P(H^0(P^n, \mathfrak{O}(H))^*)$ the identity.

Given any non-degenerate holomorphic mapping $f: M \to P^n$, we may set $L = f^*H$ and $V = f^*H^0(P^n, \mathfrak{O}(H))$ to obtain a holomorphic line bundle $L \to M$ and base point free linear system $V \subset H^0(L)$ whose corresponding map is just f. The "dictionary"

$$\left\{\begin{matrix}\text{non-degenerate}\\ \text{holomorphic mappings}\\ f: M \to P^n\end{matrix}\right\} \leftrightarrow \left\{\begin{matrix}\text{holomorphic line bundles}\\ L \to M \text{ and base point free}\\ \text{linear subspaces } V \subset H^0(L)\end{matrix}\right\}$$

will be used throughout.

In case $M = C$ is a compact Riemann surface and $f: C \to P^n$ a non-degenerate algebraic curve – we always assume that f is generically one to one – then the *degree* $\deg(C)$ is defined to be $\deg(f^*H)$. It is the number of points in which the image meets a hyperplane. The algebraic curve is normal if and only if the corresponding linear system is complete.

In order to illustrate the above dictionary, let us prove that a non-degenerate curve in P^n has degree $\geqq n$. Using the language of line bundles, we set $L = f^*H$ and let $\mathfrak{O}(L-p) \subset \mathfrak{O}(L)$ be the subsheaf of sections vanishing at a point $p \in C$. Then $\mathfrak{O}(L-p)$ is the sheaf of the line bundle $L \otimes [-p]$, and the exact sheaf sequence

$$0 \to \mathfrak{O}(L-p) \to \mathfrak{O}(L) \to L_p \to 0$$

is valid. From the exact cohomology sequence we obtain

$$h^0(L) \leqq h^0(L-p) + 1$$

where the notation

$$h^i(L) = \dim H^i(M, \mathfrak{O}(L))$$

will be used consistently. Applying this inequality recursively gives

$$h^0(L) \leqq \deg(L) + 1$$

since $\deg(L \otimes [-p]) = \deg L - 1$ and $h^0(L') = 0$ in case $\deg L' < 0$. In particular, since $h^0(f^*H) \geqq n + 1$ we obtain $\deg(C) \geqq n$.

The corresponding geometric argument is the following: Linear projection from the point $p \in f(C)$ gives

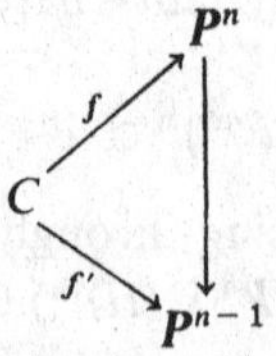

where $f'(C)$ has degree at least one less than $f(C)$. Since an algebraic curve of degree one is a straight line it follows that $\deg(C) \geqq n$.

The line bundle corresponding to f' is $L \otimes [-p]$, and the associated linear system is just $H^0(C, \mathfrak{O}(L - p)) \subset H^0(C, \mathfrak{O}(L))$. So the two arguments bounding the degree correspond under our dictionary.

The general result is this: A non-degenerate algebraic variety $V \subset \boldsymbol{P}^n$ of dimension k has degree at least $n - k + 1$. Thus a surface has degree $\geqq n - 1$, etc. Varieties of minimal degree are quite interesting, and will play an important role for us. We begin with the case of curves.

Definition. *A rational normal curve is a non-degenerate curve of degree n in* $\boldsymbol{P}^n$.

We shall prove that a rational normal curve is rational – i.e., that C is the Riemann sphere $\boldsymbol{P}^1$ – and that f is a smooth embedding. The curve is normal because it has degree n.

The second geometric argument used to bound the degree shows that the projection of a rational normal curve in $\boldsymbol{P}^n$ is a rational normal curve in $\boldsymbol{P}^{n-1}$. Iterating this we obtain

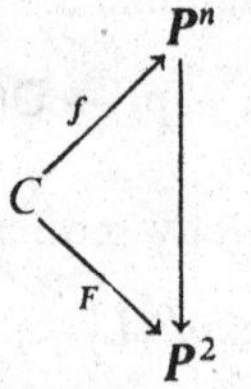

where $F(C)$ has degree two. It follows that $F(C)$ is first of all a plane conic, and secondly is smooth since otherwise it would consist of two straight lines. By projecting once more (stereographic projection), we find that $F(C)$ is isomorphic to $\boldsymbol{P}^1$. This proves that C is rational and f is a smooth embedding. The fact that *rational normal curves project onto plane conics* will be important for us.

For a rational normal curve $f : \boldsymbol{P}^1 \to \boldsymbol{P}^n$ the corresponding line bundle L has degree n. It follows that $L \cong H^n$ where $H \to \boldsymbol{P}^1$ is the hyperplane bundle. The inequalities

$$n + 1 \leqq h^0(\boldsymbol{P}^1,\mathfrak{O}(L)) \leqq \deg(L) + 1 = n + 1$$

give

$$h^0(\boldsymbol{P}^1,\mathfrak{O}(H^n)) = n + 1\,.$$

This implies:
A rational normal curve is the mapping

$$f : \boldsymbol{P}^1 \to \boldsymbol{P}^n$$

given by choosing a basis $s_0,\ldots,s_n$ *for* $H^0(\boldsymbol{P}^1,\mathfrak{O}(H^n))$ *and setting*

$$f(p) = [s_0(p),\ldots,s_n(p)]\,.$$

For example, using homogeneous coordinates $z = [z_0,z_1]$ the sections

$$z_0^n, z_0^{n-1}\, z_1,\ldots,z_0\, z_1^{n-1}, z_1^n$$

give such a basis. The image under this choice of basis will be called the *standard curve.* Setting $t = z_1/z_0$, it is given parametrically by

$$t \to [1,t,t^2,\ldots,t^n]\,.$$

We note that the two points $[1,0,\ldots,0]$ and $[0,0,\ldots,1]$ lie on the standard rational normal curve, and that the projection of this curve in $\boldsymbol{P}^{n-1}$ from either of these points is again a standard rational normal curve. This geometric property will underlie several inductive arguments later on in the paper.

Since any two bases of $H^0(\boldsymbol{P}^1,\mathfrak{O}(H^n))$ are related by a linear transformation, any two rational normal curves are projectively equivalent. This allows us to deduce properties of a general rational normal curve from those of the standard curve. Thus, e.g., distinct points $t_0,\ldots,t_N$ map to points in general position in $\boldsymbol{P}^n$ by noting the Vandermonde identity

$$\begin{vmatrix} 1 & t_0 & \ldots & t_0^k \\ 1 & t_1 & \ldots & t_1^k \\ \vdots & & & \\ 1 & t_k & \ldots & t_k^k \end{vmatrix} = n \prod_{i \neq j} (t_i - t_j)$$

to deduce that any $k + 1 \leqq n + 1$ of these points are linearly independent. Another property is that there is a 3-parameter subgroup of the projective linear group $\mathrm{PGL}_{n+1} \cong \mathrm{Aut}(\boldsymbol{P}^n)$ taking any rational normal curve into itself. For the standard curve these ∞^3 automorphisms of $\boldsymbol{P}^n$ induce the linear fractional group $t \to (at + b)/(ct + d)$ $(ad - bc = 1)$ on $\boldsymbol{P}^1$.

Using these remarks we may count the number of rational normal curves to be

$$\underbrace{(n+1)^2-1}_{\dim(PGL_{n+1})} - \underbrace{3}_{\dim(PGL_2)} = (n-1)(n+3)$$

This suggests that:

(1.1) *There is a unique rational normal curve through any $n + 3$ points in general position in $\boldsymbol{P}^n$.*

To prove this we let $p_0, \ldots, p_n, q, r$ be our $n + 3$ points in general position, and choose a homogeneous coordinate system having $p_0 = [1,0,\ldots,0], \ldots, p_n = [0,\ldots,0,1]$ as vertices of the fundamental simplex. The parametric representation

$$z_i = a_i/(t - b_i) \quad a_i, b_i \neq 0$$

gives, by finding a least common denominator, the rational normal curve

$$t \to \left[\prod_{i \neq 0} (t - b_i)a_0, \ldots, \prod_{i \neq n} (t - b_i)a_n\right]$$

sending $t = b_i$ to $p_i = [0,\ldots,1_i,\ldots,0]$. The points p and q have all coordinates non-zero, and consequently the inversion

$$z_i' = 1/z_i$$

maps p and q to distinct points p' and q' having all z_i'-coordinates non-zero. The above rational normal curve is mapped to the line

$$z_i' = (t - b_i)/a_i,$$

and we may uniquely choose the a_i and b_i so that this line passes through q' and r'. Since any rational normal curve containing the vertices of the coordinate simplex has the parametric form given above, this proves that there is a unique such curve passing through $n + 3$ points in general position.

ii) *Quadrics and rational normal curves.* Although simple to describe parametrically, the rational normal curves have a somewhat complicated set of defining equations since they are not complete intersections. The previously noted fact that rational normal curves project onto conics in the plane suggests that we study the quadrics containing them, which we now proceed to do.

A *quadric* Q in $\boldsymbol{P}^n$ is the hypersurface given by a quadratic equation

$$q(z) = \sum_{i,j} q_{ij} z_i z_j = 0 \quad (q_{ij} = q_{ji}).$$

In a suitable coordinate system, we will have

$$q(z') = \sum_{i=0}^{r(Q)} z_i'^2$$

where $r(Q)$ is defined to be the *rank* of the quadric Q. Thus $r(Q) = n$ if and only if $\det(q_{ij}) \neq 0$, or equivalently Q is non-singular. At the other extreme, if $r(Q) = 0$ then Q is a hyperplane ($z_0' = 0$) counted twice, and if $r(Q) = 1$ then

$$Q = \mathbf{P}^{n-1} + \mathbf{P}'^{n-1}$$

is the union of distinct linear spaces. In general a quadric of rank $r \geqq 2$ is the cone over a non-singular quadric in $\mathbf{P}^r$.

The quadratic polynomials $q(z)$ are just the sections of the line bundle $H^2 \to \mathbf{P}^n$, and thus

$$h^0(\mathbf{P}^n, \mathfrak{O}(H^2)) = (n+1)(n+2)/2\,.$$

A linear subspace $V \subset H^0(\mathbf{P}^n, \mathfrak{O}(H^2))$ defines a linear system $\mathbf{P}(V)$ of quadrics corresponding to sections $q \in V$. Given a rational normal curve $f: \mathbf{P}^1 \to \mathbf{P}^n$, we set $C = f(\mathbf{P}^1)$ and denote by $V(C)$ the quadratic polynomials which vanish on C. Using the notation $|V(C)|$ for the linear system $\mathbf{P}(V(C))$, we may think of $|V(C)|$ as the space of quadrics passing through C. We will now derive the classical result:

(1.2) *The linear system $|V(C)|$ has dimension given by*

$$\dim V(C) = n(n-1)/2\,.$$

The curve C is the ideal-theoretic intersection of the quadrics $Q \in |V(C)|$. Finally, the linear system $|V(C)|$ is spanned by quadrics of rank two.

Proof. Suppose that we choose coordinates such that C has a parametric representation

$$t \to [1, t, \ldots, t^n]\,.$$

Under the substitution $z_i = t^i$ each quadric $q(z)$ goes into a polynomial $g(t)$ in t of degree $\leqq 2n$, and every such $g(t)$ appears in this way. Thus

$$\dim V(C) = (n+1)(n+2)/2 - (2n+1) = n(n-1)/2\,.$$

By projecting C onto a plane conic C' and taking the cone in $\mathbf{P}^n$ over C' we may find a rank two quadric containing C and not passing through a preassigned point in $\mathbf{P}^n - C$. Thus C is the set-theoretic intersection of the rank two quadrics which contain it.

To check that the quadrics in $V(C)$ generate the ideal of C, it will suffice to do this at one point p_0 and then use the fact there is a transitive group of projective transformations acting on the configuration $C \subset \boldsymbol{P}^n$. We may suppose that C is the standard curve, $p_0 = [1,0,\ldots,0]$, and take $x = [1,x_1,\ldots,x_n]$ as affine coordinates around p_0. The $n-1$ quadrics

$$q_k(x) = x_{k+1} - x_1 x_k = 0 \quad (k = 1,\ldots,n-1)$$

all vanish on C and satisfy the Jacobian condition

$$dq_1 \wedge \cdots \wedge dq_{n-1} \neq 0$$

at p_0.

Finally, it is not difficult to write down a basis for $V(C)$ consisting of rank two quadrics. Q.E.D.

There is a lovely synthetic construction of rational normal curves and the rank two quadrics which contain them which will prove useful in a little while. Before going into it we recall that the hyperplanes in $\boldsymbol{P}^n$ which contain a fixed $\boldsymbol{P}^{n-2}$ form a *pencil* $\boldsymbol{P}^{n-1}(t)$ — here t denotes a linear coordinate on the $\boldsymbol{P}^1$ of hyperplanes containing $\boldsymbol{P}^{n-2}$. To obtain this pencil we take a $\boldsymbol{P}^2$ meeting $\boldsymbol{P}^{n-2}$ in a point p_0. The lines in $\boldsymbol{P}^2$ passing through p_0 form a pencil $\boldsymbol{P}^1(t)$ and the linear span $\boldsymbol{P}^1(t) \oplus \boldsymbol{P}^{n-2} = \boldsymbol{P}^{n-1}(t)$ traces out the pencil of hyperplanes with axis $\boldsymbol{P}^{n-2}$. Two such pencils are in *correspondence*, written

$$\boldsymbol{P}^{n-1}(t) \barwedge \boldsymbol{P}^{n-1}(t'),$$

if we are given a projective isomorphism

$$t' = (at + b)/(ct + d)$$

between their parameter spaces. Such an isomorphism is uniquely specified by giving three distinct values

$$t_i' = (at_i + b)/(ct_i + d) \quad (i = 1,2,3).$$

Now let p and p' be two points in the plane and $\boldsymbol{P}^1(t)$ and $\boldsymbol{P}^1(t')$ the respective pencils of lines through them. A correspondence $\boldsymbol{P}^1(t) \barwedge \boldsymbol{P}^1(t')$ is uniquely determined by requiring that corresponding lines pass through three non-colinear points r,r',r'' (Fig. 1).

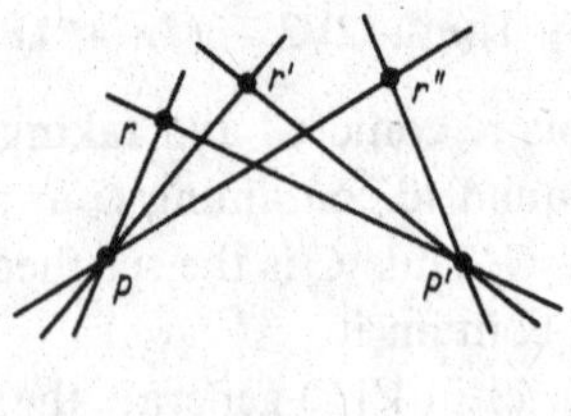

Fig. 1

The variable point of intersection $p(t) = \mathbf{P}^1(t) \cap \mathbf{P}^1(t'(t))$ will then trace out a conic passing through p, p', r, r', r''. This is Steiner's synthetic construction of a conic through five points in general position in the plane.

Before giving the generalizations of this, we need one remark concerning quadrics. Given points $p_1, p_2, \ldots, p_N$ in $\mathbf{P}^n$ we denote by $V(p_1, \ldots, p_N)$ the linear space of quadratic polynomials vanishing at the p_i. Then we have:

(1.3) *codim* $V(p_1, \ldots, p_N) = N$ *provided that* $N \leqq 2n + 1$ *and the* p_i *are in general position.*

Proof. By adding additional points it will suffice to do the case $N = 2n + 1$. We must show for each i there is a quadric Q passing through the p_j for $j \neq i$ but not containing p_i. Relabelling we may assume that $i = 2n + 1$. By general position,

$$\overline{\{p_1, \ldots, p_n\}} = \mathbf{P}^{n-1} \text{ and } \overline{\{p_{n+1}, \ldots, p_{2n}\}} = \mathbf{P}'^{n-1}.$$

We may take

$$Q = \mathbf{P}^{n-1} + \mathbf{P}'^{n-1}. \qquad \text{Q.E.D.}$$

We shall refer to what we just proved by saying that $N \leqq 2n + 1$ *points in general position impose independent conditions on the linear system of quadrics containing them.*

Steiner's construction generalizes in two ways. The first gives a synthetic method for finding rank two quadrics containing a rational normal curve $C \subset \mathbf{P}^n$. Let $p_0, \ldots, p_{n-2}$; $p'_0, \ldots, p'_{n-2}$; r, r', r'' be $2n + 1$ distinct points on our curve. Then these points are in general position, and since $\dim V(C) = 2n + 1$ and, as just proved, these points impose independent conditions on the quadrics containing them, we deduce that: *any quadric passing through* $2n + 1$ *distinct points on* C *must contain* C *entirely.* To find Q passing through these points, we set $\mathbf{P}^{n-2} = \overline{\{p_0, \ldots, p_{n-2}\}}$ and $\mathbf{P}'^{n-2} = \overline{\{p'_0, \ldots, p'_{n-2}\}}$, and consider the pencils $\mathbf{P}^{n-1}(t)$ and $\mathbf{P}^{n-1}(t')$ of hyperplanes having $\mathbf{P}^{n-2}$ and $\mathbf{P}'^{n-2}$ as respective axes. A correspondence $\mathbf{P}^{n-1}(t) \barwedge \mathbf{P}'^{n-1}(t')$ is set up by requiring that corresponding hyperplanes should pass through each of r, r', r''. Then the variable intersection

$$\mathbf{P}^{n-1}(t) \cap \mathbf{P}^{n-1}(t'(t))$$

traces out a quadric Q, which is of rank two since it contains $\mathbf{P}^{n-2}$'s but no $\mathbf{P}^{n-1}$'s, and which passes through the $2n + 1$ given points on C and hence contains the curve by our previous remark.

A second generalization of Steiner's construction gives a synthetic method for tracing out a rational normal curve through $n + 3$ points

$p_1, \ldots, p_n, r_1, r_2, r_3$ in general position in $\mathbf{P}^n$. Let

$$\mathbf{P}_i^{n-2} = \{p_1, \ldots, \hat{p}_i, \ldots, p_n\}$$

be the edges of the simplex determined by the p_i's and $\mathbf{P}^{n-1}(t_i)$ the corresponding pencil of hyperplanes with axis $\mathbf{P}_i^{n-2}$. For a fixed pencil $\mathbf{P}^{n-1}(t)$ of hyperplanes we establish a correspondence

$$\mathbf{P}^{n-1}(t) \barwedge \mathbf{P}^{n-1}(t_i)$$

by requiring that corresponding hyperplanes pass through r_1, r_2, r_3. Thus $r_i \in \mathbf{P}^{n-1}(t_i(t))\,(= 1,2,3)$ for each i. The variable point of intersection

$$p(t) = \mathbf{P}_1^{n-1}(t_1(t)) \cap \cdots \cap \mathbf{P}_n^{n-1}(t_n(t))$$

traces out a rational normal curve thorugh our given $n + 3$ points (Fig. 2).

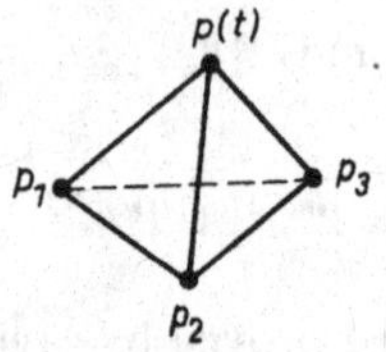

Fig. 2

We have been leading up to the following lemma, which appears implicitly in the classical literature, and will play a crucial role for us:

(1.4) *Let $p_1, \ldots, p_N$ be points in general position in $\mathbf{P}^n$ and assume that*

$$N > 2n + 2$$
$$\dim V(p_1, \ldots, p_N) \geqq n(n-1)/2\,.$$

Then these points lie on a unique rational normal curve, and the second inequality is an equality[1]).

Proof. Let $C = \bigcap_{Q \in |V(\Gamma)|} Q$ be the intersection of all the quadrics containing our set $\Gamma = p_1, \ldots, p_N$. Then C is an algebraic variety and we will first prove that

$$\dim C \leqq 1\,.$$

If, on the contrary, this dimension is $\geqq 2$ then C will meet *every* $\mathbf{P}^{n-2}$ in a non-empty set Z. We will choose $\mathbf{P}^{n-2}$ such that the quadrics in $V(\Gamma)$ remain linearly independent when restricted to this subspace.

[1]) Briefly, we may say that a set of $N > 2n + 2$ points in general position lying on $n(n-1)/2$ quadrics must lie on a rational normal curve. Since $(n+1)(n+2)/2 = 2n + 1 + n(n-1)/2$, the second inequality is equivalent to codim $V(p_1, \ldots, p_N) \leqq 2n + 1$.

For this we let $\mathbf{P}^{n-1} = \{p_1,\ldots,p_n\}$ and choose a $\mathbf{P}^{n-2} \subset \mathbf{P}^{n-1}$ not containing any point of Γ. If some $Q \in |V(\Gamma)|$ contains $\mathbf{P}^{n-2}$, then $Q|\mathbf{P}^{n-1} = \mathbf{P}^{n-2} + \mathbf{P}'^{n-2}$ since this restriction has rank one. But $Q|\mathbf{P}^{n-1}$ must pass through $p_1,\ldots,p_n$ and hence these points lie on a $\mathbf{P}'^{n-2}$ contradicting general position. This proves that the restriction

$$V(\Gamma) \to H^0(\mathbf{P}^{n-2},\mathfrak{O}(H^2))$$

is an injection. But $h^0(\mathbf{P}^{n-2},\mathfrak{O}(H^2)) = n(n-1)/2$, and so all quadrics in $\mathbf{P}^{n-2}$ must pass through Z, which is not the case. This contradiction shows that $\dim C \leqq 1$.

We now use Steiner's method to construct C. Label our points as

$$p_1,\ldots,p_{2n+1};\ r_1,\ldots,r_{N-2n-1} \quad (N - 2n - 1 \geqq 2),$$

and set

$$\mathbf{P}_i^{n-2} = \{p_1,\ldots,\hat{p}_i,\ldots,p_n\},\ \mathbf{P}^{n-2} = \{p_{n+2},\ldots,p_{2n}\}.$$

Denote by $\mathbf{P}^{n-1}(t_i)$ and $\mathbf{P}^{n-1}(t)$ the pencils of hyperplanes with axes $\mathbf{P}_i^{n-2}$ and $\mathbf{P}^{n-2}$, and establish a correspondence

$$\mathbf{P}^{n-1}(t) \barwedge \mathbf{P}^{n-1}(t_i)$$

by requiring that corresponding hyperplanes pass through p_i, p_{n+1}, p_{2n+1}. Then the quadric Q_i traced out by the intersection $\mathbf{P}^{n-1}(t) \cap \mathbf{P}^{n-1}(t_i(t))$ passes through $p_1,\ldots,p_{2n+1}$ and hence must contain the points r_α as well. In particular, for suitable values $t_{n+1}, t_{2n+1}, t_\alpha$ ($\alpha = 1,\ldots,N - 2n - 1$) of t we will have, *for all i*,

$$p_{n+1} \in \mathbf{P}_i^{n-1}(t_i(t_{n+1}))$$
$$p_{2n+1} \in \mathbf{P}_i^{n-1}(t_i(t_{2n+1}))$$
$$r_\alpha \in \mathbf{P}_i^{n-1}(t_i(t_\alpha)).$$

The rational normal curve traced out by the variable intersection

$$p(t) = \mathbf{P}_1^{n-1}(t_1(t)) \cap \cdots \cap \mathbf{P}_n^{n-1}(t_n(t))$$

then passes through $p_1,\ldots,p_n,p_{n+1},p_{2n+1},r_1,\ldots,r_{N-2n-1}$. Put another way, *the unique rational normal curve C' through $p_1,\ldots,p_{n+1},r_1,r_2$ will then contain p_{2n+1} and the remaining r_α's*. It follows that C' passes through Γ. Now every quadric in $|V(\Gamma)|$ meets C' in $\geqq 2n + 3$ points and hence contains C' entirely. Thus $V(\Gamma) \subset V(C')$, and since $\operatorname{codim} V(\Gamma) = 2n + 1 = \operatorname{codim} V(C')$ we must have $V(\Gamma) = V(C')$ and $C = C'$. Q.E.D.

B. Extremal algebraic curves of positive genus

i. *Castelnuovo's bound.* Let C be a compact Riemann surface of genus π. For a holomorphic line bundle $L \to C$ we denote by

$$h^0(L) = \dim H^0(C, \mathfrak{O}(L))$$

the number of linearly independent holomorphic sections of the bundle. Following general conventions L^* will be the line bundle dual to L, and K will denote the *canonical line bundle* whose associated sheaf $\mathfrak{O}(K) = \Omega^1$ is the 1-forms. The basic fact in the study of curves is the *Riemann-Roch theorem*

$$h^0(L) = \deg L - \pi + h^0(K \otimes L^*) + 1 . \tag{1.5}$$

Especially useful is the case when $L = [D]$ is the line bundle associated to an effective divisor $D = p_1 + \cdots + p_d$ of degree d. Then $\mathfrak{O}([D])$, or frequently just $\mathfrak{O}(D)$, is the sheaf of meromorphic functions having poles no worse than D, and one traditionally sets

$$l(D) = h^0([D]) .$$

On the other hand, $\mathfrak{O}(K \otimes [D]^*)$ is naturally identified with the sheaf $\Omega^1(-D)$ of holomorphic differentials which vanish on D, and here one traditionally sets

$$i(D) = h^0(K \otimes [D]^*)$$

and says that the divisor D is *special* in case the *index of speciality* $i(D) > 0$. With these notations the Riemann-Roch becomes

$$l(D) = d - \pi + i(D) + 1 . \tag{1.6}$$

Finally, if we let $|D|$ denote the complete linear system associated to $H^0(C, \mathfrak{O}(D))$, then $|D|$ is the projective space of all divisors $D' = p'_1 + \cdots + p'_d$ linearly equivalent to D in the sense that

$$D' - D = (\phi)$$

for some meromorphic function ϕ on C. With this notation (1.5) assumes its most symmetric form

$$\dim |D| = d - \pi + i(d) . \tag{1.7}$$

Now suppose that $f : C \to \boldsymbol{P}^n$ is a non-degenerate algebraic curve of degree d. We assume that f is generically one-to-one and set $L = f^* H$. We have proved that $d \geqq n$, and that if $d = n$ then $\pi = 0$ and C is a rational normal curve. In 1889 Castelnuovo found a general bound on π in terms of d and n, and was moreover able to determine the structure of the curves of maximum genus[2]). Because of the central role which these *extremal algebraic curves* play in the theory of webs we shall sketch Castelnuovo's argument here – a complete discussion and details may be found in Chapters II and IV of [5].

The bound is based on a lemma which will now be explained. We denote by V_k the image of the mapping

$$H^0(\boldsymbol{P}^n, \mathfrak{O}(H^k)) \to H^0(C, \mathfrak{O}(L^k)).$$

Since $H^0(\boldsymbol{P}^n, \mathfrak{O}(H^k))$ is just the vector space of homogeneous polynomials of degree k, the linear system $|V_k|$ consists of the divisors on $f(C)$ cut out by the hypersurfaces of degree k in $\boldsymbol{P}^n$. We will prove the

(1.8) **Lemma.** $$\dim V_k - \dim V_{k-1} = \begin{cases} \geqq k(n-1)+1 & \text{if } k \leqq (d-1)/(n-1) \\ = d & \text{if } k > (d-1)/(n-1) \end{cases}$$

Proof. For $k = 1$ the bound is

$$\dim V_1 \geqq n+1,$$

which is equivalent to the curve being non-degenerate. For $k > 1$ we shall use the following (loc. cit)

General position principle: A generic hyperplane $\xi \in \boldsymbol{P}^{n*}$ cuts the curve C in d points

$$\xi \cdot C = p_1 + \cdots + p_d,$$

any n of which are linearly independent.

Now assume that $k(n-1)+1 \leqq d$ and break the points up into k groups of $n-1$ each plus a remainder as follows:

$$\underbrace{p_1, \ldots, p_{n-1}}_{\displaystyle \| \atop \displaystyle D_1};\ \underbrace{p_n, \ldots, p_{2(n-1)}}_{\displaystyle \| \atop \displaystyle D_2};\ \cdots;\ \underbrace{p_{k(n-2)+1}, \ldots, p_{k(n-1)}}_{\displaystyle \| \atop \displaystyle D_k};\ \ldots, p_d.$$

By the general position principle we may find hyperplanes $\xi_1, \ldots, \xi_k$ such that each ξ_i contains D_i but not $p_{k(n-1)+1}$. The hypersurface $h = \xi_1 + \cdots + \xi_k$ in $\boldsymbol{P}^n$ has degree k and passes through $p_1, \ldots, p_{k(n-1)}$ but not $p_{k(n-1)+1}$. On the other hand, any hypersurface of the form $h' = \xi + h''$ where h'' has degree $k-1$ passes through all the p_i. There are clearly at least $k(n-1)+1+\dim V_{k-1}$ linearly independent hypersurfaces of the form

$$h + h',$$

[2]) For $n = 2$ the bound was classical. For $n = 3$ it was found by Clifford and Max Noether.

which proves the lemma in case $k(n-1)+1 \leqq d$. The other case is similar and easier. Q.E.D.

Now we set $m = [(d-1)/(n-1)]$ and successively apply the lemma for $k = 0,1,\ldots,m+m'$ to obtain

$$\dim V_{m+m'} \geq (1/2m(m+1))(n-1) + m + 1 + m'd\,.$$

On the other hand, obviously

$$h^0(L^k) \geqq \dim V_k$$

for any k, while the index of speciality

$$i(L^k) = 0$$

for large k[3]). Applying (1.5) we obtain

$$d(m+m') - \pi + 1 = h^0(L^{m+m'}) \geqq (1/2m(m+1))(n-1) + (m+1) + m'd$$

or

$$\pi \leqq \pi(d,n)$$

where

$$\begin{cases} \pi(d,n) = m[d - 1/2(m+1)(n-1) - 1] & \text{and} \\ m = [(d-1)/(n-1)] \end{cases}$$

This is Castelnuovo's bound on the genus.

We now make some brief remarks about the various cases.

Case i: For an extremal algebraic curve C of degree d with $n \leqq d \leqq 2n-1$, we have

$$\pi = d - n$$

$$h^0(L) = n + 1, \quad i(L) = 0$$

The bound on π follows immediately from Castelnuovo. In the extremal case when $\pi = \pi(d,n) = d - n$ we use the form (1.5) of Riemann-Roch to obtain

$$h^0(L) = \deg L - \pi + i(L) + 1 = n + 1 + i(L)$$

$$h^0(L) \geqq n + 1\,,$$

which together imply the second statements.

It can be shown that for any compact Riemann surface C of genus π and integers d, n with

$$\pi = d - n$$

$$n \leqq d \leqq 2n - 1\,,$$

[3]) Specifically, since $i(L^k) = h^0(K \otimes L^{-k})$ we will have $i(L^k) = 0$ as soon as $\deg(K \otimes L^k) = 2\pi - 2 - kd < 0$.

there is a holomorphic embedding $f: C \to \boldsymbol{P}^n$ of C as a curve of degree d. Consequently, there is nothing special about those Riemann surfaces which appear as extremal curves when $n \leqq d \leqq 2n - 1$.

Case ii: For an extremal algebraic curve of degree $d = 2n$, the line bundle L is the canonical bundle and C is a canonical curve. By Castelnuovo's bound

$$\pi \leqq \pi(2n, n) = n + 1\,.$$

Arguing as before, in case $\pi = n + 1$

$$h^0(L) = n + h^0(K \otimes L^*)$$
$$h^0(L) \geqq n + 1$$

which together imply that $h^0(L) = n + 1$ and $h^0(K \otimes L^*) = 1$. Now then

$$\deg(K \otimes L^*) = \deg K - \deg L = 2\pi - 2 - 2n = 0\,,$$

and $h^0(K \otimes L^*) \neq 0$ implies that $K = L$ since any section of $K \otimes L^*$ can have no zeroes.

By definition, a *canonical curve* is $f: C \to \boldsymbol{P}^n$ where f is generically one-to-one and $f^*H = K$ is the canonical line bundle (it then turns out that f is a smooth embedding). We shall say more about these curves in section I-B iii.

Case iii. These are the curves where $d > 2n$, and are the ones in which we are most interested. They will be described in more detail in the next sections.

We conclude this discussion with three observations. The first is that for line bundles $L \to C$ whose associated complete linear system gives a generically one-to-one mapping $f: C \to \boldsymbol{P}^n$ we have proved *Clifford's theorem:*

$$\begin{cases} i(L) \neq 0 \Rightarrow h^0(L) \leqq \deg L/2 + 1 \\ \text{with equality} \Leftrightarrow L = K\,. \end{cases}$$

Proof. Setting $d = \deg L$ and $h^0(L) = n + 1$, we assume

$$n + 1 > d/2 + 1$$

or $d < 2n$. Then $\pi \leqq d - n$ and

$$n + 1 = h^0(L) = d - \pi + 1 + i(L) \geqq n + 1 + i(L)\,,$$

which implies that $i(L) = 0$. If $n + 1 = d/2 + 1$ then $d = 2n$ and we have proved that $L = K$. Q.E.D.

The second observation is that in case C is extremal, then all the inequalities in the proof of Castelnuovo's bound must be equalities and so

$$h^0(L^k) = \dim V_k \quad k = 1,2,\ldots, {}^{4}).$$

In particular

$$h^0(L^2) = 3n \quad \text{in case} \quad d \geqq 2n - 1.$$

Since $h^0(P^n, \mathfrak{O}(H^2)) = (n+1)(n+2)/2$ and $(n+1)(n+2)/2 - 3n = (n-1)(n-2)/2$, this implies that:

(1.9) *An extremal algebraic curve C of degree* $d \geqq 2n - 1$ *lies on* $\infty^{(n-1)(n-2)/2}$ *quadrics in* P^n.

Recently, Joe Harris took up the question of maximizing the genus π of a curve C of degree d in P^3 which lies on a surface S of degree k but not on one of degree $k - 1$. Asymptotically he found the bound

$$\pi(C) \leqq d^2/2(k+1).$$

When $k = 3$ the precise bound is

$$\pi(C) \leqq \begin{cases} (d^2 - 3d + 2)/6 & \text{for} \quad d \equiv 1,2,(3) \\ (d^2 - 3d + 6)/6 & \text{for} \quad d \equiv 0\,(3) \end{cases}$$

As a consequence,

If $C \subset P^3$ *is a non-degenerate algebraic curve of degree d and genus* π *with*

$$\pi > (d^2 - 3d)/6 + 1,$$

then C lies on a quadric surface.

In this case Castelnuovo's bound is

$$\pi \leqq \begin{cases} (d-2)^2/4 & d \equiv 0\,(2) \\ (d-1)(d-3)/2 & d \equiv 1\,(2) \end{cases}$$

Asymptotically we have

$$\pi \leqq d^2/4 \text{ for any non-degenerate curve}$$
$$\pi > d^2/6 \Rightarrow C \text{ lies on a quadric surface.}$$

There are also results of a similar nature for non-degenerate curves in P^n.

ii. *Extremal curves of degree* $d > 2n$. We now consider an extremal algebraic curve $f: C \to P^n$ of degree d where $d > 2n$ and $n \geqq 3$. It will turn out that f is generically an embedding, and we shall identify C with its image.

---

[4]) This means that the hypersurfaces of degree k cut out *complete* linear systems on the curve for all k; such curves are called *arithmetically normal.* That canonical curves are arithmetically normal is a classical result of Max Noether.

The basic fact is

(1.10) *C lies on a surface S of degree n − 1* [5]).

Proof. Denote by $|V(C)|$ the linear system of $\infty^{(n-1)(n-2)/2}$ quadrics which contain C − cf. (1.9) above. Since C is non-degenerate, no quadric $Q \in |V(C)|$ has rank one. Consequently the restriction

$$\text{quadrics} \quad Q \in |V(C)| \to Q|\boldsymbol{P}^{n-1}$$

is injective for any hyperplane $\boldsymbol{P}^{n-1} \subset \boldsymbol{P}^n$. It follows that all the hyperplane sections $\xi \cdot C$ consists of

$$d > 2(n-1) + 2$$

points lying on $\infty^{(n-1)(n-2)/2}$ quadrics. For generic ξ these points are in general position, and we deduce from (1.4) that

$$\xi \cdot \bigcap_{Q \in |V(C)|} Q = D_\xi$$

is a rational normal curve. It follows that

$$S = \bigcap_{Q \in |V(C)|} Q$$

is a surface of degree $n-1$ containing the curve C. Q.E.D.

Now then the non-degenerate surfaces of degree $n-1$ in $\boldsymbol{P}^n$ were classified by del Pezzo and may be described as follows (cf. Chapter IV of [5] for proofs):

First, a cone over a rational normal curve is clearly a surface of minimal degree having a singularity at the vertex of the cone.

Secondly, the embedding $\boldsymbol{P}^2 \to \boldsymbol{P}^5$ defined by the complete linear system $|H^0(\boldsymbol{P}^2, \mathfrak{O}(H^2))|$ of quadrics in $\boldsymbol{P}^2$ defines a smooth surface, the *Veronese surface*. Since any two conics meet in 4 points the degree of this surface is the minimal number four. The Veronese surface is the unique non-degenerate surface in $\boldsymbol{P}^5$ whose chordal variety has dimension four, rather than five as one would generally expect.

In general a minimal surface[6]) turns out to be rational, and so $h^1(S, \mathfrak{O}) = h^2(S, \mathfrak{O}) = 0$. It follows that the group of line bundles

$$\operatorname{Pic}(S) \cong H^2(S, \boldsymbol{Z}),$$

the isomorphism being via the Chern class. The remaining minimal surfaces

$$S = S(k, l)$$

[5]) We recall that this is the minimal degree of a non-degenerate surface in $\boldsymbol{P}^n$.

[6]) Minimal here means having no exceptional curves of the 1st kind. This is the only time we shall use this terminology.

are described as follows: Over P^1 we consider the standard line bundle $H^k \to P^1$ of degree $k \geqq 0$. Adding a point at infinity to each fibre we obtain a surface $S(k)$ which is a P^1-bundle over P^1. The zero section of $H^k \to P^1$ is a curve on $S(k)$ having self-intersection number k. We denote by $e \in \mathrm{Pic}(S(k)) \cong H^2(S(k), Z)$ the class of the curve at infinity, so that

$$e \cdot e = -k .$$

We may characterize $S(k)$ as being the unique minimal rational surface having a curve of self-intersection number $-k$. Let $f \in \mathrm{Pic}(S(k))$ be the class of a fibre so that

$$\begin{cases} f \cdot f = 0 \\ f \cdot e = 1 . \end{cases}$$

The curves e and f give an integral basis for $\mathrm{Pic}(S(k))$. Therefore, any line bundle is a linear combination $\alpha e + \beta f$ $(\alpha, \beta \in Z)$. In particular, the canonical bundle is of this form and using the *adjunction formula*

$$\pi(C) = 1/2(C \cdot C + C \cdot K) + 1$$

for the genus of a smooth curve C applied to e and f we obtain

$$K = -2b + (-k-2)f .$$

We now consider the line bundle $L(k,l) \to S$ whose Chern class

$$c_1(L(k,l)) = e + (k + l + 1)f .$$

In [5] it is proved that

$$h^0(S(k), \mathfrak{O}(L(k,l))) = k + 2l + 4$$

and that the complete linear system $|H^0(S(k), \mathfrak{O}(L(k,l)))|$ gives an embedding for $l \geqq 0$. We identify $S(k)$ with its image $S(k,l)$ to obtain a surface in P^n where $n = k + 2l + 3$. The degree of $S(k,l)$ is

$$(e + (k + l + 1)f) \cdot (e + (k + l + 1)f) = -k + 2k + 2l + 2 = n - 1 .$$

We note that since

$$(e + (k + l + 1)f) \cdot f = e \cdot f = 1$$

the fibres of $S(k) \to P^1$ map into straight lines in P^n. Thus the $S(k,l)$ are *rational ruled surfaces*, sometimes referred to as *scrolls*. *Together with the cones over rational normal curves and Veronese surface, the surfaces $S(k,l)$ give all non-degenerate surfaces of degree $n-1$ in P^n.*

Moreover, the genus of a curve on $S(k,l)$ may be easily computed by the adjunction formula with the following conclusion:

(1.11) *For each $n \geqq 3$ and $d > 2n$, curves C of maximum genus $\pi(d,n)$ exist. These curves are non-singular and lie either on a cone over a rational normal curve or on a surface $S(k,l)$. In all cases the surface S is the intersection of the quadrics containing C. The hyperplane sections $\xi \cdot C(\xi \in \mathbf{P}^{n*})$ are d points on a rational normal curve $D_\xi = \xi \cdot S$. Finally, extremal curves exist over $\mathbb{R}$ as well as $\mathbb{C}$.*

Given an extremal algebraic curve C, the unique ruled surface S which contains the curve will be referred to as *Castelnuovo's ruled surface.* When $n = 3$, S is the standard doubly ruled quadric. An extremal algebraic curve has either type $(k, k-1)$ or (k,k) depending on whether its degree is odd or even. In the latter case C has degree $2k$ and is a complete intersection of S with a hypersurface of degree k; in the former case C + (line) is a complete intersection. This property of extremal algebraic curves to be as close as possible to complete intersections persists in higher dimensions.

iii. *The canonical curve and Poincaré mapping.* We have now discussed extremal algebraic curves of degree d in $\mathbf{P}^n$ when $n \leqq d \leqq 2n-1$ and $d > 2n$, and have mentioned in passing that those of degree $d = 2n$ are canonical curves. These will now be described in more detail, and then we shall relate those of degree $d > 2n$ to the canonical curve.

Suppose that C is a compact Riemann surface of genus $\pi \geqq 2$. The space $H^0(C,\mathfrak{O}(K)) = H^0(C,\Omega^1)$ of holomorphic 1-forms on C has dimension π and the associated complete linear system is base point free – i.e., for every point $p \in C$ there is an $\omega \in H^0(C,\Omega^1)$ with $\omega(p) \neq 0$. Choosing a basis $\omega_1,\ldots,\omega_\pi$ for the holomorphic differentials we obtain the *canonical mapping*

$$\varkappa : C \to \mathbf{P}^{\pi-1}$$

defined by

$$\varkappa(p) = [\omega_1(p),\ldots,\omega_\pi(p)] .$$

We note that

$$\mathbf{P}^{*\pi-1} \cong \mathbf{P}(H^0(C,\Omega^1));$$

i.e., the hyperplane sections are just the divisors of holomorphic differentials. In case C is non-hyperelliptic, which is the general case when the genus $\pi \geqq 3$, the canonical mapping is a one-to-one embedding. The image $C_\varkappa$ is a *canonical curve.* It has degree

$$2\pi - 2 = 2(\dim \mathbf{P}^{\pi-1}),$$

is intrinsically attached to the compact Riemann surface C and, as we shall now discuss, the geometry of $C_\varkappa$ reflects many properties of the special divisors on C.

The basic observation is this: Suppose that C is non-hyperelliptic and

$$D = p_1 + \cdots + p_d$$

is a divisor of degree d. Denote by $\{D\} = \{p_1, \ldots, p_d\}$ the linear subspace of $P^{\pi-1}$ spanned by the points p_i on the canonical curve. Then, since the index of speciality is just the number of linearly independent hyperplanes in $P^{\pi-1}$ which contain D, by elementary linear algebra

$$\dim\{D\} = \pi - 1 - i(D).$$

Comparing this with the third form (1.7) of the Riemann-Roch gives

$$\dim\{D\} = (d-1) - \dim|D|.$$

Consequently, $\dim|D|$ *exactly measures the extent to which the points* p_i *fail to be linearly independent on the canonical curve* [7]).

Put geometrically, suppose we say that a linear subspace $P^{d-n-1} (n > 0)$ of $P^{\pi-1}$ is a *multisecant plane* in case it is the span of d points on the canonical curve. For example, when $d = 3$ and $n = 1$ we have a trisecant line, and so forth. Then the study of special divisors is equivalent to the study of multisecant planes on the canonical curve. *This study is governed by the remarkable fact that if there is one multisecant* $P^{d-n-1} = \{p_1, \ldots, p_d\} (p_i \in C_\varkappa)$, *then there are at least* ∞^n *such multisecant planes*, as follows from the Rieman-Roch. Although it does not seem to have been made rigorous, this is perhaps the distinguishing characteristic of the canonical curve — at least provided we assume there is a non-empty set of special divisors.

Now, suppose $f: C \to P^n$ is a normal algebraic curve of degree d with typical hyperplane section

$$D_\xi = \xi \cdot f(C) = p_1(\xi) + \cdots + p_d(\xi).$$

By the Riemann-Roch

$$n + \pi = d + i(D_\xi).$$

Therefore, fixing n and d and maximizing the genus π is equivalent to maximizing the index of speciality $i(D_\xi)$. By the preceding remark this is the same as finding extremal multisecant planes, and what is suggested is that we investigate how the multisecant plane $\{D_\xi\}$ varies with $\xi \in P^{n*}$.

---

[7]) We are primarily interested in divisors of degree $d \leqq \pi - 1$. For a generic D of this degree $i(D) = \pi - d$. Consequently

$$\dim|D| = 0$$

for a generic such D, hence the source of the name *special divisors*.

So, given any normal algebraic curve $f: C \to \mathbf{P}^n$ we define the *Poincaré mapping*[8])

$$F: \mathbf{P}^{n^*} \to \mathbf{G}(d-n-1, r-1)$$

by

$$F(\xi) = \{p_1(\xi), \ldots, p_d(\xi)\}.$$

This allows us to study our original algebraic curve by investigating a mapping between two familiar spaces.

The understanding of F requires analyzing its infinitesimal structure. To see what is involved, we denote by $p_i'(\xi)$ a point on the tangent line to the canonical curve at $p_i(\xi)$. Then the infinitesimal structure of the Poincaré mapping is reflected in the linear space

$$\{p_1(\xi), \ldots, p_d(\xi); p_1'(\xi), \ldots, p_d'(\xi)\} = \{p_i(\xi); p_i'(\xi)\}$$

spanned by the points $p_i(\xi)$ and $p_i'(\xi)$. This clearly relates to the divisor $2D_\xi$, and in fact

$$\dim\{p_i(\xi); p_i'(\xi)\} = 2d - \pi + i(2D_\xi).$$

Referring to the lemma (1.8) in § I B i,

$$3n - 1 \leqq \dim|2D|$$

while

$$\dim|2D_\xi| = 2d - \pi + i(2D_\xi)$$

by the Riemann-Roch. These combine to yield

$$\dim\{p_i(\xi), p_i'(\xi)\} \leqq 2d - 3n,$$

with equality holding exactly when $f(C)$ lies on $\infty^{(n-1)(n-2)/2}$ quadrics. *In general, we may say that the infinitesimal structure of the Poincaré mapping reflects the quadrics containing the curve $f(C)$.*

Suppose in particular that $f: C \to \mathbf{P}^n$ is an extremal algebraic curve of degree $d > 2n$, $n \geqq 3$. Let S be *Castelnuovo's ruled surface* on which the curve lies. We will prove that:

(1.12) *The canonical mapping $\varkappa: C \to \mathbf{P}^{\pi-1}$ extends uniquely to a mapping $\varkappa: S \to \mathbf{P}^{\pi-1}$. Consequently the canonical curve C lies on a rational surface $S_\varkappa$ in $\mathbf{P}^{\pi-1}$.*

Proof. From surface theory we recall the adjunction formula now in the form

$$K_C = K_S \otimes [C]_C$$

[8]) In a somewhat different context this method was used by Poincaré in his paper *Sur les surfaces de translation et les fonctions Abeliennes, Bull. Soc. Math. France* vol. **29** (1901) 61–86 on the Sophus Lie theorem. We introduce the Poincaré mapping here because it will make sense in a purely web-theoretic context, and in this form will play a crucial role in our discussions.

where, as usual, K_X denotes the canonical bundle of a variety X. The adjunction formula yields the exact sheaf sequence

$$0 \to \Omega_S^2 \to \Omega_S^2(C) \to \Omega_C^1 \to 0$$

where

$$\Omega_S^2(C) \cong \mathfrak{O}(K_S \otimes [C]).$$

Since S is rational

$$h^0(S,\Omega_S^2) = 0 = h^1(S,\Omega_S^2),$$

and we obtain an isomorphism

$$H^0(S,\Omega^2(C)) \simeq H^0(C,\Omega_C^1).$$

If we think of the canonical curve $\varkappa: C \to \boldsymbol{P}^{\pi-1}$ as being given by the line bundle $K_C \to C$ together with its complete linear system, then we have just shown that there is a unique line bundle $J \to S$ with $J|C = K_C$, and that $H^0(S,(J)) \simeq H^0(C,\mathfrak{O}(K_C))$ is an isomorphism.

The linear system $|H^0(S,\mathfrak{O}(J))|$ has no base points and gives a one-to-one mapping $S \to \boldsymbol{P}^{\pi-1}$ along the curve C. By allowing C to vary on S we may conclude that the above linear system is entirely base point free and gives an embedding $\varkappa: S \to \boldsymbol{P}^{\pi-1}$. To carry this out it is only necessary to know that C varies in a large linear system on S, and this is proved in Chapter IV of [5]. Q.E.D.

We shall use this to study the Poincaré mapping associated to the extremal algebraic curve. The notation

$$\{p_1(\xi),\ldots,p_d(\xi)\} = \boldsymbol{P}^{d-n-1}(\xi)$$

will be used in lieu of $F(\xi)$ – this has the advantage of emphasizing not only the linear space $F(\xi)$, but also the set of d points $p_i(\xi)$ generating this multisecant $\boldsymbol{P}^{d-n-1}(\xi)$. A first property is

(1.13) *The points $p_i(\xi)$ lie on a rational normal curve E_ξ in $\boldsymbol{P}^{d-n-1}(\xi)$.*

Proof. Of course we will take

$$E_\xi = \varkappa(\xi \cdot S)$$

as the image under $\varkappa: S \to \boldsymbol{P}^{\pi-1}$ of a hyperplane section $\xi \cdot S$ of Castelnuovo's ruled surface. Clearly E_ξ is rational, and we must prove that

$$\deg E_\xi = d - n - 1.$$

By the intersection properties derived in the previous section

$$\begin{aligned}\deg(E_\xi) &= [\xi \cdot S] \cdot (K_S \otimes [C]) \\ &= E \cdot K_S + E \cdot C \\ &= d - n - 1.\end{aligned}$$

Q.E.D.

Now we shall prove that:

(1.14) *Any two of the rational normal curves* E_ξ *and* $E_{\xi'}$ *have* $(n-1)$ *points in common. Consequently, as* ξ *varies the* ∞^n *rational normal curves* E_ξ *trace out the surface* $S_\varkappa$, *the image under the canonical mapping of Castelnuovo's ruled surface.*

Proof. Two hyperplanes ξ and ξ' meet in a $\boldsymbol{P}^{n-2}$; we let $\sigma = \xi \circ \xi'$. Then

$$\sigma \cdot S = \xi \cdot \xi' \cdot S$$

is just $(n-1)$ points, and under $\varkappa$ these go into $E_\xi \cdot E_{\xi'}$. Q.E.D.

Note that as we move along the line[9])

$$\xi_t = \xi + t\xi'$$

in $\boldsymbol{P}^{n*}$, the corresponding points

$$\boldsymbol{P}^{d-n-1}(\xi(t))$$

will contain the $(n-2)$-fold secant $\boldsymbol{P}^{n-2}(\sigma)$ spanned by the $(n-1)$ points in $E_\xi \cdot E_{\xi'}$.

Summarizing, *we are able to reconstruct Castelnuovo's ruled surface in the space* $\boldsymbol{P}^{n-1}$ *of the canonical curve as being the locus of the rational normal curves* E_ξ *lying on the multisecant* $\boldsymbol{P}^{d-n-1}(\xi)$*'s containing a fixed* $\boldsymbol{P}^{n-2}(\sigma)$ *as* σ *runs over the hyperplanes in* ξ.

Actually, we have only proved that the intersection $\boldsymbol{P}^{d-n-1}(\xi) \cdot \boldsymbol{P}^{d-n-1}(\xi')$ contains the $\boldsymbol{P}^{n-2}(\sigma)$ spanned by the $n-1$ points on $E_\xi \cdot E_{\xi'}$. But the linear space $\{p_i(\xi), p_i'(\xi)\}$ spanned by the points $p_i(\xi)$ of $\xi \cdot C$ and the tangent lines to the canonical curve at these points is a $\boldsymbol{P}^{2d-3n}$ containing $\boldsymbol{P}^{d-n-1}(\xi)$ and $\boldsymbol{P}^{d-n-1}(\xi')$. Since $2(d-n-1) - (2d-3n) = n-2$ we have the equality $\boldsymbol{P}^{d-n-1}(\xi) \cdot \boldsymbol{P}^{d-n-1}(\xi') = \boldsymbol{P}^{n-2}(\sigma)$.

These special properties of the Poincaré map associated to an extremal algebraic curve will be important when we prove our main result characterizing the webs defined by such curves. These properties may all be stated purely in terms of the web defined by an extremal curve. This will be donne in § II A ii., and in § III we shall reprove them in a purely web-theoretic context.

II. Webs and abelian equations

A. Basic definitions and examples

i. *Definitions and the first non-trivial example.* Our study will be local, and we shall work in an open set U in $\boldsymbol{R}^n$ or $\boldsymbol{C}^n$ with coordinates denoted

[9]) This line is the pencil of hyperplanes with axis $\sigma = \xi \cdot \xi'$. We may identify the lines in $\boldsymbol{P}^{n*}$ through ξ with the dual space ξ^* of all such $\boldsymbol{P}^{n-2}$'s σ contained in ξ.

$x = (x_1, \ldots, x_n)$ in either case. The open set U may be thought of as a sufficiently small neighborhood of some point x_0, and it may be shrunk a finite number of times during the discussion. We shall consider functions $u(x)$, differential forms $\omega(x) = \sum_{\alpha=1}^{n} f_\alpha(x) dx_\alpha$, etc. In the real case these will be real-valued and C^∞; in the complex case they will be holomorphic, i.e., convergent power series in the x_α's. Since no use will be made of the conjugate variables $\bar{x}_\alpha$ or Cauchy-Riemann equations $\partial u/\partial \bar{x}_\alpha = 0$, it will not be necessary to specify whether we are in the real or complex case.

Definition. *A d-web is given by d codimension-one foliations in U. The leaves of the foliations will be called the web hypersurfaces, and we shall always assume that the tangent hyperplanes to the web hypersurfaces through a point in U are in general position.*

In general we may define a d-web of codimension k by requiring that the leaves of the d foliations should be submanifolds of codimension k whose tangent spaces are in general position. The study of such webs will almost certainly be quite interesting, but in this paper we shall be concerned exclusively with the codimension one case. Some remarks pertaining to the higher codimension case are given in § 3 of the second author's paper *On Abel's Differential Equations*, to appear in Amer. J. Math. The Blaschke-Bol book [1] contains an extensive discussion of the origins of the study of webs and of the simplest special cases, such as a 3-web of curves in plane. It will be our main reference.

In general we may say that *the study of webs is concerned with the local invariants of a set of d foliations in general position*. Invariance here means under the group of diffeomorphisms. As an example of an invariant, suppose we are given a 4-web of curves in the plane. At each point p the tangent lines to the four curves through p will define four points in the projectivized tangent space $\boldsymbol{P}(T_p) \cong \boldsymbol{P}^1$, and the *cross ratio* of these is invariant under diffeomorphisms.

One of the main interests in the theory is to find conditions under which a web may be put in a standard local form. For example, suppose we make the

Definition. *A d-web is linear in case the web hypersurfaces are (pieces of) linear hyperplanes in $\boldsymbol{R}^n$ or $\boldsymbol{C}^n$. A web is linearizable in case it is equivalent to a linear web under a change of coordinates.*

The main result of this paper is to give sufficient conditions under which a web is linearizable.

We first remark that when $d \leqq n$ any d-web is linearizable. Indeed, a general d-web may be given by the level sets

$$u_i(x) = \text{constant} \quad (i = 1,\dots,d) \tag{2.1}$$

of functions $u_i(x)$. These defining functions are unique up to a change

$$U_i(x) = v_i(u_i(x)) \tag{2.2}$$

where $v_i(u)$ is a function of a single variable with derivative $v_i'(u) \neq 0$. The non-degeneracy condition

$$du_{i_1} \wedge \cdots \wedge du_{i_k} \neq 0 \quad (k \leqq n)$$

implies that when $d \leqq n$ we may take $u_1,\dots,u_d$ as part of a coordinate system in which the web is linear[10]).

Consequently the first non-trivial case is when $d = n + 1$. Before discussing this we remark that it is frequently convenient to give the ith foliation in our web by

$$\omega^i(x) = 0 \tag{2.3}$$

where ω^i is a 1-form satisfying the *integrability condition*

$$d\omega^i \wedge \omega^i = 0\,. \tag{2.4}$$

Here the equation (2.3) means that the tangent hyperplane to the ith web hypersurface is defined by

$$\langle \omega^i(x), \xi \rangle = 0\,, \quad \xi \in T_x.$$

The form ω^i is well-defined up to a change

$$\tilde{\omega}^i = \varrho^i \omega^i$$

where ϱ^i is a non-zero function. Consequently the points

$$\omega^i(x) \in \boldsymbol{P}(T_x^*)$$

in the projectivized cotangent spaces are well-defined; these will be called the web normals. The non-degeneracy condition is just that the web normals should be d points in general position in $\boldsymbol{P}(T_x^*) \cong \boldsymbol{P}^{n-1}$.

---

[10]) This contrasts sharply with the higher codimension case. For example, suppose we have a 2-web of codimension 2 in $U \subset \boldsymbol{R}^3$. Thru each point x there will pass two curves whose normal spaces will have a line $L_x \subset T_x^*$ in common. Taking $\omega(x)$ to be a 1-form generating L_x, the condition

$$d\omega \wedge \omega \not\equiv 0$$

is independent of the choice of ω and gives the obstruction to linearization.

The relation between these two ways of defining a web is

$$du_i = \varrho^i \omega^i . \tag{2.5}$$

Now suppose that $d = n + 1$. There will be a linear relation

$$\alpha_1 \tilde{\omega}^1 + \cdots + \alpha_{n+1} \tilde{\omega}^{n+1} = 0$$

among the web normals $\tilde{\omega}^i(x)$. The coefficients are all non-zero, and this is the unique such relation up to multiplication by non-zero homogeneity factor. Setting $\omega^i = \alpha_i \tilde{\omega}^i$ we thus have

$$\omega^1 + \cdots + \omega^{n+1} = 0 \tag{2.6}$$

where the ω^i are unique up to

$$\tilde{\omega}^i = \varrho \, \omega^i$$

with the same ϱ.

The integrability condition (2.4) is equivalent to

$$d\omega^i = \pi^i \wedge \omega^i$$

where π^i is determined up to

$$\pi^i \to \pi^i + \beta^i \omega^i .$$

The form $\tilde{\pi}^i$ corresponding to $\tilde{\omega}^i = \varrho \omega^i$ is

$$\tilde{\pi}^i = \pi^i + d\varrho/\varrho$$

Consequently the condition

There exist functions β^i such that

$$\pi^i + \beta^i \omega^i = \pi \tag{2.7}$$

is the same for all i

has intrinsic meaning. We will show that

(2.8) *If $n \geqq 3$ and if (2.7) is satisfied, then the web is linearizable.*

Proof. We may assume that

$$d\omega^i = \pi \wedge \omega^i$$

with the same π for all i. Taking exterior derivatives gives

$$0 = d\pi \wedge \omega^i ,$$

so that $d\pi$ is a multiple of ω^i for all i. Since $n \geqq 3$ it follows that $d\pi = 0$. Locally we may find a function ϱ with $d\varrho = \pi$, and then

$$\tilde{\omega}^i = e^{-\varrho} \omega^i$$

satisfies $d\tilde{\omega}^i = 0$. If $u_i = \int \tilde{\omega}^i$ is a function with $du_i = \tilde{\omega}^i$, then by (2.6)

$$u_1(x) + \cdots + u_{n+1}(x) = \text{constant}.$$

Taking $u_1, \ldots, u_n$ as a coordinate system, the web is equivalent to one formed by $n + 1$ families of parallel hyperplanes. In particular, it is linearizable. Q.E.D.

Definition. *A web is octahedral*[11]) *in case it is equivalent to one formed by* $(n + 1)$ *families of parallel hyperplanes.*

An octahedral web is linearizable, but not conversely. Since an octahedral web is, in a suitable coordinate system, defined by

$$\begin{cases} x_\alpha = \text{constant} & \alpha = 1, \ldots, n \\ x_1 + \cdots + x_n = \text{constant}, \end{cases}$$

the converse of (2.8) is obvious. Thus:

(2.9) *The condition* (2.7) *is equivalent to the* $(n + 1)$*-web being octahedral.*

We now wish to reinterpret the conditions (2.6) and (2.7) for a web given by $\{u_i(x) = \text{constant}\}$. If (2.7) is satisfied then we may multiply (2.6) by a non-zero factor to have $d\omega^i = 0$. It follows that

$$\omega^i(x) = f_i(u_i(x))\, du_i(x)$$

and (2.6) assumes the form

$$\sum f_i(u_i(x))\, du_i(x) \equiv 0. \tag{2.11}$$

Definition. *For a general d-web an equation* (2.11) *is called an abelian equation. The maximum number of linearly independent abelian equations is called the rank r of the web.*

In other words, *an abelian equation is a relation among the web normals* $du_i(x)$ *whose coefficients* $f_i(u_i(x))$ *are constant along the web hypersurfaces.* The terminology abelian equation will be explained in the next section, and in § II B we will prove that the rank satisfies

$$r \leqq \pi(d, n)$$

where $\pi(d, n)$ is Castelnuovo's bound on the genus of a non-degenerate curve of degree d in $\boldsymbol{P}^n$.

For the moment we note that the rank $r = 0$ in case $d \leqq n$ and that $r \leqq 1$ for $d = n + 1$. Then (2.8) may be rephrased as:

---

[11]) For $n = 2$ we say that the web is *hexagonal*. The terminology is explained in Blaschke-Bol [1].

(2.12) *An $(n+1)$-web of maximum rank $r = 1$ is linearizable*[13]).

The main theorem of this paper will be a generalization of this to d-webs of maximal rank $r = \pi(d,n)$ when $d \geqq 2n$ and $n \geqq 3$. In fact, the result will be similar to (2.9) in that a stronger assertion than just linearizability will be proved. The problem of finding necessary and sufficient conditions for just linearizability seems difficult – cf. Blaschke-Bol [1].

ii. *Webs defined by algebraic varieties; relation to Abel's theorem.* We first recall the

Principle of projective duality: The mapping

$$\boldsymbol{P}^n \to (\boldsymbol{P}^{n*})^*$$

given by sending a point $p \in \boldsymbol{P}^n$ to the set $\boldsymbol{P}^{n-1}_p$ of hyperplanes passing through p is an isomorphism.

Now suppose that $f: C \to \boldsymbol{P}^n$ is a non-degenerate algebraic curve of degree d. For simplicity of notation we identify C with its image $f(C)$. Since the curve is non-degenerate, a general hyperplane ξ will meet C in d points in general position in $\xi \cong \boldsymbol{P}^{n-1}$. We write

$$\xi \cdot C = p_1(\xi) + \cdots + p_d(\xi).$$

In a neighborhood $U \subset \boldsymbol{P}^{n*}$ of such a general point we define a d-web by letting the ith web hypersurface passing through ξ be $\boldsymbol{P}^{n-1}_{p_i(\xi)}$ (Fig. 3).

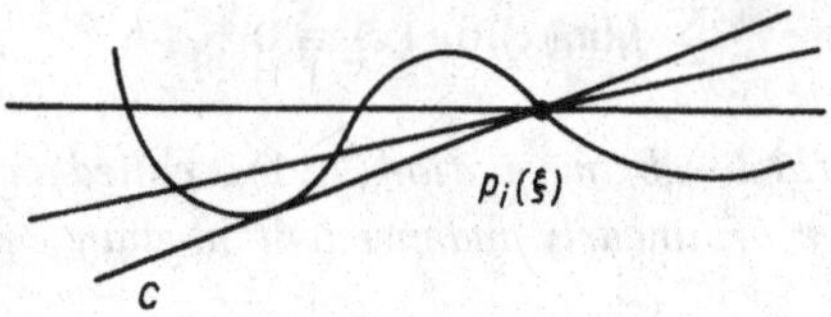

Fig. 3

Thus, a non-degenerate algebraic curve of degree d defines a linearizable d-web in sufficiently small open sets $U \subset \boldsymbol{P}^{n*}$ [14]).

For these webs defined by an algebraic curve the web normals have an especially nice description. To give it we recall that: The projectivized

[13]) Our discussion above pertained to the case $n \geqq 3$. However, it is immediate that a 3-web in the plane which satisfies one abelian equation (2.11) is linearizable.

[14]) When $d \geqq n + 2$, this web is not in general equivalent to one given by d families of parallel hyperplanes. For example, suppose that $n = 2$ and $d = 4$. Then the four lines through each point have a cross-ratio which is invariant under diffeomorphisms and which is a constant in $x \in U$ for 4 families of parallel lines. It is, however, not constant for the web defined by a quartic curve as may be seen by letting ξ tend to a tangent line to the curve.

cotangent space $P(T_\xi^*)$ at a point $\xi \in P^{n*}$ is naturally identified with ξ itself, i.e.,

$$P(T_\xi^*) \cong \xi .$$

Indeed, the projectivized tangent space is just the set of directions emanating from ξ, and this has a natural identification with the set of lines in P^{n*} passing through ξ. Such a line is a pencil $|\xi + t\xi'|_{t\in P^1}$ of hyperplanes containing ξ, and this pencil is uniquely determined by its axis $\xi \cap \xi' \cong P^{n-2}$, which is a hyperplane in $\xi \cong P^{n-1}$. Dualizing this natural identification

$$P(T_\xi) \cong \xi^*$$

gives (2.13).

Now a moment's reflection shows that:

(2.14) *Under the identification* (2.13), *the web normals* $\omega^i(\xi) \in P(T_\xi^*)$ *are just the points* $p_i(\xi) \in \xi$ [15]).

If it happens to be the case that the curve lies on a surface S in P^n, then the hyperplane section $\xi \cdot S$ will be an algebraic curve $D_\xi \subset P(T_\xi^*)$ on which the web normals $\omega^i(\xi)$ will lie. This will be the case for Castelnuovo's extremal curves of degree $d > 2n$ when $n \geqq 3$.

We now wish to relate the classical Abel theorem to the abelian equations (2.11) of a web defined by an algebraic curve. Although not strictly necessary it will simplify our explanations if we assume that $f: C \to P^n$ is a smooth embedding – cf. [4] for a discussion of Abel's theorem for possibly singular curves.

Suppose that C has genus π and let $\omega \in H^0(C, \Omega^1)$ be a holomorphic differential. The indefinite integral

(2.15) $$u(p) = \int_{p_0}^{p} \omega$$

is defined modulo the periods of ω and is called an *abelian integral.* A special case of the classical form of *Abel's theorem* states that the *abelian sum*

$$u(p_1(\xi)) + \cdots + u(p_d(\xi)) = \text{const}$$

associated to variable points of intersection of a hyperplane ξ with the curve C is constant. Differentiation of (2.15) gives the equivalent form

(2.16) $$\omega(p_1(\xi)) + \cdots + \omega(p_d(\xi)) \equiv 0$$

of Abel's theorem. Here $\omega(p_i(\xi))$ denotes the pullback to $U \subset P^{n*}$ of ω under the mapping $\xi \to p_i(\xi)$. Since the ith web hypersurface through ξ defined by

[15]) Referring to the preceding footnote, this identification makes it apparent that the cross-ratio of the four points $\omega^i(\xi) = p_i(\xi)$ is not constant in ξ.

the algebraic curve is obtained by holding $p_i(\xi)$ fixed, we see that (2.16) gives an abelian equation associated to the web. It is not difficult to verify that *any* abelian equation arises in this manner, and consequently:

For the web defined by an algebraic curve C in $\mathbf{P}^n$, the abelian equations are given by the holomorphic differentials ω via Abel's theorem in the form (2.16). *In particular, the rank of the web is equal to the genus of the curve.*

In this case maximizing the rank of the d-web is the same as maximizing the genus of the curve of fixed degree d, and this should help explain the bound mentioned in the previous section.

There is also a relationship between the linearization of a web and the classical converse to Abel's theorem on a compact Riemann surface C. Let $M = C^{(d)}$ be the *d-fold symmetric product* of C. M is a smooth compact, complex manifold of dimension d which we may think of as the set of effective divisors

$$D = p_1 + \cdots + p_d \quad (p_i \in C)$$

of degree d on C. The holomorphic differentials $\omega \in H^0(C, \Omega^1)$ induce holomorphic differentials $\boldsymbol{\omega} \in H^0(M, \Omega^1)$ by

$$\boldsymbol{\omega}(D) = \omega(p_1) + \cdots + \omega(p_d),$$

and the mapping $\omega \to \boldsymbol{\omega}$ is an isomorphism.

An *integral variety* $U \subset M$ is a local complex-analytic subvariety of M such that all

$$\boldsymbol{\omega} | U \equiv 0 .$$

If x denotes a coordinate on U and $D(x) = p_1(x) + \cdots + p_d(x)$ the corresponding point in U, then we may define a d-web in U by requiring that the ith web hypersurface passing through x_0 is given by

$$p_i(x) = p_i(x_0) .$$

The web is non-degenerate in case each $p_i(x)$ varies in an open set on C, and since

$$\boldsymbol{\omega}(x) = \omega(p_1(x)) + \cdots + \omega(p_d(x)) \equiv 0$$

for any $\omega \in H^0(C, \Omega^1)$ it has rank $r = \pi(C)$. Linearizing this web is analogous to proving that $D(x)$ varies in a linear equivalence class on C, which is a consequence of the classical converse to Abel's theorem.

B. Bound on the rank of a web

i. *Proof of the bound.* We consider a d-web $\{u_i(x) = \text{constant}\}$ and recall that the rank is the maximum number of linearly independent abelian equations

(2.11) $$\sum_i f_i(u_i(x))du_i(x) \equiv 0.$$

We set

$$m = [(d-1)/(n-1)]$$

and recall Castelnuovo's number

$$\pi(d,n) = m[d - 1/2(m+1)(n-1) - 1]$$

giving the maximum genus of a non-degenerate curve of degree d in $\boldsymbol{P}^n$. The following bound was proved for $n = 2,3$ by Blaschke-Bol [1] and for general n by Chern [3]:

(2.17) **Proposition.** *The rank r of a d-web in n-space satisfies*

$$r \leqq \pi(d,n).$$

Webs of maximum rank $r = \pi(d,n)$ exist by taking the web associated to an extremal algebraic curve.

The method of proof, which will play a central role in this paper, originated with Poincaré. Briefly stated, the idea is *to mimic for general webs the construction of the canonical curve associated to an algebraic curve in $\boldsymbol{P}^n$*. We shall now explain this. Suppose that C is a smooth non-degenerate algebraic curve of degree d in $\boldsymbol{P}^n$ and having genus π. Consider the canonical mapping

$$\varkappa : C \to \boldsymbol{P}^{\pi-1}$$

defined by

$$\varkappa(p) = [\omega_1(p), \ldots, \omega_\pi(p)]$$

where $\omega_1, \ldots, \omega_\pi$ are a basis for the holomorphic differentials on the curve. For a variable hyperplane $\xi \in \boldsymbol{P}^{n*}$ we write

$$\xi \cdot C = p_1(\xi) + \cdots + p_d(\xi)$$

and set

$$\begin{aligned} Z_i(\xi) &= \varkappa(p_i(\xi)) \\ &= [\omega_1(p_i(\xi)), \ldots, \omega_\pi(p_i(\xi))]. \end{aligned}$$

The $Z_i(\xi)$ give d points in $\boldsymbol{P}^{\pi-1}$, and Abel's theorem in differential form

$$\omega(p_1(\xi)) + \cdots + \omega(p_d(\xi)) \equiv 0, \quad \omega \in H^0(C, \Omega^1),$$

says that the $Z_i(\xi)$ span at most a $\boldsymbol{P}^{d-n-1}(\xi)$ in $\boldsymbol{P}^{\pi-1}$ [16]. A basic observation

[16]) We also encountered this statement in the discussion of the Riemann-Roch theorem in § I B i. This suggests a link between Abel's theorem and the Riemann-Roch theorem. We shall attempt to clarify this point at the end of this section.

is that *since* $Z_i(\xi)$ *depends only on the* i^{th} *web hypersurface passing through* ξ, *the Poincaré mapping*

$$\xi \to \boldsymbol{P}^{d-n-1}(\xi)$$

can be expressed purely in terms of the web defined by the given algebraic curve.

In general, suppose that we have r independent abelian equations

$$\sum_i f_i^\lambda(u_i(x))du_i(x) \equiv 0 \quad (\lambda = 1,\dots,r). \tag{2.18}$$

We may assume that the coefficient matrix has no column ${}^t(f_i^1,\dots,f_i^r)$ identically zero since otherwise we would be reduced to a $(d-1)$-web. If we set

$$Z_i(x) = [f_i^1(u_i(x)),\dots,f_i^r(u_i(x))], \tag{2.19}$$

Then the points $Z_i(x)$ *are intrinsically defined by the web and lie in a* $\boldsymbol{P}^{d-n-1}$ *in* $\boldsymbol{P}^{r-1}$. Indeed, under a change

$$\tilde{u}_i = v_i(u_i(x))$$

of defining function for the ith hypersurface

$$\tilde{f}_i^\lambda(u_i(x)) = v_i'(u_i(x))^{-1} f_i^\lambda(u_i(x))$$

so that the homogeneous coordinate vector (2.19) has intrinsic meaning. Setting $u_{i\alpha} = \partial u_i/\partial x_\alpha$ we may rewrite the abelian equations (2.18) as

$$\sum Z_i(x)u_{i\alpha}(x) \equiv 0 \tag{2.20}$$

where the coefficient matrix $(u_{i\alpha}) = \partial(u_1,\dots,u_d)/\partial(x_1,\dots,x_n)$ has rank n. Consequently the points span at most a $\boldsymbol{P}^{d-n-1}$.

Now we set out to bound the rank of a general web. The abelian equations (2.18) impose at least n independent conditions on the points $Z_i(x)$ in $\boldsymbol{P}^{r-1}$. We shall assume that these are all the conditions, and at the end of the proof it will be apparent that if there were more conditions then the estimate on the rank would be improved[17].

We note that as x varies the points $Z_i(x) = Z_i(u_i(x))$ will trace out a piece of an arc C_i in $\boldsymbol{P}^{r-1}$. Moreover, these arcs span $\boldsymbol{P}^{r-1}$ since otherwise the abelian equations (2.18) would not be independent. Each x determines d points, one $Z_i(x)$ on each arc C_i – we may say that there is a correspondence $x \to Z_1(x),\dots,Z_d(x)$ – where the linear span

$$\{Z_1(x),\dots,Z_d(x)\} = \boldsymbol{P}^{d-n-1}(x).$$

[17]) This is the web-theoretic analogue of the observation that the curves of maximal genus of fixed degree d in $\boldsymbol{P}^n$ are necessarily normal; i.e., the linear system of hyperplane sections is complete.

The mapping $x \to \{Z_1(x),\ldots,Z_d(x)\}$ is the *Poincaré mapping* for the abelian equations (2.18) associated to a general web. Here is an attempt at the picture for $n = 2$, $d = 4$, $r = 3$ (Fig. 4).

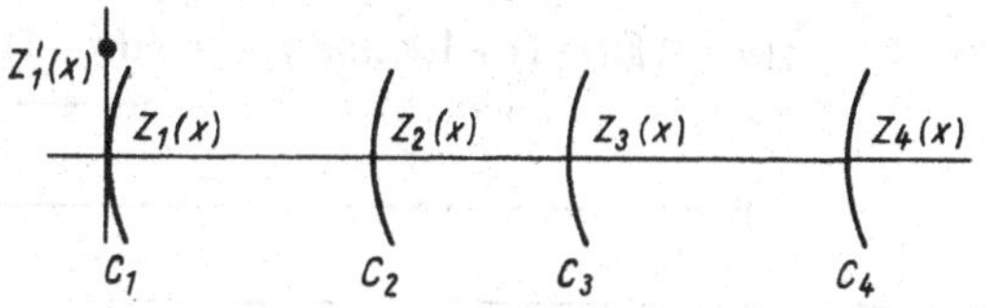

Fig. 4

We denote by $Z_i'(x)$ a point on the tangent line to C_i at $Z_i(x)$, by $Z_i''(x)$ a point such that $\{Z_i(x), Z_i'(x), Z_i''(x)\}$ is the osculating 2-plane to C_i at $Z_i(x)$, and so forth. In general we let

$$\begin{aligned} \boldsymbol{P}^{N(k)}(x) &= \{Z_1(x),\ldots,Z_d(x); Z_1'(x),\ldots,Z_d'(x);\ldots;Z_1^{(k)}(x),\ldots,Z_d^{(k)}(x)\} \\ &= \{Z_i(x); Z_i'(x);\ldots;Z_i^{(k)}(x)\} \qquad i = 1,\ldots,d \end{aligned}$$

denote the span in $\boldsymbol{P}^{r-1}$ of the kth osculating spaces to the arcs C_i at corresponding points $Z_i(x)$. As before, the rank will turn out to be maximized when the dimensions $N(k)$ are maximal subject to the conditions obtained by successively differentiating (2.20), and so we assume this to be the case. There are obvious inclusions

$$\boldsymbol{P}^{N(0)}(x) \subset \boldsymbol{P}^{N(1)}(x) \subset \cdots \subset \boldsymbol{P}^{r-1}.$$

Lemma: *If* $N(l) = N(l+1)$ *for some* l*, then* $N(l) = r - 1$.

Proof. We may assume that $\partial u_i/\partial x_\alpha \neq 0$. With the notations

$$f_i^{\lambda,(k)}(u) = \mathrm{d}^k f_i^{\lambda}(u)/\mathrm{d}u^k,$$

we have

$$\begin{aligned} \partial Z_i^{(k)}(x)/\partial x_\alpha &= [\partial f_i^{1,(k)}(u_i(x))/\partial x_\alpha,\ldots,\partial f_i^{r,(k)}(u_i(x))/\partial x_\alpha] \\ &= [f_i^{1,(k+1)}(u_i(x)),\ldots,f_i^{r,(k+1)}(u_i(x))] \\ &= Z_i^{(k+1)}(x). \end{aligned}$$

If $N(l) = N(l+1)$ then $Z_i^{(l+1)}(x)$ lies in $\boldsymbol{P}^{N(l)}(x)$, and consequently for any $Z(x) \in \boldsymbol{P}^{N(l)}(x)$ the derivative

$$\partial Z(x)/\partial x_\alpha \in \boldsymbol{P}^{N(l)}(x).$$

This says that the subspace $\boldsymbol{P}^{N(l)}(x)$ is a constant independent of x. Since this fixed subspace contains all arcs C_i we conclude that $N(l) = r - 1$.

Q.E.D.

4*

Now to the proof of the bound (2.17). We break the integers $1,\ldots,d-1$ up into $m=[(d-1)/(n-1)]$ disjoint subsets of $n-1$ elements each plus a remainder labelled as follows:

$$\overbrace{1,\ldots,n-1};\overbrace{n,\ldots,2n-2};\ldots;\underbrace{\underbrace{(m-1)(n-1)+1,\ldots,m(n-1);\underbrace{m(n-1)+1,\ldots,d-1}_{I_{m-1}}}_{I_{m-2}}}_{I_0}$$

We will inductively estimate the dimensions $N(k)$ using the following device: For any $k<m$ the functions

$$u_{k(n-1)+1},\ldots,u_{(k+1)(n-1)},u_d$$

form a coordinate system in which

$$\partial u_{k(n-1)+1}/\partial u_d=\cdots=\partial u_{(k+1)(n-1)}/\partial u_d=0\,. \tag{2.21}$$

We will successively differentiate the abelian equation (2.20) using (2.21). For the first step we take $(u_1,\ldots,u_{n-1},u_d)$ as coordinate system, and then by (2.21) for $\alpha=d$ the equation (2.20) becomes

$$Z_n\partial u_n/\partial u_d+\cdots+Z_{d-1}\partial u_{d-1}/\partial u_d+Z_d=0\,. \tag{2.22}$$

Thus Z_d is a linear combination of $Z_n,\ldots,Z_{d-1}$, and using the evident symmetry of the indices

$$P^{N(0)}=\{Z_i(x):i\in I_0\}\,.$$

Next, choose $(u_n,\ldots,u_{2n-2},u_d)$ as coordinate system and apply $\partial/\partial u_d$ to (2.22) to obtain

$$\varrho_{2n-1}Z'_{2n-1}+\cdots+\varrho_{d-1}Z'_{d-1}+Z'_d\equiv 0 \text{ modulo } P^{N(0)}(x)\,.$$

Again, by symmetry of the indices

$$P^{N(1)}(x)=\{Z'_i(x):i\in I_1\}\oplus P^{N(0)}(x)$$

where "$\oplus$" denotes the linear span of the two subspaces of P^{r-1}. Iterating the argument gives

$$P^{N(k)}(x)=\{Z_i^{(k)}(x):i\in I_k\}\oplus P^{N(k-1)}(x) \tag{2.23}$$

where $k\leq m-1$ and

(2.24) $N(k)=N(k-1)+(d-1-(k+1)(n-1)),\quad N(-1)=-1\,.$

The first two dimensions are

$$\text{(2.25)} \qquad \begin{cases} N(1) = d - n - 1 \\ N(2) = 2d - 3n . \end{cases}$$

At the last step

$$P^{N(m-1)}(x) = \{Z_i^{(m-1)}(x) : i \in I_{m-1}\} \oplus P^{N(m-2)}(x),$$

so that in particular

$$Z_d^{(m-1)}(x) \equiv \sum_{j \in I_{m-1}} \sigma_j Z_j^{(m-1)}(x) \text{ modulo } P^{N(m-2)}(x).$$

Choose $(u_i : i \in I_{m-1}, u_d)$ as part of a coordinate system and apply $\partial/\partial u_d$ to this relation to obtain

$$Z_d^{(m)}(x) \equiv 0 \text{ modulo } P^{N(m-1)}(x),$$

i.e.,

$$N(m) = N(m-1).$$

By the lemma

$$r - 1 = N(m-1) = m(d-1) - \sum_{k=1}^{m} k(n-1) - 1$$

or

$$r = m(d-1) - 1/2m(m+1)(n-1) = m[d - 1/2(m+1)(n-1) - 1]$$

as desired. Q.E.D.

We note the similarity between the argument just given and the algebro-geometric proof bounding the genus of a non-degenerate curve of degree d in P^n. Both relied on a combinatorial argument decomposing the integers $1,\dots,d$ into $m = [(d-1)/(n-1)]$ disjoint subsets, and in both instances the kth step had to do with the kth osculating curves. On the other hand, whereas Castelnuovo's argument was based on the Riemann-Roch the web-theoretic proof relied on Abel's theorem. To explain the connection we shall prove:

For an effective divisor D on a compact Riemann surface C, Abel's theorem (2.16) *implies the inequality*

$$\dim|D| \leqq \deg D - \pi + i(D)$$

in the Riemann-Roch theorem[18]).

We shall assume that C is non-hyperelliptic and that the complete linear system $|D|$ gives a projective embedding $C \to P^n$ ($n = \dim|D|$). This case

[18]) This inequality is all that is needed to establish Castelnuovo's bound. The converse of Abel's theorem yields the opposite inequality, and hence the full Riemann-Roch.

will cover the comparison we are trying to draw between the two proofs of Castelnuovo's bound and the argument we are about to give may be modified to cover the general situation. For a hyperplane $\xi \in P^{n^*}$ we write

$$\xi \cdot C = p_1(\xi) + \cdots + p_d(\xi) \in |D|$$

and set

$$Z_i(\xi) = \varkappa(p_i(\xi))$$

as above. Then by Abel's theorem (2.16) the $Z_i(\xi)$ span at most a $P^{d-n-1}(\xi)$ in the space $P^{\pi-1}$ of the canonical curve. On the other hand $\dim\{Z_1(\xi),\ldots,Z_d(\xi)\} = \pi - 1 - i(D)$. Combining we obtain

$$d - n - 1 \leqq \pi - 1 - i(D)$$

which implies (2.26).

Finally, if we go back to the proof of (2.17) we see that the first two steps (2.25) remain valid under the assumption

$$r > \mu(d,n)$$

for some $\mu(d,n)$ strictly less than Castelnuovo's number $\pi(d,n)$. For example, when $n = 3$ one finds (2.25) under the assumption

$$r > (d^2 - 3d)/6 + 1$$

in agreement with the bound obtained by Joe Harris for curves in P^3 which we mentioned at the end of § I B i. We have not determined the best exact value for $\mu(d,n)$, but the asymptotic estimates are

$$\begin{cases} r < d^2/2(n-1) \text{ for any } d\text{-web in } n\text{-space} \\ r > d^2/3(n-1) \Rightarrow \text{the bound (2.25)} \end{cases}$$

ii. *Statement and discussion of the main theorem*

Definition. *A web has maximum rank in case equality holds in* (2.17).

As mentioned before webs of maximum rank may be found in either the real or complex case, by taking the web in $U \subset P^{n^*}$ associated to an extremal algebraic curve $C \subset P^n$. By construction these webs are linear. The principal result of this paper is a converse

(2.27) **Linearization theorem.** *A d-web in n-space of maximal rank* $r = \pi(d,n)$ *is linearizable provided that*

$$\begin{cases} d = n+1, 2n & \textit{for any } n\text{[19]} \\ d > 2n & \text{for } n \geqq 3 \end{cases}$$

[19]) The index restrictions will turn out to be sharp. Cf. the discussion following Corollary (2.28).

Combining this with the main theorem proved in [4][20]) we deduce the

(2.28) **Corollary.** *Under the conditions in the main theorem, the given web is equivalent to one defined by an extremal algebraic curve C.*

We emphasize that there are *two* steps in constructing the algebraic curve C. The deeper local step consists of linearizing the web; the globalization of a linear web is of a different and more analytic character.

Regarding the index restrictions in the main theorem, recall that

$$\pi(d,n) = d - n \quad \text{for} \quad n \leqq d \leqq 2n - 1\,.$$

The case $d = n + 1$ where $\pi(d,n) = 1$ has been taken up in § II A i and the main theorem proved there – cf. (2.12). If, on the other hand $n + 2 \leqq d \leqq 2n - 1$, we may construct non-linearizable d-webs of maximal rank as follows: Define the web by level sets $\{u_i(x) = \text{constant}\}$ with

$$\text{(2.29)} \qquad \begin{cases} u_1 = x_1 \\ \vdots \\ u_n = x_n \\ u_{n+1} = U_{1,1}(x_1) + \cdots + U_{1,n}(x_n) \\ \vdots \\ u_{n+r} = U_{r,1}(x_1) + \cdots + U_{r,n}(x_n) \end{cases}$$

where the $U_{s,\alpha}(x_\alpha)$ $(1 \leqq s \leqq r = d - n,\ 1 \leqq \alpha \leqq n)$ are functions of a single variable. A typical n-fold Jacobian is

$$\begin{aligned} &du_{k+1} \wedge \cdots \wedge du_n \wedge du_{n+1} \wedge \cdots \wedge du_{n+k} \\ &= \pm \begin{vmatrix} U'_{1,1} & \cdots & U'_{1,k} \\ \vdots & & \\ U'_{k,1} & \cdots & U'_{k,k} \end{vmatrix} du_1 \wedge \cdots \wedge du_n\,, \end{aligned}$$

involving the Wronskians of the functions $U_{s,\alpha}$. For generic choice of such functions the web defined by the functions (2.29) will therefore be nondegenerate. Again, because Wronskians appear as coefficient matrices the abelian equations

$$du_{n+s} - \sum_\alpha U'_{s,\alpha}(u_\alpha)du_\alpha = 0 \quad (s = 1,\ldots,r)$$

will generically be linearly independent so that the web has maximal rank $r = d - n$. If this web were linearizable, then by the Sophus Lie theorem (cf. footnote 20) there would be an algebraic curve C in $\boldsymbol{P}^n$ of degree $n + r$ defining an equivalent web. This curve depends on only a finite number

[20]) The result referred to implies that a *linear* d-web of rank $r \geqq 1$ is defined by an algebraic curve C of degree d. We shall call this the *generalized Sophus Lie theorem.*

of constants (Chow variety), and moreover since any local diffeomorphism preserving a linear d-web for $d > n + 1$ is a projective transformation we see again that a general web (2.29) is not linearizable.

A further remark is that the proof of the linearization theorem will primarily depend on showing that under the assumptions

$$\begin{cases} r = \pi(d,n) \\ d > 2n \end{cases}$$

the web normals lie on a rational normal curve in the projectivized cotangent spaces. This follows from (2.25) and will be proved in § III A i; an outline of how this fits into the overall proof is discussed in § III A ii. On the other hand, as remarked at the end of the preceding section, the step (2.25) necessary to prove that the web normals lie on a rational normal curve will still be true provided $d > 2n$ and the rank

$$r > \mu(d,n)\,.$$

The upshot is that, by an posteriori analysis of the proof, our linearization theorem may be extended to d-webs ($d > 2n$) *whose rank*

$$r > \mu(d,n)\,,$$

where

$$\mu(d,3) = (d^2 - 3d)/6 + 1\,,$$

and in general

$$\mu(d,n) \sim d^2/3(n-1)$$

is strictly less than Castelnuovo's number $\pi(d,n) \sim d^2/2(n-1)$.

Now we shall prove the linearization theorem in the case

$$d = 2n,\ \pi(d,n) = n + 1$$

corresponding to the canonical curve. The Poincaré map is

$$x \to \{Z_1(x), \ldots, Z_{2n}(x)\} = \boldsymbol{P}^{n-1}(x) \subset \boldsymbol{P}^n\,,$$

which is consequently an *equidimensional* map

$$F : U \to \boldsymbol{P}^{n*}\,.$$

It is easy to check that as a consequence of the non-degeneracy F has non-zero Jacobian, and since the ith web hypersurface is defined by

$$Z_i(x) = \text{constant}\,,$$

this hypersurface corresponds under F to the hyperplane

$$\boldsymbol{P}^{n-1}_{Z_i(x)} \in (\boldsymbol{P}^{n*})^{*}\ {}^{21)}\,.$$

[21]) The notation is explained in the beginning of § II A ii where projective duality is formally stated.

Consequently, the Poincaré mapping F serves to linearize the web, and the theorem is proved in this case.

When $d > 2n, n \geqq 3$ the Poincaré mapping

$$x \to \{Z_1(x), \ldots, Z_d(x)\} = \mathbf{P}^{d-n-1}(x) \subset \mathbf{P}^{r-1}$$

associated to our web of maximal rank $r = \pi(d,n)$ will induce

$$F: U \to \mathbf{G}(d-n-1, r-1).$$

The Poincaré space $\mathbf{P}^{r-1}$ carries a natural linear structure, but F is no longer equidimensional and so is of a more complicated nature.

In fact, suppose for each point $Z \in \mathbf{P}^{r-1}$ we let $\mathbf{G}_Z(d-n-1, r-1)$ be the Schubert cycle in $\mathbf{G}(d-n-1, r-1)$ of all $\mathbf{P}^{d-n-1}$'s passing through Z. Then

$$\begin{cases} \dim \mathbf{G}(d-n-1, r-1) = (d-n)(r-d+n) \\ \dim \mathbf{G}_Z(d-n-1, r-1) = (d-n-1)(r-d+n) \end{cases}$$

and consequently

$$\operatorname{codim} \mathbf{G}_Z(d-n-1, r-1) = r-d+n.$$

Now the ith web hypersurface passing through $x \in U$ is, as in the equidimensional case,

$$F^{-1}(\mathbf{G}_{Z_i(x)}(d-n-1, r-1)).$$

If $F(U)$ and $\mathbf{G}_{Z_i(x)}(d-n-1, r-1)$ are in general position then they should meet in a variety of dimension $d-r$. Since for large d, $r = \pi(d,n) \sim d^2/2(n-1)$ this number will eventually be negative. In fact the right number

$$d - r = n - 1$$

occurs exactly when $d = 2n$ so that $r = n+1$ and we are in the case of the canonical curve. As a result we see that *when $d > 2n$ the Poincaré mapping F is highly non-generic, and we are in a situation somewhat analogous to an overdetermined system of equations for which the abelian relations* (2.20) *serve as compatibility conditions.*

iii. *Properties of webs defined by extremal algebraic curves.* From now on we will assume that $d > 2n$ and $n \geqq 3$. Let $C \subset \mathbf{P}^n$ be an extremal algebraic curve of degree d and genus $\pi = \pi(d,n)$. We denote by $C_\kappa \subset \mathbf{P}^{\pi-1}$ the canonical curve, and recall that the original curve C defined a d-web of maximal rank π in suitable open sets $U \subset \mathbf{P}^{n*}$ according to the prescription: *If, for a hyperplane ξ we write*

$$\xi \cdot C = p_1(\xi) + \cdots + p_d(\xi),$$

then the ith web hypersurface passing through ξ is the $\mathbf{P}^{n-1}_{p_i(\xi)}$ of hyperplanes ξ' with $p_i(\xi') = p_i(\xi)$. We restate what is perhaps the basic underlying principle in our study of maximal rank webs: *Any property of the curve C in $\mathbf{P}^n$ or of its canonical image C_κ in $\mathbf{P}^{\pi-1}$ which is expressible purely in terms of the web defined by C may be expected to hold for a general d-web of maximal rank.* For example, the Poincaré mapping given by

$$\xi \to \{Z_1(\xi),\ldots,Z_d(\xi)\} = \mathbf{P}^{d-n-1}(\xi) \subset \mathbf{P}^{\pi-1}$$

makes sense for a general web of maximal rank. Moreover, the Poincaré space $\mathbf{P}^{\pi-1}$ and projectivized cotangent spaces $\mathbf{P}(T_\xi^*)$ have intrinsic linear structures, even though this will not be true of the manifold U in which a general web is given.

We shall now list some properties of webs defined by extremal algebraic curves, and then the proof of our main theorem will proceed by showing that these same properties hold for general webs of maximal rank and may be used to interrelate the intrinsic linear structures in the Poincaré spaces and projectivized cotangent spaces in a sufficiently tight manner as to eventually yield the linearization theorem. These properties were either proved or are consequences of the discussion in § I d iii.

The first property of the web defined by C is (cf. (2.14))

(2.30) *The web-normals $\omega^i(\xi) \in \mathbf{P}(T_\xi^*)$ lie on a rational normal curve D_ξ.*

We shall be able to prove that, in general, a maximal rank web defines a field of rational normal curves D_x in the projectivized cotangent spaces $\mathbf{P}(T_x^*(U))$, and that the web normals $\omega^i(x)$ give completely integrable cross-sections of this structure.

The second property is (cf. (1.13))

(2.31) *The points $Z_i(\xi)$ lie on a rational normal curve E_ξ in $\mathbf{P}^{d-n-1}(\xi)$. The curves D_ξ and E_ξ are in projective correspondence in such a way that $\omega^i(\xi)$ corresponds to $Z_i(\xi)$.*

Again this property will be proved to remain valid for a general web of maximal rank.

The third property (cf. (1.14) and the end of § I d iii) refines the link established by the projectivity $D_\xi \barwedge E_\xi$ in (2.31). In fact it tells us how to define the straight lines in the linear structure on $U \subset \mathbf{P}^{n*}$ purely in terms of the web associated to our given curve C.

(2.32) *A tangent direction $\sigma \in \mathbf{P}(T_\xi)$ gives a hyperplane $\mathbf{P}^{n-2}(\sigma)$ in $\mathbf{P}(T_\xi^*)$. This hyperplane meets D_ξ in $n-1$ points, and the corresponding $n-1$ points on E_ξ span a secant plane $\mathbf{P}^{n-2}(\xi,\sigma)$ to C_ξ. In fact $\mathbf{P}^{n-2}(\xi,\sigma)$ is defined by the* *condition (cf. the explanation given in the list of notations)*

(2.33) $$dZ/d\sigma \in \boldsymbol{P}^{d-n-1}(\xi), \quad Z \in \boldsymbol{P}^{d-n-1}(\xi).$$

The straight line through $\xi \in U$ in the direction σ is given by all ξ' such that $\boldsymbol{P}^{d-n-1}(\xi')$ contains $\boldsymbol{P}^{n-2}(\xi,\sigma)$.

For a general web of maximal rank we will be able to show that the correspondence $D_x \barwedge E_x$ takes the hyperplanes $\boldsymbol{P}^{n-2}(\sigma)$ in $\boldsymbol{P}(T_x^*)$ bijectively onto the $(n-2)$-fold secants $\boldsymbol{P}^{n-2}(x,\sigma)$ to the rational normal curve E_x in $\boldsymbol{P}^{d-n-1}(x)$, and moreover this $\boldsymbol{P}^{n-2}(x,\sigma)$ will be defined by the equation

$$dZ/d\sigma \in \boldsymbol{P}^{d-n-1}(x)$$

analogous to (2.33). The condition

$$\boldsymbol{P}^{d-n-1}(x') \quad \text{contains} \quad \boldsymbol{P}^{n-2}(x,\sigma)$$

will then be proved to define a path $x(t)$ in U passing through x and with tangent direction σ there. In this way the maximal rank web will induce a path geometry in U such that the linearization theorem will be true if, and only if, there is a change of coordinates in U transforming the paths into straight lines.

Finally to show that there is such a change of coordinates, we will use the property for general webs of maximal rank analogous to the following property of webs defined on extremal algebraic curve C in $\boldsymbol{P}^n$:

(2.34) *Consider Castelnuovo's ruled surface S in $\boldsymbol{P}^n$ on which the curve C lies. For a hyperplane ξ the intersection $\xi \cdot S$ is the curve D_ξ under the identification $\xi \cong \boldsymbol{P}(T_\xi^*)$. Consequently, for each point $p \in \xi \cdot S$ there is a hypersurface $\boldsymbol{P}_p^{n-1}$ in U passing through ξ and whose normal corresponds to p under the above identification. In other words, the original web can be embedded in a larger family of ∞^2 hypersurfaces defined by the condition that their normals should lie on the field of rational normal curves $D_\xi \subset \boldsymbol{P}(T_\xi^*)$. Going over to the Poincaré space, if $\omega \in D_\xi$ corresponds to $Z \in E_\xi$, then the hypersurface passing through ξ and with normal ω is given by ξ' satisfying*

(2.35) $$Z \in \boldsymbol{P}^{d-n-1}(\xi').$$

Again we shall prove the analogous property for general webs of maximal rank. The ∞^2 hypersurfaces will turn out to be totally geodesic for the path geometry defined by (2.33), and the existence of this large number of totally geodesic hypersurfaces will imply projective flatness.

At this juncture it is likely that the complexity of the structure of webs arising from extremal algebraic curves is confusing, and perhaps the following correspondence table will help clarify the situation.

(1) given space U $\longleftrightarrow$ Poincaré space $P^{\pi-1}$

(2) point $\xi \in U$ $\longleftrightarrow$ subspace $P^{d-n-1}(\xi)$ in P^{r-1}

(3) $\begin{Bmatrix}\text{web normals} \\ \omega^i(\xi) \in P(T_\xi^*)\end{Bmatrix} \leftrightarrow \begin{Bmatrix}\text{points } Z_i(\xi) \text{ with} \\ \{Z_1(\xi),\ldots,Z_d(\xi)\} = P^{d-n-1}(\xi)\end{Bmatrix}$

(4) $\begin{Bmatrix}\text{rational normal} \\ \text{curve } D_\xi \subset P(T_\xi^*)\end{Bmatrix} \longleftrightarrow \begin{Bmatrix}\text{rational normal curve} \\ E_\xi \subset P^{d-n-1}(\xi)\end{Bmatrix}$

Under the projectivity $D_\xi \barwedge E_\xi$ the web normals $\omega^i(\xi)$ correspond to the $Z_i(\xi)$

(5) $\begin{Bmatrix}\text{tangent direction} \\ \sigma \in P(T_\xi)\end{Bmatrix} \longleftrightarrow \begin{Bmatrix}\text{secant plane} \\ P^{n-2}(\xi,\sigma) \text{ to } E_\xi\end{Bmatrix}$

Here, σ defines a hyperplane in $P(T_\xi^*)$ which meets D_ξ in $n-1$ points, and the secant plane $P^{n-2}(\xi,\sigma)$ is spanned by the corresponding points on E_ξ

(6) $\begin{Bmatrix}\text{line in } U \text{ through} \\ \xi \text{ and with tangent } \sigma\end{Bmatrix} \longleftrightarrow \begin{Bmatrix}\text{set of } \xi' \text{ such that} \\ P^{n-2}(\xi,\sigma) \subset P^{d-n-1}(\xi')\end{Bmatrix}$

We may think of this condition as defining a path geometry in U

(7) $\begin{Bmatrix}\text{totally geodesic} \\ \text{hypersurface through } \xi \\ \text{and with normal } \omega \in D_\xi\end{Bmatrix} \longleftrightarrow \begin{Bmatrix}\text{set of } \xi' \text{ satisfying} \\ Z \in P^{d-n-1}(\xi') \\ \text{where } Z \in E_\xi \text{ corresponds} \\ \text{to } \omega \text{ under } D_\xi \barwedge E_\xi\end{Bmatrix}$

Provided we replace the word "line" with "path" in (6), all the statements in this dictionary make sense for general webs of maximal rank, and beginning in the next section we shall prove them in this context. Once this has been done the main theorem will follow from some rather general results about projective differential geometry.

III. Path geometry associated to maximal rank webs

A. Abelian equations and rational normal curves

i. *Properties of the Poincaré map for maximal rank webs.* Let $\{u_i(x) = \text{constant}\}$ define a d-web of maximal rank r in an open set U in n-space. Suppose that

$$\sum_i f_i^\lambda(u_i(x))\,du_i(x) = 0 \quad (\lambda = 1,\ldots,r) \tag{3.1}$$

give a basis for the abelian equations associated to the web, and define

$$Z_i(x) = [f_i^1(u_i(x)),\ldots,f_i^r(u_i(x))] \in P^{r-1}.$$

As x varies the points $Z_i(x)$ trace out an arc C_i in P^{r-1}. Each $x \in U$ has associated to it points $Z_i(x) \in C_i$ – one may think of the assignment

$$x \to Z_1(x), \ldots, Z_d(x)$$

as a *correspondence*. We denote by $Z_i'(x)$ a point on the tangent line to C_i at $Z_i(x)$. From (2.25) we have:

If $d \geqq n+1$ then the $Z_i(x)$ span a $P^{d-n-1}(x)$ in P^{r-1}; If $d \geqq 2n$ then the $Z_i(x)$ and $Z_i'(x)$ together span a $P^{2d-3n}(x)$ in P^{r-1}.

We shall abbreviate these statements by writing

(3.2) $$\{Z_1(x), \ldots, Z_d(x)\} = P^{d-n-1}(x)$$

(3.3) $$\{Z_1(x), \ldots, Z_d(x); Z_1'(x), \ldots, Z_d'(x)\} = P^{2d-3n}(x).$$

The mapping (3.2)

$$F: U \to G(d-n-1, r-1)$$

is the *Poincaré mapping*. We will see that (3.3) gives the infinitesimal structure of F.

The purpose of the present section is to prove analogues for general d-webs of maximal rank of the properties (2.30)–(2.34) of the webs defined by extremal algebraic curves. We shall begin with the two statements:

(3.4) *Under the assumption $d > 2n, n \geqq 3$ the web normals $\omega^i(x)$ lie on a unique rational curve D_x in the projectivized cotangent space $P(T_x^*) \cong P^{n-1}$.*

(3.5) *With the same assumptions as in (3.4), the points $Z_i(x)$ lie on a rational normal curve E_x in $P^{d-n-1}(x)$. There is a projectivity*

$$D_x \barwedge E_x$$

taking $\omega^i(x)$ to $Z_i(x)$.

Proof of (3.4): Using the notations

$$u_{i\alpha} = \partial u_i/\partial x_\alpha, \quad u_{i\alpha\beta} = \partial^2 u_i/\partial x_\alpha \partial x_\beta$$

the abelian equations (3.1) may be written

(3.6) $$\sum_i Z_i(x) u_{i\alpha}(x) \equiv 0 \quad \alpha = 1, \ldots, n.$$

Moreover, *any linear relation among the $Z_i(x)$ is a combination of the equations* (3.6). Applying $\partial/\partial x_\beta$ to (3.6) gives

(3.7) $$\sum_i Z_i'(x) u_{i\alpha}(x) u_{i\beta}(x) + \sum_i Z_i(x) u_{i\alpha\beta}(x) = 0.$$

Taken together, (3.6) and (3.7) yield

$$n + n(n+1)/2 = n(n+3)/2$$

relations among the $Z_i(x)$ and $Z_i'(x)$. According to (3.3) only $3n - 1$ of these can be independent, and so there must be

$$n(n+3)/2 - (3n-1) = (n-1)(n-2)/2$$

relations among the relations. This implies that there will be equations

$$\sum_{\alpha,\beta} k^{\alpha\beta} u_{i\alpha}(x) u_{i\beta}(x) = 0 \tag{3.8}$$

$$\sum_{\alpha,\beta} k^{\alpha\beta} u_{i\alpha\beta}(x) = \sum_{\gamma} m^{\gamma} u_{i\gamma}(x) \tag{3.9}$$

where $k^{\alpha\beta} = k^{\beta\alpha}$ varies over an $(n-1)(n-2)/2$-dimensional linear space of quadrics.

The first equation (3.8) says that the web normals

$$\omega^i(x) = [u_{i1}(x), \ldots, u_{in}(x)] \in \boldsymbol{P}(T_x^*)$$

lie on $\infty^{(n-1)(n-2)/2}$ independent quadrics in $\boldsymbol{P}(T_x^*) \cong \boldsymbol{P}^{n-1}$. Since the $\omega^i(x)$ are in general position, if

$$d > 2(n-1) + 2 = 2n$$

then (1.4) implies that there is a unique rational normal curve D_x passing through the $\omega^i(x)$. Q.E.D. for (3.4).

We note that the restrictions $d > 2n$, $n \geqq 3$ appear naturally in this proof.

Also, (3.9) may be interpreted as stating that *the defining functions* $u_i(x)$ *for the web satisfy inhomogenous Laplace equations relative to the quadrics containing the web normals.* This will be of crucial importance in our later work.

Proof of (3.5): Choose $u_1, \ldots, u_n$ as coordinate system. Then D_x is a rational normal curve passing through the vertices $[1,0,\ldots,0], \ldots, [0,\ldots,0,1]$ of the coordinate simplex in $\boldsymbol{P}^{n-1}$. Using $\xi = [\xi_1, \ldots, \xi_n]$ as homogeneous coordinates, according to the proof of (1.1) D_x is given parametrically by

$$\varrho \xi_\alpha = a_\alpha/(t - b_\alpha) \quad \alpha = 1, \ldots, n \tag{3.10}$$

where ϱ is a homogeneity factor and a_α, b_α are functions of x. The n points $t = b_\alpha$ correspond to the vertices ω^α.

Since the web normals lie on D_x

$$\varrho_i u_{i\alpha} = a_\alpha/(t_i - b_\alpha) \quad \alpha = 1, \ldots, n$$

for $i = 1,\dots,d$. We set $\omega_{i\alpha} = \varrho_i u_{i\alpha}$ and write the abelian equations (3.6) in the form

$$Z_\alpha + \sum_{s=n+1}^{d} Z_s \omega_{s\alpha} = 0 \quad \alpha = 1,\dots,n. \tag{3.11}$$

This shows that $Z_{n+1}(x),\dots,Z_d(x)$ give a basis for $P^{d-n-1}(x)$. In terms of the homogeneous coordinates corresponding to this basis, (3.10) and (3.11) imply that

$$Z_\alpha = [1/(t_{n+1} - b_\alpha),\dots,1/(t_d - b_\alpha)] \quad \alpha = 1,\dots,n.$$

Consequently, the points $Z_i(x)$ all lie on the rational normal curve E_x in $P^{d-n-1}(x)$ given parametrically by

$$b \to [1/(t_{n+1} - b),\dots,1/(t_d - b)].$$

The rational normal curves D_x and E_x have respective linear parameters t and b. Setting $b = t$ gives a projectivity $D_x \barwedge E_x$ under which the corresponding points are

$$\begin{cases} t = b_\alpha \leftrightarrow \omega^\alpha \\ b = b_\alpha \leftrightarrow Z_\alpha \end{cases} \quad 1 \leqq \alpha \leqq n$$

$$\begin{cases} t = t_s \leftrightarrow \omega^s \\ b = t_s \leftrightarrow Z_s \end{cases} \quad n+1 \leqq s \leqq d.$$

Hence the projectivity takes ω^i to Z_i. Q.E.D. for (3.5).

When $n = 2$ the $Z_i(x)$ are d points on a $P^{d-3}(x)$, and these will always lie on a unique rational curve. Consequently this part of (3.5) remains valid for webs in the plane but imposes no restriction on the Z_i.

Next we shall prove an analogue of part of (2.32):

(3.12) *For a tangent direction $\sigma \in P(T_x)$, the set of $Z \in P^{d-n-1}(x)$ satisfying*

$$dZ/d\sigma \in P^{d-n-1}(x) \tag{3.13}$$

constitutes a $P^{n-2}(x,\sigma)$ [22]*. This $P^{n-2}(x,\sigma)$ is obtained by first considering σ as a hyperplane in $P(T_x^*)$ meeting D_x in $n-1$ points and then taking the $(n-2)$-fold secant plane spanned by the corresponding points on E_x.*

Proof. We may assume that

$$Z_{n+1}(x),\dots,Z_d(x) \quad \text{span } P^{d-n-1}(x).$$

---

[22]) Geometrically, $P^{n-2}(x,\sigma)$ is the intersection of $P^{d-n-1}(x)$ with the infinitely nearly linear space $P^{d-n-1}(x+\varepsilon\sigma)$. The notation (3.13) is explained in index of notations in the introduction.

On account of (3.3) there will be $(n-1)$ independent relations among the points $Z'_s(x)(n+1 \leqq s \leqq d)$. We write these as

$$\sum_s A_{\mu s} Z'_s \equiv 0 \text{ modulo } \boldsymbol{P}^{d-n-1} \quad (\mu = 1,\ldots,n-1). \tag{3.14}$$

A general point Z in $\boldsymbol{P}^{d-n-1}$ is written

$$Z = \sum_s p_s Z_s,$$

and

$$dZ/d\sigma \equiv \sum_s p_s du_s/d\sigma\, Z'_s \text{ modulo } \boldsymbol{P}^{d-n-1} \tag{3.15}$$

where we are thinking of σ as a non-zero tangent vector and have set $du_s/d\sigma = \langle du_s, \sigma \rangle$. Comparing (3.14) and (3.15), the condition (3.13) is equivalent to

$$p_s du_s/d\sigma = \sum_{\mu=1}^{n} c_\mu A_{\mu s}. \tag{3.16}$$

This shows that the solutions to (3.13) constitute a $\boldsymbol{P}^{n-2}(x,\sigma)$ which is the image of a standard $\boldsymbol{P}^{n-2}$ with homogeneous coordinates $[c_1,\ldots,c_{n-1}]$ under a linear map

$$\boldsymbol{P}^{n-2} \to \boldsymbol{P}^{d-n-1}(x)$$

as prescribed in (3.16)[23]).

We now want to prove that this $\boldsymbol{P}^{n-2}(x,\sigma)$ is given by the $(n-2)$-fold secant plane description. If ω lies on the rational normal curve D_x in $\boldsymbol{P}(T_x^*)$, then the tangent directions σ satisfying

$$\langle \omega, \sigma \rangle = 0$$

constitute a $\boldsymbol{P}^{n-2}(\omega)$ in $\boldsymbol{P}(T_x) \cong \boldsymbol{P}^{n-1}$. Let $\sigma_1,\ldots,\sigma_{n-1}$ be a basis for $\boldsymbol{P}^{n-2}(\omega)$, and consider the intersection

$$\boldsymbol{P}^{n-2}(x,\sigma_1) \cap \cdots \cap \boldsymbol{P}^{n-2}(x,\sigma_{n-1}) \tag{3.17}$$

in $\boldsymbol{P}^{d-n-1}(x)$. The crucial observation is that *if* (3.12) *is to be true, then this intersection must be the point* $Z \in E_x$ *which corresponds to* ω *under the projectivity* $D_x \barwedge E_x$. We shall prove – somewhat indirectly – that this is the case.

Now suppose that ω is *any* point in $\boldsymbol{P}(T_x^*)$. If $\sigma_1,\ldots,\sigma_{n-1}$ are a basis for $\boldsymbol{P}^{n-2}(\omega)$, we shall call a point Z lying on the intersection (3.17) a *knotpoint*. For example in case $\omega = \omega^i$ is a web normal, then for any vector σ which is tangent to the ith web hypersurface

$$dZ_i/d\sigma = Z'_i du_i/d\sigma = 0,$$

[23]) If $du_s/d\sigma = 0$ then we should take $p_s = 1$ in (3.16); the justification for this will emerge below. The clearest picture of the infinitely near $\boldsymbol{P}^{n-2}(x,\sigma)$ is obtained by using moving frames – cf. the discussion at the end of § III B i, especially equations (3.57) and (3.58).

and consequently Z_i is a knotpoint. In general, suppose we can prove that:

(3.18) *The set of all knotpoints – i.e., points lying on* $(n-1)$*-fold intersections* (3.17) *– is a rational normal curve in* $P^{d-n-1}(x)$.

Then we may complete the proof of (3.12) as follows: Since it contains the $Z_i(x)$, by uniqueness the rational normal curve of all knotpoints must be E_x. Moreover, since we have just proved the knotpoint corresponding to the web normal ω^i is Z_i, it follows that the projectivity

$$D_x \barwedge E_x$$

takes the normal curve D_x bijectively onto the set of knotpoints. A general tangent direction $\sigma \in P(T_x)$ defines a hyperplane in $P(T_x^*)$ meeting D_x in $(n-1)$ points $\omega_1(\sigma),\ldots,\omega_{n-1}(\sigma)$, and what we have just said proves that $P^{n-2}(x,\sigma)$ is the $(n-2)$-fold secant plane spanned by the knotpoints $Z(\omega_1(\sigma)),\ldots,Z(\omega_{n-1}(\sigma))$.

So all that remains is to prove (3.18). We set

$$\begin{cases} du_s/d\sigma = \sum_\alpha u_{s\alpha}\sigma^\alpha \\ \varrho_s = -1/p_s \end{cases}$$

and rewrite (3.16) in the form

$$\sum_\alpha u_{s\alpha}\sigma^\alpha + \sum_\mu c^\mu A_{\mu s}\varrho_s = 0 \qquad s = n+1,\ldots,d. \tag{3.19}$$

We consider (3.19) as a set of $(d-n)$ homogeneous linear equations in the $2n-1$ unknowns $(\sigma^1,\ldots,\sigma^n;c^1,\ldots,c^{n-1})$. The condition that

$$Z = \sum_s p_s Z_s$$

be a knotpoint is that (3.19) should have $(n-1)$ linearly independent solutions. If we consider the coefficient matrix

$$\begin{pmatrix} u_{n+1,1} & \cdots & u_{n+1,n} & \overbrace{A_{1,n+1}\varrho_{n+1} & \cdots & A_{n-1,n+1}\varrho_{n+1}} \\ \vdots & & & \vdots & & \\ u_{d,1} & \cdots & u_{d,n} & A_{1,d}\varrho_d & \cdots & A_{n-1,d}\varrho_d \end{pmatrix}$$

as a linear map from R^{2n-1} to R^{d-n}, the condition is that the image have dimension $\leq n+1$. Since the first n columns in this matrix are linearly independent, this in turn is equivalent to saying that the last $n-1$ columns – i.e., those with the bracket over them – be linear combinations of the first n columns. Equivalently, any $(n+1)\times(n+1)$ minor

$$\begin{vmatrix} u_{i_1,1} & \cdots & u_{i_1,n} & A_{\mu,i_1}\varrho_{i_1} \\ \vdots & & & \\ u_{i_{n+1},1} & \cdots & u_{i_{n+1},n} & A_{\mu,i_{n+1}}\varrho_{i_{n+1}} \end{vmatrix} = 0. \tag{3.20}$$

The equations (3.20) are linear in the ρ_s, and so may be given parametrically by $\varrho_s = \varrho_s(t)$ where $t = (t_1,\ldots,t_k)$ are linear parameters. It is easy to see that since any $n \times n$ minor from the Jacobian matrix $\partial(u_{n+1},\ldots,u_d)/\partial(x_1,\ldots,x_n)$ is non-zero, the number of linear parameters is *at most* one. On the other hand, it must be *at least* one since we have already found the d knotpoints Z_i. It follows that

$$\varrho_s = -(\alpha_s t + \beta_s),$$

and consequently

$$Z(t) = 1/(\alpha_s t + \beta_s) Z_s$$

gives a parametric representation of the knotpoints. It is now clear that these constitute a rational normal curve. Q.E.D. for (3.12).

The proof of (3.18) is similar to the argument in pages 266–272 in Blaschke-Bol [1], from which the word knotpoint was taken. We observe that the projectivity

$$D_x \barwedge E_x$$

is characterized by: *A point* $\omega \in D_x \subset \mathbf{P}(T_x^*)$ *corresponds to* $Z \in E_x \subset \mathbf{P}^{d-n-1}(x)$ *if and only if*

$$dZ/d\sigma \in \mathbf{P}^{d-n-1}(x)$$

for all σ *satisfying* $\langle\omega,\sigma\rangle = 0$.

ii. *Strategy of the proof.*

Definition. *A path geometry is the system of curves defined by* 2nd *order differential equations*

$$(3.21)\qquad \begin{aligned} &dx_\beta/dt(d^2x_\alpha/dt^2 + \textstyle\sum \Gamma_\alpha^{\lambda\mu}(dx_\lambda(t)/dt)(dx_\mu/dt)) \\ &= dx_\alpha/dt(d^2x_\beta/dt^2 + \textstyle\sum \Gamma_\beta^{\lambda\mu}(dx_\lambda(t)/dt)(dx_\mu(t)/dt)). \end{aligned}$$

Through each point x_0 and in each tangent direction $\sigma \in \mathbf{P}(T_{x_0})$ there is a unique solution curve $x(t)$ of the system (3.21) having the initial data $x(0) = x_0$ and $x'(0) = \sigma$ in $\mathbf{P}(T_{x_0})$ – these are the *paths*. It is important to note that there is no distinguished parameter, such as arc length, for the paths. In fact, the form of the equations (3.21) is invariant under arbitrary changes of coordinates

$$(3.22)\qquad y_\alpha = y_\alpha(x_1,\ldots,x_n)$$

and changes of parameter

$$(3.23)\qquad s = s(t).$$

This is not true of the corresponding system of homogeneous equations.

Definition. *The path geometry is flat in case there is a change of coordinates* (3.22) *and change of parameter* (3.23) *transforming* (3.21) *into the differential equations*

$$d^2 y_\alpha / ds^2 = 0$$

characteristic of straight lines in Euclidean space.

In section IV we shall discuss how a path geometry leads to an intrinsically defined *projective connection* whose associated projective curvature tensor is zero if and only if the path geometry is flat. The situation is somewhat analogous to the manner in which a Riemannian metric leads to an intrinsic Riemannian connection whose curvature is zero exactly when the original structure is equivalent to the standard Euclidean one. The difference is that the flat model space is $\boldsymbol{P}^n$ with the projective group operating in the first case and $\boldsymbol{R}^n$ with the Euclidean group in the second.

Now if our linearization theorem is true then, according to (2.33) and the analogous property (3.12) for general webs of maximal rank, the straight line through x_0 and in the direction σ is characterized as

$$\{x : \boldsymbol{P}^{n-2}(x_0, \sigma) \subset \boldsymbol{P}^{d-n-1}(x)\}. \tag{3.24}$$

Equivalently, as we move along $x(t)$ the secant plane $\boldsymbol{P}^{n-2}(x(t), x'(t))$ should be the fixed $\boldsymbol{P}^{n-2}(x_0, \sigma)$ where $x(0) = x_0, x'(0) = \sigma$, i.e., $x(t)$ is a solution curve of the differential equation

$$d/dt(\boldsymbol{P}^{n-2}(x(t), x'(t))) \subset \boldsymbol{P}^{n-2}(x(t), x'(t)). \tag{3.25}$$

Now (3.25) may be written out as a system of 2nd order equations which, for the same reasons as those discussed at the end of § II B ii, is over-determined. The main step in our proof of the linearization theorem will be to show that the compatibility conditions in (3.25) are automatically satisfied, i.e., that:

(3.26) *The equations* (3.25) *define a path geometry* (3.21).

Once (3.26) has been established, the proof of the main theorem is reduced to showing that the associated projective connection is flat. To explain how this is done, we need the

Definition. *A totally geodesic submanifold S for a path geometry given by* (3.21) *is characterized by the property that a path $x(t)$ lies entirely in S in case, for some t_0, $x(t_0) \in S$ and $x'(t_0)$ is tangent to S.*

5*

It is a classical theorem that if $n \geqq 3$, and if for every point x and normal $\omega \in P(T_x^*)$ there is a totally geodesic hypersurface passing through x with normal ω, then the path geometry is flat. We are able to refine this to

(3.27) *Given a path geometry* (3.21) *and field* $D_x \subset P(T_x^*)$ *of rational normal curves, if for every* x *and* $\omega \in P(T_x^*)$ *there is a totally geodesic hypersurface passing through* x *and with normal* ω, *then the path geometry is flat.*

Assuming (3.27) the proof of the main theorem is completed as follows: According to (3.5) and (3.12) a point $\omega \in D_x$ corresponds to $Z \in P^{d-n-1}(x)$ characterized by (c.f. the end of § III A (i)).

$$dZ/d\sigma \in P^{d-n-1}(x) \Leftrightarrow \langle \omega, \sigma \rangle = 0 . \tag{3.28}$$

The equations $\langle \omega, \sigma \rangle = 0$ define a $P^{n-2}(\omega)$ in $P(T_x)$, and the paths emanating from x and with tangent direction $\sigma \in P^{n-2}(\omega)$ fill out a hypersurface passing through x and with normal ω. Points x' in this hypersurface satisfy (cf. (2.35))

$$Z \in P^{d-n-1}(x') . \tag{3.29}$$

On the other hand, the conditions (3.29) can define at most a hypersurface in U since not all $P^{d-n-1}(x)$'s pass through any one point of P^{r-1}. Consequently, (3.29) defines a totally geodesic hypersurface for our path geometry, and in this way we have embedded our original web in a larger family of ∞^2 hypersurfaces whose normals fill out the rational normal curves $D_x \subset P(T_x^*)$. Our result then follows from (3.27).

One way of summarizing the proof is this: *The maximal rank web defines a field of rational normal curves* D_x *in the projectivized cotangent spaces* $P(T_x^*)$. *We want to construct Castelnuovo's ruled surface S, and the* D_x *constitute the disjoint union of the hyperplane sections of S. So, in order to find the identifications necessary to obtain S we go to the Poincaré space* P^{r-1} *and field of rational normal curves* $E_x \subset P^{d-n-1}(x)$ *in projective correspondence with* D_x. *Our theorem is true exactly when the* E_x *lie on a 2-dimensional surface S in* P^{r-1}, *and since there are* ∞^n *curves* E_x *this will be the case when two infinitely nearby curves* E_x *and* $E_{x+\varepsilon\sigma}$ *meet in* $(n-1)$ *points. Thinking of* σ *as a hyperplane in* $P(T_x^*)$, *these* $(n-1)$*-points are the images of the hyperplane section* $\sigma \cap D_x$ *of* D_x *under the projectivity* $D_x \barwedge E_x$. *In fact, the lines in the* P^n *in which Castelnuovo's surface S is to lie are characterized by letting* $x \in U$ *vary subject to the condition that the* $P^{d-n-1}(x)$ *have a fixed* $P^{n-2}(x_0, \sigma)$ *as axis. These lines are described by an — a priori overdetermined — system of O.D.E.'s; the necessary compatibility conditions are a reflection of* E_x *and* $E_{x+\varepsilon\sigma}$ *meeting in* $(n-1)$*-points.*

We conclude this section by discussing informally what is involved in the proof of the central result (3.26). A first remark is that the group G of projective transformations leaving fixed a rational normal curve $D \subset P^{n-1}$ induces the full transitive group of projectivities on the curve $D \cong P^1$. Since any two rational normal curves are projectively equivalent, the structure of a field $D_x \subset P(T_x^*)$ of rational curves given by (3.4) is a *G-structure.* When $n = 3$, which was the case considered by Blaschke-Bol [1], D_x is a conic in the plane $P(T_x^*) \cong P^2$ and the G-structure is a *conformal Riemannian structure.* In particular, there exist torsion-free G-connections in the tangent bundle (Weyl connections) to which one may apply existing formalism in differential geometry. It was in this setting that Bol gave his proof of the $n = 3$ case of the theorem.

Now, and this was the principal technical difficulty we encountered, when $n \geqq 4$ the group G is relatively small and there need not exist torsion-free connections leaving fixed a general field of rational normal curves [24]). However, we have two additional pieces of information:

(3.30) the field of rational normal curves has a large number d of *completely integrable cross-sections* $\omega^i(x)$; and

(3.31) the defining functions for the foliations given by $\omega^i(x) = 0$ have the *harmonic property* (3.9).

So our G-structure has rather special properties which will have to come into play if we are to be able to prove (3.26). In fact, one might hope that the properties (3.30) and (3.31) might imply the existence of a torsion-free connection in the tangent bundle which leaves invariant the curves $D_x \subset P(T_x^*)$, and for a while we thought this would be the case. However, this hope turned out to be naive, and in § III B iv we have given the structure of the "best" connection possible for the problem; it is a torsion-free connection leaving invariant the D_x only when $n = 3$.

So in this paper we show by direct computation that the desired path geometry can be introduced. This is done in § III B ii, and constitutes the essential step in the proof. Then a continuation of this computation leads to the best connection in the following section.

B. Introduction of the path geometry

i. *Structure equations for maximal rank webs.* We begin by considering a general d-web in n-space given by a Pfaffian system

$$\omega^i(x) = 0 \qquad i = 1,\dots,d$$

[24]) This reflects the overdetermined character of the equations (3.25).

satisfying the complete integrability condition

$$d\omega^i = \pi^i \wedge \omega^i \tag{3.33}$$

(no summation of the index i). The form ω^i is non-vanishing and is determined up to multiplication by a non-zero function. In particular we will have

$$\omega^i = e^{\varrho_i} du_i$$

where $u_i(x)$ is a function whose level sets define the ith foliation in the web.

Now suppose the web has maximal rank r with basis

$$\sum_i f_i^\lambda(u_i(x))\, du_i(x) \equiv 0 \qquad \lambda = 1, \ldots, r$$

for its abelian equations. The point in the Poincaré space $\boldsymbol{P}^{r-1}$ corresponding to $x \in U$ is given by

$$Z_i(x) = [f_i^1(u_i(x)), \ldots, f_i^r(u_i(x))] .$$

In this section we shall slightly abuse notation and denote by

$$Z_i(x) = e^{-\varrho_i}(f_i^1(u_i(x)), \ldots, f_i^r(u_i(x)))$$

the designated vector lying over the point in $\boldsymbol{P}^{r-1}$. When this is done the condition that $Z_i(x) \in \boldsymbol{P}^{r-1}$ should depend only on the ith web hypersurface through x is expressed by

$$d(Z_i \omega^i) = 0$$

(no summation again). This inplies

$$(dZ_i + \pi^i Z_i) \wedge \omega^i = 0 ,$$

and so we may define the point Z_i' on the tangent line to the arc C_i traced out at Z_i by

$$dZ_i + \pi^i Z_i = Z_i' \omega^i . \tag{3.34}$$

Now we suppose that $d > 2n$ and $n \geqq 3$. Then the web normals $\omega^i(x)$ lie on a rational normal curve D_x in the projectivized cotangent space $\boldsymbol{P}(T_x^*)$. This in turn defines a G-structure, to which one may seek to apply standard techniques in differential geometry – in particular the method of moving frames – to find the invariants. Carrying this out to arrive at the path geometry will be the backbone of the proof of our main theorem.

We recall that for any closed subgroup $G \subset \mathrm{GL}_n$ a G-structure on a manifold M is given relative to an open covering $U, V, \ldots$ of M by bases $\omega_U^\alpha, \omega_V^\beta, \ldots$ for the 1-forms in the respective open sets such that in intersections $U \cap V$

$$\omega_U^\alpha = \sum_\beta \omega_V^\beta (g_{UV})_\beta^\alpha$$

where $g_{UV} = \{(g_{UV})^{\alpha}_{\beta}\}$ is a G-valued matrix. In our case we may use the fact that any two rational normal curves in $\boldsymbol{P}^{n-1}$ are projectively equivalent and that such a curve has a transitive group of automorphisms induced by projective transformations of the ambient space to deduce the existence of a G-structure where G is the group leaving fixed the standard rational normal curve

$$t \rightarrow [t, t^2, \ldots, t^n]\ ^{25)}.$$

Definition. *A moving frame is given by a coframe* $\{\phi^{\alpha}\}$, *or basis for the cotangent bundle, such that the rational normal curve* C_x *is given parametrically by*

$$e^{\varrho(x)}\Phi(x,t) = \sum_{\alpha=1}^{n} t^{\alpha}\phi^{\alpha}(x). \tag{3.35}$$

In other words a moving frame gives a homogeneous coordinate system for $\boldsymbol{P}(T_x^*) \cong \boldsymbol{P}^{n-1}$ in which C_x is the standard rational normal curve. To determine the group G, it is more convenient to set $t = t_1/t_0$ and use the homogeneous coordinate representation

$$e^{\varrho}\Phi = \sum_{\alpha=1}^{n} t_0^{\alpha-1} t_1^{n-\alpha}\phi^{\alpha} \tag{3.36}$$

for the standard rational normal curve. If we make a change of homogeneous variables

$$\begin{cases} t_0 = a_{00}t_0^* + a_{01}t_1^* \\ t_1 = a_{10}t_0^* + a_{11}t_1^* \end{cases}$$

and substitute in (3.36), then

$$e^{\varrho}\Phi = \sum_{\alpha} t_0^{*\alpha-1} t_1^{*n-\alpha}\phi^{*\alpha}$$

where

$$\phi^{*\alpha} = \sum_{\beta}\phi^{\beta} g^{\alpha}_{\beta}$$

and $g(x) = (g^{\alpha}_{\beta}(x))$ describes a general element in the subgroup $G \subset \mathrm{GL}_n$. We note that G has 4 parameters rather than the usual 3 when GL_n is considered as acting on $\boldsymbol{P}^{n-1}$. This is because we are in effect considering the *cone* in T_x^* lying over the rational normal curve C_x in $\boldsymbol{P}(T_x^*)$.

Now the equations of G are somewhat messy, but those for the Lie algebra of $\mathfrak{g} \subset \mathfrak{gl}(n)$ are controllable. To derive them we set

$$\begin{aligned} t_0 &= (1 + E_{00})t_0^* + E_{01}t_1^* \\ t_1 &= E_{10}t_0^* + (1 + E_{11})t_1^*. \end{aligned}$$

[25]) We shall use this parametric representation rather than the usual $t \rightarrow [1, t, \ldots, t^{n-1}]$ so as to allow all indices to run from 1 to n.

Then, modulo higher order terms in the $E_{\mu\nu}$,

$$t_0^{\alpha-1}t_1^{n-\alpha} = (1 + (\alpha - 1)E_{00} + (n - \alpha)E_{11})t_0^{*\alpha-1}t_1^{*n-\alpha} + E_{10}(n-2)t_0^{*\alpha}t_1^{*n-\alpha-1} + E_{01}(\alpha - 1)t_0^{*\alpha-2}t_1^{*n-\alpha+1}.$$

It follows that $\mathfrak{g}$ is generated by the transformations

$$\begin{cases} \phi^{\alpha *} = (\alpha - 1)\phi^\alpha \\ \phi^{\alpha *} = (n - \alpha)\phi^\alpha \\ \phi^{\alpha *} = (n - \alpha + 1)\phi^{\alpha-1} \\ \phi^{\alpha *} = \alpha\phi^{\alpha+1} \end{cases}$$

corresponding to the matrices

$$h_{11} = \begin{pmatrix} n-1 & & 0 \\ & \ddots & \\ & & 1 \\ 0 & & 0 \end{pmatrix} \qquad h_{22} = \begin{pmatrix} 0 & & 0 \\ & 1 & \\ & & \ddots \\ 0 & & n-1 \end{pmatrix}$$

(3.37)

$$h_{12} = \begin{pmatrix} 0 & n-1 & 0 \\ & \ddots & \ddots \\ & & \ddots & 1 \\ 0 & & 0 \end{pmatrix}, \qquad h_{21} = \begin{pmatrix} 0 & & 0 \\ 1 & \ddots & \\ & \ddots & \ddots \\ 0 & n-1 & 0 \end{pmatrix}$$

The indices are meant to suggest that G is the image of GL_2 under the standard representation $\varrho : GL_2 \to GL_n$ corresponding to the $(n-1)^{\text{st}}$ symmetric power.

The property which is special about our G-structure in the presence of the large number of completely integrable cross-sections $\omega^i(x)$. If $\{\phi^\alpha\}$ is a moving frame, then after multiplying $\omega^i(x)$ by a scalar factor if necessary, we will have an equation

(3.38) $$\omega^i(x) = \sum_\alpha t_i^\alpha(x)\phi^\alpha(x).$$

In order to conveniently express the complete integrability condition (3.33) we need to have a formula for $d\phi^\alpha$; i.e., a connection in the tangent bundle. Moreover, since equality of mixed partials will be involved it is desirable that this connection be *symmetric*[26]). Therefore, we recall that such a symmetric connection is given by a matrix $\{\phi_\beta^\alpha\}$ of 1-forms satisfying

(3.39) $$d\phi^\alpha = \sum_\beta \phi^\beta \wedge \phi_\beta^\alpha.$$

[26]) We are using symmetric rather than the more common *torsion free*, since the G-structure in question has its own well-defined torsion.

Once we have chosen a symmetric connection, any other one is given by

(3.40) $$\phi_\beta^{*\alpha} = \phi_\beta^\alpha + \sum_\gamma h_{\alpha\beta\gamma}\phi^\gamma, \quad h_{\alpha\beta\gamma} = h_{\alpha\gamma\beta}\,.$$

Under a change of coframe

$$\phi^{\alpha *} = \sum_\beta \phi^\beta g_\beta^\alpha$$

the connection matrix transforms according to the usual equation

(3.41) $$\phi^* = g^{-1}\phi g - g^{-1}dg\,.$$

Definition. *The G-structure is torsion-free is there exists a symmetric connection* $\{\phi_\beta^\alpha\}$ *with values in the Lie algebra* $\mathfrak{g} \subset \mathfrak{gl}(n)$.

Note that this definition makes sense for any G-structure, and that by (3.41) the condition $\{\phi_\beta^\alpha\} \in \mathfrak{g}$ is independent of the choice of moving frame.

For example, suppose that we have a Riemannian structure corresponding to $G = 0(n)$. A moving coframe is characterized by

$$ds^2 = \sum_\alpha (\phi^\alpha)^2\,.$$

Proving that this G-structure has no torsion is equivalent to first choosing an arbitrary symmetric connection $\{\phi_\beta^\alpha\}$ and then modifying it according to (3.40) so that

(3.42) $$\phi_\beta^{*\alpha} + \phi_\alpha^{*\beta} = 0 \quad (\alpha < \beta)\,.$$

The number of functions $h_{\alpha\beta\gamma}$ is $1/2(n^2(n+1))$, and the number of equations (3.42) is $1/2(n^2(n-1))$. Since $\{\phi_\beta^\alpha\}$ has n^3 entries and

$$n^3 - 1/2(n^2(n+1)) = 1/2(n^2(n-1))\,,$$

what is suggested is that a Riemannian structure is *uniquely* torsion free. This is well known to be the case (Levi-Civita connection).

We shall now derive the equations analogous to (3.42) that $\{\phi_\beta^\alpha\}$ have values in $\mathfrak{g} \subset \mathfrak{gl}(n)$ for the G-structure defined by a field of rational normal curves. Referring to (3.37) a $\mathfrak{g}$-valued 1-form may be written as

(3.43) $$h_{11}\omega_{11} + h_{22}\omega_{22} + h_{12}\omega_{12} + h_{21}\omega_{21}$$

where the $h_{\mu\nu}$ are as in (3.37) and the $\omega_{\mu\nu}$ are 1-forms. The conditions that $\{\phi_\beta^\alpha\}$ have the form (3.43) are

(3.44) $$\begin{cases} \phi_\beta^\alpha = 0 & |\alpha - \beta| \geqq 2 \\ \phi_{\alpha+1}^\alpha = (n-\alpha)\omega_{12} \\ \phi_{\alpha-1}^\alpha = (\alpha-1)\omega_{21} \\ \phi_\alpha^\alpha = (n-\alpha)\omega_{11} + (\alpha-1)\omega_{22}\,. \end{cases}$$

If we eliminate the $\omega_{\mu\nu}$'s we obtain

$$(3.45)\quad \begin{cases} \phi^\alpha_\beta = 0 \quad |\alpha - \beta| \geqq 2 \\ (n-\alpha-1)\phi^\alpha_{\alpha+1} = (n-\alpha)\phi^{\alpha+1}_{\alpha+2}, \\ (\alpha+1)\phi^{\alpha+1}_\alpha = \alpha\phi^{\alpha+2}_{\alpha+1} \\ \phi^{\alpha+1}_{\alpha+1} + \phi^{\alpha-1}_{\alpha-1} = 2\phi^\alpha_\alpha . \end{cases}$$

The number of these equations is

$$2((n-1)(n-2)/2) + 2(n-2) + (n-2) = n^2 - 4,$$

which is correct since the codimension of $\mathfrak{g}$ in $\mathfrak{gl}(n)$ is $n^2 - 4$. Now $\{\phi^\alpha_\beta\}$ has n^3 entries and may be modified by the $1/2n^2(n+1)$ functions $h_{\alpha\beta\gamma}$. To be in the Lie algebra $\mathfrak{g}$ imposes $n^3 - 4n$ conditions, so that in general our G-structure will be torsion-free only when

$$n^3 - 1/2n^2(n+1) \geqq n^3 - 4n .$$

This is turn occurs only when $n = 3$. In this case the G-structure is equivalent to a *conformal pseudo-Riemannian structure*, and one has the Weyl geometry. Bol's proof of the theorem when $n = 3$ was based on this fortunate occurrence.

When $n \geqq 4$ the G-structure given by a field of rational normal curves will in general *not* be torsion-free. This caused us considerable difficulty, and meant that if our theorem were true then the restrictions on the G-structure imposed by the completely integrable cross-sections $\omega^i(x)$ would have to play a crucial role – cf. (3.30) and (3.31). This turns out to be so, and also the "harmonic" property (3.9) of the defining equations of the web comes essentially into the picture. Before taking up these matters, we want to give in a convenient manner the analogue of the equations (3.42) for our G-structure.

Recall that the algebro-geometric study of rational normal curves in $\boldsymbol{P}^n$ could, by successively projecting from points on the curve, be reduced to the study of conics in the plane. For somewhat similar reasons there will turn out to be an asymmetry involving the indices 1 and 2 in the structure equations of a field of rational normal curves. This will now appear when we write the connection matrix $\phi = \{\phi^\alpha_\beta\}$ in the form

$$\phi = \eta + \psi$$

where η has values in the Lie algebra $\mathfrak{g}$, and where the equation

$$\psi = 0$$

expresses the condition that ϕ have values in $\mathfrak{g}$. To do this we observe that, if this latter is the case, then the whole matrix ϕ is uniquely determined by the four entries $\phi_1^1, \phi_2^2, \phi_2^1, \phi_1^2$. Indeed, according to (3.44)

$$\begin{aligned} \phi_1^1 &= (n-1)\omega_{11} & \phi_2^1 &= (n-1)\omega_{12} \\ \phi_2^2 &= (n-2)\omega_{11} + \omega_{22} & \phi_1^2 &= \omega_{21} \end{aligned}$$

uniquely determines the $\omega_{\mu\nu}$ in terms of the ϕ_ν^μ. We then write

$$\phi_\beta^\alpha = \eta_\beta^\alpha + \psi_\beta^\alpha \tag{3.46}$$

where

$$\begin{aligned} \eta_\beta^\alpha = {} & \delta_\beta^{\alpha+1}((n-\alpha)/(n-1))\phi_2^1 + \delta_\beta^{\alpha-1}(\alpha-1)\phi_1^2 \\ & + \delta_\beta^\alpha[(\alpha-1)\phi_2^2 - (\alpha-2)\phi_1^1] \end{aligned} \tag{3.47}$$

has values in $\mathfrak{g}$. The forms ψ_β^α defined by (3.46) and (3.47) have the properties:

i. $\psi_1^1 = \psi_2^1 = \phi_1^2 = \psi_2^2 = 0$;
ii. the $n^2 - 4$ equations $\psi_\beta^\alpha = 0$ express the condition that ϕ have values in $\mathfrak{g}$; and
iii. the forms ψ_β^α are *horizontal* for any choice of symmetric connection ϕ_β^α.

Here, horizontal means that if we consider the $(n+4)$-dimensional principal bundle B_G of all coframes for the G-structure, then the ϕ^α are intrinsically defined on B_G and

$$\psi_\beta^\alpha \equiv 0 \text{ modulo } \{\phi^1, \ldots, \phi^n\}.$$

The special role played by the indices 1 and 2 will be particularly apparent in the next section.

We now come to expressing the complete integrability property (3.33) for the web normal $\omega^i(x)$ given by (3.38). By (3.39)

$$d\omega^i = \sum_\alpha dt_i^\alpha \wedge \phi^\alpha + \sum_{\alpha,\beta} t_i^\beta \phi^\alpha \wedge \phi_\alpha^\beta.$$

Comparison with (3.33) yields

$$\sum_\alpha (dt_i^\alpha - \pi^i t_i^\alpha - \sum_\beta t_i^\beta \phi_\alpha^\beta) \wedge \phi^\alpha = 0.$$

By the *Cartan lemma*

$$dt_i^\alpha - \pi^i t_i^\alpha - \sum_\beta t_i^\beta \phi_\alpha^\beta = \sum_\beta t_{i\alpha\beta}\phi^\beta \quad \text{where} \quad t_{i\alpha\beta} = t_{i\beta\alpha}. \tag{3.48}$$

The symmetry $t_{i\alpha\beta} = t_{i\beta\alpha}$ exactly expresses the complete integrability condition (3.33).

In the notation of moving frames the abelian equation (3.1) becomes

$$\sum_i Z_i t_i^\alpha = 0 \quad (\alpha = 1, \ldots, n). \tag{3.49}$$

Taking the exterior derivative we obtain

$$0 = \sum_i (dZ_i t_i^\alpha + Z_i dt_i^\alpha) = \sum_\beta \left(\sum_i Z_i' t_i^{\alpha+\beta} + Z_i t_{i\alpha\beta} \right) \phi^\beta$$

by (3.34), (3.48), and (3.49). This implies that

$$\sum_i Z_i' t_i^{\alpha+\beta} + Z_i t_{i\alpha\beta} = 0, \quad 1 \leqq \alpha, \beta \leqq n. \tag{3.50}$$

This beautiful relation is the intrinsic analogue of (3.7). One advantage of using moving frames, as we shall now see, is that this method renders most visible the quadrics containing the rational normal curves $D_x \subset \boldsymbol{P}(T_x^*)$.

The condition that a general quadric $\sum_{\alpha,\beta} k^{\alpha\beta} \zeta^\alpha \zeta^\beta = 0$ in $\boldsymbol{P}^{n-1}$ should pass through the standard rational normal curve $\zeta^\alpha = t^\alpha$ is clearly

$$\sum_{\alpha+\beta=\lambda} k^{\alpha\beta} = 0 \qquad \lambda = 2, \ldots, 2n. \tag{3.51}$$

This gives $2n - 1$ independent equations, and so, as previously noted, there are

$$1/2(n(n+1)) - (2n-1) = (n-1)(n-2)/2$$

independent quadrics containing a rational normal curve. If we multiply (3.50) by $k^{\alpha\beta}$ satisfying (3.51) and sum over α and β we obtain

$$\sum_i \left(\sum_{\alpha,\beta} k^{\alpha\beta} t_{i\alpha\beta} \right) Z_i = 0.$$

Since the equations (3.49) are a basis, we will obtain the intrinsic analogue of (3.9)

$$\sum_{\alpha,\beta} k^{\alpha\beta} t_{i\alpha\beta} = \sum_\gamma m_\gamma t_i^\gamma \tag{3.52}$$

whenever $k^{\alpha\beta}$ satisfies (3.51). As mentioned before, these are sort of inhomogeneous Laplace equations, and the simplest situation would be if the right hand side of (3.52) were zero.

Definition. *We shall call the symmetric connection* $\{\phi_\beta^\alpha\}$ *harmonic in case*

$$\begin{cases} \sum_{\alpha,\beta} k^{\alpha\beta} t_{i\alpha\beta} = 0 & \textit{whenever} \\ \sum_{\alpha+\beta=\lambda} k^{\alpha\beta} = 0. & \end{cases} \tag{3.53}$$

Lemma. *The connection is harmonic if, and only if,*

$$t_{i\alpha\beta} = t_{i\alpha-1,\beta+1} = \cdots \tag{3.54}$$

depends only on the sum $\alpha + \beta$. *Harmonic connections exist, and once* ϕ_β^α *gives a harmonic connection then any other is of the form* (3.40) *where*

$$h_{\alpha\beta\gamma} = h_{\alpha,\beta+1,\gamma-1} \tag{3.55}$$

depends only on the sum $\beta + \gamma$ *and on* α.

Before proving the lemma we remark that we may now define

$$t_{i\alpha\beta} = t_{i,\alpha+\beta}$$

for any α, β with $2 \leqq \alpha + \beta \leqq 2n$. When this is done, equations (3.49) and (3.50) assume the particularly symmetric form

$$\begin{cases} \sum_i Z_i t_i^\alpha = 0 & \alpha = 1, \ldots, n \\ \sum_i Z_i' t_i^\varrho + Z_i t_{i,\varrho} = 0 & \varrho = 2, \ldots, 2n\,. \end{cases} \tag{3.56}$$

We note that there are here just the correct number $n + (2n - 1) = 3n - 1$ of these equations.

Proof of lemma (3.54): We let

$$\phi_\beta^{*\alpha} = \phi_\beta^\alpha + \sum_\gamma h_{\alpha\beta\gamma} \phi^\gamma$$

where

$$h_{\alpha\beta\gamma} = h_{\alpha\gamma\beta}\,.$$

Referring to (3.48)

$$t_{i\alpha\beta}^* = t_{i\alpha\beta} - \sum_\gamma h_{\gamma\alpha\beta} t_i^\gamma\,.$$

For each quadric $k^{\alpha\beta}$ satisfying

$$\sum_{\alpha+\beta=\lambda} k^{\alpha\beta} = 0, \quad 2 \leqq \lambda \leqq 2n\,,$$

we have for all i

$$\sum_{\alpha,\beta} k^{\alpha\beta} t_{i\alpha\beta} = \sum_\gamma m_\gamma(k) t_i^\gamma$$

where $m_\gamma(k)$ depends on the quadric. It is required to determine $h_{\gamma\alpha\beta}$ so that

$$\sum_{\alpha,\beta} k^{\alpha\beta} t_{i\alpha\beta}^* = 0\,.$$

This is equivalent to

$$\sum_{\alpha,\beta} k^{\alpha\beta} h_{\gamma\alpha\beta} = m_\gamma(k)\,.$$

Fix λ with $4 \leqq \lambda \leqq 2n - 2$. We suppose first that $\lambda = 2\mu$ is even, and let $\mu + \eta = \min(\lambda - 1, n)$. We consider quadrics $k^{\alpha\beta}$ whose only non-zero entries are when $\alpha + \beta = \lambda$ and which satisfy $\sum k^{\alpha\beta} = 0$. We represent such a quadric by the quadratic form

$$\sum_{\alpha,\beta} k^{\alpha\beta} \xi^\alpha \xi^\beta = 0 .$$

A basis for these is

$$\begin{aligned} &\xi^{\mu+1}\xi^{\mu-1} - (\xi^\mu)^2 = 0 \\ &\xi^{\mu+2}\xi^{\mu-2} - \xi^{\mu+1}\xi^{\mu-1} = 0 \\ &\vdots \\ &\xi^{\mu+\eta}\xi^{\mu-\eta} - \xi^{\mu+\eta-1}\xi^{\mu-\eta+1} = 0 . \end{aligned}$$

We label the corresponding quadrics as $k_1, \ldots, k_\eta$. The equations we must solve are

$$\begin{aligned} h_{\gamma,\mu+1,\mu-1} &= h_{\gamma,\mu,\mu} + m_\gamma(k_1) \\ h_{\gamma,\mu+2,\mu-2} &= h_{\gamma,\mu+1,\mu-1} + m_\gamma(k_2) \\ &\vdots \\ h_{\gamma,\mu+\eta,\mu-\eta} &= h_{\gamma,\mu+\eta-1,\mu-\eta+1} + m_\gamma(k_\eta) . \end{aligned}$$

Setting $h_{\gamma,\mu,\mu} = 0$ for instance, the recursive solution

$$h_{\gamma,\mu+\nu,\mu-\nu} = m_\gamma(k_1) + \cdots + m_\gamma(k_\nu)$$

established the lemma for λ even. The case λ odd is similar. Q.E.D.

As another illustration of the use of moving frames, and also as preparation for the computation in the next section, let us re-examine the proof of (3.12) – especially the equation (3.13). A point $Z \in \boldsymbol{P}^{d-n-1}(x)$ is written

$$Z = \sum_i p^i Z_i .$$

For a non-zero tangent vector σ we set $\sigma^\alpha = \langle \phi^\alpha, \sigma \rangle$, and use the notation "$\equiv$" to mean "congruent modulo $\boldsymbol{P}^{d-n-1}(x)$". By definition

$$\boldsymbol{P}^{n-2}(x,\sigma) = \{Z : \mathrm{d}Z/\mathrm{d}\sigma \equiv 0\} .$$

Using (3.34) and (3.38)

$$\begin{aligned} \mathrm{d}Z/\mathrm{d}\sigma &\equiv \sum_i p^i \mathrm{d}Z_i/\mathrm{d}\sigma \\ &\equiv \sum_i p^i \langle \omega^i, \sigma \rangle Z_i' \\ &\equiv \sum_{i,\alpha} p^i t_i^\alpha \sigma^\alpha Z_i' . \end{aligned}$$

According to (3.50),

$$Z \in \boldsymbol{P}^{n-2}(x,\sigma) \Leftrightarrow p^i = \left(\sum_{\varrho=2}^{2n} k^\varrho t_i^\varrho\right) \Big/ \left(\sum_{\alpha=1}^{n} t_i^\alpha \sigma^\alpha\right). \tag{3.57}$$

The number of parameters k^ϱ is $2n-1$. However, if p^i is given by (3.57) then p^i is only determined modulo the abelian equations (3.49). In other words the k^ϱ of the form

$$k^\varrho = \sum_{\alpha+\beta=\varrho} l^\alpha \sigma^\beta \tag{3.58}$$

give trivial solutions to the left hand side of (3.57), and so the number of essential parameters k^ϱ is

$$2n - 1 - n = n - 1,$$

which by homogeneity gives a $\boldsymbol{P}^{n-2}$. The equations (3.58) illustrate quite clearly how $\boldsymbol{P}^{n-2}(x,\sigma)$ varies with σ, and one may use them to give an alternate proof of (3.12).

ii. *The main computation.* We want to prove (3.26). Changing the notation for the path parameter from t to s, this amounts to showing that the condition

$$\mathrm{d}/\mathrm{d}s(\boldsymbol{P}^{n-2}(x(s),x'(s))) \subset \boldsymbol{P}^{n-2}(x(s),x'(s)) \tag{3.59}$$

is expressed by an equation of the type (3.21). Along the path $x(s)$ we set

$$\begin{cases} \phi^\alpha = y^\alpha \mathrm{d}s \\ \mathrm{d}y^\alpha + y^\beta \phi^\alpha_\beta = z^\alpha \mathrm{d}s \\ x = x(s) \quad \text{and} \quad y = x'(s). \end{cases}$$

Here, as throughout this section, repeated Greek indices are summed and we shall use the index ranges

$$\begin{cases} 1 \leqq \alpha,\beta,\gamma,\delta \leqq n \\ 2 \leqq \varrho,\sigma,\tau \leqq 2n. \end{cases}$$

With these notations (3.21) is a system of second order O.D.E.'s

$$y^\beta(z^\alpha + \Gamma^\alpha_{\gamma\delta} y^\gamma y^\delta) = y^\alpha(z^\beta + \Gamma^\beta_{\gamma\delta} y^\gamma y^\delta) \tag{3.60}$$

for all α and β and where $\Gamma^\alpha_{\gamma\delta} = \Gamma^\alpha_{\delta\gamma}$.

Now, according to (3.57), points in $\boldsymbol{P}^{n-2}(x,y)$ are given parametrically by

$$Z = \sum_i (k^\varrho t_i^\varrho / y^\alpha t_i^\alpha) Z_i. \tag{3.61}$$

For such a Z (3.59) is equivalent to

$$\mathrm{d}Z/\mathrm{d}s = \sum_i (K^\varrho t_i^\varrho / y^\alpha t_i^\alpha) Z_i. \tag{3.62}$$

where

$$-\alpha = \sum_i (-1)^{i-1} d(a_1, x) \wedge \cdots \wedge d(a_{i-1}, x) \wedge (a_i, e_{m+1}) \wedge d(a_{i+1}, x) \wedge \cdots \wedge d(a_m, x). \tag{20}$$

By an elementary calculation, using the fact that an element of a properly orthogonal matrix is euqal to its cofactor, we can write

$$+\alpha = \sum_i (-1)^{i-1} (a_{m+1}, e_i)\, \omega_1 \wedge \cdots \wedge \omega_{i-1} \wedge \omega_{i+1} \wedge \cdots \wedge \omega_m. \tag{21}$$

Suppose M be a compact piece of hypersurface with smooth boundary ∂M. Application of STOKES Theorem then gives

$$m \int_M t\sigma_1 \Phi = \int_{\partial M} \alpha. \tag{22}$$

This formula pertains to the hypersurface $x(M)$ and a fixed unit vector a_{m+1}. For simplicity write

$$a = a_{m+1}, \quad v_k = (a, e_k). \tag{23}$$

We project $x(M)$ orthogonally into the hyperplane through the origin and orthogonal to a. If $x'(p)$ is the image point of $x(p)$, $p \in M$, under this orthogonal projection, we have

$$x' = x - (a, x) a, \tag{24}$$

and hence

$$dx' = \sum_i \omega_i e_i - (\sum v_k \omega_k) a. \tag{25}$$

Over the submanifold ∂M there are now defined three quantities: the $(m-1)$-dimensional volumes of $x(\partial M)$ and $x'(\partial M)$, and the integral of α. There is an important relation between them, which we wish to establish. In fact, let Ψ be a non-zero differential form on ∂M, defined locally. Let the elements of volume of $x'(\partial M)$, $x(\partial M)$ be respectively $P\Psi$, $Q\Psi$, $P \geqq 0$, $Q \geqq 0$, Let $\alpha = R\Psi$. Then we claim that the following relation is valid:

$$P^2 = t^2 Q^2 + R^2. \tag{26}$$

Clearly this relation is independent of the choice of Ψ.

We proceed to prove (26). On ∂M let

$$\omega_{i_1} \wedge \cdots \wedge \omega_{i_{m-1}} = p_{i_1, \ldots, i_{m-1}} \Psi, \tag{27}$$

so that the p's are anti-symmetric in any two of the indices and

$$R = \sum_i (-1)^{i-1} v_i p_{1, \ldots, i-1, i+1, \ldots, m}.$$

Now the term containing Dk^{ϱ} has the desired form (3.62). In fact, along the paths $x(s)$ whose equation we shall derive we may determine $k^{\varrho}(x(s))$ with given initial value and satisfying $Dk^{\varrho}(x(s)) = 0$. So we may as well set $Dk^{\varrho} = 0$, and then (3.64) becomes

(3.65) $$\mathrm{d}Z/\mathrm{d}s = \sum_i (-k^{\varrho} t_{i,\varrho} + 1/\Delta\{k^{\alpha+\beta}(t_i^{\beta} t_{i\alpha\gamma} y^{\gamma} + t_i^{\alpha} t_{i\beta\gamma} y^{\gamma})\} \\ - 1/\Delta^2\{k^{\varrho} t_i^{\varrho}(z^{\alpha} t_i^{\alpha} + t_{i\alpha\beta} y^{\alpha} y^{\beta})\}) Z_i .$$

There are four terms involving the 2nd derivatives $t_{i\alpha\beta} = t_{i,\alpha+\beta}$ (harmonic property of the connection). *Of these, two have plus and two have minus signs.* It turns out that if our connection matrix $\{\phi_{\beta}^{\alpha}\}$ had values in the Lie algebra $\mathfrak{g}$, then these four terms cancel out and the paths are given by

$$z^{\alpha} = 0 ,$$

i.e., they are the geodesics for this connection. Since we don't know this good property of the connection we must set about simplifying the terms containing the $t_{i\alpha\beta}$, essentially by expressing them in powers of t_i.

As a preliminary we will prove the

(3.66) **Lemma.** *Suppose that, for some σ_i, l, m,*

$$t_{i\alpha\beta} = \sigma_i(\alpha + \beta - l) t_i^{\alpha+\beta-m} + T_{i\alpha\beta}$$

then $\mathrm{d}Z/\mathrm{d}s$ *is given by the same formula* (3.65) *with $T_{i\alpha\beta}$ replacing $t_{i\alpha\beta}$.*

Proof. Moving minus signs across we must prove that

$$k^{\varrho}(\rho - l) t_i^{\varrho-m} \Delta^2 + k^{\varrho} t_i^{\varrho}(\alpha + \beta - l) t_i^{\alpha+\beta-m} y^{\alpha} y^{\beta} \\ = \Delta\{k^{\alpha+\beta}((\alpha + \gamma - l) t_i^{\alpha+\beta+\gamma-m} y^{\alpha} + (\beta + \gamma - l) t_i^{\alpha+\beta+\gamma-m} y^{\gamma})\}.$$

The left hand side is

(3.67) $$k^{\varrho}(\varrho + \alpha + \beta - 2l) t_i^{\varrho+\alpha+\beta-m} y^{\alpha} y^{\beta} .$$

The right hand side is

$$\Delta\{k^{\alpha+\beta}((\alpha + \beta + 2\gamma - 2l) t_i^{\alpha+\beta+\gamma-m} y^{\gamma})\} \\ = \Delta\{k^{\varrho}((\varrho + \alpha - 2l) t_i^{\varrho+\alpha-m} y^{\alpha} + \beta t_i^{\varrho+\beta-m} y^{\beta}) \\ = k^{\varrho}(\varrho + \alpha + \beta - 2l) t_i^{\varrho+\alpha+\beta-m} y^{\alpha} y^{\beta} \\ = (3.67) .$$

Q.E.D.

We now turn to the problem of simplifying the $t_{i\alpha\beta}$. Recall the horizontal forms ψ_{β}^{α} defined by (3.46). We first will establish the formula:

(3.68) $$t_i^{\beta} \phi_{\alpha}^{\beta} - (\alpha - 1) t_i^{\alpha+\beta-2} \phi_2^{\beta} + (\alpha - 2) t_i^{\alpha+\beta-1} \phi_1^{\beta} \\ = t_i^{\beta} \psi_{\alpha}^{\beta} - (\alpha - 1) t_i^{\alpha+\beta-2} \psi_2^{\beta} + (\alpha - 2) t_i^{\alpha+\beta-1} \psi_1^{\beta} .$$

Proof. Using (3.47) we must show that the η^α_β terms drop out of the left hand side of (3.66). This is equivalent to the vanishing of the coefficients of $\phi^1_1, \phi^1_2, \phi^2_1, \phi^2_2$ under the substitution (3.47). The coefficient of ϕ^1_1 is

$$-(\alpha-2)t_i^\alpha + (\alpha-2)t_i^\alpha = 0.$$

The coefficient of ϕ_{12} is

$$((n-\alpha)/(n-1))t_i^{\alpha+1} - (\alpha-1)((n-2)(n-1))t_i^{\alpha+1} + (\alpha-2)t_i^{\alpha+1} = 0.$$

Similarly the coefficients of ϕ^2_1 and ϕ^2_2 are zero. Q.E.D.

The meaning of this relation is that for any choice of symmetric connection $\{\phi^\alpha_\beta\}$ the combinations on the left hand side of (3.68) are horizontal. Moreover, they are zero in case $\{\phi^\alpha_\beta\}$ has values in the Lie algebra g [27]). The symmetry arising from the complete integrability of $\omega^i = t_i^\alpha \phi^\alpha$ is not present in (3.68), but does appear in the following alternate expression for the left hand side of (3.68):

$$\begin{aligned} t_i^\beta \phi^\beta_\alpha - (\alpha-1)t_i^{\alpha+\beta-2}\phi^\beta_2 + (\alpha-2)t_i^{\alpha+\beta-1}\phi^\beta_1 \\ = -t_{i\alpha\gamma}\phi^\gamma + (\alpha-1)t_i^{\alpha-2}t_{i2\gamma}\phi^\gamma - (\alpha-2)t_i^{\alpha-1}t_{i1\gamma}\phi^\gamma. \end{aligned} \tag{3.69}$$

Proof. Write down the equations (3.48) for 1, 2, and α to obtain

$$dt_i - \pi^i t_i - t_i^\beta \phi^\beta_1 = t_{i1\gamma}\phi^\gamma \tag{1}$$

$$2t_i dt_i - \pi^i t_i^2 - t_i^\beta \phi^\beta_2 = t_{i2\gamma}\phi^\gamma \tag{2}$$

$$\alpha t_i^{\alpha-1} dt_i - \pi^i t_i^\alpha - t_i^\beta \phi^\beta_\alpha = t_{i\alpha\gamma}\phi^\gamma. \tag{3}$$

Then the linear combination

$$-((\alpha-2)t_i^{\alpha-1})(1) + ((\alpha-1)t_i^{\alpha-2})(2) - (3)$$

eliminates the coefficients of dt_i and π^i, and gives our desired formula. Q.E.D.

The left hand side of (3.69) is a polynomial in t_i of degree $\leqq \alpha + n - 1$. We set it equal to

$$\sum_{1 \leqq A \leqq \alpha+n-1} t_i^A a_{A\alpha\gamma}\phi^\gamma = A_{i\alpha\gamma}\phi^\gamma \tag{3.70}$$

and note that

$$A_{i1\gamma} = A_{i2\gamma} = 0. \tag{3.71}$$

The symmetry

$$t_{i\alpha\gamma} = t_{i\gamma\alpha} \tag{3.72}$$

[27]) The converse is also true, but we won't need this here.

and harmonic property

$$t_{i\alpha\gamma} = t_{i\alpha-1,\gamma+1} = \cdots \tag{3.73}$$

are by no means apparent from the left hand side of (3.69), and by (3.68) this imposes conditions on the ψ_β^α's. We proceed to exploit this, first by showing that

$$t_{i\alpha\gamma} = t_{i11}(-\alpha-\gamma+3)t_i^{\alpha+\gamma-2} + t_{i12}(\alpha+\gamma-2)t_i^{\alpha+\gamma-3} + T_{i\alpha\gamma} \tag{3.74}$$

where

$$\begin{aligned} T_{i\alpha\gamma} = &-(\alpha-1)(\gamma-1)t_i^{\alpha+\gamma-4}A_{i31} - A_{i\alpha\gamma} \\ &-(\alpha-1)t_i^{\alpha-2}A_{i\gamma 2} + (\alpha-2)t_i^{\alpha-1}A_{i\gamma 1} \end{aligned} \tag{3.75}$$ [28].

Proof. By (3.69) and (3.70)

$$t_{i\alpha\gamma} - (\alpha-1)t_i^{\alpha-2}t_{i2\gamma} + (\alpha-2)t_i^{\alpha-1}t_{i1\gamma} + A_{i\alpha\gamma} = 0 \tag{3.76}$$

$$t_{i\gamma\alpha} - (\gamma-1)t_i^{\gamma-2}t_{i2\alpha} + (\gamma-2)t_i^{\gamma-1}t_{i1\alpha} + A_{i\gamma\alpha} = 0\,. \tag{3.77}$$

We put $\alpha = 1, 2$ in the second equation and use (3.72) to obtain

$$\begin{cases} t_{i1\gamma} - (\gamma-1)t_i^{\gamma-2}t_{i21} + (\gamma-2)t_i^{\gamma-1}t_{i11} + A_{i\gamma 1} = 0 \\ t_{i2\gamma} - (\gamma-1)t_i^{\gamma-2}t_{i22} + (\gamma-2)t_i^{\gamma-1}t_{i12} + A_{i\gamma 2} = 0\,. \end{cases} \tag{3.78}$$

By (3.73) and (3.69)

$$t_{i22} = t_{i13} = -t_i^2 t_{i11} + 2t_i t_{i12} - A_{i31}\,.$$

If we substitute this in the second equation in (3.78), then we will have expressed $t_{i1\gamma}$ and $t_{i2\gamma}$ in terms of $t_{i11}, t_{i12}, A_{i\gamma 1}, A_{i\gamma 2}$, and A_{i31}. Plugging these into the first equation in (3.76) gives (3.74) and (3.75). Q.E.D.

According to lemma (3.66), dZ/ds is given by (3.65) with $T_{i\alpha\beta}$ in (3.75) replacing $t_{i\alpha\beta}$. We note that by (3.72), (3.73), and (3.74)

$$\begin{aligned} T_{i\alpha\gamma} &= T_{i\gamma\alpha} \\ T_{i\alpha\gamma} &= T_{i\alpha-1,\gamma+1} = \cdots = T_{i\varrho}, \quad \varrho = \alpha+\gamma\,, \end{aligned}$$

and shall now use these relations to express $T_{i\alpha\gamma}$ entirely in terms of $A_{i3\beta}$'s. For $\alpha + \gamma \leqq n+1$ the desired formula is

$$T_{i\varrho} = -\sum_{3\leqq\sigma\leqq\varrho-1} (\varrho-\sigma)t_i^{\varrho-\sigma-1}A_{i3,\sigma-2} \qquad 4 \leqq \varrho \leqq n+1\,. \tag{3.79}$$

---

[28]) Here we can make an interesting observation. If ϕ_β^α is a symmetric connection with values on the Lie algebra g, then by (3.46), (3.68), and the definition (3.70) $A_{i\alpha\gamma} = 0$. Consequently, $T_{i\alpha\gamma} = 0$ and $t_{i\alpha\gamma}$ is a linear combination of terms $\sigma_i(\alpha+\gamma-m)t_i^{\alpha+\gamma-l}$. According to lemma (3.66)

$$dZ/ds = -\sum_i ((k^\varrho t_i^\varrho)(t_i^\alpha z^\alpha)/\Delta^2)Z_i\,,$$

and so the paths are defined by $z^\alpha = 0$.

6*

Proof. By (3.71)

$$T_{i2} = T_{i3} = 0\,, \quad T_{i4} = -A_{i31}\,.$$

For $\gamma \geqq 4$

$$\begin{aligned} T_{i1\gamma} &= -A_{i\gamma 1} \\ T_{i2,\gamma-1} &= -(\gamma-2)t_i^{\gamma-3}A_{i31} - A_{i\gamma-1,2} \\ T_{i3,\gamma-2} &= -2(\gamma-3)t_i^{\gamma-3}A_{i31} - A_{i3,\gamma-2} \\ &\quad -2t_iA_{i\gamma-2,2} + t_i^2A_{i\gamma-2,1}\,. \end{aligned} \tag{3.80}$$

These expressions are all equal. Thus equating the second and third

$$\begin{aligned} &-(\gamma-2)t_i^{\gamma-3}A_{i31} - A_{i\gamma-1,2} \\ &= -2(\gamma-3)t_i^{\gamma-3}A_{i31} - A_{i3,\gamma-2} - 2t_iA_{i\gamma-2,2} \\ &\quad + t_i^2\{(\gamma-4)t_i^{\gamma-5}A_{i31} + A_{i\gamma-3,2}\}\,, \end{aligned}$$

where the term in curly brackets came from equating the first two equations for $\gamma - 2$. The terms involving A_{i31} cancel and we obtain

$$-A_{i3,\gamma-2} + A_{i\gamma-1,2} - 2t_iA_{i\gamma-2,2} + t_i^2A_{i\gamma-3,2} = 0\,.$$

We multiply by $(\alpha-\gamma+1)t_i^{\alpha-\gamma}$ and sum

$$\begin{aligned} \sum_{4\leqq\gamma\leqq\alpha}(\alpha-\gamma+1)t_i^{\alpha-\gamma}A_{i3,\gamma-2} &= \sum_{4\leqq\gamma\leqq\alpha}(\alpha-\gamma+1)t_i^{\alpha-\gamma}A_{i\gamma-1,2} \\ &\quad - \sum_{4\leqq\gamma\leqq\alpha}2(\alpha-\gamma+1)t_i^{\alpha-\gamma+1}A_{i\gamma-2,2} \\ &\quad + \sum_{4\leqq\gamma\leqq\alpha}(\alpha-\gamma+1)t_i^{\alpha-\gamma+2}A_{i\gamma-3,2}\,. \end{aligned}$$

Telescoping occurs with the result that

$$A_{i\alpha-1,2} = \sum_{4\leqq\gamma\leqq\alpha}(\alpha-\gamma+1)t_i^{\alpha-\gamma}A_{i3,\gamma-2}\,.$$

Similarly

$$A_{i\alpha 1} = \sum_{3\leqq\gamma\leqq\alpha}(\alpha-\gamma+1)t_i^{\alpha-\gamma}A_{i3,\gamma-2}\,.$$

By the first of these and middle equation in (3.80), for $4 \leqq \varrho \leqq n+1$

$$\begin{aligned} T_{i\varrho} = T_{i2,\varrho-2} &= -(\varrho-3)t_i^{\varrho-4}A_{i31} \\ &\quad - \sum_{4\leqq\sigma\leqq\varrho-1}(\varrho-\sigma)t_i^{\varrho-\sigma-1}A_{i3,\sigma-2} \\ &= (3.79)\,. \end{aligned}$$

Q.E.D.

By using

$$T_{i,n+2} = T_{i3,n-1}\,, \quad T_{i,n+3} = T_{i3,n}$$

this formula holds for $\rho = n + 2, n + 3$. For $\alpha \geqq 4$

$$\begin{aligned} T_{i\alpha n} &= -(\alpha - 1)(n - 1)t_i^{\alpha+n-4}A_{i31} - A_{i\alpha n} \\ &\quad -(\alpha - 1)t_i^{\alpha-2}\left(\sum_{4 \leqq \sigma \leqq n+1}(n - \sigma + 2)t_i^{n-\sigma+1}A_{i3,\sigma-2}\right) \\ &\quad +(\alpha - 2)t_i^{\alpha-1}\left(\sum_{3 \leqq \sigma \leqq n}(n - \sigma + 1)t_i^{n-\sigma}A_{i3,\sigma-2}\right) \\ &= -\sum_{3 \leqq \sigma \leqq n+1}(n + \alpha - \sigma)t_i^{n+\alpha-\sigma-1}A_{i3,\sigma-2} - A_{i\alpha n}. \end{aligned}$$

We may then introduce $A_{i3,n+1}, \ldots, A_{i3,2n-3}$ so that (3.79) holds for all ϱ; i.e.,

$$T_{i\varrho} = -\sum_{3 \leqq \sigma \leqq \varrho-1}(\varrho - \sigma)t_i^{\varrho-\sigma-1}A_{i3,\sigma-2}, \quad 4 \leqq \varrho \leqq 2n. \tag{3.81}$$

Now we may complete the proof. By (3.65) and (3.66)

$$\begin{aligned} dZ/ds = \sum_i(&-k^{\varrho}T_{i,\varrho} + 1/\Delta\{k^{\alpha+\beta}(t_i^{\beta}T_{i\alpha\gamma}y^{\gamma} + t_i^{\alpha}T_{i\beta\gamma}y^{\gamma}) \\ &-1/\Delta^2 k^{\varrho}t_i^{\varrho}(z^{\alpha}t_i^{\alpha} + T_{i\alpha\beta}y^{\alpha}y^{\beta})\})Z_i. \end{aligned} \tag{3.82}$$

Since we have now expressed $T_{i\alpha\beta} = T_{i,\alpha+\beta}$ as a polynomial in t_i, the last term may be simplified. Specifically, we have the

Delected due to error.

Lemma. *We have*

$$\left\{\begin{aligned} A_{i\alpha\gamma} &= \sum_{\gamma+2 \leqq \sigma \leqq \lambda-1}(\lambda - \sigma)t_i^{\lambda-\sigma-1}A_{i3,\sigma-2}, \quad \lambda = \alpha + \gamma \\ &= \sum_{\substack{\gamma+2 \leqq \sigma \leqq \lambda-1 \\ 1 \leqq D \leqq n+2}}(\lambda - \sigma)t_i^{\lambda-\sigma-1+D}a_{D,3,\sigma-2}. \end{aligned}\right. \tag{3.90}$$

Proof. Combining equation (3.75) with (3.79), we obtain

$$\begin{aligned} A_{i\alpha\gamma} &= -(\alpha - 1)(\gamma - 1)t_i^{\alpha+\gamma-4}A_{i31} + \sum_{3 \leqq \sigma \leqq \alpha+\gamma-1}(\alpha + \gamma - \sigma)A_{i3,\sigma-2} \\ &\quad -(\alpha - 1)t_i^{\alpha-2}\left(\sum_{4 \leqq \sigma \leqq \gamma+1}(\gamma - \sigma + 2)t_i^{\gamma-\sigma+1}A_{i3,\sigma-2}\right) \\ &\quad +(\alpha - 2)t_i^{\alpha-1}\left(\sum_{3 \leqq \sigma \leqq \gamma+1}(\gamma - \sigma + 1)t_i^{\gamma-\sigma}A_{i3,\sigma-2}\right). \end{aligned}$$

Here we have substituted for $A_{i\gamma 2}$ and $A_{i\gamma 1}$ the expressions appearing just at the end of the proof of (3.79). The coefficient of A_{i31} is

$$t_i^{\alpha+\gamma-4}(-(\alpha - 1)(\gamma - 1) + (\alpha + \gamma - 3) + (\alpha - 2)(\gamma - 2)) = 0.$$

This follows from the observation that the $(m-2)$-form

$$(68)\qquad \lambda = \sum_{\alpha} (-1)^{\alpha-1}\theta_1 \wedge \cdots \wedge \theta_{\alpha-1} \wedge (z_\alpha z_m) \wedge \theta_{\alpha+1} \wedge \cdots \wedge \theta_{m-1}$$

is globally defined on Y_u, so that $\int_{Y_u} d\lambda = 0$. By using (59) and (60), we get (on Y_u),

$$d\lambda = \{\sum z_{\alpha\alpha} \cdot z_m + \sum z_\alpha z_{m\alpha} + z_m^2 \sum A_{\alpha m\alpha} - \sum A_{\alpha m\beta} z_\alpha z_\beta\}\, \theta_1 \wedge \cdots \wedge \theta_{m-1},$$

where the product of the θ's is equal to dV. Now the lefthand side of (67) is equal to

$$LHS = \int_{Y_u} \left\{ -\frac{1}{2} \frac{\partial}{\partial u} (\sum z_\alpha^2 + z_m^2) + z_m (\sum_i z_{ii}) \right\} dV,$$

and the right-hand side of (67) is

$$RHS = \int_{Y_u} \left\{ -\frac{\partial}{\partial u} \sum z_\alpha^2 - z_m^2 \sum A_{\alpha m\alpha} + \sum A_{\alpha m\beta} z_\alpha z_\beta \right\} dV.$$

It follows that their difference

$$LHS - RHS = \int_{Y_u} d\lambda = 0.$$

This proves (67).

We now apply (67) to the family of spheres $\partial \sum_r$ in the space $(x_1, \ldots, x_m)$. Then $u = r$ and

$$(69)\qquad \theta_{\alpha m} = -\frac{1}{r}\theta_\alpha,$$

or

$$(69\text{a})\qquad A_{\alpha m\beta} = -\frac{1}{r}\delta_{\alpha\beta}.$$

From (54) and (67) we get

$$(70)\qquad (m-2)! \int_{\partial\Sigma r} \beta = \int_{\partial\Sigma r} \left\{ -\frac{\partial}{\partial r} (\sum z_\alpha^2) + \frac{m-1}{r} z_r^2 - \frac{1}{r} \sum z_\alpha^2 \right\} \Theta_r.$$

For $x \in \partial \sum_1$, the map $x \to rx$ establishes a diffeomorphism between $\partial \sum_1$ and $\partial \sum_r$. Under this diffeomorphism we have

$$\Theta_r = r^{m-1} \Theta_1.$$

It follows that

$$(71)\qquad \begin{aligned} (m-2)! \int_{\Sigma_r} B_2 &= (m-2)! \int_{\partial\Sigma_r} \beta \\ &= \int_{\partial\Sigma_1} \left\{ -\frac{\partial}{\partial r} (r^{m-1} \sum z_\alpha^2) + (m-1) r^{m-2} z_r^2 + (m-2)\, r^{m-2} \sum z_\alpha^2 \right\} \Theta_1. \end{aligned}$$

For $m = 2$ this formula is due to S. Bernstein (cf. [1]).

Equating to zero the powers of t_i gives $n(n+2)$ equations which clearly determine uniquely the $h_{\beta,\lambda}$ for $1 \leqq \beta \leqq n$, $2 \leqq \lambda \leqq n+3$. Explicitly, we obtain

(3.95a) $$-h_{\beta,\lambda} = \sum_{0 \leqq \sigma \leqq \beta-1} (\beta - \sigma) a_{\sigma+1,3,\lambda-\beta-2+\sigma}, \quad \beta+3 \leqq \lambda \leqq n+3$$

(3.95b) $$-h_{\beta,\lambda} = \sum_{0 \leqq \sigma \leqq n-\beta} (n+1-\beta-\sigma) a_{n+2-\sigma,3,\lambda+n-1-\beta-\sigma}, \quad \lambda \leqq \beta+1$$

(3.95c) $$-h_{\beta,\beta+2} = -\beta h_{1,3} + \sum_{2 \leqq \sigma \leqq \beta} (\beta+1-\sigma) a_{\sigma,3,\sigma-1}$$

(3.95d) $$(n+1) h_{1,3} = \sum_{\beta} (n-\beta+1) a_{\beta+1,3,\beta}.$$ Q.E.D.

According to the expansion of $A_{i\alpha\gamma}$ in terms of $A_{i3\gamma}$, the equations (3.93) will imply relations on $A^*_{i\alpha\gamma}$. Moreover, the $n(n-3)$ functions $h_{\beta,\lambda}$, $n+4 \leqq \lambda \leqq 2n$ are still at our disposal. To see what happens we set

(3.96) $$\begin{aligned} B_{i\alpha\gamma} &= A_{i\alpha\gamma} - (\alpha-1) t_i^{\alpha+\beta-2} h_{\beta,2+\gamma} + (\alpha-2) t_i^{\alpha+\beta-1} h_{\beta,1+\gamma} \\ &= \sum_{1 \leqq \nu \leqq n+\alpha-1} p_{\alpha\gamma\nu} t_i^{\nu} \end{aligned}$$

where the second equation defines the $p_{\alpha\gamma\nu}$. The reason for considering this $B_{i\alpha\gamma}$ is that the $h_{\beta,\lambda}$ appearing on the right are already determined by (3.95). Moreover, for $\lambda = \alpha + \gamma \geqq n+4$

(3.97) $$A^*_{i\alpha\gamma} = t_i^{\beta} h_{\beta,\lambda} + B_{i\alpha\gamma}.$$

Now the $p_{\alpha\gamma\nu}$ may according to (3.92) and (3.95) be expressed in terms of the $a_{\varrho 3\sigma}$. To do this we separate into cases.

Case i) $\nu \geqq \lambda - 1$ where $\lambda = \alpha + \gamma$. Then

$$\begin{aligned} p_{\alpha\gamma\nu} = &- \sum_{0 \leqq \varrho \leqq n-\nu+\alpha-1} (-n+\nu+\varrho-1) a_{n+2-\varrho,3,n-\nu-1+\lambda-\varrho} \\ &+ \sum_{\gamma+2 \leqq \varrho \leqq \lambda-1} (\lambda-\sigma) a_{\nu-\lambda+1+\sigma,3,\sigma-2}. \end{aligned}$$

When $n+2-\varrho = \nu - \lambda + 1 + \sigma$ we have

$$\begin{cases} -n+\nu+\varrho-1 = \lambda-\sigma \\ n-\nu-1+\lambda-\varrho = \sigma-2 \end{cases}$$

and so $p_{\alpha\gamma\nu} = 0$ for $\nu \geqq n+2$ since $\lambda - 1 \geqq \sigma \Leftrightarrow \nu + \varrho - n - 2 \geqq 0$. When $\nu = n+1$ we also find $p_{\alpha\gamma\nu} = 0$. If $\nu \leqq n$, say $\nu = \beta$, then

$$p_{\alpha\gamma\beta} = a_{\beta+2,3,\lambda-1} + 2a_{\beta+3,3,\lambda} + \cdots + (n+1-\beta) a_{n+2,3,\beta-\lambda+1-n}.$$

By (3.97) and (3.95b)

$$\begin{aligned} A^*_{i\alpha\gamma} &= t_i^{\beta} h_{\beta\lambda} + B_{i\alpha\gamma} \\ &= 0. \end{aligned}$$

We conclude that:

A_i^* is a linear combination of $t_i^{\lambda-2}, t_i^{\lambda-3}, \ldots, t_i$, which we may describe as saying that the equations (3.93) eliminate the high powers of t_i from $A_{i\alpha\gamma}^*$ in (3.92). The remaining choices for $h_{\beta\lambda}$ will eliminate low powers, as will now be seen

Case ii) $v = \lambda - 2$. Then by (3.91) and (3.95)

$$\begin{aligned} p_{\alpha\gamma v} &= -(\lambda-2)h_{13} + (\alpha-1)\Big(\sum_{2\leqq\varrho\leqq\gamma}(\gamma+1-\varrho)a_{\varrho,3,\varrho-1}\Big) \\ &\quad -(\alpha-2)\Big(\sum_{2\leqq\varrho\leqq\gamma-1}(\gamma-\varrho)a_{\varrho,3,\varrho-1}\Big) + \sum_{\gamma+2\leqq\sigma\leqq\lambda-1}(\lambda-\sigma)a_{\sigma-1,3,\sigma-2} \\ &= -(\lambda-2)h_{13} + \sum_{2\leqq\varrho\leqq\lambda-2}(\lambda-\varrho-1)a_{\varrho,3,\varrho-1}\,. \end{aligned}$$

Case iii) $v \leqq \lambda - 3$. Then, as before

$$\begin{aligned} p_{\alpha\gamma v} &= (\alpha-1)\Big(\sum_{0\leqq\varrho\leqq v-\alpha+1}(v-\alpha+2-\varrho)a_{\varrho+1,3,\lambda-v-2+\varrho} \\ &\quad -(\alpha-2)\Big(\sum_{0\leqq\varrho\leqq v-\alpha}(v-\alpha+1-\varrho)a_{\varrho+1,3,\lambda-v-2+\varrho}\Big) \\ &\quad + \sum_{\gamma+2\leqq\sigma\leqq\lambda-1}(\lambda-\sigma)a_{v-\lambda+1+\sigma,3,\sigma-2} \\ &= \sum_{0\leqq\varrho\leqq v-\alpha+1}(v-\varrho)a_{\varrho+1,3,\lambda-v-2+\varrho} + \sum_{\gamma+2\leqq\sigma\leqq\lambda-1}(\lambda-\sigma)a_{v-\lambda+1+\sigma,3,\sigma-2} \\ &= \sum_{0\leqq\varrho\leqq v-1}(v-\varrho)a_{\varrho+1,3,\lambda-v-2+\varrho}\,. \end{aligned}$$

Now if $v = \beta \leqq n$ we set

$$A_{i\alpha\gamma}^* = t_i^\beta h_{\beta\lambda} + B_{i\alpha\gamma} \equiv 0 \text{ modulo } t_i^{n+1}$$

by choosing

$$\text{(3.95e)} \quad -h_{\beta\lambda} = \sum_{0\leqq\varrho\leqq\beta-1}(\beta-\varrho)a_{\varrho+1,3,\lambda-\beta-2+\varrho}, \quad \beta+3 \leqq \lambda \leqq 2n$$

as prescribed by cases ii) and iii). We note that (3.95e) are the same equations as (3.95a) only extended now to the full index range. Continuing this with case i) we deduce that *$A_{i\alpha\gamma}^*$ depends only on i, $\alpha + \gamma = \lambda$, and is a linear combination of $t_i^{\lambda-2}, \ldots, t_i^{n+1}$. In particular,*

$$A_{i\alpha\gamma}^* = 0 \quad \text{for} \quad \alpha + \gamma \leqq n + 2\,.$$

Moreover, at this point the connection $\phi_\beta^{*\alpha}$ is uniquely determined, and it remains only to examine the case $\lambda = n + 3$. By (3.98) $A_{i\alpha\gamma}^*(\alpha + \gamma = n + 3)$ is a multiple of t_i^{n+1}. By case ii) when $v = n + 1$ the coefficient is

$$-(n+1)h_{13} + \sum_{2\leqq\varrho\leqq n+1}(n+2-\varrho)a_{\varrho,3,\varrho-1} = 0$$

by (3.95d).

IV. Projective differential geometry and completion of the proof

A. Projective connections and path geometry

i. *Basic definitions.* We first give the structure equations for $\boldsymbol{P}^n$, as this provides the model space for projective differential geometry. The index ranges

$$\begin{cases} 0 \leqq a, b, c \leqq n \\ 1 \leqq \alpha, \beta, \gamma \leqq n \end{cases}$$

will be used, and repeated indices are summed. We shall work with real projective space – the discussion carries over with the same notation to the complex case.

Definition. *A frame for $\boldsymbol{P}^n$ is given by*

$$F = \{Z_0, Z_1, \ldots, Z_n\}$$

where the Z_a form a basis for $\boldsymbol{R}^{n+1}$.

The manifold of all frames will be denoted by $\mathscr{F}(\boldsymbol{R}^n)$. It may be identified with GL_{n+1}. Occasionally, we shall speak of *normalized frames* defined by

$$Z_0 \wedge Z_1 \wedge \cdots \wedge Z_n = 1$$

the set of which may be identified with SL_{n+1}. There is a projection $\pi : \mathscr{F}(\boldsymbol{P}^n) \to \boldsymbol{P}^n$ given by

$$\pi(F) = Z_0 \,^{29)},$$

and the fibre $\pi^{-1}(Z_0)$ consists of all frames $F^* = \{Z_0^*, Z_1^*, \ldots, Z_n^*\}$ where

$$(4.1) \qquad \begin{cases} Z_0^* = A_0^0 Z_0 \\ Z_1^* = A_1^0 Z_0 + A_1^1 Z_1 + \cdots + A_1^n Z_n \\ \vdots \\ Z_n^* = A_n^0 Z_0 + A_n^1 Z_1 + \cdots + A_n^n Z_n \,. \end{cases}$$

This equation may be abbreviated by writing F as a column vector and (A_b^a) as a matrix A whereby (4.1) becomes

$$(4.2) \qquad F^* = A \cdot F \,.$$

Each frame F gives a coordinate simplex (Fig. 5) in $\boldsymbol{P}_n$, and up to the homogeneity factor A_0^0, $\pi^{-1}(Z_0)$ consists of all coordinate simplices whose first point is Z_0. The linear structure on $\boldsymbol{P}^n$ is given by identifying each

[29]) We shall frequently abuse notation and denote by Z_0 the point in $\boldsymbol{P}^n$ defined by the non-zero vector Z_0 in $\boldsymbol{R}^{n+1}$.

line $\overrightarrow{Z_0 Z}$ with a tangent vector to $\boldsymbol{P}^n$ at Z_0, and this should be kept in mind in the following discussion.

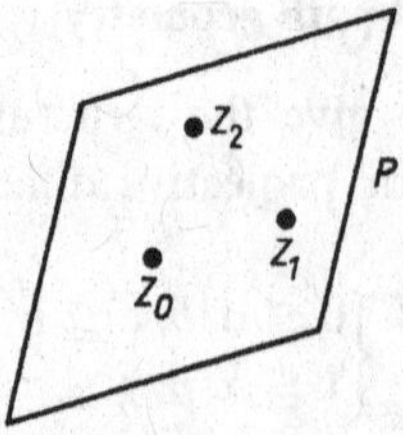

Fig. 5

The frame entries Z_a may be considered as vector-valued functions

$$Z_a : \mathscr{F}(\boldsymbol{P}^n) \to \boldsymbol{R}^{n+1}.$$

Expanding the exterior derivative dZ_a at $F \in \mathscr{F}(\boldsymbol{P}^n)$ in terms of the basis determined by the frame F we obtain

$$dZ_a = \theta_a^b Z_b. \tag{4.3}$$

The matrix $\theta = (\theta_a^b)$ gives the coefficients of infinitesimal displacement of the moving frame F. It is the Maurer-Cartan matrix on GL_{n+1}. Equation (4.3) may be abbreviated as

$$dF = \theta \cdot F. \tag{4.4}$$

Under a change of frame (4.2)

$$\theta^* = dA \cdot A^{-1} + A\theta A^{-1}. \tag{4.5}$$

The integrability relation, or Maurer Cartan equation,

$$d\theta = \theta \wedge \theta \tag{4.6}$$

follows by taking the exterior derivative of (4.4).

It will be useful to write (4.5) out making use of the special block from (4.1) of the matrix A. For this we introduce the notations

$$\begin{cases} \theta^\alpha = \theta_0^\alpha \\ \phi_\beta^\alpha = \theta_\beta^\alpha - \delta_\beta^\alpha \theta_0^0, \end{cases}$$

which will be motivated in a little while. Then (4.5) becomes

$$\begin{cases} \theta^{*\alpha} = A_0^0 \theta^\beta (A^{-1})_\beta^\alpha \\ \phi^{*\alpha}_{\beta} = dA_\beta^\gamma (A^{-1})_\gamma^\alpha - \delta_\beta^\alpha d\log A_0^0 + A_\beta^\gamma \phi_\gamma^\lambda (A^{-1})_\lambda^\alpha \\ \qquad + A_\beta^0 \theta^\lambda (A^{-1})_\lambda^\alpha \\ \theta^{*0}_{\alpha} = dA_\alpha^0 (A^{-1})_0^0 + dA_\alpha^\beta (A^{-1})_\beta^0 + A_\alpha^\beta \theta_\beta^0 (A^{-1})_0^0 \\ \qquad + A_\alpha^\gamma \theta_\gamma^\lambda (A^{-1})_\lambda^0 + A_\alpha^0 \theta^\beta (A^{-1})_\beta^0. \end{cases} \tag{4.7}$$

The integrability conditions (4.6) are

$$(4.8)\qquad \begin{cases} d\theta^\gamma = \theta^\beta \wedge \phi^\gamma_\beta \\ d\phi^\alpha_\beta = \phi^\gamma_\beta \wedge \phi^\alpha_\gamma + \theta^0_\beta \wedge \theta^\alpha - \delta^\alpha_\beta \theta^\gamma \wedge \theta^0_\gamma \\ d\theta^0_\beta = \phi^\alpha_\beta \wedge \theta^0_\alpha . \end{cases}$$

From the first equations in (4.7) and (4.8) it follows that on $\mathscr{F}(\boldsymbol{P}^n)$

$$\theta^\alpha = 0$$

defines a completely integrable Pfaffian system, one whose integral manifolds are the fibres of $\mathscr{F}(\boldsymbol{P}^n) \xrightarrow{\pi} \boldsymbol{P}^n$.

To interpret the second and third equations, we denote by $H^* \to \boldsymbol{P}^n$ the universal bundle whose fibre over $Z_0 \in \boldsymbol{P}^n$ is the line $\lambda \cdot Z_0$ in $\boldsymbol{R}^{n+1}$. Letting

$$Q = \boldsymbol{P}^n \times \boldsymbol{R}^{n+1}/H^*$$

be the quotient of the trivial bundle by H^*, the identification of lines with tangent vectors gives an isomorphism

$$T \cong \mathrm{Hom}(H^*, Q) = H \otimes Q .$$

More precisely, sections of $H \otimes \boldsymbol{R}^{n+1}$ may be interpreted as vector fields of the form

$$\theta = \sum_{a,b} \theta^a_b(z_1/z_0, \dots, z_n/z_0) z^b \partial/\partial z^a .$$

The inclusion $H^* \to \boldsymbol{P}^n \times \boldsymbol{R}^{n+1}$ induces an inclusion of the trivial bundle in $H \otimes \boldsymbol{R}^{n+1}$ with the generator 1 going into $\sum_a z^a \partial/\partial z^a$. By Euler's theorem $\pi(\theta) = 0$ in T_{Z_0} if and only if

$$\theta = \lambda\Big(\sum_a z^a \partial/\partial z^a\Big) \quad (Z_0 = [1, z_1/z_0, \dots, z_n/z_0]) .$$

Summarizing, the exact sequence

$$(4.9)\qquad 0 \to H^* \to \boldsymbol{P}^n \times \boldsymbol{R}^{n+1} \to Q \to 0$$

and isomorphism

$$(4.10)\qquad T \cong H \otimes Q$$

embody the linear structure on $\boldsymbol{P}^n$. $\mathscr{F}(\boldsymbol{P}^n)$ is the manifold of frames for the trivial bundle in the middle of (4.9), and $\phi^\alpha_\beta = \theta^\alpha_\beta - \delta^\alpha_\beta \theta^0_0$ looks something like a connection in $T \cong Q \otimes H$. This is not quite correct, since the flat connection on $\boldsymbol{P}^n \times \boldsymbol{R}^{n+1}$ does not leave H^* invariant, or equivalently since the extension in (4.9) is non-trivial relative to the connection in $\boldsymbol{P}^n \times \boldsymbol{R}^{n+1}$.

In general, a projective connection on a manifold will be given by a vector bundle sequence

$$0 \to L^* \to E \to Q \to 0$$

with isomorphism

$$T(M) \cong L \otimes Q$$

and with a connection in E. Rather than try to formalize this, we shall adopt a "working definition". So, given a covering $\{U, V, \ldots, \ldots\}$ of M we assume given in each open set U an $(n+1) \times (n+1)$ matrix of 1-forms θ_U satisfying

$$\theta_U = dA_{UV} A_{UV}^{-1} + A_{UV} \theta_V A_{UV}^{-1}$$

in $U \cap V$ where A_{UV} has the block form

$$A_{UV} = \begin{pmatrix} * & 0 & \cdots & 0 \\ * & * & \cdots & * \\ \vdots & & & \\ * & * & \cdots & * \end{pmatrix}.$$

We assume that the entries

$$(\theta_U)_0^1, \ldots, (\theta_U)_0^n$$

are linearly independent, and shall admit over U all connection matrices

$$\theta = dA \cdot A^{-1} + A \theta_U A^{-1}$$

where A has the block form given above. This will be our definition of a *projective connection.*

For example, suppose that $M_n \subset \boldsymbol{P}^{n+k}$ is a submanifold and consider the *Darboux frames* $F = \{Z_0; Z_1, \ldots, Z_n; Z_{n+1}, \ldots, Z_{n+k}\} \in \mathscr{F}(\boldsymbol{P}^n)$ where $Z_0 \in M$ and $Z_0, Z_1, \ldots, Z_n$ span the projective tangent space to M at Z_0. The connection matrix for $\boldsymbol{P}^{n+k}$ has the block form

$$\theta_{\boldsymbol{P}^{n+k}} = \left(\begin{array}{ccc|ccc} * & \cdots & * & 0 & \cdots & 0 \\ \vdots & & \vdots & * & \cdots & * \\ \vdots & & \vdots & \vdots & & \vdots \\ * & \cdots & * & * & \cdots & * \\ \hline * & \cdots & * & * & \cdots & * \\ \vdots & & \vdots & \vdots & & \vdots \\ * & \cdots & * & * & \cdots & * \end{array}\right) \begin{array}{l} \left.\vphantom{\begin{array}{c} a\\a\\a\\a \end{array}}\right\} n+1 \\ \left.\vphantom{\begin{array}{c} a\\a\\a \end{array}}\right\} k \end{array}$$

on the Darboux frames, and the upper left hand block defines a projective connection on M.

At this point, since our discussion is local it will simplify matters if we assume that our projective connection is *special* in the sense that the trace

(4.11) $$\sum_a (\theta_{UV})_a^a = 0.$$

It follows that $\det(A_{UV}) = \text{constant}$, and we shall assume this constant is $+1$. Although somewhat unnatural, the use of special projective connections will alleviate the necessity of saying when two projective connections are equivalent, a notion which takes into account the fact that GL_{n+1} does not act effectively on $\boldsymbol{P}^n$. There is a natural special projective connection on the manifold of normalized frames on $\boldsymbol{P}^n$. There are $(n+1)^2 - 1 = n(n+2)$ forms in the connection matrix of a special projective connection, and θ_0^0 is then determined by the ϕ_α^α according to

(4.12) $$-(n+1)\theta_0^0 = \phi_1^1 + \cdots + \phi_n^n.$$

Given a special projective connection with connection matrix $\theta = (\theta_a^b)$ (we omit reference to the open set U), set

(4.13) $$\begin{cases} \theta^\alpha = \theta_0^\alpha \\ \phi_\beta^\alpha = \theta_\beta^\alpha - \delta_\beta^\alpha \theta_0^0. \end{cases}$$

The *projective torsion* Θ^α and *projective curvature* Φ_β^α are defined by

(4.14) $$\Theta^\alpha = d\theta^\alpha - \theta^\beta \wedge \phi_\beta^\alpha$$

(4.15) $$\Phi_\beta^\alpha = d\phi_\beta^\alpha - \phi_\beta^\gamma \wedge \phi_\gamma^\alpha - \theta_\beta^0 \wedge \theta^\alpha + \delta_\beta^\alpha \theta^\gamma \wedge \theta_\gamma^0.$$

By the same computations as led to (4.7) and (4.8) it follows that Θ^α and Φ_β^α are tensors. In particular, they are horizontal and we set

$$\Phi_\beta^\alpha = 1/2(R_{\beta\gamma\lambda}^\alpha \theta^\gamma \wedge \theta^\lambda), \quad R_{\beta\gamma\lambda}^\alpha + R_{\beta\lambda\gamma}^\alpha = 0.$$

Definition. *The special projective connection is normal in case*

(4.16) $$\begin{cases} \Theta^\alpha = 0 \\ \sum_\alpha R_{\beta\gamma\alpha}^\alpha = 0. \end{cases}$$

For a torsion-free projective connection there is the *Bianchi identity*

(4.17) $$R_{\beta\gamma\lambda}^\alpha + R_{\lambda\beta\gamma}^\alpha + R_{\gamma\lambda\beta}^\alpha = 0.$$

Proof. By (4.14)

$$\begin{aligned} 0 = d^2\theta^\alpha &= d\theta^\gamma \wedge \phi_\gamma^\alpha - \phi^\beta \wedge d\phi_\beta^\alpha \\ &= -\theta^\beta \wedge (d\phi_\beta^\alpha - \phi_\beta^\gamma \wedge \phi_\gamma^\alpha) \\ &= -\theta^\beta \wedge (\Phi_\beta^\alpha + \theta_\beta^0 \wedge \theta^\alpha - \delta_\beta^\alpha \theta^\gamma \wedge \theta_\gamma^0) \end{aligned}$$

by (4.15)

$$= -\theta^\beta \wedge \Phi^\alpha_\beta ,$$

which implies (4.17). Q.E.D.

We note that (4.17) is vacuous unless $n \geqq 3$. Somewhat deeper is the following: Define (cf. the 3rd equation in (4.8))

$$\Theta^0_\beta = d\theta^0_\beta - \phi^\alpha_\beta \wedge \phi^0_\alpha . \tag{4.18}$$

Then, in case $n \geqq 3$,

$$\Theta^\alpha = 0 \quad \text{and} \quad \Phi^\alpha_\beta = 0 \Rightarrow \Theta^0_\beta = 0 . \tag{4.19}$$

Proof. Applying exterior differentiation to (4.15) gives

$$\begin{aligned} 0 = & - d\phi^\gamma_\beta \wedge \phi^\alpha_\gamma + \phi^\gamma_\beta \wedge d\phi^\alpha_\gamma - d\theta^0_\beta \wedge \theta^\alpha + \theta^0_\beta \wedge d\theta^\alpha \\ & + \delta^\alpha_\beta d\theta^\gamma \wedge \theta^0_\gamma - \delta^\alpha_\beta \theta^\gamma \wedge d\theta^0_\gamma \\ = & -(\phi^\lambda_\beta \wedge \phi^\gamma_\lambda + \theta^0_\beta \wedge \theta^\gamma - \delta^\gamma_\beta \theta^\lambda \wedge \theta^0_\lambda) \wedge \theta^\alpha_\gamma \\ & + \phi^\gamma_\beta \wedge (\phi^\lambda_\gamma \wedge \phi^\alpha_\lambda + \theta^0_\gamma \wedge \theta^\alpha - \delta^\alpha_\gamma \theta^\lambda \wedge \theta^0_\lambda) \\ & - (\Theta^0_\beta + \phi^\gamma_\beta \wedge \theta^0_\gamma) \wedge \theta^\alpha + \theta^0_\beta \wedge \theta^\gamma \wedge \phi^\alpha_\gamma \\ & + \delta^\alpha_\beta \theta^\lambda \wedge \phi^\gamma_\lambda \wedge \theta^0_\gamma - \delta^\alpha_\beta \theta^\gamma \wedge (\Theta^0_\gamma + \phi^\lambda_\gamma \wedge \theta^0_\lambda) \\ = & -\Theta^0_\beta \wedge \theta^\alpha - \delta^\alpha_\beta \theta^\gamma \wedge \Theta^0_\gamma . \end{aligned}$$

Suppose that $n = 3$ and take $\beta = 1$ and $\alpha = 2,3$ to obtain

$$\Theta^0_1 \wedge \theta^2 = 0 = \Theta^0_1 \wedge \theta^3 .$$

This implies that

$$\Theta^0_1 = \varrho_1 \theta^2 \wedge \theta^3 .$$

Similarly

$$\Theta^0_2 = -\varrho_2 \theta^1 \wedge \theta^3 , \quad \Theta^0_3 = \varrho_3 \theta^1 \wedge \theta^2 .$$

Taking $\alpha = \beta = 1$ gives

$$(-2\varrho_1 - \varrho_2 - \varrho_3)\theta^1 \wedge \theta^2 \wedge \theta^3 = 0 ,$$

and consequently

$$\begin{aligned} -2\varrho_1 &= \varrho_2 + \varrho_3 \\ -2\varrho_2 &= \varrho_1 + \varrho_3 \\ -2\varrho_3 &= \varrho_1 + \varrho_2 \end{aligned}$$

which implies

$$-2(\varrho_1 + \varrho_2 + \varrho_3) = 2(\varrho_1 + \varrho_2 + \varrho_3)$$

or $\varrho_1 + \varrho_2 + \varrho_3 = 0$. But then $\varrho_1 = \varrho_2 = \varrho_3 = 0$.

The case $n \geqq 4$ is even easier. Q.E.D.

Definition. *The connection is projectively flat in case* $\Theta^\alpha = \Phi^\alpha_\beta = \Theta^0_\beta = 0$.

Assuming projective flatness, the projective connection matrix $\theta = (\theta^a_b)$ satisfies

$$\begin{cases} d\theta = \theta \wedge \theta \\ \sum_a \theta^a_a = 0 \,. \end{cases}$$

According to a standard application of the Frobenius theorem there are locally maps $f: M \to \mathrm{SL}_{n+1}$ inducing θ from the Maurer-Cartan matrix on SL_{n+1}. In this sense connections which are projectively flat are equivalent to the standard projective connection on $\boldsymbol{P}^n$. We shall say more about this in the next section.

In concluding this section, we assume that the projective torsion $\Theta^\alpha = 0$ so that the Bianchi identity (4.17) holds. Then normality implies

$$\sum_\alpha R^\alpha_{\alpha\beta\gamma} = 0 \,. \tag{4.20}$$

ii. *Fundamental theorem in local projective differential geometry; the Beltrami theorem.* We shall discuss the relationship between projective connections and path geometry as defined in § III A ii, beginning with the case of the straight lines in the model space $\boldsymbol{P}^n$. Following the notations in the preceding section, let $\{Z_0(t), Z_1(t), \dots, Z_n(t)\}$ be a curve in the frame manifold lying over the curve $Z_0(t)$ in $\boldsymbol{P}^n$. We denote t-derivatives by a dot, and for a 1-form ψ the notation $\dot{\psi}$ will be used for ψ/dt along the curve. Thus, by (4.3)

$$\dot{Z}_0 = \dot{\theta}^a_0 Z_a$$
$$\ddot{Z}_0 = (\ddot{\theta}^a_0 + \dot{\theta}^b_0 \dot{\theta}^a_b) Z_0$$

and

$$Z_0 \wedge \dot{Z}_0 \wedge \ddot{Z}_0 = \sum_{\alpha<\beta} \{\dot{\theta}^\alpha(\ddot{\theta}^\beta + \dot{\theta}^\gamma \dot{\phi}^\beta_\gamma + 2\dot{\theta}^\beta \dot{\theta}^0_0) - \dot{\theta}^\beta(\ddot{\theta}^\alpha + \dot{\theta}^\gamma \dot{\phi}^\alpha_\gamma + 2\dot{\theta}^\alpha \dot{\theta}^0_0)\} Z_0 \wedge Z_\alpha \wedge Z_\beta \,.$$

Since the equation

$$Z_0 \wedge \dot{Z}_0 \wedge \ddot{Z}_0 = 0$$

characterizes the straight lines, it follows that these lines are given in terms of the standard projective connection on $\boldsymbol{P}^n$ by the O.D.E. system

$$(\ddot{\theta}^\beta + \dot{\theta}^\gamma \dot{\phi}^\beta_\gamma)/\dot{\theta}^\beta = (\ddot{\theta}^\alpha + \dot{\theta}^\gamma \dot{\phi}^\alpha_\gamma)/\dot{\theta}^\alpha \,. \tag{4.21}$$

We recognize (4.21) as being of the form (3.21), and the resulting path geometry is the geometry of lines in $\boldsymbol{P}^n$.

Now suppose that we are given a special projective connection on a manifold M. If θ_U is the connection matrix in the open set U, then we are permitted to use any other connection matrix

$$\theta = dA \cdot A^{-1} + A \cdot \theta_U A^{-1}$$

where A has the block form given by (4.1) and $\det A = +1$. In particular, since $(\theta_U)_0^1 \wedge \cdots \wedge (\theta_U)_0^n \neq 0$ we may choose a coordinate system $x = (x^1,\ldots,x^n)$ and A such that for the new connection matrix

$$\theta^\alpha = dx^\alpha \text{ [30]}. \tag{4.22}$$

If we think of the various choices of A as giving the possible frames for the projective connection, then we may refer to a frame which satisfies (4.22) as a *coordinate frame*. We write

$$\phi^\alpha_\beta = \Gamma^\alpha_{\beta\lambda} dx^\lambda$$
$$\theta^0_\beta = \Gamma^0_{\beta\lambda} dx^\lambda,$$

and determine θ^0_0 by (4.12) so that $\theta^\alpha_\beta = \phi^\alpha_\beta + \delta^\alpha_\beta \theta^0_0$. By the definition (4.14), the torsion

$$\Theta^\alpha = -1/2\left\{\sum_{\beta,\lambda}(\Gamma^\alpha_{\beta\lambda} - \Gamma^\alpha_{\lambda\beta})dx^\beta \wedge dx^\lambda\right\},$$

so that for a coordinate frame the torsion being zero is equivalent to the symmetry

$$\Gamma^\alpha_{\beta\lambda} = \Gamma^\alpha_{\lambda\beta}. \tag{4.23}$$

Henceforth, although not strictly necessary, we shall suppose that the torsion is zero. Equations (4.21) are of the form

$$(\ddot{x}^\beta + \Gamma^\beta_{\gamma\lambda}\dot{x}^\gamma\dot{x}^\lambda)/\dot{x}^\beta = (\ddot{x}^\alpha + \Gamma^\alpha_{\gamma\lambda}\dot{x}^\gamma\dot{x}^\lambda)/\dot{x}^\alpha,\ \Gamma^\beta_{\gamma\lambda} = \Gamma^\beta_{\lambda\gamma}, \tag{4.24}$$

and define a path geometry in the sense of § III A ii. A basic remark is that *the form of the equations* (4.24) *is invariant under an arbitrary change of coordinates* $x^\alpha = x^\alpha(y^1,\ldots,y^n)$ *and change of parameter* $t = t(s)$.

For example, suppose we change parameter and denote by a prime the s-derivatives. Then

$$\begin{cases} x^{\alpha\prime}(t(s)) = \dot{x}^\alpha t' \\ x^{\alpha\prime\prime}(t(s)) = \ddot{x}^\alpha t'^2 + \dot{x}^\alpha t'' \end{cases}$$

and so

$$\begin{aligned}(\ddot{x}^\alpha + \Gamma^\gamma_{\gamma\lambda}\dot{x}^\gamma\dot{x}^\lambda)/\dot{x}^\alpha &= ((x^{\alpha\prime\prime}/t'^2) - (\dot{x}^\alpha t''/t'^2) \\ &\quad + (\Gamma^\alpha_{\gamma\lambda}x^{\gamma\prime}x^{\lambda\prime}/t'^2))/(x^{\alpha\prime}/t') \\ &= 1/t'((x^{\alpha\prime\prime} + \Gamma^\alpha_{\gamma\lambda}x^{\gamma\prime}x^{\lambda\prime})/x^{\alpha\prime}) - 1/t'^2,\end{aligned}$$

which implies the invariance of (4.24) under a change of parameter. The behaviour of (4.24) under a change of coordinates is done by a similar

---

[30] Recall that under the above transformation $\theta^\alpha = A^0_0(\theta_U)^\beta(A^{-1})^\alpha_\beta$.

calculation, and taken together these show that *the path geometry defined by* (4.24) *is intrinsically attached to the projective connection.*

The fundamental theorem of local projective differential geometry is the converse:

(4.25) **Proposition.** *A path geometry* (4.24) *intrinsically defines a unique special, normal projective connection.*

Proof. We write (4.24) in the form

$$(4.26) \qquad (d^2x^\beta + \Gamma^\beta_{\gamma\lambda}dx^\gamma dx^\lambda)/dx^\beta = (d^2x^\alpha + \Gamma^\alpha_{\gamma\lambda}dx^\gamma dx^\lambda)/dx^\alpha\,,$$

and set

$$Q^\alpha(dx) = \Gamma^\alpha_{\gamma\lambda}dx^\gamma dx^\lambda$$

considered as a quadratic polynomial in the dx^γ's. The most general change preserving (4.26) is

$$\tilde{Q}^\alpha(dx) = Q^\alpha(dx) + dx^\alpha P(dx)$$

where P is linear. We will uniquely determine P so that

$$(4.27) \qquad \sum_\alpha \partial\tilde{Q}^\alpha(\xi)/\partial\xi^\alpha = 0\,.$$

This is equivalent to

$$\begin{aligned} 0 &= \sum_\alpha \partial/\partial\xi^\alpha(Q^\alpha(\xi) + \xi^\alpha P(\xi)) \\ &= \sum_\alpha \partial Q^\alpha(\xi)/\partial\xi^\alpha + 2P(\xi) \end{aligned}$$

so that $P = -1/2(\sum_\alpha \partial Q^\alpha/\partial\xi^\alpha)$ gives the unique solution. Changing notation, we assume (4.27) for the $Q^\alpha(\xi)$, or equivalently that

$$\sum_\alpha \Gamma^\alpha_{\alpha\lambda} = 0\,.$$

This equation is the same as

$$(4.28) \qquad \sum_\alpha \phi^\alpha_\alpha = 0\,.$$

Now to prove (4.25) we must show how to uniquely determine the connection matrix (θ^a_b) subject to certain properties. We set

$$\begin{cases} \theta^\alpha_0 = \theta^\alpha = dx^\alpha \\ \theta^\alpha_\beta = \phi^\alpha_\beta = \Gamma^\alpha_{\beta\lambda}dx^\lambda\,, \quad \Gamma^\alpha_{\beta\lambda} = \Gamma^\alpha_{\lambda\beta}\,. \end{cases}$$

Then by (4.12) and (4.28) we must have $\theta^0_0 = 0$, and it remains to determine

$$\theta^0_\alpha = \Gamma^0_{\alpha\beta}dx^\beta$$

7*

by the normality condition (cf. (4.16))

$$\sum_{\alpha} R^{\alpha}_{\beta\gamma\alpha} = 0 . \tag{4.29}$$

These are n^2 equations in the n^2 unknowns $\Gamma^0_{\alpha\beta}$, and it must be proved that they are linear equations whose coefficient matrix is non-singular.

We define $S^{\alpha}_{\beta\gamma\lambda} = -S^{\alpha}_{\beta\lambda\gamma}$ by

$$-1/2\left\{\sum_{\gamma,\lambda} S^{\alpha}_{\beta\gamma\lambda} dx^{\gamma} \wedge dx^{\lambda}\right\} = d\phi^{\alpha}_{\beta} - \phi^{\gamma}_{\beta} \wedge \phi^{\alpha}_{\gamma} .$$

These are known functions in terms of the $\Gamma^{\alpha}_{\gamma\lambda}$'s and their 1st derivatives. By the definition (4.15) of the projective curvature tensor

$$\begin{aligned} 1/2\{R^{\alpha}_{\beta\gamma\lambda} dx^{\gamma} \wedge dx^{\lambda}\} + 1/2\{S^{\alpha}_{\beta\gamma\lambda} dx^{\gamma} \wedge dx^{\lambda}\} &= -\theta^0_{\beta} \wedge \theta^{\alpha} + \delta^{\alpha}_{\beta}\theta^{\gamma} \wedge \theta^0_{\gamma} \\ &= (\delta^{\alpha}_{\gamma}\Gamma^0_{\beta\lambda} + \delta^{\alpha}_{\beta}\Gamma^0_{\gamma\lambda}) dx^{\gamma} \wedge dx^{\lambda} . \end{aligned}$$

Consequently

$$R^{\alpha}_{\beta\gamma\lambda} + S^{\alpha}_{\beta\gamma\lambda} = (\delta^{\alpha}_{\gamma}\Gamma^0_{\beta\lambda} - \delta^{\alpha}_{\lambda}\Gamma^0_{\beta\gamma}) + \delta^{\alpha}_{\beta}(\Gamma^0_{\gamma\lambda} - \Gamma^0_{\lambda\gamma}) .$$

Setting $\alpha = \lambda$ and summing gives, by (4.29),

$$\sum_{\alpha} S^{\alpha}_{\beta\gamma\alpha} = S_{\beta\gamma} = -n\Gamma^0_{\beta\gamma} + \Gamma^0_{\gamma\beta} .$$

When $n > 1$ the coefficient matrix of this linear system of n^2 equations in n^2 unknowns has a non-zero determinant. Q.E.D.

We recall from section III A ii that the path geometry defined by (4.24) is *flat* in case there is locally a diffeomorphism taking the paths onto straight lines in $\boldsymbol{P}^n$. Combining (4.25) with (4.19) we obtain the

(4.30) **Corollary.** *When* $n \geqq 3$ *the path geometry* (4.24) *is flat if, and only if, the associated projective connection has projective curvature*

$$\Phi^{\alpha}_{\beta} = 0 .$$

While we are discussing flatness we want to prove the beautiful

Beltrami theorem. *The path geometry defined by the geodesics of a Riemannian metric is flat if, and only if, the metric has constant sectional curvatures.*

Proof. We choose an orthonormal coframe ω^{α} for the metric – thus

$$ds^2 = \sum_{\alpha} (\omega^{\alpha})^2 .$$

The structure equations for the Riemannian connection matrix ω^{α}_{β} and curvature matrix $\Omega^{\alpha}_{\beta} = 1/2\{T^{\alpha}_{\beta\gamma\lambda}\omega^{\gamma} \wedge \omega^{\lambda}\}$ are

$$\begin{cases} d\omega^{\alpha} = \omega^{\beta} \wedge \omega^{\alpha}_{\beta}, \quad \omega^{\alpha}_{\beta} + \omega^{\beta}_{\alpha} = 0 \\ \Omega^{\alpha}_{\beta} = d\omega^{\alpha}_{\beta} - \omega^{\gamma}_{\beta} \wedge \omega^{\alpha}_{\gamma} . \end{cases} \tag{4.31}$$

Taking exterior derivatives gives the two Bianchi identities

$$\text{(4.32)}\qquad \begin{cases} T^{\alpha}_{\beta\gamma\lambda}\omega^{\beta} \wedge \omega^{\gamma} \wedge \omega^{\lambda} = 0 \\ T^{\alpha}_{\beta\gamma\lambda|\mu}\omega^{\gamma} \wedge \omega^{\lambda} \wedge \omega^{\mu} = 0 . \end{cases}$$

The geodesics of the Riemannian metric are the solution curves to the O.D.E. system

$$\text{(4.33)}\qquad \ddot{\omega}^{\alpha} + \dot{\omega}^{\beta}\dot{\omega}^{\alpha}_{\beta} = 0 .$$

According to the proof of (4.25), the projective connection matrix $\{\theta^{\alpha}, \phi^{\alpha}_{\beta}, \theta^{0}_{\beta}\}$ associated to the path geometry given by the geodesics (4.33) is uniquely determined by the equations

$$\begin{cases} \theta^{\alpha} = \omega^{\alpha} \\ \phi^{\alpha}_{\beta} = \omega^{\alpha}_{\beta} \quad {}^{31)} \\ \theta^{0}_{\beta} = \Gamma^{0}_{\beta\gamma}\omega^{\gamma} \end{cases}$$

where

$$\text{(4.33)}\qquad \sum_{\alpha} T^{\alpha}_{\beta\gamma\alpha} = T_{\beta\gamma} = n\Gamma^{0}_{\beta\gamma} - \Gamma^{0}_{\gamma\beta} .$$

Consider first the case $n = 2$. The only non-zero component of the Riemannian curvature tensor is

$$K = T^{1}_{212}\,{}^{32)} .$$

Taking $\beta = \gamma = 1$ in (4.33) gives

$$-K = T^{2}_{112} = T_{11} = 2\Gamma^{0}_{11} - \Gamma^{0}_{11} = \Gamma^{0}_{11} ,$$

and similarly $\Gamma^{0}_{22} = -K$. Taking $\beta = 1$, $\gamma = 2$ and then interchanging β and γ gives

$$\begin{aligned} 0 &= T_{12} = 2\Gamma^{0}_{12} - \Gamma^{0}_{21} \\ 0 &= T_{21} = 2\Gamma^{0}_{21} - \Gamma^{0}_{12} , \end{aligned}$$

which implies that

$$\begin{aligned} \Gamma^{0}_{12} &= \Gamma^{0}_{21} = 0 , \\ \Gamma^{0}_{\beta\gamma} &= -\delta^{\gamma}_{\beta}K . \end{aligned}$$

It follows that the projective curvature tensor

$$R^{\alpha}_{\beta\gamma\lambda} = T^{\alpha}_{\beta\gamma\lambda} - K(\delta^{\alpha}_{\gamma}\delta^{\beta}_{\lambda} - \delta^{\alpha}_{\lambda}\delta^{\beta}_{\gamma}) .$$

---

31) Note that trivially $\sum_{\alpha} \phi^{\alpha}_{\alpha} = 0$, so that $\theta^{0}_{0} = 0$.

32) Up to the symmetries $T^{\alpha}_{\beta\gamma\lambda} = -T^{\beta}_{\alpha\gamma\lambda} = -T^{\alpha}_{\beta\lambda\gamma}$.

Since $R^\alpha_{\beta\gamma\lambda} = 0$ if $\alpha = \beta$, the only possible non-zero components of this tensor are

$$R^1_{212} = K - K = 0$$
$$R^2_{112} = -K - K(-1) = 0\,.$$

So, when $n = 2$ the projective torsion and curvature are always zero, and the obstruction to projective flatness is entirely measured by

$$\begin{aligned}\Theta^0_\beta &= d\theta^0_\beta - \theta^\alpha_\beta \wedge \theta^0_\alpha \\ &= -dK \wedge \omega^\beta\end{aligned}$$

since
$$\theta^0_\beta = -K\omega^\beta\,.$$

It follows that

$$\Theta^0_\beta = 0 \Leftrightarrow dK = 0\,,$$

which proves the result in this case.

Now suppose that $n \geqq 3$. According to (4.30) the geodesic path geometry is flat if, and only if, $R^\alpha_{\beta\gamma\lambda} = 0$. By the symmetries of $T^\alpha_{\beta\gamma\lambda}$,

$$T_{\beta\gamma} = T_{\gamma\beta}$$

in (4.33). We may choose the coframe ω^α to diagonalize $T_{\beta\gamma}$ at a point x_0. Call the eigenvalues $(n-1)T_\beta$. Then, by (4.33)

$$n\Gamma^0_{\beta\gamma} - \Gamma^0_{\gamma\beta} = (n-1)\delta^\beta_\gamma T_\beta$$
$$n\Gamma^0_{\gamma\beta} - \Gamma^0_{\beta\gamma} = (n-1)\delta^\gamma_\beta T_\gamma\,,$$

which gives

$$\Gamma^0_{\beta\gamma} = \Gamma^0_{\gamma\beta} = \delta^\beta_\gamma T_\beta\,.$$

Suppose that $R^\alpha_{\beta\gamma\lambda} = 0$. Then, by (4.15), at x_0

$$\begin{cases} 0 = T^\alpha_{\beta\gamma\lambda} + (\delta^\alpha_\gamma\delta^\beta_\lambda - \delta^\alpha_\lambda\delta^\beta_\gamma)T_\beta \\ 0 = T^\beta_{\alpha\gamma\lambda} + (\delta^\beta_\gamma\delta^\alpha_\lambda - \delta^\beta_\lambda\delta^\alpha_\gamma)T_\alpha\,. \end{cases}$$

Adding these equations and using $T^\alpha_{\beta\gamma\lambda} + T^\beta_{\alpha\gamma\lambda} = 0$ gives

$$(T_\alpha - T_\beta)\delta^\alpha_\lambda\delta^\beta_\gamma = (T_\alpha - T_\beta)\delta^\alpha_\gamma\delta^\beta_\lambda\,.$$

Taking $\alpha = \lambda$, $\beta = \gamma$ but $\alpha \neq \beta$ we find

$$T_\alpha - T_\beta = 0\,.$$

It follows that, at x_0, all T_α are the same. Since x_0 was any point,

$$T^\alpha_{\beta\gamma\lambda}(x) = (\delta^\alpha_\gamma\delta^\beta_\lambda - \delta^\alpha_\lambda\delta^\beta_\gamma)K(x)\,.$$

By the second Bianchi identity in (4.32),

$$K_\mu\omega^\alpha \wedge \omega^\beta \wedge \omega^\mu = 0\,.$$

Thus $K_\mu = 0$ for $\mu \neq \alpha, \beta$. Since $n \geqq 3$ this implies that all $K_\mu = 0$, and so ds^2 has constant sectional curvature.

If, conversely, the sectional curvatures are constant, then we may reverse the calculation to conclude that $R^\alpha_{\beta\gamma\lambda} = 0$.

When the sectional curvatures are a constant $K > 0$, we may argue geometrically as follows: Represent the manifold as a portion of the sphere S^n in $\boldsymbol{R}^{n+1}$. The geodesics are the intersections of planes through the origin with S^n. Intersecting these planes with the $\boldsymbol{P}^n$ at infinity in $\boldsymbol{R}^{n+1}$ maps the geodesics to lines.

Deleted due to error.

References

[1] Blaschke, W.; Bol, G.: Geometrie der Gewebe. Berlin 1938
[2] Cartan, E.: Leçons sur la théorie des espaces à connexion projective. Paris 1937
[3] Chern, S. S.: Abzählungen für Gewebe. Abh. Math. Sem. Hamb. **11** (1936) 163–170
[4] Griffiths, P.: Variations on a Theorem of Abel. Invent. Math. **35** (1976) 321–390
[5] Griffiths, P.; Harris, J.: Algebraic Geometry. To be published

(Eingegangen: 20. 2. 1977)

University of California
Dept. of Mathematics
Berkeley, California 94720
USA

Harvard University
Dept. of Mathematics
Cambridge, Mass. 02138
USA

Jber. d. Dt. Math.-Verein.
83 (1981) 78–83

Corrections and Addenda to Our Paper: Abel's Theorem and Webs[1])

S. S. Chern*, Berkeley, and P. Griffiths**, Cambridge, MA

We regret to report that our paper in question contains two essential errors:

1. The lemmas on pp. 85, 86 are incorrect, because of a faulty elementary argument. The divisibility in question is true only under an additional hypothesis, which we will call normality and which will be defined below.

2. The proof of (4.34) contains a gap, which we are not able to fill. Following Bol, we will complete the proof of our main theorem by an algebraic-geometric argument, which will be given below. Thus our main theorem is valid with the additional condition of normality. For completeness we will state it as follows (cf. p. 18):

Consider a normal d-*web of codimension* 1 *in* $\mathbf{R}^n$ *of maximum rank* $\pi(d, n)$. *Suppose that* $n \geqslant 3$, $d \geqslant 2n$. *Then the web is linearizable, i.e., equivalent to a web whose leaves are hyperplanes.*

Our paper is in order with the deletion of the lemmas on pp. 85, 86 and §IV B. We wish to thank C. Im Hof for detecting the second error, which led us to a reexamination of the paper and to the need of the condition of normality. In the following we will follow the structure of the original paper and the notations:

III B. iv. *Divisibility for normal webs.* By (3.81) and (3.90) we introduced $A_{i3,n+1}, \ldots, A_{i3,2n-3}$ (the inequality $1 \leqslant D \leqslant n+2$ under the summation sign should be deleted.). By (3.90) we derive immediately

$$(3.101) \qquad A_{i3,n+\rho} = A_{i,\rho+3,n} - 2\,t_i\,A_{i,\rho+2,n} + t_i^2\,A_{i,\rho+1,n}, \qquad 1 \leqslant \rho \leqslant n-3\,.$$

Since by (3.70) $A_{i\alpha\gamma}$ is of degree $\alpha + n - 1$ in t_i, it follows from the above formula (3.101) that $A_{i3,n+\rho}$ is of degree $n + \rho + 2$ in t_i.

Under a change of the harmonic connection (3.86) the $A_{i\alpha\gamma}$ are transformed according to (3.92), which gives

$$(3.102) \qquad A^*_{i\alpha n} = A_{i\alpha n} + t_i^\beta h_{\beta,\alpha+n} - (\alpha-1)\,t_i^{\alpha+\beta-2}\,h_{\beta,n+2} + (\alpha-2)\,t_i^{\alpha+\beta-1}\,h_{\beta,n+1}\,.$$

[1]) Jber. d. Dt. Math.-Verein. **80** (1978) 13–110.

* Research partially supported by NSF Grant MCS 77-23579.
** Research partially supported by NSF Grant MCS 78-07348.

0012–0456/81/02 0078–06$01.20/0

It follows that

$$(3.103)\quad A^*_{i3,n+\rho} = A_{i3,n+\rho} + t_i^\beta h_{\beta,n+\rho+3} - 2\,t_i^{\beta+1} h_{\beta,n+\rho+2} + t_i^{\beta+2} h_{\beta,n+\rho+1},\quad 1 \leqslant \rho \leqslant n-3 .$$

Thus only the terms of degree $\leqslant n+2$ in t_i can be modified by a change of the harmonic connection.

Definition *The web is called normal, if all* A_{i3s}, $1 \leqslant s \leqslant 2n-3$, *are of degrees* $\leqslant n+2$ *in* t_i. *From now on we suppose our web normal.*

Before we prove the divisibility on normal webs, we wish to make a remark on the best harmonic connection. We think there is a harmonic connection simpler than the one described in Proposition (3.87). It is given by

(3.104) **Proposition** (New Best Harmonic Connection) *There is a unique symmetric, harmonic connection such that*

$$(3.105a)\quad A_{i3\gamma} = 0,$$

$$(3.105b)\quad A_{i3,n+\rho} = p_\rho\, t_i^{n+1} + q_\rho\, t_i^{n+2},\qquad 1 \leqslant \rho \leqslant n-3 .$$

In fact, by Lemma (3.94), conditions (3.105a) leave only the $h_{\beta\lambda}$, $n+4 \leqslant \lambda \leqslant 2n$, undetermined. By keeping (3.105a), we can suppose

$$h_{\beta\lambda} = 0,\qquad 2 \leqslant \lambda \leqslant n+3 .$$

The lemma then follows from (3.103).

By (3.57) and (3.61) the intersection $\mathbf{P}^{n-2}(x,\sigma)$ of two neighboring $\mathbf{P}^{\lambda-n-1}(x)$'s is spanned by

$$(3.106)\quad Y_\lambda = \sum_i \frac{t_i^\lambda}{\Delta_i} Z_i,\qquad 2 \leqslant \lambda \leqslant 2n,$$

where

$$(3.107)\quad \Delta_i = \Sigma\, y^\alpha t_i^\alpha$$

was previously denoted by Δ (cf. (3.63)). We wish to express the condition that $\mathbf{P}^{n-2}(x,\sigma)$ is fixed, i.e.,

$$(3.108)\quad dY_\lambda \equiv 0,\quad \text{mod } Y_2,\ldots,Y_{2n} .$$

The left-hand side is given by

$$(3.82)\quad \frac{dZ}{ds} = \sum_i \Big\{ -k^\rho T_{i\rho} + \frac{1}{\Delta_i} k^{\alpha+\beta}(t_i^\beta T_{i\alpha\gamma} Y^\gamma + t_i^\alpha T_{i\beta\gamma} Y^\gamma) - \frac{1}{\Delta_i^2} k^\rho t_i^\rho (z^\alpha t_i^\alpha + T_{i\alpha\beta}\, y^\alpha y^\beta) \Big\} Z_i$$

with

$$(3.109)\quad k^\lambda = 1,\quad k^\rho = 0,\quad \rho \neq \lambda,\qquad 2 \leqslant \lambda, \rho \leqslant 2n .$$

With the values (3.109) for k^ρ the first two terms of (3.82) can be written

(3.110) $\Sigma \frac{1}{\Delta_i} E_{\lambda\alpha\gamma} y^\gamma Z_i$,

where

(3.111) $E_{\lambda\alpha\gamma} = -T_{i\lambda} t_i^\gamma + T_{i,\alpha+\gamma} t_i^{\lambda-\alpha} + T_{i,\lambda-\alpha+\gamma} t_i^\alpha$.

As we are expressing the condition (3.108), it suffices, by (3.106), to evaluate $E_{\lambda\alpha\gamma} \bmod t_i^2, \ldots, t_i^{2n}$. Since our web is normal, $T_{i\rho}$ is of degree $n + \rho - 2$ in t_i, so that $E_{\lambda\alpha\gamma}$ is of degree $\lambda + \gamma + n - 2$ in t_i.

For this purpose we set

(3.112) $P_\alpha = -\sum_{3 \leqslant \rho \leqslant \alpha+n} (2\alpha - \rho)(t_i^{\alpha-\rho-1} A_{i3,\rho-2})^+$,

where $(\ )^+$ denotes the sum of terms of positive degree in t_i. (This expression differs from the same on p. 85 in that the ranges of summation are different. Notice also that the power $\alpha - \rho - 1$ of t_i could be negative.) The terms in (3.112) are of decreasing degrees and P_α is of degree $\leqslant n + \alpha - 2$ in t_i. The first step in the reduction is achieved by the lemma:

(3.113) $E_{\lambda\alpha\gamma} - t_i^\lambda P_\gamma \equiv 0, \quad \bmod t_i^2, \ldots, t_i^{2n}$.

This is shown by direct inspection. For simplicity we will write

(3.114) $A_{i3s} = A_s$.

Since α is a summand of λ, there are two cases: 1) $2 \leqslant \lambda \leqslant n + 1$, in which case we suppose $\lambda = \alpha + 1$; 2) $n + 1 \leqslant \lambda \leqslant 2n$, where we suppose $\lambda = n + \alpha$.

Consider first the case $\lambda = \alpha + 1$. The left-hand side of (3.113) is of degree $\alpha + \gamma + n - 1$. If the latter is $\leqslant 2n$, there is nothing to prove. Hence we suppose

$$\alpha + \gamma \geqslant n + 2.$$

Each term of P_γ is a numerical multiple of

$$(t_i^{\gamma-\rho-1} A_{\rho-2})^+.$$

Its degree is $\leqslant n + \gamma - \rho + 1$, for, the web being normal, $A_{\rho-2}$ is of degree $\leqslant n + 2$ in t_i. To prove (3.113) it suffices to restrict to the ρ such that

$$\alpha + 1 + n + \gamma - \rho + 1 \geqslant 2n + 1$$

or

$$\rho \leqslant \alpha + \gamma - n + 1.$$

It follows that

$$t_i^{\alpha+1} P_\gamma \equiv t_i^{1+\alpha} \{(3 - 2\gamma)(t_i^{\gamma-4} A_1)^+ + \ldots + (\alpha + \gamma - n + 1 - 2\gamma)(t_i^{n-2-\alpha} A_{\alpha+\gamma-n-1})^+\}$$

On the other hand, we have, by (3.111) and (3.81),

$$E_{\alpha+1,\alpha\gamma} = (3 - 2\gamma) t_i^{\alpha+\gamma-3} A_1 + (4 - 2\gamma) t_i^{\alpha+\gamma-4} A_2 + \ldots + (\alpha + \gamma - n + 1 - 2\gamma) t_i^{n-1} A_{\alpha+\gamma-n-1}.$$

Clearly this expression is congruent to $t_i^{\alpha+1} P_\gamma$.

The second case $\lambda = n + \alpha$ can be treated in a similar way. However, it will not be necessary, for $Y_{1+\alpha}$ already span $P^{n-2}(x, \sigma)$, the $Y_2, \ldots, Y_{2n}$ being linearly dependent.

Using (3.113) and (3.82), the condition (3.108) becomes

$$(3.115) \quad \sum_i \frac{t_i^\lambda}{\Delta_i^2} \{z^\alpha t_i^\alpha + T_{i\alpha\beta} y^\alpha y^\beta - (P_\alpha y^\alpha)(t_i^\beta y^\beta)\} Z_i = 0 .$$

The sum of the second and third terms in this expression is of the form $\widetilde{Q}_{i\alpha\beta} y^\alpha y^\beta$, where

$$(3.116) \quad 2\,\widetilde{Q}_{i\alpha\beta} = 2\,T_{i\alpha\beta} - P_\alpha t_i^\beta - P_\beta t_i^\alpha .$$

We wish to prove the lemma:

(3.117) **Lemma** $\widetilde{Q}_{i\alpha\beta}$ *is a linear combination of* t_i^γ.

For this purpose we make use of the new best harmonic connection in (3.104). Then

$$T_{i\lambda} = 0, \qquad 2 \leqslant \lambda \leqslant n+3,$$

$$T_{i\lambda} = -(\lambda - n - 3)\, q_1 t_i^{\lambda-2} - \{(\lambda - n - 3)\, p_1 + (\lambda - n - 4)\, q_2\}\, t_i^{\lambda-3} - \ldots - p_{\lambda-n-3} t_i^{n+1}, \qquad \lambda \geqslant n+4.$$

By (3.112),

$$P_1 = P_2 = 0,$$

$$t_i^\beta P_\alpha \equiv (n+3-2\alpha)\, q_1 t_i^{\alpha+\beta-2} + \{(n+3-2\alpha)\, p_1 + (n+4-2\alpha) q_2\} t_i^{\alpha+\beta-3} + \ldots + \{(\alpha+\beta-1-2\alpha)\, p_{\alpha+\beta-n-3} + (\alpha+\beta-2\alpha)\, q_{\alpha+\beta-n-2}\}\, t_i^{n+1},$$
$$\bmod t_i^\gamma.$$

From these expressions the statement on $\widetilde{Q}_{i\alpha\beta}$ follows immediately.

From the lemma (3.117) we get the equation of paths following the formula (3.85).

IV B′. Completion of the proof of linearization

This section replaces IV B. To complete the proof of our main theorem we will need some results on algebraic geometry. We begin with the lemma:

(4.101) **Lemma** *Suppose that in* $\mathbf{P}^N$ *we have a family* $\{C\}$ *of* ∞^n *rational curves, any pair of which meet in* $n-1$ *points. Then these curves lie on an algebraic surface.*

P r o o f. We assume that a general curve C has degree δ. Since the normal bundle $N \to C$ is a quotient of the ample bundle $T(\mathbf{P}^N)|C$, its Grothendieck decomposition has only factors $O_C(k)$ where $k > 0$. It follows that $H^1(C, N) = 0$, so that the Hilbert scheme of curves of degree δ is reduced and smooth at C. We denote by Ξ the irreducible component of the Hilbert scheme containing the curves of our family, and for each curve C we let $\Xi_C \subset \Xi$ be the set of curves that meet C in at least n – 1 points. Clearly Ξ_C is an algebraic subvariety of Ξ, and therefore so is the intersection

$$\Xi_0 = \bigcap_C \Xi_C$$

By assumption dim $\Xi_0 \geqslant n$. We choose a general irreducible algebraic curve T in Ξ_0 and denote by $\{C_t\}_{t\in T}$ the corresponding family of ∞^1 rational curves in P^N. Then

$$S = \bigcup_{t\in T} C_t$$

is an irreducible algebraic surface, and each of our given curves C meets S in infinitely many points. Hence they all lie on S. ■

We will now describe the algebraic surface S.

(4.102) S *is a regular algebraic surface.*

Proof. Since the tangents to the ∞^n curves {C} span the tangent plane to S at a general point, any holomorphic differential would have to restrict to a nonzero holomorphic differential on $C \cong P^1$.

(4.103) *The* ∞^n *rational curves are contained in a linear system* |L|.

Proof. This is a property of an algebraic family of curves on any regular algebraic surface.

(4.104) *The linear system* |L| *has dimension equal to* n.

Proof. Let $C \cong \mathbf{P}^1$ be a curve in our system. Since any two curves meet in n–1 points, we have

$$L|C \cong O_C(n-1).$$

From the exact cohomology sequence

$$0 \to O_S \to O_S(L) \to O_C(n-1) \to 0$$

and $h^1(O_S) = 0$ we obtain

$$0 \to C \to H^0(O_L(L)) \to C^n \to 0.$$

This implies (4.104).

The linear system |L| clearly has no fixed component, and it induces a rational mapping

(4.105) $\varphi_L : S \to \mathbf{P}^n$.

(4.106) *The mapping* (4.105) *is birational onto its image.*

Proof. The proof of (4.104) shows that |L| has no base points. Hence φ_L is a holomorphic mapping of S onto a non-degenerate surface S_0 of some degree δ in $\mathbf{P}^n$. Since

$$\delta \cdot (\text{degree of } \varphi_L) = n - 1, \qquad \delta \geqslant n - 1,$$

we infer that φ_L is birational.

We now apply the above results to the family $\{E(x)\}$ of rational curves in the Poincare space $\mathbf{P}^{\pi-1}$. Indeed, the condition "$K^{\rho} = 0$ in (3.62)" on line 7 of page 80 means that the axis $\mathbf{P}^{n-2}$ is pointwise fixed along a path $\{x(t)\}$. Since any pair of points x,x' may be joined by a path, the condition in the lemma (4.101) is fulfilled. This gives

(4.107) *The* ∞^n *rational curves* $E(x)$ *lie on a two-dimensional surface* $S \subset \mathbf{P}^{\pi-1}$. *The paths* $\{x(t)\}$ *are characterized by the property: As we move along* $x(t)$, $n-1$ *points of* $E(x(t))$ *remain fixed.*

We also have:

(4.108) *The surface* $S \subset \mathbf{P}^{\pi-1}$ *is birational to a surface* $S_0 \subset \mathbf{P}^n$ *of minimal degree* $n-1$. *Under this mapping the rational curves* $E(x)$ *are in a one-to-one correspondence with the hyperplane sections* $H(x) \cap S_0$, $H(x) \in P^{n*}$.

By (4.107) and (4.108) the proof of the main theorem can be completed as follows: The mapping

(4.109) $U \to \mathbf{P}^{n*}$

given by

$$x \to H(x)$$

has the defining property

$$E(x) = H(x) \cap S_0.$$

It follows that the paths $\{x(t)\}$ are mapped to pencils of hyperplanes in such a way that the $n-1$ fixed points correspond to the intersection $B \cap S_0$ where $B \cong \mathbf{P}^{n-2}$ is the base of the pencil. Thus the mapping (4.109) maps our path geometry to the flat one on $\mathbf{P}^{n*}$, and this completes the proof of our main theorem.

Before concluding we cannot refrain from mentioning what we consider to be the fundamental problem on the subject, which is to determine the maximum rank non-linearizable webs. The strong conditions must imply that there are not many. It may not be unreasonable to compare the situation with the exceptional simple Lie groups. Perhaps the error in 1. makes our subject more interesting.

Prof. S. S. Chern
Department of Mathematics
University of California
Berkeley, CA 94720, USA

Prof. Phillip Griffiths
Harvard University
Cambridge, MA 02138, USA

(*Eingegangen: 2. 5. 1980*)

Annali Scuola Normale Superiore - Pisa
Classe di Scienze
Serie IV - Vol. V, n. 3 (1978)

An Inequality for the Rank of a Web and Webs of Maximum Rank (*).

SHIING-SHEN CHERN (**) - PHILLIP A. GRIFFITHS (***)

dedicated to Hans Lewy

1. - Statement of results.

A *web* is given in a neighborhood $U \subset R^N$ by a set of codimension k foliations in general position, a notion we shall make precise in a moment. *Web geometry* is the study of local diffeomorphism invariants of a web (¹); for example, we may ask if it is equivalent to a standard web whose foliations consist of parallel linear spaces of dimension $N-k$. An invariant arises from the consideration of the *abelian q-equations* $(1 \leqslant q \leqslant k)$ (²) associated to the web. In this paper we will be concerned with the abelian equations when $q = k$. Specifically, we will find a bound on the *rank* or maximum number of linearly independent abelian k-equations, and will show that webs of maximal rank give a very special G-structure in the projectivized tangent spaces PT_x $(x \in U)$. In a future paper we hope to use this to show that such webs have a standard local form, generalizing our previous result in the codimension-one case (³).

For simplicity of notation we will carry out our study in detail only in the case $k = 2$. Therefore we now agree, until specified otherwise, that a

(*) Research partially supported by NSF Grants MCS 74-23180, A01 and 72-07782.

(**) Department of Mathematics, University of California, Berkeley 94720.
(***) Department of Mathematics, Harvard University, Cambridge 02138.

(¹) The basic reference is Blaschke-Bol [1].

(²) These are defined in general in [4]. The definitions relevant to our present discussion will be given below.

(³) Cf. [2], and also [3] for an outline of the main result from [2].

Pervenuto alla Redazione il 27 Giugno 1977.

web will be given in an open set $U \subset R^{2n}$ (4) by d foliations by codimension-two submanifolds. The leaves of the i-th foliation will be taken as level sets

$$u_i(x) = \text{const}, \qquad v_i(x) = \text{const};$$

these functions are defined up to a local diffeomorphism in (u_i, v_i)-space. The i-th *web normal* is

$$\Omega_i(x) = du_i(x) \wedge dv_i(x).$$

Under a diffeomorphism of (u_i, v_i), Ω_i is multiplied by a non-vanishing factor so that what is intrinsic is the point

$$\Omega_i(x) \in P(\Lambda^2 T^*),$$

the latter being the projective space associated to the vector space of 2-forms at $x \in U$.

We want to say what it means for the web to be in general position. For this some linear algebra preliminaries are required. It will be convenient not to distinguish between a point $u \in P^N = P(R^{N+1})$ and its homogeneous coordinate vector $u \in R^{N+1} - \{0\}$. A set of points $u_1, \ldots, u_d \in P^N$ is in general position in case any $k \leqslant N+1$ of them span a P^{k-1}; i.e. $u_{i_1} \wedge \ldots \wedge u_{i_k} \neq 0$ for $1 \leqslant i_1 < \ldots < i_k \leqslant d$, $k \leqslant N+1$.

When we come to the notion of general position of lines some care is necessary. Denote by $G(1, 2n-1)$ the Grassmannian of lines in $P^{2n-1} = P(R^{2n})$. We will identify $G(1, 2n-1)$ with its image under the *Plücker embedding*

$$G(1, 2n-1) \hookrightarrow P^{\binom{2n}{2}-1} = P(\Lambda^2 R^{2n})$$

given by sending the line spanned by points $u, v \in P^{2n-1}$ into $u \wedge v \in P(\Lambda^2 R^{2n})$. A first guess is that a set of lines

$$\Omega_1, \ldots, \Omega_d \in G(1, 2n-1)$$

should be said to be in general position if any $k \leqslant n$ of them span a P^{2k-1}; i.e. if all

$$\Omega_{i_1} \wedge \ldots \wedge \Omega_{i_k} \neq 0 \qquad 1 \leqslant i_1 < \ldots < i_k \leqslant d, \; k \leqslant n. \tag{1.1}$$

This condition is certainly necessary, but for some purposes may not be sufficient.

(4) The reason for taking the dimension $N = 2n$ will appear in § 4.

For example (5), consider a set of four lines $\Omega_1, \Omega_2, \Omega_3, \Omega_4$ in P^3. The condition (1.1) is equivalent to those lines being pairwise skew. Now it is well-known that there are « in general » two lines meeting each of four skew lines in P^3. To see better what « in general » means recall that a non-singular quadric surface S in P^3 is doubly ruled by two families of lines, called the A-lines and B-lines. The A-lines (resp. B-lines) are pairwise skew, and any A-line meets any B-line exactly once. All of these facts follow easily by representing S as the image under the Segre embedding

$$P^1 \times P^1 \to P^3$$

given by

$$(s, t) \to [1, s, t, st] .$$

The A-lines and B-lines are given by $s = \text{const}$ and $t = \text{const}$ respectively. Now there is a unique non-singular quadric surface S containing $\Omega_1, \Omega_2, \Omega_3$ as A-lines.

For the remaining line Ω_4 there are the three possibilities:

Ω_4 meets S in distinct points u_1, u_2;

Ω_4 is tangent to S at u;

Ω_4 is an A-line lying in S.

In the first case each of the B-lines through u_1, u_2 meets all four Ω_i once, and the second possibility is the limiting case of the first when $u_1 = u_2$. But in the third case there are infinitely many lines meeting the four skew lines Ω_i.

In our study we will give a definition of general position motivated by webs arising from non-degenerate algebraic surfaces in P^{n+1}. For this we assume first the condition (1.1). Given any $n-1$ of the Ω_i, say $\Omega_1, \ldots, \Omega_{n-1}$, spanning a P^{2n-3} we consider any P^{2n-5} contained in this P^{2n-3} and the linear projection

$$\pi : P^{2n-1} - P^{2n-5} \to P^3 .$$

Our second requirement is:

(1.2) the lines $\pi(\Omega_n), \ldots, \pi(\Omega_d)$ do not all pass through a common point.

(5) This observation is due to Ran Donagi.

A set of lines satisfying (1.1) and (1.2) will be said to be in *general position.* It is *not* the case that lines in general position have as Plücker images points in general position in $P^{\binom{2n}{2}-1}$. The linear algebra subtlety here is crucial in our study.

A set of foliations is in general position in case the normals $\Omega_i(x)$ are lines in general position in PT^*_x.

An *abelian equation* is a relation

$$\sum_i f_i\big(u_i(x), v_i(x)\big)\,\Omega_i(x) = 0\,, \tag{1.3}$$

and the *rank* r of the web is the maximum number of linearly independent abelian equations. Our first result is a bound on this rank. Namely, define the integer t uniquely by the conditions

$$t \equiv -d + 1 \bmod n-1\,, \qquad 0 \leqslant t \leqslant n-2 \tag{1.4}$$

and set

$$p_g(d, n) = \frac{1}{6(n-1)^2}(d-2n+1+t)(d-n+t)(d-1-2t)\,. \tag{1.5}$$

THEOREM I. *The rank of a d-web in $U \subset R^{2n}$ satisfies*

$$r \leqslant p_g(d, n) \tag{1.6}$$

In particular, $r = 0$ when $d < 2n$.

This bound may be seen to be sharp. Webs for which equality holds in (1.6) will be said to be of *maximal rank.* Our remaining results will, in this case, give a particular type of G-structure in the projectivized cotangent spaces PT^*_x.

Before stating the next theorem we recall a little algebraic geometry. A *ruled surface* S in P^N may be constructed by taking two skew subspaces P^m, $P^{m'}$ $(m + m' = N-1)$ spanning P^N together with rational curves $C \subset P^m$, $C' \subset P^{m'}$ in projective correspondence and letting S be the surface of ∞^1 lines obtained by joining corresponding points. In case $N = 2n-1$, $m = m' = n-1$, and C together with C' are rational normal curves we obtain what will be termed a *special ruled surface.* In suitable coordinates it is the image of $P^1 \times P^1$ under the map

$$(t, s) \to [t, \ldots, t^n; st, \ldots, st^n]\,. \tag{1.7}$$

The lines $\Omega(t)$ on S are obtained by holding t fixed. They all have a common linear parameter s, and are therefore all in projective correspondence with corresponding points spanning a $P^{n-1}(s)$ (where $P^{n-1}(0) = P^{n-1}$, $P^{n-1} \cdot (\infty) = P'^{n-1}$).

THEOREM II. *Assume given a d-web of maximal rank in* $U \subset R^{2n}$ *where*

$$d \geqslant 2n + 1, \qquad n \geqslant 3 \ (^6)$$

Then there is defined a field of special ruled surfaces

$$S(x) \subset PT^*_x$$

such that the web normals are lines belonging to this surface. In particular the web normals are all in projective correspondence, written

$$\Omega_i(x) \overline{\wedge} \Omega_j(x) \qquad 1 \leqslant i, j \leqslant d \tag{1.9}$$

with corresponding points spanning a P^{n-1} *in* PT^*_x.

2. – Proof of the bound on the rank.

Suppose that

$$\sum_{i=1}^{d} f_i^\lambda(u_i(x), v_i(x))\, du_i(x) \wedge dv_i(x) = 0 \qquad \lambda = 1, \ldots, r \tag{2.1}$$

are r linearly independent abelian equations.

Set $\Omega_i = du_i \wedge dv_i$ and

$$Z_i(x) = [\ldots, f_i^\lambda(u_i(x), v_i(x)), \ldots] \in P^{r-1}.$$

The abelian equations (2.1) become

$$\sum_i Z_i(x) \otimes \Omega_i(x) = 0. \tag{2.2}$$

As x varies over $U \subset R^{2n}$, $Z_i(x)$ traces out a piece of two-dimensional surface S_i in P^{r-1}. We may take (u_i, v_i) as local coordinates on S_i. There

(6) The rank is zero when $d < 2n$ and is $\leqslant 1$ when $d = 2n$. These cases are excluded.

is a 1-to-d correspondence

$$x \to Z_1(x), \dots, Z_d(x),$$

but because of (2.2) corresponding points $Z_i(x)$ are not in general position. At first glance it might appear that there are in fact $\binom{2n}{2}$ independent linear relations among the Z_i, but this is not so. Letting $\{Z_1, \dots, Z_d\}$ denote the linear span in P^{r-1} of $Z_1(x), \dots, Z_d(x)$ we shall prove that

$$\dim\{Z_1, \dots, Z_d\} \leqslant d - 2n, \tag{2.3}$$

an estimate which will turn out to be sharp.

To see this we note that since the lines $\Omega_1(x), \dots, \Omega_d(x) \in G_x(1, 2n-1)$ are in general position we may choose points

$$\omega_1 = du_1 + \lambda_1 dv_1 \in \Omega_1, \dots, \omega_{2n-2} = du_{2n-2} + \lambda_{2n-2} dv_{2n-2} \in \Omega_{2n-2}$$

such that

$$\omega_1 \wedge \dots \wedge \omega_{2n-2} \wedge \Omega_{2n-1} \neq 0.$$

If we multiply (2.2) by $\omega_1 \wedge \dots \wedge \omega_{2n-2}$ the first $2n-2$ terms drop out and Z_{2n-1} appears with a non-zero coefficient, i.e.

$$Z_{2n-1} \in \{Z_{2n}, \dots, Z_d\}. \tag{2.4}$$

By symmetry it follows that at most $d - 2n + 1$ of $Z_i(x)$ are linearly independent, which proves (2.3).

Now the argument proceeds as in the codimension-one case. By (2.4)

$$Z_{2n-1} = \varrho_{2n} Z_{2n} + \dots + \varrho_d Z_d. \tag{2.5}$$

Choose $(u_{2n-1}, v_{2n-1}, u_{2n}, v_{2n}, \dots, u_{3n-2}, v_{3n-2})$ as coordinates on U and differentiate (2.5) to obtain

$$\frac{\partial Z_{2n-1}}{\partial u_{2n-1}}, \frac{\partial Z_{2n-1}}{\partial v_{2n-1}} \equiv \frac{\partial Z_{3n-1}}{\partial u_{3n-1}}, \frac{\partial Z_{3n-1}}{\partial v_{3n-1}}, \dots, \frac{\partial Z_d}{\partial u_d}, \frac{\partial Z_d}{\partial v_d} \mod Z_1, \dots, Z_d. \tag{2.6}$$

Let

$$P^{k(0)} = \{Z_1, \dots, Z_d\}, \qquad P^{k(1)} = \left\{Z_1, \frac{\partial Z_1}{\partial u_1}, \frac{\partial Z_1}{\partial v_1}, \dots, Z_d, \frac{\partial Z_d}{\partial u_d}, \frac{\partial Z_d}{\partial v_d}\right\}, \dots,$$

$P^{k(\mu)}=$ span of the μ-th osculating spaces to the surfaces S_i at corresponding points. By (2.4) and (2.6)

$$\dim P^{k(0)} \leqslant (d-2n+1)-1\,,$$

$$\dim P^{k(1)} \leqslant (d-2n+1)+2(d-3n+2)-1\,,$$

and in general

$$\dim P^{k(\mu)} \leqslant (d-2n+1)+2(d-3n+2)+\ldots+ + (\mu+1)(d-(\mu+2)n+\mu+1)-1\,. \tag{2.7}$$

Here we agree that zero is put in whenever one of the first $\mu+1$ terms on the right becomes $\leqslant 0$, which obviously happens for large μ.

The $P^{k(\mu)}(x)$ give an increasing sequence of linear subspaces of P^{r-1}, which eventually terminates at say $P^{k(\infty)}(x)$. Since $P^{k(\infty)}(x)$ does not change by differentiation, it must be a constant linear subspace, and hence is all of P^{r-1} since the equations (2.1) were assumed linearly independent. By (2.7)

$$r \leqslant \sum_{\mu\geqslant 0} \max\,[(\mu+1)\{d-n(\mu+2)+(\mu+1)\},0]\,. \tag{2.8}$$

It remains to identify this sum with the expression (1.5). Write

$$\tau=\frac{d-2n+1}{n-1}+\frac{t}{n-1}$$

where t is determined by (1.4); τ is an integer. Put

$$a=d-2n+1\,,\qquad s=n-1\,.$$

Then the R.H.S. of (2.8) is

$$a+2(a-s)+3(a-2s)+\ldots+\tau(a-(\tau-1)s)=$$
$$=\tfrac{1}{2}\tau(\tau+1)a-\tfrac{1}{3}(\tau-1)\tau(\tau+1)s=\tfrac{1}{6}\tau(\tau+1)\{3a-2(\tau-1)s\}\,.$$

Now

$$(n-1)\tau=d-2n+1+t\,,$$

$$(n-1)(\tau+1)=d-n+t\,,\;(n-1)(\tau-1)=d-3n+2+t\,,$$

$$3a-2(\tau-1)s=3d-6n+3-2(d-3n+2+t)=d-1-2t\,,$$

so that the R.H.S. of (2.8) is

$$\frac{1}{6(n-1)^2}(d-2n+1+t)(d-n+t)(d-1-2t)=p_g(d,n)$$

according to (1.5). This completes the proof of Theorem I.

When $n=2$, we have $t=0$ and

$$p_g(d,n)=\tfrac{1}{6}(d-1)(d-2)(d-3)= \tag{2.9}$$
$$= \text{geometric genus of a smooth surface of degree } d \text{ in } P^3 .$$

Exactly the same considerations can be carried out for a d-web of codimension k in a neighborhood $U\subset R^{kn}$, $k\leqslant n$. Let r_k be its rank, i.e., the maximum number of linearly independent abelian k-equations. Then we have

$$r_k\leqslant\pi(d,n,k) \tag{2.10}$$

where

$$\pi(d,n,k)=\sum_{\mu\geqslant 0}\max\left(\binom{k+\mu-1}{\mu}\{d-(k+\mu)n+k-1+\mu\},0\right). \tag{2.11}$$

The first term in $\pi(d,n,k)$ is

$$d-kn+k-1 .$$

Hence we have

$$\pi(d,n,k)=0 \text{ when and only when } d\leqslant kn-k+1 . \tag{2.12}$$

The first two terms in $\pi(d,n,k)$ are

$$d-kn+k-1+k\{d-(k+1)n+k\}=$$
$$=(k+1)d-k(k+2)n+k^2+k-1 .$$

Hence we have

$$\pi(d,n,k)=n+k , \tag{2.13a}$$

when and only when

$$d=(k+1)n-(k-1) . \tag{2.13b}$$

3. - Proof of Theorem II.

By the assumption of maximal rank

$$\{Z_1, \ldots, Z_d\} = \{Z_{2n}, \ldots, Z_d\} = P^{d-2n}(x)\,, \tag{3.1}$$

i.e. there are exactly $(2n-1)$ independent relations among the $Z_i(x)$. On the other hand, (2.2) gives what appears to be $\binom{2n}{2}$ relations, and consequently some of these must be dependent. We will see that the geometrical consequence of this is the presence in $P(T^*_x)$, $x \in U \subset R^{2n}$, of a field of special ruled surfaces. An intermediate step is the normal form (3.16), which we will derive first.

To carry this out we choose $u_1, v_1, \ldots, u_n, v_n$ as coordinate system and write, at a fixed point $x_0 \in U$,

$$\left|\begin{aligned} du_s &= du_1 + \sum_{\lambda=2}^{n} A_{s\lambda}\, du_\lambda + \sum_{\lambda=2}^{n} B_{s\lambda}\, dv_\lambda\,, \\ dv_s &= dv_1 + \sum_{\lambda=2}^{n} C_{s\lambda}\, du_\lambda + \sum_{\lambda=2}^{n} D_{s\lambda}\, dv_\lambda\,, \qquad s = n+1, \ldots, d\,. \end{aligned}\right. \tag{3.2}$$

This is possible since by general position all

$$\frac{\partial(u_s, v_s)}{\partial(u_1, v_1)} \neq 0\,.$$

We set

$$A_\lambda = (A_{n+1,\lambda}, \ldots, A_{d,\lambda}) \in R^{d-n}$$

and similarly for B_λ, C_λ, D_λ.

(3.3) Lemma. *The vectors A_λ, B_λ are multiples of a vector E_λ, and the E_λ are linearly independent (here $\lambda = 2, \ldots, n$).*

Proof. The abelian equations (2.2) are

$$\sum_{\alpha=1}^{n} Z_\alpha\, du_\alpha \wedge dv_\alpha + \sum_{s=n+1}^{d} Z_s\, du_s \wedge dv_s = 0\,. \tag{3.4}$$

The coefficient of $du_\alpha \wedge dv_\alpha$ gives Z_α, $\alpha = 1, \ldots, n$, as a linear combination of Z_s, $s = n+1, \ldots, d$, so that by (3.1) there are at most $n-1$ independent relations among Z_s. In particular, the coefficient of $du_1 \wedge dv_1$ gives

$$Z_1 + \sum_{s=n+1}^{d} Z_s = 0 , \tag{3.5}$$

and the coefficients of $dv_1 \wedge du_\lambda$ and $dv_1 \wedge dv_\lambda$ give

$$(3.6)_\lambda \quad \left| \begin{array}{l} \sum_{s=n+1}^{d} A_{s\lambda} Z_s = 0 , \\ \sum_{s=n+1}^{d} B_{s\lambda} Z_s = 0 , \qquad \lambda = 2, \ldots, n . \end{array} \right.$$

By the above remark at most $(n-1)$ of the $2(n-1)$ equations $(3.6)_\lambda$ can be independent. In other words, if $R_\lambda \subset R^{d-n}$ is the span of A_λ, B_λ then

$$\dim\left(\sum_{\lambda=2}^{n} R_\lambda\right) \leq n-1 .$$

The lemma amounts to

$$\sum_{\lambda=2}^{n} R_\lambda = \bigoplus_{\lambda=2}^{n} R_\lambda \cong R^{n-1} ,$$

which is implied by

$$R_\lambda \not\subset \sum_{\lambda \neq \gamma} R_\gamma$$

for fixed λ. If, on the contrary, the equations $(3.6)_\lambda$ are linear combinations of $(3.6)_\gamma$ for $\gamma \neq \lambda$, then taking $\lambda = n$ we will have

$$A_n = \sum_{\gamma=2}^{n-1} a_\gamma A_\gamma + b_\gamma B_\gamma ,$$

$$B_n = \sum_{\gamma=2}^{n-1} c_\gamma A_\gamma + d_\gamma B_\gamma .$$

By (3.2)

$$du_s = du_1 + \sum_{\gamma=2}^{1-n} A_{s\gamma}(du_\gamma + a_\gamma\, du_n + c_\gamma\, dv_n) + \sum_{\gamma=2}^{n} B_{s\gamma}(dv_\gamma + b_\gamma\, du_n + d_\gamma\, dv_n) .$$

In the R^4 defined by

$$(3.7) \quad \left\{ \begin{array}{l} du_\gamma + a_\gamma\, du_n + c_\gamma\, dv_n = 0 , \\ dv_\gamma + b_\gamma\, du_n + d_\gamma dv_n = 0 , \qquad \gamma = 2, \ldots, n-1 , \end{array} \right.$$

we have

$$du_s = du_1, \qquad s = n+1, \ldots, d, \tag{3.8}$$

contradicting general position. More precisely, the $2(n-2)$ one-forms (3.7) span a P^{2n-5} in $P^{2n-1} = P(T_x^*)$. This P^{2n-5}, does not meet any of the web normals $\Omega_1, \Omega_{n+1}, \ldots, \Omega_d$. Under the linear projection

$$P^{2n-1} - P^{2n-5} \xrightarrow{\pi} P^3$$

(3.8) says exactly the lines $\pi(\Omega_1), \pi(\Omega_{n+1}), \ldots, \pi(\Omega_d)$ all pass through a common point, and this contradicts general position. Thus Lemma (3.3) is proved.

By the lemma we have $A_\lambda = \alpha_\lambda E_\lambda$, $B_\lambda = \beta_\lambda E_\lambda$, α_λ, β_λ not both zero. Replacing du_λ by $\alpha_\lambda du_\lambda + \beta_\lambda dv_\lambda$ we obtain

$$A_\lambda = E_\lambda, \qquad B_\lambda = 0$$

in (3.2). After a similar argument applied to C_λ, D_λ we may assume

$$C_\lambda = 0, \qquad D_\lambda = F_\lambda,$$

so that (3.2) is now

$$\left\{\begin{aligned} du_s &= du_1 + \sum_{\lambda=2}^{n} E_{s\lambda}\, du_\lambda, \\ dv_s &= dv_1 + \sum_{\lambda=2}^{n} F_{s\lambda}\, dv_\lambda. \end{aligned}\right. \tag{3.9}$$

(3.10) LEMMA. *E_λ is a non-zero multiple of F_λ.*

PROOF. By the proof of Lemma (3.3) the $2(n-1)$ vectors E_γ, F_γ span an R^{n-1} in R^{d-n}. Thus, if R_λ is the span of E_λ, F_λ

$$\dim \sum_{\lambda=2}^{n} R_\lambda = n-1.$$

If some E_λ and F_λ are linearly independent, i.e.

$$\dim R_\lambda = 2,$$

then for some other γ we must have

$$R_\gamma \subset \sum_{\lambda \neq \gamma} R_\lambda .$$

Taking $\gamma = n$ we obtain a relation

$$\text{(3.11)} \qquad \begin{cases} E_n = \sum_{\gamma=2}^{n-1} a_\gamma E_\gamma + b_\gamma F_\gamma , \\ F_n = \sum_{\gamma=2}^{n-1} c_\gamma E_\gamma + d_\gamma F_\gamma , \end{cases}$$

which we will show leads to a contradiction.

Using (3.9) the coefficients of $du_\gamma \wedge dv_n$ and $du_n \wedge dv_\gamma$ in (3.4) give

$$\text{(3.12)} \qquad \begin{cases} \sum_{s=n+1}^{d} (E_{s\gamma} \; F_{sn}) \; Z_s = 0 , \\ \sum_{s=n+1}^{d} (E_{sn} F_{s\gamma}) Z_s = 0 , \qquad 2 \leqslant \gamma \leqslant n-1 . \end{cases}$$

The coefficient of $du_\lambda \wedge dv_\gamma$ gives

$$\text{(3.13)} \qquad \delta_\gamma^\lambda Z_\gamma + \sum_{s=n+1}^{d} (E_{s\gamma} F_{s\lambda}) Z_s = 0 , \qquad 2 \leqslant \gamma , \; \lambda \leqslant n-1 .$$

Finally the coefficient of $du_n \wedge dv_n$ gives, after we plug in (3.11),

$$\text{(3.14)} \qquad Z_n + \sum_{\lambda,\gamma=2}^{n-1} \sum_{s=n+1}^{d} (a_\lambda E_{s\lambda} + b_\lambda F_{s\lambda})(c_\gamma E_{s\gamma} + d_\gamma F_{s\gamma}) Z_s = 0 .$$

Substituting (3.11) into (3.12) and using (3.13) gives

$$\text{(3.15)} \qquad \begin{cases} \sum_{\lambda=2}^{n-1} \sum_{s=n+1}^{d} (c_\lambda E_{s\gamma} E_{s\lambda}) Z_s = - d_\gamma Z_\gamma , \\ \sum_{\lambda=2}^{n-1} \sum_{s=n+1}^{d} (b_\lambda F_{s\lambda} F_{s\gamma}) Z_s = - a_\gamma Z_\gamma , \qquad 2 \leqslant \gamma \leqslant n-1 . \end{cases}$$

Expanding out (3.14) and plugging in (3.15) and (3.13) we obtain

$$Z_n - \sum_{\gamma=2}^{n-1} (3a_\gamma d_\gamma + b_\gamma c_\gamma) Z_\gamma = 0 ,$$

which contradicts the maximal rank assumption (3.1). This proves Lemma (3.10).

We now arrive at our *normal form* for the du_i, dv_i. Namely, we may multiply by a scale factor and assume $E_\lambda = F_\lambda$. If we relabel and define $A_{i\alpha}$ by

$$\begin{aligned} &A_{s\lambda} = E_{s\lambda} , && n+1 \leqslant s \leqslant d ,\ 2 \leqslant \lambda \leqslant n , \\ &A_{s1} = 1 , && n+1 \leqslant s \leqslant d , \\ &A_{\alpha\beta} = \delta_{\alpha\beta} , && 1 \leqslant \alpha ,\ \beta \leqslant n , \end{aligned}$$

then (3.9) becomes

$$\text{(3.16)} \qquad \left| \begin{aligned} du_i &= \sum_{\alpha=1}^{n} A_{i\alpha} du_\alpha , \\ dv_i &= \sum_{\alpha=1}^{n} A_{i\alpha} dv_\alpha , \qquad 1 \leqslant i \leqslant d . \end{aligned} \right.$$

(3.17) LEMMA. *The vectors* $A_i = [\ldots, A_{i\alpha}, \ldots] \in P^{n-1}$ *lie on* $\infty^{(n-1)(n-2)/2}$ *linearly independent quadrics.*

PROOF. The basic abelian equation (3.4) gives upon substituting in (3.16)

$$\text{(3.18)} \qquad \sum_{i=1}^{d} A_{i\alpha} A_{i\beta} Z_i = 0 , \qquad 1 \leqslant \alpha ,\ \beta \leqslant n .$$

These are $n(n+1)/2$ relations among the Z_i, and by (3.1) only $2n-1$ of the equations (3.18) can be independent. In other words we have

$$\frac{n(n+1)}{2} - (2n-1) = \frac{(n-1)(n-2)}{2}$$

linearly independent relations

$$\sum_{\alpha,\beta=1}^{n} k_{\alpha\beta} A_{i\alpha} A_{i\beta} = 0 , \qquad k_{\alpha\beta} = k_{\beta\alpha} ,$$

among the coefficients in (3.18), and this gives the lemma.

Now we can complete the proof of Theorem II. Namely, by (3.17) the A_i lie on a rational normal curve C in P^{n-1}. After a linear change of coordinates we may assume that C is given parametrically by

$$t \to [t, t^2, \ldots, t^n]. \tag{3.19}$$

According to (3.16) we have now written $P^{2n-1} = P(T^*_x)$ as the span of the P^{n-1} determined by the du_α and P'^{n-1} determined by the dv_α, and in each of P^{n-1}, P'^{n-1} we have the rational normal curve (3.19) such that setting $t = t_i$ gives $du_i \in P^{n-1}$ and $dv_i \in P'^{n-1}$ respectively. The i-th web normal Ω_i is the line $du_i + s\, dv_i$, which is just the line $t = t_i$ on the standard ruled surface given parametrically by

$$(s, t) \to [t, t^2, \ldots, t^n; st, st^2, \ldots, st^n].$$

4. – Webs defined by algebraic varieties.

A projective algebraic variety of dimension k, $V_k \subset P^m$ is non-degenerate in case it does not lie in a P^{m-1}. The degree d is the number of intersections with a generic P^{m-k}, written

$$V \cdot P^{m-k} = p_1 + \ldots + p_d. \tag{4.1}$$

(Here and in what follows, we frequently omit the index of V_k.) For non-degenerate V, which we will always assume, the $p_i \in P^{m-k}$ are in general position (c.f. Lemma 1.8 in [2]).

We continue to denote by $G(m-k, m)$ the Grassmannian of P^{m-k}'s in P^m, and for fixed $p \in P^m$ we let $\Sigma(p)$ designate the Schubert variety of all P^{m-k}'s which pass through the point p. Note that $\Sigma(p) \cong G(m-k-1, m-1)$ and has codimension k in $G(m-k, m)$. The algebraic variety V defines a web in open sets $U \subset G(m-k, m)$ by specifying the i-th web leaf through P^{m-k} to be $\Sigma(p_i)$ where the p_i are given by (4.1). The basic geometric object here is the incidence correspondence

$$I_V \subset V \times G(m-k, m) \tag{4.2}$$

defined by V, where $I_V = \{(p, A): p \in V,\ A \in G(m-k, m),\ p \in A \cap V\}$. By taking V and the p_i to be defined over the real numbers, we have associated to a projective variety $V_k \subset P^m$ a d ($=$ degree V) web of codimension k ($= \dim V$) submanifolds in $U \subset R^{k(m-k+1)}$.

We now wish to verify that the web defined by a non-degerate algebraic variety is non-degenerate according to our definition, which we shall do for a surface $S \subset P^{n+1}$. For this consider the linear projection

$$\pi: P^{n+1} - P^{n-3} \to P^3 \tag{4.3}$$

with center P^{n-3} defined by

$$\pi(p) = (p \wedge P^{n-3}) \cdot P^3$$

where $p \wedge P^{n-3}$ is the P^{n-2} spanned by $p \in P^{n+1} - P^{n-3}$ and the center. Under such a projection, $\pi(S) = S'$ is a non-degenerate surface in P^3 of degree

$$d' = d - \# \text{ of points in } S \cap P^{n-3}.$$

The projection induces an inclusion

$$\pi^{-1}: G(1, 3) \to G(n-1, n+1) \tag{4.4}$$

whose image is the Schubert cycle of all P^{n-1}'s containing the center P^{n-3}. Our first observation is that the *web in* $G(1, 3)$ *defined by* S' *is the intersection of* $\pi^{-1}G(1, 3)$ *with the web in* $G(n-1, n+1)$ *defined by* S, even in case there are finitely many points in $S \cap P^{n-3}$.

Now consider the web in $G(1, 3)$ defined by a non-degenerate surface $S' \subset P^3$. For a generic line P^1 in P^3 the intersection

$$P^1 \cdot S' = p_1 + \dots + p_{d'}$$

where the p_i are distinct. The Schubert cycle $\Sigma(p)$ consists of all lines passing through $p \in P^3$, and under the Plücker embedding

$$G(1, 3) \to P^5$$

$\Sigma(p)$ is a plane. If $\Sigma(p)$ and $\Sigma(q)$ fail to intersect transversely, then they must have in common a line in P^5. Any line on $G(1, 3)$ is the $\Sigma(p, P^2)$ $(p \in P^2 \subset P^3)$ of lines in P^3 passing through p and contained in P^2. Consequently $\Sigma(p)$ and $\Sigma(q)$ meet transversely unless $p = q$. From this we deduce that the normals to the web defined by $S' \subset P^3$ are skew lines in the projectivized cotangent spaces to $G(1, 3)$, these being P^3's.

Finally, for the web defined by a non-degenerate surface S in P^{n+1}, the projection

$$P^{2n-1} - P^{2n-5} \to P^3 \qquad (P^{2n-1} = P(T_x^*))$$

in the definition of web non-degeneracy corresponds to the transposed differential of the inclusion (4.4) induced by the linear projection (4.3) whose center P^{n-3} contains $p_1, \dots, p_{n-1}$. But since $S' = \pi(S)$ is still non-degenerate we deduce that the web defined by S in the neighborhood of a generic $P^{n-1} \in G(n-1, n+1)$ is non-degenerate according to the definition used in Theorems I and II.

Given $V_k \subset P^m$ we consider a meromorphic k-form ω on V and define the *trace* $\boldsymbol{\omega}$, a meromorphic k-form on the Grassmannian $G(m-k, m)$, by

$$\boldsymbol{\omega}(A) = \sum_{i=1}^{d} \omega(p_i(A)) , \qquad A \in G(m-k, m) ,$$

where the intersection

$$A \cdot V = p_1(A) + \dots + p_d(A)$$

for a variable $(m-k)$-plane A. In terms of the diagram (4.2)

$$\boldsymbol{\omega} = (\pi_2)_* \pi_1^* \omega$$

where π_1, π_2 are respectively the projections $I_V \to V$, $I_V \to G(m-k, m)$. The form ω is a *differential of the first kind* (d.f.k.) if $\boldsymbol{\omega}$ is holomorphic (cf. § II of [4]). The space of d.f.k. will be denoted by $H^{k,0}(V)$ and its dimension by $h^{k,0}(V)$. In case V is non-singular $H^{k,0}(V)$ are just the holomorphic k-forms and $h^{k,0}(V)$ is the usual Hodge number.

Since there are no holomorphic forms on $G(m-k, m)$, for ω a d.f.k. we have $\boldsymbol{\omega} = 0$, which is *Abel's theorem*

$$\sum_{i=1}^{d} \boldsymbol{\omega}(p_i(A)) = 0 . \tag{4.5}$$

Clearly (4.5) gives an abelian k-equation on the web defined by V. Conversely, it is not difficult to see that every abelian k-equation is of this form, and consequently the rank of the web is equal to $h^{k,0}(V)$. From Theorem I we deduce the bound

$$h^{k,0}(V) \leqslant \pi(d, m-k+1, k) \tag{4.6}$$

on the number of linearly independent d.f.k. of a non-degenerate $V_k \subset P^m$.

In case $k=1$ and $m=n$ we obtain *Castelnuovo's bound* (cf. [2])

$$\pi(C) \leqslant \pi(d, n) = \pi(d, n, 1) \tag{4.7}$$

on the genus of a curve of degree d in P^n. The curves for which equality holds in (4.1) were extensively discussed in our previous paper [2], where in fact we proved that their properties could be deduced by web-theoretic methods.

When $k=2$ we set $m=n+1$ so that our variety is a surface $S \subset P^{n+1}$ corresponding to a codimension-2 web in $U \subset R^{2n}$. We denote by $p_g(S)$ the number $h^{2,0}(S)$ of d.f.k.; for smooth S this is the *geometric genus*. Theorem I gives the bound

(4.8) $$p_g(S) \leqslant p_g(d, n) = \pi\,(d, n, 2)\,.$$

This inequality has been proved algebro-geometrically by Joe Harris in his Harvard thesis, which contains general methods of estimating the superabundance (= « number of relations among conditions imposed by ») of linear systems with base conditions imposed.

A special case of (4.7) is (7)

$$p_g(S) = 0 \qquad \text{for degree } S < 2n\,.$$

The general statement for a non-degenerate $V_k \subset P^m$ is, by (2.12) and with $n = m-k+1$,

(4.8) $$h^{k,0}(V) = 0 \qquad \text{for } d < k(m-k)+2\,.$$

These bounds are sharp. For example, for each $n \geqslant 2$ there are $K3$ surfaces $S \subset P^{n+1}$ of degree $2n$, characterized by having as hyperplane sections canonical curves of genus $n+1$. In general

$$h^{k,0}(V) \leqslant 1 \qquad \text{for } d = k(m-k)+2\,,$$

and if V is smooth and if $h^{k,0}(V)=1$, then V is simply-connected (for $k \geqslant 2$) with trivial canonical bundle.

To give another application we first observe that, by (2.13*a*) and (2.13*b*), there is, for each k a unique function $m \to d(m)$ satisfying

$$\pi(d(m), m-k+1, k) = m+1\,.$$

(7) After we mentioned this result to R. Hartshorne, he showed us an algebraic-geometric proof, together with the result that S must be a $K3$-surface, if deg $S = 2n$, $p_g(S) = 1$.

(Notice that $n = m - k + 1$). For example we have

$$(4.9)\qquad \begin{cases} d(m) = 2m & \text{when } k = 1\,, \\ d(m) = 3m - 4 & \text{when } k = 2\,, \end{cases}$$

and in general

$$(4.10)\qquad d(m) = (k+1)m - (k-1)(k+2)\,.$$

Next we remark that (4.6) can be inverted to

$$(4.11)\qquad d \geqslant d(h^{k,0}, m)$$

bounding from below the degree of a non-degenerate $V_k \subset P^m$ with fixed $h^{k,0}$.

In particular we consider *canonical algebraic varieties*, defined by the property that their canonical linear system $|K|$ gives a birational and biregular mapping of the abstract variety onto its image in P^m $(m + 1 = h^{k,0})$. For such varieties the degree of the canonical image is

$$(-1)^k c_1^k$$

where c_1 is the 1-st Chern class, and by combining (4.10) and (4.11) we deduce the bound

$$(4.12)\qquad h^{k,0}(V) \leqslant \frac{1}{k+1}\left[(-1)^k c_1^k + k^2 + 2k - 1\right]$$

on the Hodge number of a canonical variety. For $k = 1, 2$ we may use (4.9) to obtain

$$\pi(C) \leqslant \frac{-c_1}{2} + 1\,, \quad p_g(S) \leqslant \frac{c_1^2}{3} + \frac{7}{3}\,.$$

The first is an equality due to $c_1 = 2 - 2\pi$, but the second is in general an inequality. It may be compared with Max Noether's estimate

$$(4.13)\qquad p_g(S) \leqslant \frac{c_1^2}{2} + 2$$

valid for any surface. We remark that here the factor $\frac{1}{2}$ ultimately comes from the 2 in

CLIFFORD'S THEOREM:

$$\dim |L| \leqslant \frac{\deg L}{2}$$

for a special linear series $|L|$ *on a curve,* and consequently the generalization of (4.13) to higher dimension is

$$h^{k,0} \leqslant \frac{(-1)^k c_1^k}{2} + \text{const},$$

which is sharp for suitable double coverings of P^k.

The estimates (4.6)-(4.8), (4.11)-(4.13) were consequences of Theorem I. It is of course, interesting to ask whether or not these bounds are sharp, and if so to determine the structure of the *extremal varieties* defined as those for which equality holds. Now Theorem II gives at least the infinitesimal structure of extremal surfaces $S \subset P^{n+1}$ where the degree $d > 2n$. By continuing the reasoning in the proof of that result we may show that an extremal surface lies in a very special way as a divisor on a threefold $V \subset P^{n+1}$ of minimal degree $n-1$, and this leads to an effective determination of all extremal surfaces. These matters will be taken up in a future paper.

REFERENCES

[1] W. BLASCHKE - G. BOL, *Geometrie der Gewebe,* Springer, Berlin (1938).

[2] S. CHERN - P. A. GRIFFITHS, *Abel's theorem and webs,* to appear in Jahresberichte der deutschen Mathematiker Vereinigung (1978).

[3] S. CHERN - P. A. GRIFFITHS, *Linearization of webs of codimension one and maximum rank,* to appear in Proc. of International Symposium on Algebraic Geometry, Kyoto 1977.

[4] P. A. GRIFFITHS, *Variations on a theorem of Abel,* Inventiones Math., **35** (1976), pp. 321-390.

Minimal Submanifolds and Geodesics
Kaigai Publications, Tokyo, 1978, 17–30

AFFINE MINIMAL HYPERSURFACES

SHIING-SHEN CHERN[1)]

Introduction

In this paper I wish to call attention to a class of variational problems which arise naturally in affine geometry. In recent years there have been important developments on minimal varieties and variational problems with more general integrands, such as Jean Taylor's crystalline integrands. These integrals involve first partial derivatives. In the affine differential geometry of hypersurfaces there is an invariant volume element which depends on second partial derivatives and the corresponding Euler-Lagrange equation is a partial differential equation of the fourth order. Because of their simple geometrical origin affine minimal hypersurfaces have many remarkable properties, some of which will be discussed below. On the other hand, fundamental problems, such as the analogues of Plateau and Bernstein problems, have not been touched.

1. Hypersurfaces in unimodular affine space (cf. [1], [2], [3], [5])

Let A^{n+1} be the unimodular affine space of dimension $n+1$, i.e., the space with the real coordinates $x^1, \cdots, x^{n+1}$, provided with the group G of transformations (called the "unimodular affine group")

$$x^{*\alpha} = \sum_\beta c^\alpha_\beta x^\beta + d^\alpha\,, \qquad 1 \leqslant \alpha, \beta, \gamma \leqslant n+1\,, \tag{1}$$

where

$$\det(c^\alpha_\beta) = +1\,. \tag{2}$$

In this space distance and angle have no meaning, but there is the notion of parallelism. Vectors

$$V = (v^1, \cdots, v^{n+1}) \tag{3}$$

are transformed according to the equations

$$v^{*\alpha} = \sum c^\alpha_\beta v^\beta\,. \tag{4}$$

1) Work done under partial support of NSF grant MCS 74-23180.

As a consequence of (2) the determinant of $n+1$ vectors, e.g.,

$$(V_1, \cdots, V_{n+1}) = \det(v^\alpha_\beta)\,, \tag{5}$$

where

$$V_\beta = (v^1_\beta, \cdots, v^{n+1}_\beta)\,, \tag{5a}$$

is an invariant.

A hypersurface consists of an n-dimensional manifold M and an immersion

$$x : M \to A^{n+1}\,. \tag{6}$$

We will study its properties invariant under the group (1), and we will apply the method of moving frames.

By a *unimodular affine frame*, or simply a *frame*, is meant a point x and $n+1$ vectors e_α, satisfying the condition

$$(e_1, \cdots, e_{n+1}) = 1\,. \tag{7}$$

The importance of frames in affine geometry lies in the fact that there is exactly one unimodular affine transformation carrying one frame into another. In the space of all frames, which is the same as the group manifold G, we can write

$$\begin{aligned} dx &= \sum \omega^\alpha e_\alpha\,, \\ de_\alpha &= \sum \omega^\beta_\alpha e_\beta\,. \end{aligned} \tag{8}$$

The $\omega^\alpha, \omega^\beta_\alpha$ are the Maurer-Cartan forms of G. Differentiating (7) and using (8), we get

$$\sum_\alpha \omega^\alpha_\alpha = 0\,. \tag{9}$$

Exterior differentiation of (8) gives the *structure equations* of A^{n+1} or the *Maurer-Cartan equations* of G:

$$\begin{aligned} d\omega^\alpha &= \sum_\beta \omega^\beta \wedge \omega^\alpha_\beta\,, \\ d\omega^\beta_\alpha &= \sum_\gamma \omega^\gamma_\alpha \wedge \omega^\beta_\gamma\,, \end{aligned} \tag{10}$$

We restrict to the submanifold of frames such that x lies on the hypersurface and $e_1, \cdots, e_n$ span the tangent hyperplane at x. Then

$$\omega^{n+1} = 0 \tag{11}$$

and the first equation of (10) gives

$$d\omega^{n+1} = \sum_i \omega^i \wedge \omega_i^{n+1} = 0 , \tag{12}$$

where, as later, we agree on the index range

$$1 \leqslant i, j, k \leqslant n . \tag{13}$$

By Cartan's lemma we have

$$\omega_i^{n+1} = \sum_k h_{ik}\omega^k , \tag{14}$$

where

$$h_{ik} = h_{ki} . \tag{15}$$

The quadratic differential form

$$II = \sum_i \omega^i \omega_i^{n+1} = \sum_{i,k} h_{ik}\omega^i\omega^k \tag{16}$$

is called the *second fundamental form* of M. In our terminology there is no first fundamental form.

Let $xe_1^* \cdots e_{n+1}^*$ be a frame satisfying also the condition that $e_1^*, \cdots, e_n^*$ span the tangent hyperplane to the hypersurface $x(M)$ at x. Then we have

$$\begin{aligned} e_i^* &= \sum a_i^k e_k , \\ e_{n+1}^* &= A^{-1} e_{n+1} + \sum a_{n+1}^i e_i , \qquad A = \det (a_i^k) \end{aligned} \tag{17}$$

or

$$\begin{aligned} e_i &= \sum b_i^k e_k^* , \\ e_{n+1} &= A e_{n+1}^* + \sum_i b_{n+1}^i e_i^* , \end{aligned} \tag{17a}$$

where (b_i^k) is the inverse matrix of (a_i^k). Relative to the new frames $xe_1^* \cdots e_{n+1}^*$ let

$$\begin{aligned} dx &= \sum \omega^{*i} e_i^* , \\ de_\alpha^* &= \sum_\beta \omega_\alpha^{*\beta} e_\beta^* . \end{aligned} \tag{18}$$

Then we have

$$\begin{aligned} \omega^k &= \sum a_i^k \omega^{*i} , \\ \omega_i^{n+1} &= (e_1, \cdots, e_n, de_i) = A^{-1} \sum b_i^k \omega_k^{*n+1} . \end{aligned} \tag{19}$$

and

$$\sum \omega^i \omega_i^{n+1} = A^{-1} \sum \omega^{*i} \omega_i^{*n+1} . \tag{20}$$

This leads to the important conclusion that the second fundamental form is defined by the hypersurface up to a non-zero factor. Its rank is an affine invariant. We will restrict ourselves to the study of *non-degenerate hypersurfaces* for which the rank of II is equal to n.

For a non-degenerate hypersurface we have thus

$$H \underset{\text{def}}{=} \det (h_{jk}) \neq 0 . \tag{21}$$

Under a change of frames (17) we have, by (20),

$$H^* = HA^{n+2} . \tag{22}$$

It follows that for even n the sign of H is an affine invariant.

Suppose M be oriented. We shall restrict to frames such that $e_1, \cdots, e_n$ define a coherent orientation. Then $A > 0$ and

$$|H^*|^{1/(n+2)} = |H|^{1/(n+2)} A . \tag{22a}$$

The normalized second fundamental form

$$\hat{II} = |H|^{-1/(n+2)} II \tag{23}$$

is affinely invariant. Since it is of rank n, it defines a pseudo-Riemannian structure on M, to which the methods of Riemannian geometry can be applied. In particular, M has the volume element

$$dV = |H|^{1/(n+2)} \omega^1 \wedge \cdots \wedge \omega^n . \tag{24}$$

If $\hat{II}$ is a definite form, M is locally convex.

Taking the exterior derivative of (14) and using (10), we get

$$\sum_k \left(dh_{ik} + h_{ik}\omega_{n+1}^{n+1} - \sum_j h_{ij}\omega_k^j - \sum_j h_{kj}\omega_i^j \right) \wedge \omega^k = 0 .$$

This gives

$$dh_{ik} + h_{ik}\omega_{n+1}^{n+1} - \sum_j h_{ij}\omega_k^j - \sum_j h_{kj}\omega_i^j = \sum_j h_{ikj}\omega^j , \tag{25}$$

where h_{ikj} is symmetric in all the indices i, k, j. Let (H^{ik}) be the adjoint matrix of (h_{ij}), so that

$$\sum_j H^{ij} h_{jk} = \delta_k^i H . \tag{26}$$

By (25) and (9)

(27) $$dH = \sum_{i,j} H^{ij} dh_{ij} = -(n+2)H\omega_{n+1}^{n+1} + \sum_{i,j,k} H^{ij} h_{ijk}\omega^k .$$

Differentiating the second equation of (17), we get

$$de_{n+1}^* \equiv -A^{-2}dAe_{n+1} + A^{-1}\omega_{n+1}^{n+1}e_{n+1} + \sum_i a_{n+1}^i \omega_i^{n+1} e_{n+1}, \text{ mod } e_k .$$

It follows that

(28) $$\omega_{n+1}^{*n+1} = \omega_{n+1}^{n+1} - d\log A + A\sum a_{n+1}^i \omega_i^{n+1} .$$

On the other hand, by (22a),

(29) $$d\log|H^*| = d\log|H| + (n+2)d\log A .$$

Combination of (28) and (29) gives

(30) $$\begin{aligned} &\omega_{n+1}^{*n+1} + \frac{1}{n+2} d\log|H^*| \\ &= \omega_{n+1}^{n+1} + \frac{1}{n+2} d\log|H| + A\sum a_{n+1}^i \omega_i^{n+1} . \end{aligned}$$

Since M is non-degenerate, we can choose a_{n+1}^i to annihilate this expression. Suppose this be done, i.e., suppose

(31) $$(n+2)\omega_{n+1}^{n+1} + d\log|H| = 0 ,$$

or

(31a) $$\sum H^{ij} h_{ijk} = 0 .$$

By (17) the vector e_{n+1} is defined up to a factor. The line through x in the direction of e_{n+1} is called the *affine normal* at x. The vector

(32) $$\nu = |H|^{1/(n+2)} e_{n+1} ,$$

affinely defined, is called the *affine normal vector*. The affine normal has interesting geometrical properties; we refer to [1] for a beautiful treatment.

Differentiating (31), we get

(33) $$\sum_i \omega_{n+1}^i \wedge \omega_i^{n+1} = 0 ,$$

which gives

(34) $$\omega_{n+1}^i = \sum l^{ik} \omega_k^{n+1} ,$$

where

$$l^{ik} = l^{ki}. \tag{35}$$

Equation (34) can be written

$$\omega_{n+1}^{i} = \sum l_k^i \omega^k , \tag{36}$$

where

$$l_k^i = \sum_j l^{ij} h_{jk} \tag{37}$$

The quadratic differential form

$$\begin{aligned} III = \sum \omega_{n+1}^{i} \omega_i^{n+1} &= \sum l^{ik} \omega_i^{n+1} \omega_k^{n+1} \\ &= \sum l^{ik} h_{ij} h_{kl} \omega^j \omega^l \end{aligned} \tag{38}$$

is called the *third fundamental form.*

With xe_{n+1} as the affine normal the allowable change of frame (17) becomes

$$\begin{aligned} e_i^* &= \sum a_i^k e_k , \\ e_{n+1}^* &= A^{-1} e_{n+1} , \qquad A = \det (a_i^k) , \end{aligned} \tag{39}$$

whence

$$\omega_i^{*n+1} = (e_1^*, \cdots, e_n^*, de_i^*) = A \sum a_i^k \omega_k^{n+1} . \tag{40}$$

On the other hand, by (39),

$$de_{n+1}^* \equiv A^{-1} de_{n+1} , \qquad \mod e_{n+1} ,$$

while

$$\begin{aligned} de_{n+1}^* &\equiv \sum \omega_{n+1}^{*i} e_i^* , \qquad \mod e_{n+1} , \\ de_{n+1} &\equiv \sum \omega_{n+1}^{i} e_i , \qquad \mod e_{n+1} . \end{aligned}$$

It follows that

$$\omega_{n+1}^{i} = A \sum a_k^i \omega_{n+1}^{*k} . \tag{41}$$

By (40) and (41), we have

$$III = \sum \omega_{n+1}^{i} \omega_i^{n+1} = A \sum a_k^i \omega_{n+1}^{*k} \omega_i^{n+1} = \sum \omega_{n+1}^{*k} \omega_k^{*n+1} .$$

Therefore the third fundamental form is invariant under a change of frame keeping the affine normal xe_{n+1} fixed.

The elementary symmetric functions of the eigenvalues of III relative to $\hat{II}$ give the affine curvatures of M, which are scalar invariants.

In particular, the trace

$$L = \frac{1}{n}|H|^{1/(n+2)} \sum l_i^i \tag{42}$$

is called the *affine mean curvature*. A hypersurface with $L = 0$ is called *affine minimal.*

We consider the case when the hypersurface is given in the "Monge form":

$$x^{n+1} = F(x^1, \cdots, x^n) . \tag{43}$$

Let

$$x = (x^1, \cdots, x^n, F(x^1, \cdots, x^n)) . \tag{44}$$

The first equation of (8) holds, if we set

$$\begin{aligned} &\omega^i = dx^i , \\ &e_i = \Big(0, \cdots, 0, \underset{i\text{th}}{1}, 0, \cdots, 0, \frac{\partial F}{\partial x^i}\Big) . \end{aligned} \tag{45}$$

Putting

$$e_{n+1} = (0, \cdots, 0, 1) . \tag{46}$$

we get

$$\omega_i^{n+1} = \sum_j F_{ij} dx^j , \qquad F_{ij} = \frac{\partial^2 F}{\partial x^i \partial x^j} \tag{47}$$

and

$$II = \sum_{i,j} F_{ij} dx^i dx^j , \tag{48}$$

so that

$$h_{ij} = F_{ij} , \qquad H = \det(F_{ij}) . \tag{49}$$

We let (F^{ij}) be the inverse matrix of (F_{ik}), so that

$$\sum F_{ij} F^{jk} = \delta_i^k . \tag{50}$$

To find the affine normal we let $a_i^k = \delta_i^k$ in (17). Then it is along the vector e_{n+1}^* where, by (30), a_{n+1}^i is determined by

$$d \log |H| + (n+2) \sum a^i_{n+1} h_{ik} dx^k = 0 .$$

It follows that the affine normal vector has the expression

$$\nu = |H|^{1/(n+2)} \left\{ e_{n+1} - \frac{1}{n+2} \sum_{i,k} \frac{\partial}{\partial x^i} (\log |H|) F^{ik} e_k \right\} . \tag{51}$$

2. Affine minimal hypersurfaces

Let D be a sufficiently small domain of M, with boundary ∂D. Its volume is

$$V(D) = \int_D dV , \tag{52}$$

where dV is given by (24). We wish to compute the first variation $\delta V(D)$ under an infinitisimal displacement of D, with ∂D kept fixed. To express this situation analytically, let I be the interval $-\frac{1}{2} < t < \frac{1}{2}$. Let $f: M \times I \to A^{n+1}$ be a smooth mapping such that its restriction to $M \times t$, $t \in I$, is an immersion and that $f(m, 0) = x(m)$, $m \in M$. We consider a frame field $e_\alpha(m, t)$ over $M \times I$ such that for every $t \in I$, $e_i(m, t)$ are tangent vectors and $e_{n+1}(m, t)$ is along the affine normal to $f(M \times t)$ at (m, t).

The analytical development parallels that of § 1, in the sense that we are studying, instead of one hypersurface, a family of hypersurfaces. Pulling the forms ω^α, ω^β_α in the frame manifold back to $M \times I$, we have, since e_i span the tangent hyperplane at $f(m, t)$,

$$\omega^{n+1} = a dt . \tag{53}$$

Its exterior differentiation gives

$$\sum \omega^i \wedge \omega^{n+1}_i + dt \wedge (a\omega^{n+1}_{n+1} + da) = 0 . \tag{54}$$

It follows that we can set

$$\begin{aligned} \omega^{n+1}_i &= \sum h_{ij} \omega^j + h_i dt , \\ a\omega^{n+1}_{n+1} + da &= \sum h_i \omega^i + h dt , \end{aligned} \tag{55}$$

where

$$h_{ij} = h_{ji} . \tag{56}$$

Exterior differentiation of (55_1) then gives

$$\begin{aligned} &\sum (dh_{ij} - \sum h_{ik}\omega^k_j - \sum h_{jk}\omega^k_i + h_{ij}\omega^{n+1}_{n+1}) \wedge \omega^j \\ &\quad + \sum (dh_i - \sum h_k \omega^k_i + h_i \omega^{n+1}_{n+1} - a \sum h_{ij} \omega^j_{n+1}) \wedge dt = 0 . \end{aligned} \tag{57}$$

Hence we can set

(58)
$$\begin{aligned} dh_{ij} &= \sum h_{ik}\omega_j^k + \sum h_{jk}\omega_i^k - h_{ij}\omega_{n+1}^{n+1} + \sum h_{ijk}\omega^k + p_{ij}dt \,, \\ dh_i &= \sum h_k\omega_i^k - h_i\omega_{n+1}^{n+1} + a\sum h_{ij}\omega_{n+1}^j + \sum p_{ij}\omega^j + q_i dt \,, \end{aligned}$$

where h_{ijk} is symmetric in all three indices, and

(59)
$$p_{ij} = p_{ji} \,.$$

We introduce H^{ik} as in (26), and find

(60)
$$\begin{aligned} dH &= \sum H^{ij}dh_{ij} \\ &= -(n+2)H\omega_{n+1}^{n+1} + \sum H^{ij}h_{ijk}\omega^k + \sum H^{ij}p_{ij}dt \,. \end{aligned}$$

As in § 1, by the change of frame (17), we can make

(61)
$$\sum H^{ij}h_{ijk} = 0 \,.$$

Geometrically this means that e_{n+1} is in the direction of the affine normal to the hypersurface $f(M \times t)$ at $f(m, t)$. The resulting equation (60) can be written as

(62)
$$f^*\omega_{n+1}^{n+1} + \frac{1}{n+2} d\log|H| = bdt \,,$$

where

(63)
$$b = \frac{1}{(n+2)H} \sum H^{ij}p_{ij} \,.$$

For later application we differentiate (26) and use (58), obtaining

(64)
$$\begin{aligned} dH^{ij} &= -\sum H^{ik}\omega_k^j - \sum H^{jk}\omega_k^i + H^{ij}(\omega_{n+1}^{n+1} + d\log|H|) \\ &\quad -\frac{1}{H}\sum H^{ik}H^{jl}h_{klr}\omega^r - \frac{1}{H}\sum H^{ik}H^{jl}p_{kl}dt \,. \end{aligned}$$

We introduce

(65)
$$h^i = \sum H^{ij}h_j \,.$$

By (58) and (64), we find

(66)
$$\begin{aligned} dh^i &= -\sum h^k\omega_k^i + h^i d\log|H| + aH\omega_{n+1}^i \\ &\quad -\frac{1}{H}\sum H^{ik}h^l h_{klr}\omega^r + \sum H^{ij}p_{jk}\omega^k \end{aligned}$$

$$-\frac{1}{H}\sum H^{ik}h^l p_{kl}dt + \sum H^{ij}q_j dt .$$

We use the structure equations (10), (9), and obtain

$$\begin{aligned} &d(\omega^1 \wedge \cdots \wedge \omega^n) \\ &\quad = \sum_i (-1)^{i-1}\omega^1 \wedge \cdots \wedge \omega^{i-1} \wedge (\omega^i \wedge \omega_i^i + \omega^{n+1} \wedge \omega_{n+1}^i) \\ &\qquad \wedge \omega^{i+1} \wedge \cdots \omega^n \\ &\quad = \omega_{n+1}^{n+1} \wedge \omega^1 \wedge \cdots \wedge \omega^n \\ &\qquad + \omega^{n+1} \wedge \sum_i \omega^1 \wedge \cdots \wedge \omega^{i-1} \wedge \omega_{n+1}^i \wedge \omega^{i+1} \wedge \cdots \wedge \omega^n . \end{aligned} \tag{67}$$

Pulling back under f, we get (cf. (42))

$$\begin{aligned} &|H|^{-1/(n+2)} d(|H|^{1/(n+2)} f^*(\omega^1 \wedge \cdots \wedge \omega^n)) \\ &\quad = \left(f^*\omega_{n+1}^{n+1} + \frac{1}{n+2} d\log|H| + n|H|^{-1/(n+2)} La dt\right) \\ &\qquad \wedge f^*(\omega^1 \wedge \cdots \wedge \omega^n) \\ &\quad = (b + n|H|^{-1/(n+2)} La) dt \wedge \tilde{\omega}^1 \wedge \cdots \wedge \tilde{\omega}^n , \end{aligned}$$

where $\tilde{\omega}^i$ is defined from $f^*\omega^i$ by "splitting off" the term in dt:

$$f^*\omega^i = \tilde{\omega}^i + a^i dt . \tag{68}$$

The operator d on $M \times I$ can also be decomposed as

$$d = d_M + dt\frac{\partial}{\partial t} . \tag{69}$$

Equating the terms in dt in the above equation, we get

$$\begin{aligned} &\frac{\partial}{\partial t}(|H|^{1/(n+2)}\tilde{\omega}^1 \wedge \cdots \wedge \tilde{\omega}^n) \\ &\qquad + d_M\Big\{|H|^{1/(n+2)} \sum_i (-1)^i a^i \tilde{\omega}^1 \wedge \cdots \wedge \tilde{\omega}^{i-1} \\ &\qquad\qquad \wedge \tilde{\omega}^{i+1} \wedge \cdots \tilde{\omega}^n\Big\} \\ &\quad = |H|^{1/(n+2)}(b + n|H|^{-1/(n+2)} La)\tilde{\omega}^1 \wedge \cdots \wedge \tilde{\omega}^n . \end{aligned} \tag{70}$$

On ∂D we have $a^i = 0$. Integrating over D and setting $t = 0$, we find the first variation of volume

$$V'(0) = \frac{\partial}{\partial t}\int_D |H|^{1/(n+2)}\tilde{\omega}^1 \wedge \cdots \tilde{\omega}^n\Big|_{t=0} \tag{71}$$
$$= \int_D (b + n|H|^{-1/(n+2)}La)dV\Big|_{t=0} .$$

The last expression can be simplified, on account of the following lemma:

Lemma. *For* $t =$ const. *the form*

$$\left(b + \frac{n}{n+2}|H|^{-1/(n+2)}La\right)dV \tag{72}$$
$$= \left(b\,|H|^{1/(n+2)} + \frac{n}{n+2}La\right)\omega^1 \wedge \cdots \wedge \omega^n$$

is exact and its integral over D *for* $t = 0$ *is zero.*

Proof. We introduce the form

$$\Omega = \frac{1}{(n-1)!}\sum \varepsilon_{i_1\cdots i_n} h^{i_1}\omega^{i_2} \wedge \cdots \wedge \omega^{i_n} \tag{73}$$
$$= \sum_i (-1)^{i-1}\omega^1 \wedge \cdots \wedge \omega^{i-1}h^i \wedge \omega^{i+1} \wedge \cdots \wedge \omega^n ,$$

where $\varepsilon_{i_1\cdots i_n}$ is $+1$ or -1 according as $i_1, \cdots, i_n$ is an even or odd permutation of $1, \cdots, n$, and is otherwise zero. By using (10) and (66), we find

$$d\Omega = \left(\omega_{n+1}^{n+1} + \frac{dH}{H}\right) \wedge \Omega \tag{74}$$
$$+ \{a \sum l_i^i + (n+2)b\}H\omega^1 \wedge \cdots \wedge \omega^n .$$

It follows that

$$d(|H|^{-(n+1)/(n+2)}\Omega) = |H|^{-(n+1)/(n+2)}\left(d\Omega - \frac{n+1}{n+2}d\log|H| \wedge \Omega\right) .$$

By (62) we have, for $t =$ const.,

$$d(|H|^{-(n+1)/(n+2)}\Omega) \tag{75}$$
$$= (\operatorname{sgn} H)\{a \sum l_i^i + (n+2)b\}\,|H|^{1/(n+2)}\omega^1 \wedge \cdots \wedge \omega^n ,$$

We suppose the variation such that $h_i(m, 0) = 0$ for $m \in \partial D$. Hence the lemma follows.

It follows from (71) that

(71a) $$V'(0) = \frac{n(n+1)}{n+2} \int_D |H|^{-1/(n+2)} L a dV \Big|_{t=0} .$$

If this is zero for arbitrary functions $a(m,t)$, $m \in D$, $t \in I$. satisfying $a(m,0) = 0$, $h_i(m,0) = 0$, $m \in \partial D$, we must have $L = 0$, i.e., M be an affine minimal hypersurface.

To better understand the analytical significance of an affine minimal hypersurface we suppose the latter be given in the Monge form (43). Then the affine normal vector ν is given by (51), with e_α defined by (45), (46). Computing mod ν, we find

$$\begin{aligned} |H|^{-1/(n+2)} d\nu \equiv & -\frac{1}{n+2} \sum_k d\Big(\sum_i (\log|H|)_i F^{ik}\Big) e_k \\ & -\frac{1}{n+2} \sum_{i,k} (\log|H|)_i F^{ik} dF_k e_{n+1} , \end{aligned}$$

the last term being congruent to

$$-\frac{1}{(n+2)^2} d \log|H| \sum_{i,k} (\log|H|)_i F^{ik} e_k , \qquad \text{mod } \nu ,$$

where the subscript i means partial differentiation with respect to x^i. From this it follows that the condition for the hypersurface (43) to be an affine minimal hypersurface is

(76) $$(n+2) \sum_{i,k} ((\log|H|)_i F^{ik})_k + \sum_{i,k} F^{ik} (\log|H|)_i (\log|H|)_k = 0 ,$$

which is a partial differential equation of the fourth order in the unknown function $F(x^1, \cdots, x^n)$.

Using the pseudo-Riemannian structure defined by $\hat{II}$ in (23), equation (76) can also be written

(77) $$\Delta |H|^{-1/(n+2)} = 0^{(1)} ,$$

where Δ is the Beltrami-Laplace operator relative to $\hat{II}$.

We have the following theorem:

Theorem. *There is no closed affine minimal hypersurface in* A^{n+1}.

Proof. We have the formula

(78) $$d(x, \nu, \underbrace{dx, \cdots, dx}_{n-1}) = -(\nu, \underbrace{dx, \cdots, dx}_{n}) + (x, d\nu, \underbrace{dx, \cdots, dx}_{n-1}) ,$$

where the determinant have entries which are vector-valued one-forms. By (32) and (31) we have

(1) This expression was communicated to me by S. Y. Cheng.

$$d\nu = |H|^{1/(n+2)}(\omega^1_{n+1}e_1 + \cdots + \omega^n_{n+1}e_n) .$$

It follows that

$$\begin{aligned}(x, d\nu, dx, \cdots, dx) &= (n-1)!\,|H|^{1/(n+2)}(x, e_1, \cdots, e_n) \\ &\quad \times (\omega^1_{n+1} \wedge \omega^2 \wedge \cdots \wedge \omega^n + \cdots \\ &\quad + \omega^1 \wedge \cdots \wedge \omega^{n-1} \wedge \omega^n_{n+1}) \\ &= (n-1)!\,|H|^{1/(n+2)}(x, e_1, \cdots, e_n) \\ &\quad \times (\textstyle\sum l^i_i)\omega^1 \wedge \cdots \wedge \omega^n ,\end{aligned} \tag{79}$$

which is zero if M is affine minimal. On the other hand,

$$\begin{aligned}(\nu, dx, \cdots, dx) &= n!\,(\nu, e_1, \cdots, e_n)\omega^1 \wedge \cdots \wedge \omega^n \\ &= (-1)^n n!\,|H|^{-1/(n+2)} dV .\end{aligned}$$

If M is a closed hypersurface without boundary, the integral over M of the left-hand side of (78) is zero. The same is therefore true of the integral of its right-hand side. By (79) the second term is zero if M is affine minimal. Hence

$$\int_M (\nu, dx, \cdots, dx) = 0 ,$$

which is clearly a contradiction.

3. Surfaces in A^3

Minimal surfaces in the three-dimensional affine space A^3 have many interesting properties. The classical account is [1]; cf. also [6]. For a recent result see [4].

On account of the great interest in classical minimal surfaces we wish to state the following theorem of G. Thomsen:

A minimal surface of the three-dimensional Euclidean space E^3 is at the same time an affine minimal surface, if and only if the image under the Gauss map of every asymptotic curve is a circle on the unit sphere.

I wish next to state a theorem on affine minimal surfaces proved recently by Chuu-lian Terng and myself [7]. It concerns with the analogue of Bäcklund's theorem in affine surface theory.

The classical Bäcklund's theorem, which is the basis of the theory of Bäcklund transformations and which has received much attention in recent studies of the soliton solutions of the sine-Gordon equation, says the following: *Let the surfaces M, M^* in E^3 be the focal surfaces of a line congruence, so that the lines of the congruence are the common tangent lines* *of M and M^*. To every line let $x \in M$, $x^* \in M^*$ be the points of contact and v, v^* be the unit normal vectors to M, M^* at x, x^* respectively. If* $r = \text{dist}(x, x^*) = \text{const.}$ *and the angle θ between v, v^* is constant, then both M and M^* have the Gaussian curvature* $-(\sin\theta/r)^2$.

The congruence in the theorem is a W-congruence, i.e., the asymptotic curves of M and M^* correspond under the mapping which sends x to x^*.

Our theorem is the following:

Let the surfaces M, M^* in A^3 be the focal surfaces of a W-congruence, such that the affine normals to M and M^* at corresponding points are parallel. Then both M and M^* are affine minimal surfaces.

Before concluding this paper I wish to state a conjecture:

Among affine minimal surfaces are the elliptic paraboloids

$$x^3 = (x^1)^2 + (x^2)^2 , \tag{80}$$

whose affine normals are parallel. The following conjecture is an analogue to Bernstein's theorem: In A^3 consider the surface

$$x^3 = F(x^1, x^2) \tag{81}$$

defined for all x^1, x^2. If it consists entirely of elliptic points, and is affine minimal, then it is affinely equivalent to the paraboloid (80).

References

[1] W. Blaschke, Vorlesungen über Differentialgeometrie, II, Berlin, 1923.

[2] E. Calabi, *Improper affine hyperspheres of convex type and a generalization of a theorem by K. Jörgens,* Michigan Math. J. **5** (1958), 105–126.

[3] H. Flanders, *Local theory of affine hypersurfaces,* J. d'Analyse Math. **15** (1965), 353–387.

[4] E. Glässner, *Ein Affinanalogon zu den Scherkschen Minimalflächen,* Archiv der Math. **28** (1977), 436–439.

[5] H. W. Guggenheimer, Differential Geometry, New York, 1963.

[6] P. A. Schirokow and A. P. Schirokow, Affine Differentialgeometrie, Leipzig, 1962 (translated from Russian).

[7] S. S. Chern and Chuu-Lian Terng, *An analogue of Bäcklund's theorem in affine geometry,* to appear in Rocky Mountain J. Math.

University of California
Berkeley, CA 94720
U.S.A.

ROCKY MOUNTAIN
JOURNAL OF MATHEMATICS
Volume 10, Number 1, Winter 1980

AN ANALOGUE OF BÄCKLUND'S THEOREM IN AFFINE GEOMETRY

SHIING-SHEN CHERN* AND CHUU-LIAN TERNG**

ABSTRACT. It is well-known that there is a correspondence between solutions of the Sine-Gordon equation (SGE)

$$\frac{\partial^2\phi}{\partial x^2} - \frac{\partial^2\phi}{\partial t^2} = \sin\phi$$

and the surfaces of constant curvature -1 in $\mathbf{R}^3$ (see below). The classical Bäcklund transformation of such surfaces furnishes a way to generate new solutions of the SGE from a given solution. This has received much attention in recent studies of the soliton solutions of the SGE, and the technique has been used successfully in the study of other non-linear evolution equations. In the first section of this paper we present a simple derivation of the classical Bäcklund theorem and its applications by using the method of moving frames.

Our main result concerns affine minimal surfaces. They arise as the solution of the variation problem for affine area. The corresponding Euler-Lagrange equation is a fourth order partial differential equation. In §2, we develop the basic properties of affine minimal surfaces. In §3 we study the transformation of affine surfaces by realizing them as the focal surfaces of a line congruence. The natural conditions that the congruence be a W-congruence and that the affine normals at corresponding points be parallel lead to the conclusion that both surfaces are affine minimal. This is the content of Theorem 4, the main result of our paper. As in the classical case, the Theorem leads to the construction of new affine minimal surfaces from a given one by the solution of a completely integrable system of first order partial differential equations.

1. **The classical Bäcklund theorem and its consequences.** Let M be a surface in $\mathbf{R}^3$. We choose a local field of orthonormal frames v_1, v_2, v_3 with origin X in $\mathbf{R}^3$ such that X is a point of M and the vectors v_1, v_2 are tangent to M at X. Let $\theta_1, \theta_2, \theta_3$ be the dual coframe of v_1, v_2, v_3. We can write

$$(1.1)\qquad \begin{aligned} dx &= \sum_\alpha \theta_\alpha v_\alpha \\ dv_\alpha &= \sum_\beta \theta_{\alpha\beta} v_\beta, \end{aligned}$$

Here and throughout this paper we shall agree on the index ranges

$$(1.2)\qquad 1 \leqq i, j, k \leqq 2, \quad 1 \leqq \alpha, \beta, \gamma \leqq 3.$$

The structure equations of $\mathbf{R}^3$ are

*Work done under partial support of NSF grant MCS74–23180.
**Work done under partial support of NSF Grant MCS76–01692.

Received by the editors on April 13, 1978.

$$d\theta_\alpha = \sum_\beta \theta_\beta \wedge \theta_{\beta\alpha}, \theta_{\alpha\beta} + \theta_{\beta\alpha} = 0 \tag{1.3}$$
$$d\theta_{\alpha\beta} = \sum_\beta \theta_{\alpha\gamma} \wedge \theta_{\gamma\beta}.$$

Restricting these forms to the frames defined above, we have

$$\theta_3 = 0 \tag{1.4}$$

and hence

$$0 = d\theta_3 = \sum_i \theta_i \wedge \theta_{i3}. \tag{1.5}$$

By Cartan's lemma we may write

$$\theta_{i3} = \sum_j h_{ij}\theta_j, \quad h_{ij} = h_{ji}. \tag{1.6}$$

The first equation of (1.3) gives

$$d\theta_i = \sum_j \theta_j \wedge \theta_{ji} \tag{1.7}$$

(where θ_{12} is the Levi-Civita connection form on M which is uniquely determined by these two equations).

The Gauss equation is

$$d\theta_{ij} = \sum_k \theta_{ik} \wedge \theta_{kj} + \Theta_{ij} \tag{1.8}$$

where

$$\Theta_{12} = -\theta_{13} \wedge \theta_{23} = -(\det(h_{ij}))\theta_1 \wedge \theta_2, \tag{1.9}$$
$$K \overset{\text{def}}{=} \det(h_{ij}) = \text{the Gaussian curvature of } M.$$

The Codazzi equations are

$$d\theta_{i3} = \sum_j \theta_{ij} \wedge \theta_{j3}. \tag{1.10}$$

The first fundamental form of the surface is

$$\mathrm{I} = (\theta_1)^2 + (\theta_2)^2, \tag{1.11}$$

and the second fundamental form is

$$\mathrm{II} = \sum_i \theta_i\theta_{i3} = \sum_{ij} h_{ij}\theta_i\theta_j. \tag{1.12}$$

Near a non-umbilical point M can be parametrized by its lines of curvature, i.e., there are coordinates u_i in which I and II are both diagonalized. Explicitly we write

$$\mathrm{I} = (a_1)^2(du_1)^2 + (a_2)^2(du_2)^2 \tag{1.13}$$
$$\mathrm{II} = b_1(a_1)^2(du_1)^2 + b_2(a_2)^2(du_2)^2,$$

so that $h_{ii} = b_i$ and $h_{12} = 0$. Observe that b_i are the two principal curvatures. The Riemannian connection is given by

$$\theta_{12} = \frac{1}{a_1}\frac{\partial a_2}{\partial u_1} du_2 - \frac{1}{a_2}\frac{\partial a_1}{\partial u_j} du_1, \tag{1.14}$$

and the Codazzi equation can be written

$$\frac{1}{b_i - b_j}\frac{\partial b_i}{\partial u_j} = -\frac{\partial(\log a_i)}{\partial u_j}, \quad i \neq j. \tag{1.15}$$

Now suppose M has constant negative curvature $K \equiv -1$; then $b_1 b_2 = -1$. By (1.15)

$$\frac{b_i}{b_i(b_i - b_j)}\frac{\partial b_i}{\partial u_j} = -\frac{\partial(\log a_i)}{\partial u_j}, \quad i \neq j, \tag{1.16}$$

or

$$\frac{1}{2}\frac{\partial(\log(b_i^2 + 1))}{\partial u_j} = -\frac{\partial(\log a_i)}{\partial u_j}, \quad i \neq j. \tag{1.17}$$

Therefore there exist two positive valued functions $c_1(u_1)$, $c_2(u_2)$ such that

$$b_i^2 + 1 = \frac{c_i(u_i)}{a_i^2}. \tag{1.18}$$

Making a change in each coordinate separately, we may assume $c_i = 1$. Writing $b_1 = \tan\psi$, $b_2 = -\cot\psi$, then $a_1 = \cos\psi$ and $a_2 = \sin\psi$, i.e.,

$$\begin{aligned} \mathrm{I} &= \cos^2\psi(du_1)^2 + \sin^2\psi(du_2)^2 \\ \mathrm{II} &= \sin\psi\cos\psi((du_1)^2 - (du_2)^2), \end{aligned} \tag{1.19}$$

so 2ψ is the angle between the two asymptotic directions. The connection form (1.14) becomes

$$\theta_{12} = \frac{\partial\psi}{\partial u_1} du_2 + \frac{\partial\psi}{\partial u_2} du_1. \tag{1.20}$$

The Gauss equation (1.8) then gives

$$\frac{\partial^2\psi}{\partial(u_1)^2} - \frac{\partial^2\psi}{\partial(u_2)^2} = \sin\psi\cos\psi, \tag{1.21}$$

i.e., $\phi = 2\psi$ satisfies the Sine-Gordon equation (SGE)

$$\frac{\partial^2\phi}{\partial(u_1)^2} - \frac{\partial^2\phi}{\partial(u_2)^2} = \sin\phi. \tag{1.22}$$

The converse is also true by the existence and uniqueness theorem on surfaces. Therefore we have proved: There is a one to one correspondence between solutions ϕ of SGE with $0 < \phi < \pi$ and the local surfaces of constant Gaussian curvature $K \equiv -1$ in $\mathbf{R}^3$ up to rigid motion.

A *congruence of lines* is an immersed surface in the Grassmann manifold Gr of all lines in $\mathbf{R}^3$. Locally we can suppose the lines be oriented, with their points given by

$$Y = X(u, v) + \lambda\xi(u, v), \quad \xi^2 = 1,$$

λ being a parameter on each line. The equations

$$u = u(t), v = v(t), u'^2 + v'^2 \neq 0$$

define a ruled surface belonging to the congruence. It is a developable if and only if the determinant

$$(\xi, dX, d\xi) = 0.$$

This is a quadratic equation in du, dv. Suppose that it has two real and distinct roots. There are then two families of developables, each of which (in the generic case) consists of the tangent lines of a surface. It follows that the lines of the congruence are the common tangent lines of two surfaces M and M^*, to be called the *focal surfaces*. There results a mapping $\ell: M \to M^*$ such that the congruence consists of the lines joining $P \in M$ to $\ell(P) \in M^*$. This construction, of great geometrical simplicity, plays a fundamental role in the theory of transformation of surfaces.

DEFINITION 1. Consider a line congruence with the focal surfaces M, M^* such that its lines are the common tangents at $P \in M$ and $P^* = \ell(P) \in M^*$. The congruence is called *pseudo-spherical* (p.s.) if
(1) $\|\mathbf{PP}^*\| = r$, which is a constant independent of P.
(2) The angle between the two normals ν_P and ν_{P^*} at P and P^* is equal to a constant τ independent of P.

We can now state the classical Bäcklund theorem and give a simple proof by the method of moving frames.

THEOREM 1. *Suppose there is a* p.s. *congruence in* $\mathbf{R}^3$ *with the focal surfaces M and M^* such that the distance r between corresponding points and the angle τ between corresponding normals are constants. Then both M and M^* have constant negative Gaussian curvature equal to* $-\sin^2\tau/r^2$.

PROOF. We choose an orthonormal frame v_1, v_2, v_3 on M such that v_1 is the unit vector in the direction of $\mathbf{PP}^*$ (where P and P^* are the corresponding points) and v_3 is the normal to M. From the definition of a p.s. congruence, there is an orthonormal frame v_1^*, v_2^*, v_3^* on M^* given by

$$\begin{aligned} &v_1^* = v_1 \\ (1.23) \qquad &v_2^* = \cos\tau v_2 + \sin\tau v_3 \\ &v_3^* = -\sin\tau v_2 + \operatorname{con}\tau v_3 = \text{the normal to } M^*. \end{aligned}$$

Suppose locally M is given by an immersion $X: U \to \mathbf{R}^3$, where U is an open subset of $\mathbf{R}^2$, then M^* is given by

$$X^* = X + rv_1 \tag{1.24}$$

Taking the differential of (1.24) gives

$$\begin{aligned} dX^* &= dX + rdv_1 \\ &= \sum_i \theta_i v_i + r\sum_\alpha \theta_{1\alpha} v_\alpha \\ &= \theta_1 v_1 + (\theta_2 + r\theta_{12})v_2 + r\theta_{13}v_3. \end{aligned} \tag{1.25}$$

On the other hand, let θ_i^* be the dual coframe of v_i^* we have

$$\begin{aligned} dX^* &= \theta_1^* v_1^* + \theta_2^* v_2^* \\ &= \theta^*{}_1 v_1 + \theta_2^* \cos\tau v_2 + \theta_2^* \sin\tau v_3. \end{aligned} \tag{1.26}$$

Comparing coefficients of v_i in (1.25) and (1.26), we get

$$\begin{aligned} \theta_1^* &= \theta_1 \\ \theta_2^* \cos\tau &= \theta_2 + r\theta_{12} \\ \theta_2^* \sin\tau &= r\theta_{13} \end{aligned} \tag{1.27}$$

This gives

$$\theta_2 + r\theta_{12} = r\cot\tau\theta_{13}. \tag{1.28}$$

Since θ_1^*, θ_2^* are linearly independent, by (1.27) h_{12} never vanishes. In order to compute the curvature of M^*, we have, by using (1.23), (1.28),

$$\begin{aligned} \theta_{13}^* &= dv_1^* \cdot v_3^* \\ &= -\sin\tau\,\theta_{12} + \cos\tau\theta_{13} \\ &= \frac{\sin\tau}{r}\theta_2, \\ \theta_{23}^* &= dv_2^* \cdot v_3^* = \theta_{23}. \end{aligned} \tag{1.29}$$

Then

$$\begin{aligned} \Theta_{12}^* &= -\theta_{13}^* \wedge \theta_{23}^* \\ &= -\frac{\sin\tau}{r}\theta_2 \wedge \theta_{23} \\ &= \frac{\sin\tau}{r} h_{12}\theta_1 \wedge \theta_2. \end{aligned} \tag{1.30}$$

On the other hand, using (1.27),

$$\Theta_{12}^* = -K^*\theta_1^* \wedge \theta_2^*$$

$$= -K^* \frac{r}{\sin\tau} h_{12}\theta_1 \wedge \theta_2. \tag{1.31}$$

Comparing coefficients in (1 30) and (1.31), we have

$$-K^* \frac{rh_{12}}{\sin\tau} = \frac{h_{12}\sin\tau}{r} \tag{1.32}$$

Since h_{12} never vanishes,

$$K^* = -\frac{\sin^2\tau}{r^2}. \tag{1.33}$$

By symmetry,

$$K = -\frac{\sin^2\tau}{r^2}. \tag{1.34}$$

The differential form equation (1.28) is called the *Bäcklund transformation.* We can write it as a system of partial differential equations. For simplicity, we may assume $K \equiv -1$, so that $r = \sin\tau$. Let α be the angle between the u_1-curves in the coordinate system (1.19) and the vector $\mathbf{PP^*}$. Then (1.28) becomes

$$\begin{aligned} \frac{\partial(\alpha+\psi)}{\partial x} &= a\sin(\alpha-\psi) \\ \frac{\partial(\alpha-\psi)}{\partial y} &= \frac{1}{a}\sin(\alpha+\psi), \end{aligned} \tag{1.35}$$

where $x = 1/2\,(u_1 + u_2)$, $y = 1/2\,(u_1 - u_2)$ are the asymptotic coordinates, and $a = \csc\tau - \cot\tau =$ constant. In the asymptotic coordinates, the SGE becomes

$$\frac{\partial^2\phi}{\partial x\partial y} = \sin\phi. \tag{1.36}$$

By equating the cross derivatives of (1.35), it follows that 2α is also a solution of SGE.

Note that the geometric derivation of (1.35) is local and the range of α and ψ are restricted to lie in $(0, \pi/2)$. However, from the analytic point of view, if 2ψ is any solution of SGE and α, ψ satisfy (1.35) then 2α is also a solution.

Next we discuss the complete integrability of (1.28) or (1.35).

THEOREM 2. *Suppose M is a surface of constant negative curvature, $K = -\sin^2\tau/r^2$, where $r > 0$ and τ are constants. Given any unit vector $v_0 \in T_{P_0}(M)$, which is not in a principal direction, there exists a unique surface M^* and a p.s. congruence with M and M^* as focal surfaces such that if*

$P_0^* \in M^*$ is the point corresponding to P_0, we have $\mathbf{P}_0\mathbf{P}_0^* = rv_0$ and τ is the angle between the normals at P_0, P_0^*.

PROOF. The differential form equation (1.28) is completely integrable, since

$$\begin{aligned} &d(\theta_2 + r\theta_{12} - r\cot\tau\theta_{13}) \\ &\quad = \theta_1 \wedge \theta_{12} + r\theta_{13} \wedge \theta_{32} - r\cot\tau\theta_{12} \wedge \theta_{23}, \\ &\quad = \left(-\frac{1}{r} - rK\csc^2\tau\right)\theta_1 \wedge \theta_2, \text{ using (1.28)} \\ &\quad = 0, \text{ since } K = -\frac{\sin^2\tau}{r^2}. \end{aligned}$$

Then the theorem follows directly from Frobenius Theorem.

DEFINITION 2. A line congruence $\ell: M \to M^*$ is called a Weingarten congruence (or W-congruence) if ℓ maps the asymptotic curves of M to the asymptotic curves of M^*.

COROLLARY 1. *A* p.s. *congruence is a W-congruence.*

PROOF. $\mathrm{II}^* = \theta_1^*\theta_{13}^* + \theta_2^*\theta_{23}^*$

$$= \frac{\sin\tau}{r}\theta_1\theta_2 + \frac{r}{\sin\tau}\theta_{13}\theta_{23}$$

$$= \frac{r\,h_{12}}{\sin\tau}[h_{11}(\theta_1)^2 + 2h_{12}\theta_1\theta_2 + h_{22}(\theta_2)^2], \text{ using } K = \det(h_{ij}) = \frac{-\sin^2\tau}{r^2}$$

$$= \frac{rh_{12}}{\sin\tau}\,\mathrm{II}_\ell$$

Since the second fundamental forms are proportional, the asymptotic curves correspond under ℓ.

The above results generalize to the case of n-dimensional submanifolds of constant negative curvature in $\mathbf{R}^{2n-1}$. This generalization will be the subject of a paper by K. Tenenblat and C. L. Terng.

In what follows we will study the analogue of the geometric Bäcklund transformation in affine differential geometry.

2. **Affine surfaces.** Let A^3 be the unimodular affine space of dimension 3, i.e., the space with real coordinates x^1, x^2, x^3 and volume element $dV = dx^1 \wedge dx^2 \wedge dx^3$. The linear group G which preserves the volume form is the unimodular affine group, i.e.,

$$x^{*\alpha} = \sum_\beta c_\beta^\alpha x^\beta + d^\alpha \tag{2.1}$$

where

$$\det(c^\alpha_\beta) = 1 \tag{2.2}$$

Following §1 we shall continue to adopt the range 1, 2, 3, for small Greek indices and the range 1, 2 for small Latin indices. In the space A^3 distance and angle have no meaning, but there are the notions of vectors and parallelism.

Let x be the position vector of the surface M sitting in A^3. Let x, e_1, e_2, e_3 be an affine frame on M such that e_1, e_2 are tangent to M at x, and

$$(e_1, e_2, e_3) = \det(e_1, e_2, e_3) = 1. \tag{2.3}$$

We can write

$$\begin{aligned} dx &= \sum_\alpha \omega^\alpha e_\alpha \\ de_\alpha &= \sum_\beta \omega^\beta_\alpha e_\beta. \end{aligned} \tag{2.4}$$

The ω^α, ω^β_α are the Maurer-Cartan forms of G. Differentiating (2.3) and using (2.4), we get

$$\sum_\alpha \omega^\alpha_\alpha = 0. \tag{2.5}$$

The structure equation of A^3 gives

$$\begin{aligned} d\omega^\alpha &= \sum_\beta \omega^\beta \wedge \omega^\alpha_\beta \\ d\omega^\beta_\alpha &= \sum_\tau \omega^\tau_\alpha \wedge \omega^\beta_\tau. \end{aligned} \tag{2.6}$$

If we restrict the forms to the surface M as defined above, we have

$$\omega^3 = 0 \tag{2.7}$$

and the first equation of (2.6) gives

$$\sum_i \omega^i \omega^3_i = 0. \tag{2.8}$$

By Cartan's lemma, we have

$$\omega^3_i = \sum_k h_{ik}\omega^k, \; h_{ik} = h_{ki}. \tag{2.9}$$

From now on *we assume that M is non-degenerate,* i.e., the rank of (h_{ij}) is equal to 2. Let $H = \det(h_{ij})$. Then it follows that the quadratic differential form

$$\mathrm{II} = |H|^{-\frac{1}{4}} \sum_{i,k} h_{ik}\omega^i\omega^k \tag{2.10}$$

is affinely invariant. Since it is of rank 2, it defines a pseudo-Riemannian structure on M, which is called the affine metric of M, to which the method of Riemannian geometry can be applied. In particular, the Gaussian curvature of II is defined.

Since M is non-degenerate, we can choose e_3 suitably such that

$$\omega_3^3 + \frac{1}{4} d \log |H| = 0. \tag{2.11}$$

Under such a choice the line through x in the direction of e_3 is called the *affine normal* at x. Moreover, the vector $|H|^{1/4} e_3$ is affinely invariant, and is called the *affine normal vector*. The affine normal has interesting geometrical properties; see **[1, 3]**.

Differentiating (2.11), we get

$$\sum_i \omega_3^i \wedge \omega_i^3 = 0, \tag{2.12}$$

which gives

$$\omega_3^i = \sum_k \ell^{ik} \omega_k^3, \quad \ell^{ik} = \ell^{ki}. \tag{2.13}$$

Equation (2.13) can be written as

$$\omega_3^i = \sum_k \ell_k^i \omega^k, \tag{2.14}$$

where

$$\ell_k^i = \sum_j \ell^{ij} h_{jk}. \tag{2.15}$$

Then we can verify easily that the quadratic differential form

$$\mathrm{III} = \sum_i \omega_3^i \omega_i^3 \tag{2.17}$$

is invariant under any change of frame keeping the affine normal e_3 fixed. Therefore the trace of III relative to II, i.e.,

$$L = \frac{1}{2} |H|^{\frac{1}{4}} \sum_i \ell_i^i \tag{2.18}$$

is an affine invariant. It is called the *affine mean curvature. A* surface is called *affine minimal* if $L = 0$. These are the critical points for the variational problem for the affine area defined by II. (See **[2]**). Formula (2.18) can also be put in the form

$$L\omega^1 \wedge \omega^2 = \frac{1}{2} |H|^{\frac{1}{4}} (\omega^1 \wedge \omega_3^2 + \omega_3^1 \wedge \omega^2). \tag{2.18a}$$

Next we define the Fubini-Pick cubic form of M. Taking the exterior differential of (2.9), we get

$$\sum_k \Big(dh_{ik} - \sum_j h_{ij}\omega_k^j - \sum_j h_{jk}\omega_i^j\Big) \wedge \omega^k = 0. \tag{2.19}$$

We define

$$Dh_{ik} = \sum_j h_{ikj}\omega^j = dh_{ik} - \sum_j h_{ij}\omega^j{}_k - \sum_j h_{jk}\omega^j_i. \tag{2.20}$$

It follows from (2.9) and (2.19) that h_{ijk} are symmetric in all the indices. The cubic differential form

$$P = \sum_{i,j,k} h_{ijk}\omega^i\omega^j\omega^k \tag{2.21}$$

is called the *Fubini-Pick form* on M. It measures the difference between the affine connection

$$De_i = \omega_i^j e_j \tag{2.22}$$

and the Levi-Civita connection of II, which we write as

$$\tilde{D}e_i = \bar{\omega}_i^j e_j. \tag{2.23}$$

In fact, we find

$$\bar{\omega}_i^j = \omega_i^j + \frac{1}{2}\sum_{k,\ell} h^{jk}h_{ik\ell}\omega^\ell, \quad (h^{jk}) = \text{inverse matrix of } (h_{ik}). \tag{2.24}$$

Therefore the curvature tensor of the affine metric II can be computed from III and the Fubini-Pick form.

For later use, we will now develop the local theory for hyperbolic surfaces (i.e., $H < 0$).

If M is hyperbolic, we may assume $X(u, v)$ to be parametrized by its asymptotic curves. Choose $e_1 = \partial x/\partial u$, $e_2 = \partial x/\partial v$, and e_3 to be in the affine normal direction such that

$$(e_1, e_2, e_3) = 1$$

Then

$$\omega^1 = du,\ \omega^2 = dv,\ \omega_1^3 = h_{12}dv,\ \omega_2^3 = h_{12}du,$$

and by supposing $h_{12} > 0$,

$$\omega_3^3 = -\frac{1}{2}d\log h_{12}.$$

The affine metric is

$$\text{II} = 2\,F du\,dv \tag{2.25}$$

where $F = (h_{12})^{1/2}$.

Then a standard computation implies that the Gaussian curvature of II, which is called the *affine curvature* is given by

$$\tilde{K} = -\frac{1}{F}\frac{\partial^2 \log F}{\partial u \partial v}. \tag{2.26}$$

By (2.20)

$$
\begin{aligned}
&h_{12;j}\omega^j = dh_{12} - h_{12}\omega_2^2 - h_{12}\omega_1^1 + h_{12}\omega_3^3 = 0 \\
(2.27) \qquad &h_{11;j}\omega^j = -2h_{12}\omega_1^2, \\
&h_{22;j}\omega^j = -2h_{12}\omega_2^1.
\end{aligned}
$$

So we have

$$
\begin{aligned}
&h_{12j} = 0, \\
(2.28) \qquad &\omega_1^2 = -\frac{h_{111}}{2h_{12}}\, du, \\
&\omega_2^1 = -\frac{h_{222}}{2h_{12}}\, dv.
\end{aligned}
$$

Therefore the Fubini-Pick form is

$$(2.29) \qquad P = h_{111}(du)^3 + h_{222}(dv)^3.$$

Using the structure equation (2.6), we get

$$
\begin{aligned}
&\omega_1^1 = \frac{\partial}{\partial u}(\log F)\, du \\
(2.30) \qquad &\omega_2^2 = \frac{\partial}{\partial v}(\log F)\, dv \\
&\omega_3^1 = \ell\, du + \ell_2^1\, dv \\
&\omega_3^2 = \ell_1^2 du + \ell\, dv, \qquad \ell = \ell_1^1 = \ell_2^2
\end{aligned}
$$

where

$$
\begin{aligned}
&\ell_1^2 = -\frac{1}{F^3}\frac{\partial}{\partial v}\left(\frac{h_{111}}{2F}\right), \\
(2.31) \qquad &\ell_2^1 = -\frac{1}{F^3}\frac{\partial}{\partial u}\left(\frac{h_{222}}{2F}\right),
\end{aligned}
$$

$$(2.32) \qquad L = F\ell = J - K,$$

$$(2.33) \qquad J = \frac{h_{111}h_{222}}{4F^5}.$$

Since $\tilde{K}$, L are affine invariants, J is also an affine invariant.

Next we develop a necessary and sufficient condition for a graph to be affine minimal. Let a surface be locally given by

$$(2.34) \qquad x^3 = f(x^1, x^2)$$

So $x = (x^1, x^2, f(x^1, x^2))$ is the position vector. Then equations (2.4), (2.5) hold if we set

$$\omega^i = dx^i$$

$$\omega^3 = dx^3 - \frac{\partial f}{\partial x^1} dx^1 - \frac{\partial f}{\partial x^2} dx^2$$

$$e_1 = \left(1, 0, \frac{\partial f}{\partial x^1}\right) \tag{2.35}$$

$$e_2 = \left(0, 1, \frac{\partial f}{\partial x^2}\right)$$

$$e_3 = (0, 0, 1)$$

with

$$\omega_i^j = 0$$

$$\omega_i^3 = \sum_j \frac{\partial^2 f}{\partial x^i \partial x^j} \omega^j. \tag{2.36}$$

Hence $h_{ij} = \partial^2 f/\partial x^i \partial x^j$ and $H =$ Hessian of f. To find the affine normal, we let

$$e_i^* = e_i$$

$$e_3^* = e_3 + a^1 e_1 + a^2 e_2 \tag{2.37}$$

where e_3^* is in the affine normal direction. Then a_i's are determined by

$$d \log|H| + 4 \sum_{i,k} a^i h_{ik} dx^k = 0. \tag{2.38}$$

Hence

$$a^i = - \sum_j \frac{h^{ij}}{4} \frac{\partial}{\partial x^j} (\log |H|) \tag{2.39}$$

where

$$(h^{ij}) = (h_{ij})^{-1}. \tag{2.40}$$

We compute

$$\omega_3^{*1} = (de_3^*, e_2^*, e_3^*) = da^1 + \frac{a^1}{4} d \log |H|$$

$$\omega_3^{*2} = (e_1^*, de_3^*, e_3^*) = da^2 + \frac{a^2}{4} d \log |H|. \tag{2.41}$$

Therefore the affine mean curvature is

$$L = \frac{1}{2} |H|^{\frac{1}{4}} \sum_{i,j} \left\{ \frac{3h^{ij}}{16} \frac{\partial}{\partial x^i} (\log |H|) \frac{\partial}{\partial x^j} (\log |H|) - \frac{h^{ij}}{4} \frac{\partial^2}{\partial x^i \partial x^j} (\log |H|) \right\}. \tag{2.42}$$

We note that the equation for affine minimal surfaces is a fourth order equation in f.

If f is a non-degenerate quadratic polynomial, then $H =$ constant. Hence the elliptic paraboloid $x^3 = (x^1)^2 + (x^2)^2$ and the hyperbolic paraboloid $x^3 = (x^1)^2 - (x^2)^2$ are affine minimal surfaces.

Our next result is a formula for affine mean curvature in terms of Riemannian geometry.

Let e_1, e_2, e_3 be a local orthonormal frame field on M such that e_1, e_2 are tangent to M, θ_i is the dual coframe and $\theta_{\alpha\beta}$ are defined by

$$de_\alpha = \sum_\beta \theta_{\alpha\beta} e_\beta, \quad \theta_{\alpha\beta} + \theta_{\beta\alpha} = 0. \tag{2.43}$$

Then we have equations (1.1) and (1.4) as in section 1, and

$$H = \det(h_{ij}) = K = \text{Gaussian curvature of } M. \tag{2.44}$$

To find the affine normal direction, we let

$$\begin{aligned} e_i^* &= e_i \\ e_3^* &= e_3 + a^1 e_1 + a^2 e_2, \end{aligned} \tag{2.45}$$

where e_3^* is in the affine normal direction, then a^i's are determined by

$$d \log |K| + 4 \sum_i a^i h_{ik} \omega^k = 0, \tag{2.46}$$

Hence

$$a^i = -\sum \frac{h^{ij}}{4} (\log |K|)_j, \tag{2.47}$$

where $(f)_j$ denotes the covariant derivative of f with respect to e_j. In fact

$$\begin{aligned} Df &= df = \sum_i f_i \omega^i \\ Df_i &= \sum_j f_{ij} \omega^j = df_i - \sum_j f_j \omega_i^j, \end{aligned} \tag{2.48}$$

We compute

$$\begin{aligned} \omega_3^{*1} &= (de_3^*, e_2^*, e_3^*) = \omega_3^1 + Da^1 + \frac{a^1}{4} d\log|K|, \\ \omega_3^{*2} &= (e_1^*, de_3^*, e_3^*) = \omega_3^2 + Da^2 + \frac{a^2}{4} d\log|K|, \end{aligned} \tag{2.49}$$

$$\begin{aligned} L = \frac{1}{2}|K|^{\frac{1}{4}}\Big\{&-(h_{11} + h_{22}) \\ &- \sum \frac{h^{ij}}{4} (\log|K|)_{ij} + \frac{3}{16}\sum h^{ij} (\log|K|)_i (\log|K|)_j\Big\}. \end{aligned} \tag{2.50}$$

One immediate application of this formula is the following theorem

THEOREM 3. *Suppose M is a surface in* $\mathbf{R}^3$ *which is isometric to a piece of the elliptic paraboloid with its induced Riemannian metric. Then as an affine surface M is affine minimal.*

PROOF. Rewrite (2.50) as follows

$$\begin{aligned} 2|K|^{-\frac{1}{4}} L = {} & h_{11}\left\{-K - \frac{1}{4}(\log|K|)_{22} + \frac{3}{16}(\log|K|)_2^2\right\} \\ & - 2h_{12}\left\{-\frac{1}{4}(\log|K|)_{12} + \frac{3}{16}(\log|K|)_1(\log|K|)_2\right\} \\ & + h_{22}\left\{-K - \frac{1}{4}(\log|K|)_{11} + \frac{3}{16}(\log|K|)_1^2\right\}. \end{aligned} \tag{2.51}$$

(In this formula the superscript 2 means square.)

For the surface $x^3 = (x^1)^2 + (x^2)^2$, we choose coordinates

$$X(u, v) = (v \cos u, v \sin u, v^2). \tag{2.52}$$

Then the coefficients of h_{ij} in (2.51) vanish identically, and the theorem follows from the fact that these coefficients only depend on the first fundamental form of the surface.

3. **Bäcklund theorem for affine surfaces.** In this section we are going to prove our main theorem:

THEOREM 4. *Let M and M* be the focal surfaces of a W*-congruence *in* A^3, *with the correspondence denoted by* $\ell: M \to M^*$ *such that the affine normals at P and* $P^* = \ell(P)$ *are parallel. Then both M and M* are affine minimal surfaces.*

PROOF. Choose an affine frame e_1, e_2, e_3 such that

$$\begin{aligned} & e_1 = \overrightarrow{\mathbf{PP}^*} \\ & e_2 \text{ is tangent to } M \text{ at } P \\ & e_3 \text{ is in the affine normal direction.} \end{aligned} \tag{3.1}$$

Suppose the position vector for M is given by X. Then the position vector for M^* is given by

$$X^* = X + e_1. \tag{3.2}$$

There exists a function k such that

$$\begin{aligned} e_1^* &= -e_1 \\ e_2^* &= e_2 + \frac{1}{k} e_3 \\ e_3^* &= -e_3 \end{aligned} \tag{3.3}$$

is an affine frame on M^*, with e_2^* tangent to M^* at X^*. Let ω^{*i} be the dual frame of (3.3). Then

$$\begin{aligned} dX^* &= \omega^{*1}e_1^* + \omega^{*2}e_2^* \\ &= -\omega^{*1}e_1 + \omega^{*2}\left(e_2 + \frac{1}{k}e_3\right). \end{aligned} \tag{3.4}$$

However, differentiating (3.2) we get

$$\begin{aligned} dX^* &= dX + de_1 \\ &= (\omega^1 + \omega_1^1)e_1 + (\omega^2 + \omega_1^2)e_2 + \omega_1^3 e_3. \end{aligned} \tag{3.5}$$

Comparing coefficients of (3.4) and (3.5), we get

$$\begin{aligned} \omega^{*1} &= -(\omega^1 + \omega_1^1) \\ \omega^{*2} &= \omega^2 + \omega_1^2 \\ \frac{1}{k}\omega^{*2} &= \omega_1^3. \end{aligned} \tag{3.6}$$

Hence

$$\omega^2 + \omega_1^2 = k\omega_1^3. \tag{3.7}$$

Let

$$\begin{aligned} \alpha &= \omega^1 + \omega_1^1 + k\omega_2^3 \\ \beta &= -\omega_3^2 - dk + 2k\omega_3^3 \\ \gamma &= \omega^2 + \omega_1^2 - k\omega_1^3 = 0. \end{aligned} \tag{3.8}$$

Then we have

$$\begin{aligned} \omega^{*1} &= k\omega_3^3 - \alpha \\ \omega^{*2} &= k\omega_1^3. \end{aligned} \tag{3.9a}$$

From (2.4) we have

$$\begin{aligned} \omega_1^{*3} &= (e_1^*, e_2^*, de_1^*) \\ &= \left(-e_1, e_2 + \frac{1}{k}e_3, -de_1\right) \\ &= \omega_1^3 - \frac{1}{k}\omega_1^2 \\ &= \frac{1}{k}\omega^2, \qquad \text{using (3.7).} \end{aligned} \tag{3.9b}$$

Similarly, we have

$$\omega_2^{*3} = -\frac{1}{k^2}(k\alpha + \beta) + \frac{1}{k}\omega^1$$

$$\begin{aligned} \omega_3^{*3} &= \omega_3^3 - \frac{1}{k}\,\omega_3^2 \\ \omega_3^{*1} &= \omega_3^1 \\ \omega_3^{*2} &= -\omega_3^2. \end{aligned} \tag{3.9c}$$

It follows from (2.18a), (3.9a), (3.9b) and (3.9c) that

$$\begin{aligned} L^*\omega^{*1} \wedge \omega^{*2} &= \frac{1}{2}\,|H^*|^{\frac{1}{4}}\,(\omega^{*1} \wedge \omega_3^{*2} + \omega_3^{*1} \wedge \omega^{*2}) \\ &= \frac{1}{2}\,|H^*|^{\frac{1}{4}}\,[(-k\omega_2^3 + \alpha) \wedge \omega_3^2 + k\omega_3^1 \wedge \omega_1^3] \\ &= \frac{1}{2}\,|H^*|^{\frac{1}{4}}\,[k(\omega_3^1 \wedge \omega_1^3 + \omega_3^2 \wedge \omega_2^3) + \alpha \wedge \omega_3^2] \\ &= \frac{1}{2}\,|H^*|^{\frac{1}{4}}\,[k\, d\omega_3^3 + \alpha \wedge \omega_3^2]. \end{aligned}$$

Since e_3 is in the affine normal direction, $d\omega_3^3 = 0$ by (2.11). So we have

$$L^*\omega^{*1} \wedge \omega^{*2} = \frac{1}{2}\,|H^*|^{\frac{1}{4}}\,\alpha \wedge \omega_3^2. \tag{3.10}$$

By hypothesis, e_3 and e_3^* are in the affine normal directions of M and M^* respectively; we rewrite (2.11) as

$$\begin{aligned} \omega_3^3 &= -\frac{1}{4}\,d\log|H| \\ \omega_3^{*3} &= -\frac{1}{4}\,d\log|H^*| \end{aligned} \tag{3.11}$$

Next we compute the following tensor by using (3.9a), (3.9b) and (3.9c) getting,

$$\begin{aligned} \sum_{i,j} h_{ij}^*\omega^{*i} \otimes \omega^{*j} &= \sum_i \omega^{*i} \otimes \omega_i^{*3} \\ &= (k\omega_2^3 - \alpha) \otimes \left(\frac{1}{k}\omega^2\right) + k\omega_1^3 \otimes \left[-\frac{1}{k^2}(k\alpha + \beta) + \frac{1}{k}\omega^1\right] \\ &= \omega_2^3 \otimes \omega^2 + \omega_1^3 \otimes \omega^1 - \frac{1}{k}\,[\alpha \otimes \omega^2 + \omega_1^3 \otimes (k\alpha + \beta)] \\ &= \sum_{i,j} h_{ij}\omega^i \otimes \omega^j - \frac{1}{k}\,[\alpha \otimes \omega^2 + \omega_1^3 \otimes (k\alpha + \beta)]. \end{aligned} \tag{3.12}$$

We note that the tensors $\sum_{i,j} h_{ij}^*\omega^{*i} \otimes \omega^{*j}$ and $\sum_{i,j} h_{ij}\omega^i \otimes \omega^j$ are symmetric and the same must be true of their difference. Because ℓ is a W-congruence (i.e., II* is a multiple of II), these two tensors are proportional in the tensor space. Hence there exists a function b such that

(3.13) $$\alpha \otimes \omega^2 + \omega_1^3 \otimes (k\alpha + \beta) = b \sum_{i,j} h_{ij}\omega^i \otimes \omega^j.$$

This $b \neq k$, for otherwise $\mathrm{II}^* = 0$, contradicting the non-degeneracy of M^*.

Suppose

(3.14) $$\begin{aligned} \alpha &= a_1\omega^1 + a_2\omega^2 \\ k\alpha + \beta &= b_1\omega^1 + b_2\omega^2. \end{aligned}$$

Comparing the coefficients of $\omega^i \otimes \omega^j$ in (3.13), we get

(3.15) $$a_1 + h_{11}b_2 = h_{12}b_1 = h_{12}b$$

(3.16) $$h_{11}b_1 = h_{11}b$$

(3.17) $$a_2 + h_{12}b_2 = h_{22}b.$$

Since M is non-degenerate, h_{11} and h_{12} cannot vanish simultaneously, and (3.15), (3.16) imply that $b_1 = b$. It also follows from (3.15) and (3 17) that

(3.18) $$a_1h_{12} - a_2h_{11} = -bH.$$

Using (3.9) and (3.18), we get

(3.19) $$\omega^{*1} \wedge \omega^{*2} = kH(b - k)\omega^1 \wedge \omega^2$$

(3.20) $$\omega_1^{*3} \wedge \omega_2^{*3} = \frac{1}{k^3}(b - k)\omega^1 \wedge \omega^2.$$

However,

(3.21) $$\omega_1^{*3} \wedge \omega_2^{*3} = H^*\omega^{*1} \wedge \omega^{*2}$$

and $(b - k)$ never vanishes, so we have

(3.22) $$k^4H^*H = 1.$$

Taking $1/4d\log$ of (3.22) and using (3.11), we obtain

$$dk - \omega_3^{*3} - \omega_3^3 = 0.$$

Then (3.9) implies that

(3.23) $$\beta = 0.$$

Therefore by (3.14)

(3.24) $$b_i = ka_i.$$

Substituting (3.24) in (3.15) and (3.17), we obtain

(3.25) $$\begin{aligned} (1 - kh_{12})a_1 + h_{11}ka_2 &= 0 \\ -h_{22}a_1 + (1 + kh_{12})a_2 &= 0. \end{aligned}$$

The determinant of (3.25) is

$$k^2H + 1.$$

If $k^2H + 1 \neq 0$ then $a_i = 0$, so $\alpha = 0$. And if $k^2H + 1 \equiv 0$, then

$$\frac{dk}{2k} - \omega_3^3 = 0. \tag{3.26}$$

But $\beta = 0$, so $\omega_3^2 = 0$. Therefore we have shown that either α or ω_3^2 is zero, so by (3.10) $L^* = 0$. Then by symmetry $L = 0$, i.e., both M and M^* are affine minimal.

We use the same notations as in the proof of the above theorem. We claim that if $\omega_3^2 = 0$ then $\overrightarrow{\mathbf{PP}}^*$ is an asymptotic vector. Indeed using (3.9), $\omega_3^2 = 0$, $L = 0$, and $L^* = 0$, we have

$$\begin{aligned}\omega_3^{*1} &= \ell_2^{*1}\omega^{*2} = k\ell_2^{*1}\omega_1^3 \\ &= \omega_3^1 = \ell_2^1\omega^2.\end{aligned} \tag{3.27}$$

Therefore $h_{11} \equiv 0$, i.e., $e_1 = \overrightarrow{\mathbf{PP}}^*$ is an asymptotic vector.

Suppose $\overrightarrow{\mathbf{PP}}^*$ is an asymptotic vector for all P, then $h_{11} \equiv 0$. By using the local theory for hyperbolic affine surfaces in section 2 and (3.7), we can conclude that $\omega_1^2 = 0$ and $k = 1/h_{12}$. By (2.28), we have $h_{111} = 0$. So $J = 0$. But we have already shown that $L = 0$, hence $\tilde{K} = 0$. Therefore we have proved the following two corollaries.

COROLLARY 1. *Assumptions as in Theorem 4. If $\overrightarrow{\mathbf{PP}}^*$ is not in the asymptotic direction for all $P \in M$, then*

$$\alpha = 0, \beta = 0, \gamma = 0. \tag{3.28}$$

COROLLARY 2. *Assumptions as in Theorem 4. If $\overrightarrow{\mathbf{PP}}^*$ is an asymptotic vector for all $P \in M$, then both M and M^* are affine minimal and affinely flat (i.e., the affine curvature is zero).*

Now we wish to prove the integrability theorem.

THEOREM 5. *Suppose M is an affine minimal surface in A^3. Given $v_0 \in T_{P_0}(M)$ which is not an asymptotic vector, then there exist a surface M^* and a W-congruence ℓ: $M \to M^*$ with parallel affine normals at $P \in M$ and $P^* = \ell(P) \in M^*$ and $\overrightarrow{\mathbf{P_0P_0}^*} = v_0$.*

PROOF. Taking the differential of the system (3.28), we have

$$\begin{aligned}d\alpha &= \gamma \wedge \omega_2^1 - \beta \wedge \omega_2^3 + \alpha \wedge \omega_1^1 \\ d\beta &= -\omega_3^1 \wedge \gamma + \omega_3^2 \wedge \alpha + 2\beta \wedge \omega_1^3 - 2|H|^{-\frac{1}{4}} L\omega^1 \wedge \omega^2 \\ d\gamma &= \alpha \wedge \omega_1^2 - \omega_1^3 \wedge \beta.\end{aligned} \tag{3.29}$$

That the system (3.28) is completely integrable follows from the fact that M is affine minimal. So there exist a function k and an affine frame e_1, e_2, e_3 with e_3 in the affine normal direction and $e_1(P_0) = v_0$ such that $\alpha = 0$, $\beta = 0$, $\gamma = 0$.

Let X be the position vector of M in A^3, and $X^* = X + e_1$. Using $\gamma = 0$, we have

$$\begin{aligned} dX^* &= dX + de_1 \\ &= (\omega^1 + \omega_1^1)e_1 + k\omega_1^3\left(e_2 + \frac{1}{k}e_3\right). \end{aligned} \tag{3.30}$$

Since $\alpha = 0$,

$$\begin{aligned} (\omega^1 + \omega_1^1) \wedge k\omega_1^3 &= k^2\omega_1^3 \wedge \omega_2^3 \\ &= k^2 H\omega^1 \wedge \omega^2. \end{aligned} \tag{3.31}$$

Since M is non-degenerate, $\omega^1 + \omega_1^1$ and $k\omega_1^3$ are linearly independent. Hence X^* defines a surface M^* having e_1, $e_2 + 1/ke_3$ as tangent at X^*. Therefore we can choose an affine frame on M^* as follows

$$\begin{aligned} e_1^* &= -e_1 \\ e_2^* &= e_2 + \frac{1}{k}e_3 \\ e_3^* &= -e_3. \end{aligned} \tag{3.32}$$

Then we have (3.9). Since $\alpha = \beta = 0$, (3.12) implies that

$$|H^*|^{\frac{1}{4}}\,\mathrm{II}^* = |H|^{\frac{1}{4}}\,\mathrm{II}, \tag{3.33}$$

i.e., $\ell : X \mapsto X^*$ is a W-congruence.

Next we want to show that e_3^* is in the direction of affine normal of M^*. By (3.9),

$$\omega_1^{*3} \wedge \omega_2^{*3} = \frac{-1}{k^2}\,\omega^1 \wedge \omega^2. \tag{3.34}$$

However,

$$\begin{aligned} \omega_1^{*3} \wedge \omega_2^{*3} &= H^*\omega^{*1} \wedge \omega^{*2}, \\ &= -k^2 H^* H\omega^1 \wedge \omega^2 \text{ using (3.9)}, \end{aligned} \tag{3.35}$$

so

$$k^4 H^* H = 1. \tag{3.36}$$

Since e_3 is in the affine normal direction, $\omega_3^3 = -1/4\, d \log |H|$. By (3.36), we have

$$\frac{dk}{k} + \frac{d\log|H^*|}{4} - \omega_3^3 = 0. \tag{3.37}$$

Using (3.9), (3.37) and $\beta = 0$, we get

$$\begin{aligned}\omega_3^{*3} &= \omega_3^3 - \frac{1}{k}\omega_3^2 \\ &= \frac{dk}{k} - \omega_3^3 \\ &= -\frac{1}{4}d\log|H^*|,\end{aligned}$$

i.e., e_3^* is in the affine normal direction of M^*.

We note that if M is affinely flat and affine minimal in A^3 with position vector X, then given any asymptotic vector field e_1 on M such that $X^* = X + e_1$ defines a surface in A^3, it follows from the local theory for hyperbolic surfaces in section 2 that $\ell: X \to X^*$ defines a W-congruence with parallel affine normals.

The significance of the above theorem in geometry is that we can construct new affine minimal surfaces by solving the completely integrable system (3.28) on a given affine minimal surface. This fact seems to be of some geometric interest. We do not know, however, whether it has any physical applications.

Our theorem can most likely be generalized to higher dimensions.

References

1. W. Blaschke, *Vorlesungen über Differentialgeometrie* II, Berlin 1923.
2. S. S. Chern, *Affine minimal hypersurfaces*, Proceedings of US-Japan Seminar on Minimal Submanifolds, Tokyo 1978, 1–14.
3. P. A. Schirokow and A. P. Schirokow, *Affine Differentialgeometrie*, Leipzig 1962 (translated from Russian).

UNIVERSITY OF CALIFORNIA, BERKELEY CA 94720.

FROM TRIANGLES TO MANIFOLDS

SHIING-SHEN CHERN

1. Geometry. I believe I am expected to tell you all about geometry; what it is, its developments through the centuries, its current issues and problems, and, if possible, a peep into the future. The first question does not have a clear-cut answer. The meaning of the word *geometry* changes with time and with the speaker. With Euclid, geometry consists of the logical conclusions drawn from a set of axioms. This is clearly not sufficient with the horizons of geometry ever widening. Thus in 1932 the great geometers O. Veblen and J. H. C. Whitehead said, "A branch of mathematics is called geometry, because the name seems good on emotional and traditional grounds to a sufficiently large number of competent people" [1]. This opinion was enthusiastically seconded by the great French geometer Elie Cartan [2]. Being an analyst himself, the great American mathematician George Birkhoff mentioned a "disturbing secret fear that geometry may ultimately turn out to be no more than the glittering intuitional trappings of analysis" [3]. Recently my friend André Weil said: "The psychological aspects of true geometric intuition will perhaps never be cleared up. At one time it implied primarily the power of visualization in three-dimensional space. Now that higher-dimensional spaces have mostly driven out the more elementary problems, visualization can at best be partial or symbolic. Some degree of tactile imagination seems also to be involved" [4].

At this point it is perhaps better to let things stand and turn to some concrete topics.

2. Triangles. Among the simplest geometrical figures is the triangle, which has many beautiful properties. For example, it has one and only one inscribed circle and also one and only one circumscribed circle. At the beginning of this century the nine-point circle theorem was known to almost every educated mathematician. But its most intriguing property concerns the sum of its angles. Euclid says that it is equal to 180°, or π by radian measure, and deduces this from a sophisticated axiom, the so-called *parallel axiom*. Efforts to avoid this axiom failed. The result was the discovery of non-Euclidean geometries in which the sum of angles of a triangle is less or greater than π, according as the geometry is hyperbolic or elliptic. The discovery of hyperbolic non-Euclidean geometry, in the eighteenth century by Gauss, John Bolyai, and Lobatchevsky, was one of the most brilliant chapters in human intellectual history.

The generalization of a triangle is an n-gon, a polygon with n sides. By cutting the n-gon into $n-2$ triangles, one sees that the sum of its angles is $(n-2)\pi$. It is better to measure the sum of the exterior angles! The latter is equal to 2π, for all n-gons, including triangles.

3. Curves in the plane; rotation index and regular homotopy. By applying calculus we can consider smooth curves and closed smooth curves in the plane, i.e., curves with a tangent line everywhere and varying continuously. As a point moves along a closed smooth (oriented) curve C once, the lines through a fixed point O and parallel to the tangent lines of C rotate through an angle $2n\pi$ or rotate n times about O. This integer n is called the rotation index of C. (See Fig. 1.) A famous theorem in differential geometry says that if C is a simple curve, i.e., if C does not intersect itself, $n = \pm 1$.

Clearly, there should be a theorem combining the theorem on the sum of exterior angles of an n-gon and the rotation index theorem of a simple closed smooth curve. This is achieved by considering the wider class of simple closed sectionally smooth curves. The rotation index of

The author received his D.Sc. from the University of Hamburg (Germany). He has taught at Tsinghua University, Academia Sinica, and the University of Chicago, and is now at the University of California, Berkeley; he has also held visiting positions at numerous universities around the world. He is a member of the National Academy of Sciences, received the National Medal of Science in 1976, and was awarded the Chauvenet Prize in 1970. His main research interests are in differential geometry, integral geometry, and topology.—*Editors*

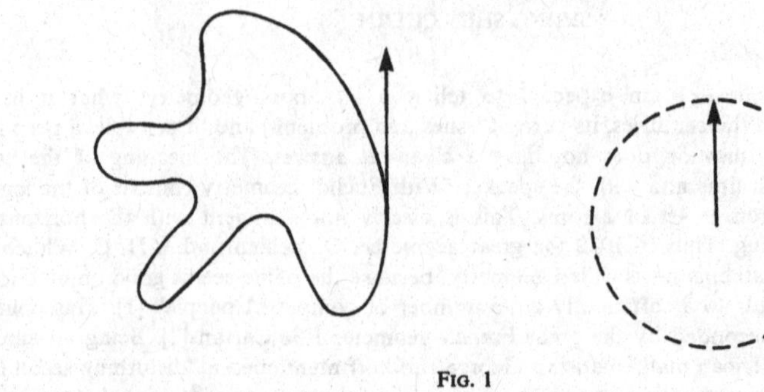

FIG. 1

such a curve can be defined in a natural way by turning the tangent at a corner an amount equal to the exterior angle. (See Fig. 2.) Then the rotation index theorem above remains valid for simple closed sectionally smooth curves. In the particular case of an n-gon formed by straight segments, this reduces to the statement that the sum of its exterior angles is 2π.

This theorem can be further generalized. Instead of simple closed curves we can allow closed curves to intersect themselves. A generic self-intersection can be assigned a sign. Then, if the curve is properly oriented, the rotation index is equal to one plus the algebraic sum of the number of self-intersections. (See Fig. 3.) For example, the figure 8 has the rotation index zero.

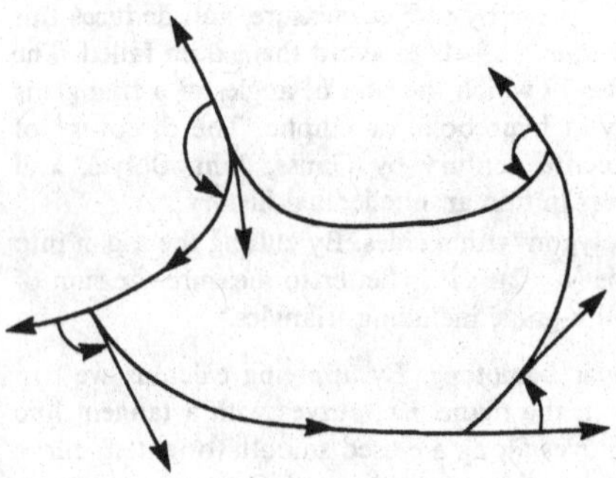

FIG. 2

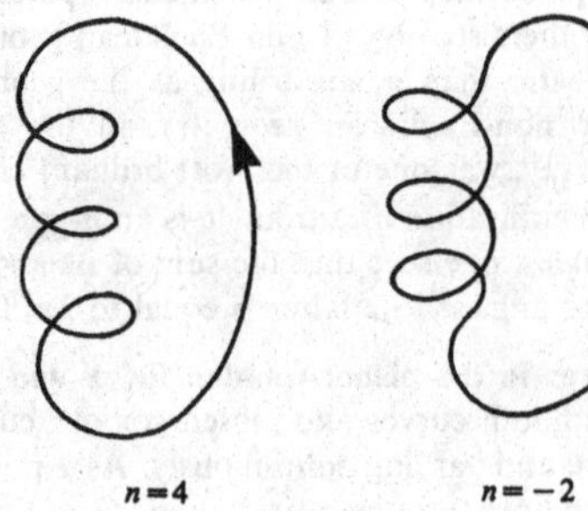

$n=4$ $n=-2$

FIG. 3

A fundamental notion in geometry, or in mathematics in general, is *deformation* or *homotopy*. Two closed smooth curves are said to be *regularly homotopic* if one can be deformed to the other through a family of closed smooth curves. Since the rotation index is an integer and varies continuously in the family, it must remain a constant; i.e., it keeps the same value when the curve is regularly deformed. A remarkable theorem of Graustein-Whitney says that the converse is true [5]: Two closed smooth curves with the same rotation index are regularly homotopic.

It is a standard practice in mathematics that in order to study closed smooth curves in the plane it is more profitable to look at all curves and to put them into classes, the regular homotopy classes in this case being an example. This may be one of the essential methodological differences between theoretical science and experimental science, where such a procedure is impractical. The Graustein-Whitney theorem says that the only invariant of a regular homotopy class is the rotation index.

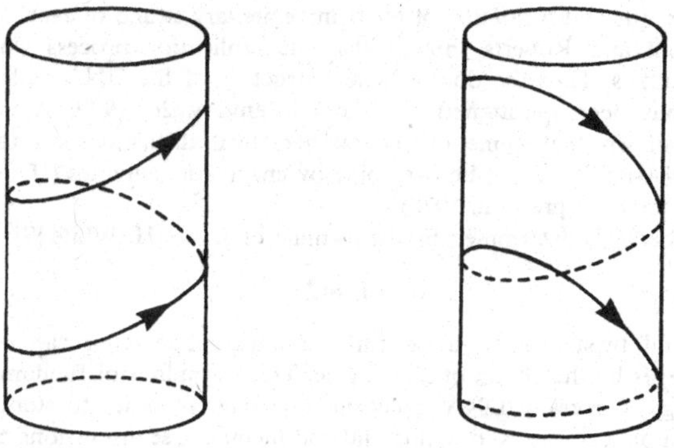

FIG. 4

4. Euclidean three-space. From the plane we pass to the three-dimensional Euclidean space where the geometry is richer and has distinct features. Perhaps the nicest space curve which does not lie in a plane is a circular helix. It has constant curvature and constant torsion and is the only curve admitting ∞^1 rigid motions. There is an essential difference between right-handed and left-handed helices (See Fig. 4), depending on the sign of the torsion; a right-handed helix cannot be congruent to a left-handed one, except by a mirror reflection. Helices play an important role in mechanics. From a geometrical viewpoint it may not be an entire coincidence that the Crick-Watson model of a DNA-molecule is double-helical. A double helix has interesting geometrical properties. In particular, by joining the end points of the helices by segments or arcs, we get two closed curves. In three-dimensional space they have a linking number. (See Fig. 5.)

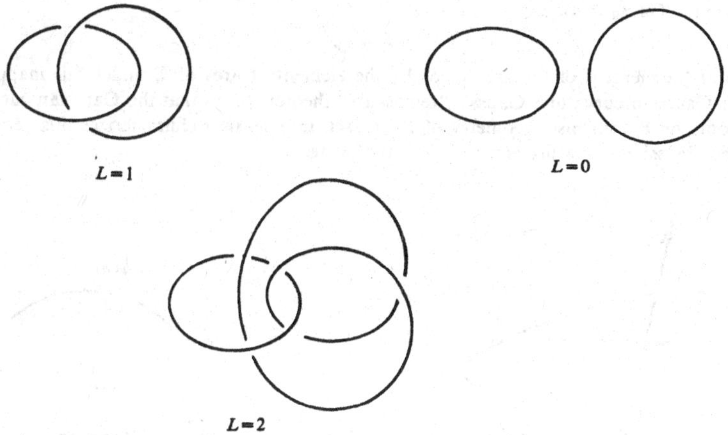

FIG. 5

A recent controversial issue in biochemistry, raised by the mathematicians William Pohl and George Roberts, is whether the chromosomal DNA is double-helical. In fact, if it is, it will have two closed strands with a linking number of the order of 300,000. The molecule is replicated by separation of the strands and formation of the complementary strand of each. With such a large linking number Pohl and Roberts showed that the replication process would have severe mathematical difficulties. Thus the double-helical structure of the DNA molecule, at least for the chromosome, has been questioned [6]. (Added January 26, 1979: A number of recent experiments have shown that some of the mathematical difficulties for the double helical structure of the DNA-molecule can be overcome by enzymatic activities (cf. F. H. C. Crick, Is DNA really a double helix? preprint, 1978).)

The linking number L is determined by the formula of James H. White [7]:

$$T + W = L, \tag{1}$$

where T is the total twist and W the writhing number. The latter can be experimentally measured and changes by the action of an enzyme. This formula is of fundamental importance in molecular biology. Generally DNA molecules are long. In order to store them in limited space, the most economical way is to writhe and coil them. These discussions could indicate the beginning of a stochastic geometry, with the main examples drawn from biology.

In a three-dimensional space surfaces have far more important properties than curves. Gauss's fundamental work elevated differential geometry from a chapter of calculus to an independent discipline. His *Disquisitiones generales circa superficies curvas* (1827) is the birth certificate of differential geometry. The main idea is that a surface has an intrinsic geometry based on the measure of arc length alone. From the element of arc other geometric notions, such as the angle between curves and the area of a piece of surface, can be defined. Plane geometry is thus generalized to any surface Σ based only on the local properties of the element of arc. This localization of geometry is both original and revolutionary. In place of the straight lines are the geodesics, the "shortest" curves between any two points (sufficiently close). More generally, a curve on Σ has a "geodesic curvature" generalizing the curvature of a plane curve and geodesics are the curves whose geodesic curvature vanishes identically.

Let the surface Σ be smooth and oriented. At every point p of Σ there is a unit normal vector $\nu(p)$ which is perpendicular to the tangent plane to Σ at p. (See Fig. 6.) The vector $\nu(p)$ can be viewed as a point of the unit sphere S_0 with center at the origin of the space. By sending p to $\nu(p)$ we get the Gauss mapping

$$g : \Sigma \to S_0. \tag{2}$$

The ratio of the element of the area of S_0 by the element of area of Σ under this mapping is called the Gaussian curvature. Gauss's "remarkable theorem" says that the Gaussian curvature depends only on the intrinsic geometry of Σ. In fact, in a sense it characterizes this geometry. Clearly the Gaussian curvature is zero if Σ is the plane.

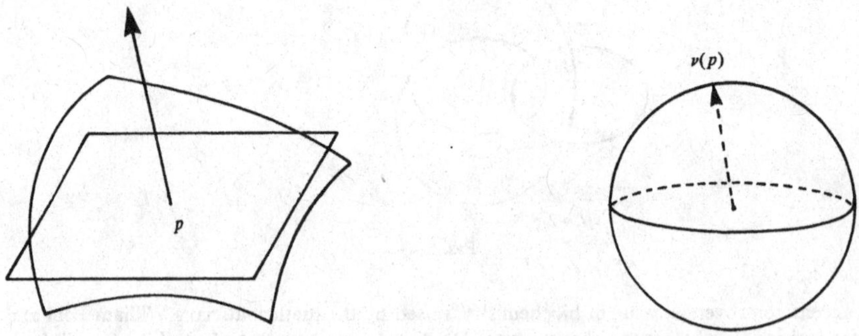

FIG. 6

As in plane geometry we consider on Σ a domain D bounded by one or more sectionally smooth curves. D has an important topological invariant $\chi(D)$, called its Euler characteristic, which is most easily defined as follows: Cut D into polygons in a "proper way" and denote by v, e, and f the number of vertices, edges, and faces, respectively. Then

$$\chi(D)=v-e+f. \tag{3}$$

(Euler's polyhedral theorem was known before Euler, but Euler seems to have been the first one to recognize explicitly the importance of the "alternating sum.")

The Gauss-Bonnet formula in surface theory is

$$\Sigma\,\text{ext angles}+\int_{\partial D}\text{geod curv}+\int\int_{D}\text{Gaussian curv}=2\pi\chi(D), \tag{4}$$

where ∂D is the boundary of D. For a plane domain the Gaussian curvature is zero. If in addition the domain is simply connected, we have $\chi(D)=1$. Then this formula reduces to the rotation index theorem discussed in §3. We are indeed a long way from the sum of angles of a triangle.

Generalizing the geometry of closed plane curves we can consider closed oriented surfaces in space. The generalization of the rotation index is the degree of the Gauss mapping g in (2). The precise definition of the degree is sophisticated. Intuitively it is the number of times that the image $g(\Sigma)$ covers S_0, counted with sign. Unlike the plane, where the rotation index can be any integer, the degree d is completely determined by the topology of Σ; it is equal to

$$d=\frac{1}{2}\chi(\Sigma). \tag{5}$$

For the imbedded unit sphere this degree is $+1$ independently of its orientation. A surprising result of S. Smale [8] says that the two oppositely oriented unit spheres are indeed regularly homotopic or, more intuitively, that the unit sphere can be turned inside out through a regular homotopy. It is essential that at each stage of the homotopy the surface has a tangent plane everywhere, but is allowed to intersect itself.

5. From coordinate spaces to manifolds. It was Descartes who in the seventeenth century revolutionized geometry by using coordinates. Quoting Hermann Weyl, "The introduction of numbers as coordinates was an act of violence" [9]. From now on, paraphrasing Weyl, figure and number, like angel and devil, fight for the soul of every geometer. In the plane the Cartesian coordinates of a point are its distances, with signs, from two fixed perpendicular lines, the coordinate axes. A straight line is the locus of all points whose coordinates x,y satisfy a linear equation

$$ax+by+c=0. \tag{6}$$

The result is the translation of geometry into algebra.

Once the door was opened for analytic geometry, other coordinate systems came into play. Among them are polar coordinates in the plane and spherical coordinates, cylindrical coordinates in space, and elliptic coordinates in the plane and in space. The latter are adapted to the confocal quadrics and are particularly suited to the study of the ellipsoids, which include our earth.

There is also a need for higher dimensions. For even if we start with a three-dimensional space, the theory of relativity calls for the inclusion of time as a fourth dimension. On a more elementary level, to record the motion of a particle, including its velocity, requires six coordinates (the hodograph). All the continuous functions in one variable form an infinite-dimensional space. Those which are square-integrable form a Hilbert space, which can be coordinatized by an infinite sequence of coordinates. Such a viewpoint, of considering all functions with prescribed properties, is fundamental in mathematics.

From the proliferation of coordinate systems it is natural to have a theory of coordinates.

General coordinates need only the property that they can be identified with points; i.e., there is a one-to-one correspondence between points and their coordinates—their origin and meaning are inessential.

If you find it difficult to accept general coordinates, you will be in good company. It took Einstein seven years to pass from his special relativity in 1908 to his general relativity in 1915. He explained the long delay in the following words: "Why were another seven years required for the construction of the general theory of relativity? The main reason lies in the fact that it is not so easy to free oneself from the idea that coordinates must have an immediate metrical meaning" [10].

After being served by coordinates in the study of geometry, we now wish to be free from their bond. This leads to the fundamental notion of a *manifold*. A manifold is described locally by coordinates, but the latter are subject to arbitrary transformations. In other words, it is a space with transient or relative coordinates (principle of relativity). I would compare the concept with the introduction of clothing to human life. It was a historical event of the utmost importance that human beings began to clothe themselves. No less significant was the ability of human beings to change their clothing. If geometry is the human body and coordinates are clothing, then the evolution of geometry has the following comparison.

Synthetic geometry	Naked man
Coordinate geometry	Primitive man
Manifolds	Modern man

A manifold is a sophisticated concept even for mathematicians. For example, a great mathematician such as Jacques Hadamard "felt insuperable difficulty . . . in maintaining more than a rather elementary and superficial knowledge of the theory of Lie groups" [11], a notion based on that of a manifold.

6. Manifolds; local tools. With coordinates practically meaningless there is a need for new tools in studying manifolds. The key word is invariance. Invariants are of two kinds: local and global. The former refer to the behavior under a change of the local coordinates, while the latter are global invariants of the manifold, examples being the topological invariants. Two of the most important local tools are the exterior differential calculus and Ricci's tensor analysis.

An exterior differential form is the integrand of a multiple integral, such as

$$\iint_D Pdydz + Qdzdx + Rdxdy, \tag{7}$$

in (x,y,z)-space, where P, Q, R are functions in x,y,z and D is a two-dimensional domain. It is observed that a change of variables in D (supposed to be oriented) will be taken care of automatically if the multiplication of differentials is anti-symmetric, i.e.,

$$dy \wedge dz = -dz \wedge dy, \text{ etc.}, \tag{8}$$

where the symbol $\wedge$ is used to denote exterior multiplication. It is also more suggestive to introduce the exterior two-form

$$\omega = Pdy \wedge dz + Qdz \wedge dx + Rdx \wedge dy \tag{9}$$

and to write the integral (7) as a pairing (D,ω) of the domain D and the form ω.

For if the same is done in n-space, then Stokes's theorem can be written

$$(D, d\omega) = (\partial D, \omega), \tag{10}$$

where D is an r-dimensional domain and ω is an exterior $(r-1)$-form; ∂D is the boundary of D and $d\omega$ is the exterior derivative of ω and is an r-form. Formula (10), the fundamental formula in multi-variable calculus, shows that ∂ and d are adjoint operators. The remarkable fact is that, while the boundary operator ∂ on domains is global, the exterior differentiation operator d on forms is local. This makes d a powerful tool. When applied to a function (=0-form) and a

1-form, it gives the gradient and the curl, respectively. All the smooth forms, of all degrees ($\leq$ dim of manifold), of a differentiable manifold constitute a ring with the exterior differentiation operator d. Elie Cartan used the exterior differential calculus most efficiently in local problems of differential geometry and partial differential equations. The global theory was founded by G. de Rham, after initial work of Poincaré. This will be discussed in the next section.

In spite of its importance the exterior differential calculus is inadequate in describing the geometrical and analytical phenomena on a manifold. A broader concept is Ricci's tensor analysis. Tensors are based on the fact that a manifold, being smooth, can be approximated at every point by a linear space, called its tangent space. The tangent space at a point leads to associated tensor spaces. Differentiation of tensor fields needs an additional structure, called an affine connection. If the manifold has a Riemannian or Lorentzian structure, the corresponding Levi-Civita connection will serve the purpose.

7. Homology. Historically a systematic study of the global invariants of a manifold began with combinatorial topology. The idea is to decompose the manifold into cells and see how they fit together. (The decomposition satisfies some mild conditions, which we will not specify.) In particular, if M is a closed manifold of dimension n and α_k denotes the number of k cells of the decomposition, $k=0,1,\ldots,n$, then, as a generalization of (3), the Euler-Poincaré characteristic of M is defined by

$$\chi(M)=\alpha_0-\alpha_1+\cdots+(-1)^n\alpha_n. \tag{11}$$

The basic notion in homology theory is that of a boundary. A chain is a sum of cells with multiplicities. It is called a cycle if it has no boundary, i.e., if its boundary is zero. The boundary of a chain is a cycle (see Fig. 7). The number of linearly independent k-dimensional cycles modulo k-dimensional boundaries is a finite integer b_k, called the kth Betti number. The Euler-Poincaré formula says

$$\chi(M)=b_0-b_1+\cdots+(-1)^n b_n. \tag{12}$$

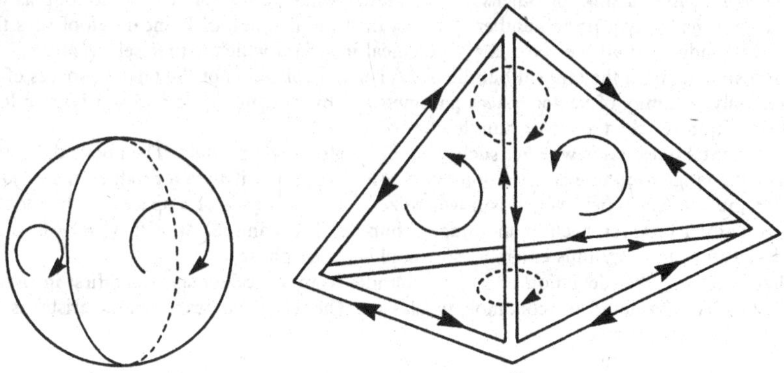

(a) (b)

FIG. 7

The Betti numbers b_k, and hence $\chi(M)$ itself, are topological invariants of M, that is, they are independent of the decomposition and remain invariant under a topological transformation of M. This and more general statements could be considered the fundamental theorems of combinatorial topology. After the path-breaking works of Poincaré and L. E. J. Brouwer, combinatorial topology blossomed in the U.S. in the 1920's under the leadership of Veblen, Alexander, and Lefschetz.

While this is an effective way in deriving topological invariants, the danger in cutting a manifold is that it might be "killed." Precisely, this means that by using a combinatorial approach we may lose sight of the relations of the topological invariants with local geometrical properties. It turns out that, while homology theory depends on the boundary operator ∂, there is a dual cohomology theory based on the exterior differentiation operator d, the latter being a local operator.

The resulting de Rham cohomology theory can be summarized as follows: The operator d has the fundamental property that, when applied repeatedly it gives the zero form; that is, for any k-form α, the exterior derivative of the $(k+1)$-form $d\alpha$ is zero. This corresponds to the geometrical fact that the boundary of any chain (or domain) has no boundary. (See (10).) A form α is called closed, if $d\alpha=0$. It is called a derived form, if there exists a form β, of degree $k-1$, such that it can be written $\alpha=d\beta$. Thus a derived form is always closed. Two closed forms are called cohomologous if they differ by a derived form. All the closed k-forms which are cohomologous to each other constitute the k-dimensional cohomology class. The remarkable fact is that, while the families of k-forms, closed k-forms, derived k-forms are immensely large, the k-dimensional cohomology classes constitute a finite-dimensional linear space whose dimension is the kth Betti number b_k.

De Rham cohomology is the forerunner of sheaf cohomology, which was founded by J. Leray [12] and perfected and applied with great success by H. Cartan and J.-P. Serre.

8. Vector fields and generalizations. On a manifold M it is natural to consider continuous vector fields, i.e., the attachment of a tangent vector to each point, varying in a continuous manner. If the Euler-Poincaré characteristic $\chi(M)$ is not zero, there is at least one point of M at which the vector vanishes. In other words, when the wind blows there is at least one spot on earth with no wind (for the Euler characteristic of the two-dimensional sphere is equal to 2). More precisely, at an isolated zero of a continuous vector field, an integer, called the index, can be defined, which describes to a certain extent the behavior of the vector field at the zero, i.e., whether it is a source, a sink, or otherwise. No matter what the vector field is, so long as it is continuous and has only a finite number of zeros, then the theorem of Poincaré-Hopf says that the sum of its indices at all the zeros is a topological invariant which is precisely $\chi(M)$.

This is a statement on the tangent bundle of M, i.e., the collection of the tangent spaces of M. More generally, a family of vector spaces parametrized by a manifold M and satisfying a local product condition is called a vector bundle over M.

A fundamental question is whether such a bundle is globally a product. The above discussion shows that the tangent bundle is not a product if $\chi(M)\neq 0$; for if it were a product, there would exist a continuous vector field which is nowhere zero. The existence of a space which is locally but not globally a product, such as the tangent bundle of a manifold M with $\chi(M)\neq 0$, is not easy to visualize; geometry thus enters a more sophisticated phase.

To describe the global deviation of a vector bundle from a product space the first invariants are the so-called characteristic cohomology classes. The Euler-Poincaré characteristic is the simplest of the characteristic classes.

The Gauss-Bonnet formula (4) in §4 takes the particularly simple form

$$\iint K\,dA = 2\pi\chi(\Sigma) \tag{4a}$$

when the surface Σ has no boundary. In this formula K is the Gaussian curvature and dA is the element of area. Formula (4a) is of paramount importance because it expresses the global invariant $\chi(\Sigma)$ as the integral of a local invariant, which is perhaps the most desirable relationship between local and global properties. This result has a wide generalization.

Let

$$\pi: E \rightarrow M \tag{13}$$

be a vector bundle. The generalization of a tangent vector field on M is a section of the bundle, i.e., a smooth mapping $s: M \rightarrow E$, such that the composition $\pi \circ s$ is the identity. Since E is only locally a product, the differentiation of s needs an additional structure, usually called a connection. The resulting differentiation, called covariant differentiation, is generally not commutative. The notion of curvature is a measure of the noncommutativity of covariant differentiation. Suitable combinations of the curvature give rise to differential forms which represent characteristic cohomology classes in the sense of the de Rham theory, of which the Gauss-Bonnet formula (4a) is the simplest example [13]. I believe that the concepts of vector bundles, connections, and curvature are so fundamental and so simple that they should be included in any introductory course on multivariable calculus.

9. Elliptic differential equations. When M has a Riemannian metric, there is an operator $*$ sending a k-form α to the $(n-k)$-form $*\alpha, n = \dim M$. It corresponds to the geometrical construction of taking the orthogonal complement of a linear subspace of the tangent space. With $*$ and the differential d we introduce the codifferential

$$\delta = (-1)^{nk+n+1} * d * \tag{14}$$

and the Laplacian

$$\Delta = d\delta + \delta d. \tag{15}$$

Then the operator δ sends a k-form to a $(k-1)$-form and Δ sends a k-form to a k-form. A form α satisfying

$$\Delta\alpha = 0 \tag{16}$$

is called *harmonic*. A harmonic form of degree 0 is a harmonic function in the usual sense.

The equation (16) is an elliptic partial differential equation of the second order. If M is closed, all its solutions form a finite dimensional vector space. By a classical theorem of Hodge this dimension is exactly the kth Betti number b_k. It follows by (12) that the Euler characteristic can be written

$$\chi(M) = d_e - d_0, \tag{17}$$

where d_e (respectively, d_0) is the dimension of the space of harmonic forms of even (respectively odd) degree. The exterior derivative d is itself an elliptic operator and (17) can be regarded as expressing $\chi(M)$ as the index of an elliptic operator. The latter is, for any linear elliptic operator, equal to the dimension of the space of solutions minus the dimension of the space of solutions of the adjoint operator.

The expression of the index of an elliptic operator as the integral of a local invariant culminates in the Atiyah-Singer index theorem. It includes as special cases many famous theorems, such as the Hodge signature theorem, the Hirzebruch signature theorem, and the Riemann-Roch theorem for complex manifolds. An important by-product of this study is the recognition of the need to consider pseudo-differential operators on manifolds, which are more general than differential operators.

Elliptic differential equations and systems are closely enmeshed with geometry. The Cauchy-Riemann differential equations, in one or more complex variables, are at the foundation of complex geometry. Minimal varieties are solutions of the Euler-Lagrange equations of the variational problem minimizing the area. These equations are quasi-linear. Perhaps the "most" non-linear equations are the Monge-Ampère equations, which are of importance in several geometrical problems. Great progress has been made in these areas in recent years [14]. With this heavy intrusion of analysis George Birkhoff's remark quoted above sounds even more disturbing. However, while analysis maps a whole mine, geometry looks out for the beautiful stones. Geometry is based on the principle that not all structures are equal and not all equations are equal.

10. Euler characteristic as a source of global invariants. To summarize, the Euler characteristic is the source and common cause of a large number of geometrical disciplines. I will illustrate this relationship by a diagram. (See Fig. 8.)

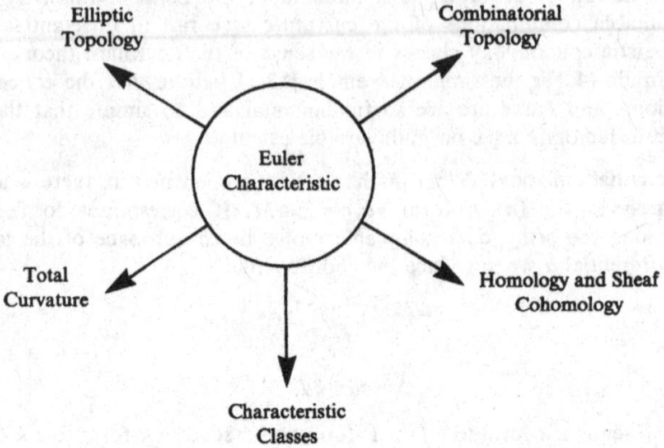

FIG. 8

11. Gauge field theory. At the beginning of this century differential geometry got the spotlight through Einstein's theory of relativity. Einstein's idea was to interpret physical phenomena as geometrical phenomena and to construct a space which would fit the physical world. It was a gigantic task and it is not clear whether he said the last word on a unified field theory of gravitational and electromagnetic fields. The introduction of vector bundles described above, and particularly the connections in them with their characteristic classes and their relations to curvature, widened the horizon of geometry. The case of a line bundle (i.e., when the fiber is a complex line) furnishes the mathematical basis of Weyl's gauge theory of an electromagnetic field. The Yang-Mills theory, based on an understanding of the isotopic spin, is the first example of a nonabelian gauge theory. Its geometrical foundation is a complex plane bundle with a unitary connection. Attempts to unify all field theories, including strong and weak interactions, have recently focused on a gauge theory, i.e., a geometrical model based on bundles and connections. It is with great satisfaction to see geometry and physics united again.

Bundles, connections, cohomology, characteristic classes are sophisticated concepts which crystallized after long years of search and experimentation in geometry. The physicist C. N. Yang wrote **[15]**: "That nonabelian gauge fields are conceptually identical to ideas in the beautiful theory of fiber bundles, developed by mathematicians without reference to the physical world, was a great marvel to me." In 1975 he mentioned to me: "This is both thrilling and puzzling, since you mathematicians dreamed up these concepts out of nowhere." This puzzling is mutual. In fact, referring to the role of mathematics in physics, Eugene Wigner spoke about the unreasonable effectiveness of mathematics **[16]**. If one has to find a reason, it might be expressed in the vague term "unity of science." Fundamental concepts are always rare.

12. Concluding remarks. Modern differential geometry is a young subject. Not counting the strong impetus it received from relativity and topology, its developments have been continuous.

I am glad that we do not know what it is and, unlike many other mathematical disciplines, I hope it will not be axiomatized. With its contact with other domains in and outside of mathematics and with its spirit of relating the local and the global, it will remain a fertile area for years to come.

It may be interesting to characterize a period of mathematics by the number of variables in the functions or the dimension of the spaces it deals with. In this sense nineteenth century mathematics is one-dimensional and twentieth century mathematics is n-dimensional. It is because of the multi-variables that algebra acquires paramount importance. So far most of the global results on manifolds are concerned with even-dimensional ones. In particular, all complex algebraic varieties are of even real dimension. Odd-dimensional manifolds are still very mysterious. I venture to hope that they will receive more attention and substantial clarification in the twenty-first century. Recent works on hyperbolic 3-manifolds by W. Thurston [17] and on closed minimal surfaces in a 3-manifold by S. T. Yau, W. Meeks, and R. Schoen have thrown considerable light on 3-manifolds and their geometry. Perhaps the problem of problems in geometry is still the so-called Poincaré conjecture which says that a closed simply connected 3-dimensional manifold is homeomorphic to the 3-sphere. Topological and algebraic methods have so far not led to a clarification of this problem. It is conceivable that tools in geometry and analysis will be found useful.

This paper, written with partial support from NSF Grant MCS77-23579, was delivered as a Faculty Research Lecture at Berkeley, California, on April 27, 1978.

References

1. O. Veblen and J. H. C. Whitehead, Foundations of Differential Geometry, Cambridge, England, 1932, p. 17.

2. Elie Cartan, Le rôle de la théorie des groupes de Lie dans l'évolution de la géométrie moderne, Congrès Inter. Math., Oslo, 1936, Tome I, p. 96.

3. George D. Birkhoff, Fifty years of American mathematics, Semicentennial Addresses of Amer. Math. Soc., 1938, p. 307.

4. A. Weil, S. S. Chern as friend and geometer, Chern, Selected Papers, Springer Verlag, New York, 1978, p. xii.

5. H. Whitney, On regular closed curves in the plane, Comp. Math. 4 (1937) 276–284.

6. William F. Pohl and George W. Roberts, Topological considerations in the theory of replication of DNA, Journal of Mathematical Biology, 6 (1978) 383–386, 402.

7. James H. White, Self-linking and the Gauss integral in higher dimensions, American J. of Math., 91 (1969), 693–728; B. Fuller, The writhing number of a space curve, Proc. Nat. Acad. Sci., 68 (1971) 815–819; F. Crick, Linking numbers and nucleosomes, Proc. Nat. Acad. Sci., 73 (1976) 2639–2643.

8. S. Smale, A classification of immersions of the two-sphere, Transactions AMS, 90 (1959) 281–290; cf. also A. Phillips, Turning a surface inside out, Scientific American, 214 (May 1966) 112–120. A film of the process, by N. L. Max, is distributed by International Film Bureau, Chicago, Ill.

9. H. Weyl, Philosophy of Mathematics and Science, 1949, p. 90.

10. A. Einstein, Library of Living Philosophers, vol. 1, p. 67.

11. J. Hadamard, Psychology of Invention in the Mathematical Field, Princeton, 1945, p. 115.

12. R. Godement, Topologie algébrique et théorie des faisceaux, Hermann, Paris, 1958.

13. S. Chern, Geometry of characteristic classes, Proc. 13th Biennial Sem. Canadian Math. Congress, 1–40 (1972).

14. S. T. Yau, The role of partial differential equations in differential geometry, Int. Congress of Math., Helsinki, 1978.

15. C. N. Yang, Magnetic monopoles, fiber bundles, and gauge fields, Annals of the New York Academy of Sciences, 294 (1977) 86–97.

16. E. Wigner, The unreasonable effectiveness of mathematics in the natural sciences, Communications on Pure and Applied Math., 13 (1960) 1–14.

17. W. Thurston, Geometry and topology in dimension three, Int. Congress of Math., Helsinki, 1978.

Department of Mathematics, University of California, Berkeley, CA 94720.

manuscripta math. 28, 207 - 217 (1979)

manuscripta
mathematica

LIE GROUPS AND KdV EQUATIONS

Shiing-shen Chern* and Chia-kuei Peng

Dedicated to Hans Lewy and Charles B. Morrey, Jr.

1. Introduction

In recent years there have been extensive studies of evolution equations with soliton solutions, among which the most important ones are the Korteweg-deVries and sine-Gordon equations. We will show that the algebraic basis of these mathematical phenomena lies in Lie groups and their structure equations; their explicit solutions with special properties often give the evolution equations. The process is thus similar to the introduction of a "potential". In fact, from SL(2; R), the special linear group of all (2×2)-real unimodular matrices, one is led naturally to the KdV and MKdV (= modified Korteweg-deVries) equations of higher order. A Miura transformation exists between them. Following H.H. Chen, [2] this leads to the Bäcklund transformations of the KdV equation.

2. KdV equations

Let

$$(1) \qquad SL(2;R) = \left\{ X = \begin{pmatrix} a & b \\ c & d \end{pmatrix} \middle| ad-bc = 1 \right\}$$

*Work done under partial support of NSF grant MCS77-23579.

0025-2611/79/0028/0207/$02.20

be the group of all (2×2)-real unimodular matrices. Its right-invariant Maurer-Cartan form is

$$\omega = dXX^{-1} = \begin{pmatrix} \omega_1^1 & \omega_1^2 \\ \omega_2^1 & \omega_2^2 \end{pmatrix}, \tag{2}$$

where

$$\omega_1^1 + \omega_2^2 = 0. \tag{3}$$

The structure equation of SL(2; R), or the Maurer-Cartan equation, is

$$d\omega = \omega \wedge \omega , \tag{4}$$

or, written explicitly,

$$\begin{aligned} d\omega_1^1 &= \omega_1^2 \wedge \omega_2^1 \\ d\omega_1^2 &= 2\omega_1^1 \wedge \omega_1^2 , \\ d\omega_2^1 &= 2\omega_2^1 \wedge \omega_1^1 . \end{aligned} \tag{4a}$$

Let U be a neighborhood in the (x,t)-plane and consider the smooth mapping

$$f : U \rightarrow SL(2; R) . \tag{5}$$

The pull-backs of the Maurer-Cartan forms can be written

$$\begin{aligned} \omega_1^1 &= \eta dx + Adt , \\ \omega_1^2 &= qdx + Bdt , \\ \omega_2^1 &= rdx + Cdt , \end{aligned} \tag{6}$$

where the coefficients are functions of x, t.

The forms in (6) satisfy the equations (4a). This gives

$$(7)\quad \begin{aligned} -\eta_t + A_x - qC + rB &= 0 , \\ -q_t + B_x - 2\eta B + 2qA &= 0 , \\ -r_t + C_x - 2rA + 2\eta C &= 0 . \end{aligned}$$

We consider the special case that $r = +1$ and η is a parameter independent of x, t. Writing

$$(8)\qquad q = u(x,t) ,$$

we get from (7) ,

$$(9)\quad \begin{aligned} A &= + \eta C + \frac{1}{2} C_x , \\ B &= uC - \eta C_x - \frac{1}{2} C_{xx} . \end{aligned}$$

Substitution into the second equation of (7) gives

$$(10)\qquad u_t = K(u) ,$$

where

$$(11)\qquad K(u) = u_x C + 2uC_x + 2\eta^2 C_x - \frac{1}{2} C_{xxx}$$

As an example we take

$$(12)\qquad C = \eta^2 - \frac{1}{2} u$$

Then (10) becomes

$$(13)\qquad u_t = \frac{1}{4} u_{xxx} - \frac{3}{2} uu_x ,$$

which is the well-known KdV (= Korteweg deVries) equation.

It is therefore natural to take C to be an arbitrary polynomial in η. Since the expression in (11) involves explicitly only η^2, we shall suppose C to be a polynomial in η^2, i.e., precisely,

$$(14)\qquad C = \sum_{0 \leqq j \leqq n} C_j(x,t)\, \eta^{2(n-j)} ,$$

where $C_j(x,t)$ are functions of x,t. Substituting (14) into (11) and equating to zero the powers of η^2, we get

(15) $$C_0 = \text{const.},$$

(16) $$C_{j+1,x} = -\frac{1}{2}u_x C_j - uC_{j,x} + \frac{1}{4}C_{j,xxx}, \quad 0 \leqq j \leqq n-1 .$$

It is interesting to note that the latter is exactly the recursion formula for the conserved densities of the KdV equation! [4] We write the right-hand number of (10) as

(17) $$K_n(x) = u_x C_n + 2uC_{n,x} - \frac{1}{2}C_{n,xxx} \underset{\text{def}}{=} -2C_{n+1,x}.$$

The last quantity is introduced by definition. Furthermore, introducing an infinite sequence of C_j we can suppose (16) be valid for all j, $0 \leqq j < +\infty$. The equation

(18) $$u_t = K_n(u)$$

is called the KdV equation of the nth order.

<u>The C_j's are polynomials in u and its successive derivatives with respect to x.</u> To prove this, we put

(19) $$\tilde{C} = \sum_{0 \leqq j} C_j \eta^{-2j}$$

There is no loss of generality in assuming

(20) $$C_0 = 1.$$

Equation (16) can be written

(21) $$\eta^2 \tilde{C}_x = -\frac{1}{2}u_x\tilde{C} - u\tilde{C}_x + \frac{1}{4}\tilde{C}_{xxx},$$

When multiplied by $2\tilde{C}$, this becomes

$$\eta^2(\tilde{C}^2)_x = -(u\tilde{C}^2)_x + \frac{1}{2}(\tilde{C}\tilde{C}_{xx} - \frac{1}{2}\tilde{C}_x^2)_x .$$

Integrating with respect to x, we have

$$\eta^2(\tilde{C}^2-1) = -u\tilde{C}^2 + \frac{1}{2}\tilde{C}\tilde{C}_{xx} - \frac{1}{4}\tilde{C}_x^2 . \tag{22}$$

Equating the coefficients of η^{-2j}, $j \geqq 0$, in both sides we get

$$2C_1 = -u ,$$

$$\begin{aligned} 2C_{j+1} = &- \sum_{1 \leqq k \leqq j} C_k C_{j+1-k} - u \sum_{0 \leqq k \leqq j} C_k C_{j-k} \\ &+ \frac{1}{2} \sum_{0 \leqq k \leqq j-1} C_k C_{j-k,xx} \\ &- \frac{1}{4} \sum_{1 \leqq k \leqq j} C_{k,x} C_{j-k,x} \qquad j = 1,2, \ldots . \end{aligned} \tag{23}$$

The second equation determines C_{j+1} as a polynomial of $C_1, \ldots, C_j$ and their x-derivatives. Hence the statement to be proved follows by induction.

We assign to u the weight 1 and to the subscript x the weight $\frac{1}{2}$. Then C_j is isobaric of weight j, as seen by induction.

From C_{n+1} one gets the explicit form of the nth order KdV equation by (17); $K_n(u)$ is a polynomial of u and its x-derivatives, and is isobaric of weight $n + \frac{3}{2}$.

The next two C_j's are given by

$$\begin{aligned} 2C_2 &= \frac{3}{4} u^2 - \frac{1}{4} u_{xx} , \\ 2C_3 &= -\frac{5}{8} u^3 + \frac{5}{16} u_x^2 + \frac{5}{8} uu_{xx} - \frac{1}{16} u_{xxxx} . \end{aligned} \tag{23a}$$

In our terminology the classical KdV equation is of the

first order; cf (12), (13).

3. MKdV equations

In the same way we set

(24) $$q = r = v(x,t)$$

in (7) and consider η to be a parameter independent of x,t. Then equations (7) become

(25) $$\begin{aligned} A_x &= v(C-B) , \\ v_t &= B_x - 2\eta B + 2vA , \\ v_t &= C_x + 2\eta C - 2vA , \end{aligned}$$

of which the last two can be written

(26) $$\begin{aligned} (C-B)_x &= 4vA - 2\eta(B + C) , \\ v_t &= \frac{1}{2}(B + C)_x + \eta(C - B) . \end{aligned}$$

Let

(27) $$C - B = \eta P, \qquad C + B = Q, \qquad A = \eta R .$$

The above equations become

(28) $$\begin{aligned} R_x &= vP , \\ P_x &= 4vR - 2Q , \\ v_t &= \frac{1}{2} Q_x + \eta^2 P . \end{aligned}$$

Eliminating P, Q, we get

(29) $$v_t = M(v) ,$$

where

(30) $$M(v) = \eta^2 \frac{R_x}{v} + (vR)_x - \frac{1}{4}\left(\frac{R_x}{v}\right)_{xx}$$

By taking

(31) $$R = \eta^2 - \frac{1}{2} v^2 ,$$

equation (29) becomes

(32) $$v_t = \frac{1}{4} v_{xxx} - \frac{3}{2} v^2 v_x ,$$

which is the well-known modified KdV equation.

In general, we set

(33) $$R = \sum_{0 \leqq j \leqq n} R_j(x,t)\eta^{2(n-j)}$$

In order that $M(v)$ be independent of η we get the conditions

$$R_0 = \text{const}$$

(34) $$v^{-1}R_{j+1,x} = \frac{1}{4}(v^{-1}R_{j,x})_{xx} - (vR_j)_x ,$$

$$0 \leqq j \leqq n-1 .$$

By using the second equation to define R_{j+1}, we can suppose (34) to be valid for all $j \geqq 0$. The right-hand side of (29), with R given by (33), will be denoted by $M_n(v)$. Then we have

(35) $$M_n(v) = -v^{-1}R_{n+1,x}(v)$$

From (34) we immediately observe that R_j are even, i.e.,

(36) $$R_j(-v) = R_j(v) .$$

It follows from (35) that $M_n(v)$ are odd, i.e.,

(37) $$M_n(-v) = -M_n(v).$$

There is no loss of generality in supposing $R_0 = 1$, and the first R_j's are found to be

$$R_0 = 1 ,$$

(38) $$R_1 = -\frac{1}{2}v^2 ,$$

$$R_2 = \frac{3}{8} v^4 + \frac{1}{8} v_x^2 - \frac{1}{4} vv_{xx} .$$

4. Miura transformation

The connection between the KdV and MKdV equations is furnished by the Miura transformation. To define it we observe that (17) can be written formally as

(39) $$K_{n+1}(u) = TK_n(u) ,$$

where

(40) $$T = \frac{1}{4} D^2 - u - \frac{1}{2} u' D^{-1}, \quad D = \frac{d}{dx}, \ u' = u_x .$$

Similarly, we write (35) as

(41) $$M_{n+1}(v) = SM_n(v) ,$$

where

(42) $$S = \frac{1}{4} D^2 - v^2 - v' D^{-1} v .$$

By an easy computation the following commutativity relations can be verified:

(43) $$\begin{aligned} (2v + D)S(v) &= T(v' + v^2)(2v + D) , \\ (2v - D)S(v) &= T(-v' + v^2)(2v - D) . \end{aligned}$$

It follows that

(44) $$\begin{aligned} K_n(v' + v^2) &= (2v + D)M_n(v) , \\ K_n(-v' + v^2) &= (2v - D)M_n(v) . \end{aligned}$$

In fact, for $n = 1$, this follows directly from (13) and (32), their right-hand sides being $K_1(u)$ and $M_1(v)$ respectively. The general case follows from induction by applying (43) to $M_{n-1}(v)$.

These results were derived by a different method

in [1], in which they, or at least a part of them, were attributed to P. Olver. Formula (44) gives a fundamental relation between the KdV and MKdV equations and is at the basis of the Miura transformation. Of importance is the relation

$$u = v_x + v^2 . \tag{45}$$

If u and v are so related and if $R_j(v)$ satisfy (34), then, by a straightforward computation, we find that

$$C_j(u) = -\frac{1}{2} M_{j-1}(v) + R_j(v)$$

satisfy the recurrent relation (16). It follows that

$$-\frac{1}{2} M_{j-1}(v) + R_j(v) = C_j(v_x + v^2) . \tag{46a}$$

Similarly, we have

$$\frac{1}{2} M_{j-1}(v) + R_j(v) = C_j(-v_x + v^2) . \tag{46b}$$

These relations can also be written

$$\begin{aligned} M_{j-1}(v) &= -C_j(v_x + v^2) + C_j(-v_x + v^2), \\ 2R_j(v) &= C_j(v_x + v^2) + C_j(-v_x + v^2) . \end{aligned} \tag{47}$$

In particular, we draw from (47) the conclusion that $M_j(v)$, $R_j(v)$ are polynomials of v and its x-derivatives. Moreover, if v and the subscript x are each given the weight 1, then $M_{j-1}(v)$ and $R_j(v)$ are isobaric of weight $2j$.

From (44) we get

$$\frac{\partial u}{\partial t} - K_n(u) = (D + 2v)\left(\frac{\partial v}{\partial t} - M_n(v)\right) , \tag{48}$$

where u is given by (45). Thus, if v is a solution of MKdV, then $u(x,t)$, given by (45), is a solution of KdV. But then $-v$ is also a solution of MKdV, so that the KdV has a new solution given by

$$\tilde{u} = -v_x + v^2 \quad . \tag{49}$$

This passage from v to u is called a Miura transformation, and that from u to $\tilde{u}$ a Bäcklund transformation, following an approach of H.H. Chen [3].

To pass from u to $\tilde{u}$ we set

$$u = w_x, \qquad \tilde{u} = \tilde{w}_x \quad . \tag{50}$$

Then

$$u - \tilde{u} = (w - \tilde{w})_x = 2v_x \ ,$$

and we can suppose

$$2v = w - \tilde{w} \ . \tag{51}$$

It follows that

$$\begin{aligned} (w + \tilde{w})_x &= \frac{1}{2} (w - \tilde{w})^2 \ , \\ (w - \tilde{w})_t &= 2M_n\left(\frac{1}{2} (w - \tilde{w})\right) \ . \end{aligned} \tag{52}$$

With w given, such that $u = w_x$ is a solution of (18), the system (52) is completely integrable. A solution $\tilde{w}$ of (52) gives a new solution $\tilde{u} = \tilde{w}_x$ of (18); $\tilde{u}$ is a Bäcklund transform of u.

From the MKdV equation one can pass to a twice modified KdV equation by a similar procedure. This and other results will be reported later.

References

1. M. Adler and J. Moser, On a class of polynomials connected with the Korteweg-deVries equation, Communications in Math Physics 61, 1-30 (1978)

2. Hsing-Hen Chen, Relation between Bäcklund transformations and inverse scattering problems, Lecture Notes in Math, no. 515, 241-252, Springer 1976

3. M. Crampin, F.A.E. Pirani, D.C. Robinson, The soliton connection, Lett. Math. Phys. 2, 15-19 (1977)

4. C. S. Gardner, J.M. Greene, M.D. Kruskal, R.M. Miura, Korteweg-deVries equation and generalizations VI, Methods for exact solution, Comm. Pure and Appl. Math. 27, 97-133 (1974)

University of California
Berkeley, California 94720, USA

and
University of Science and Technology of China
Hofei, Anhwei
People's Republic of China

(Received February 23, 1979)

16. General Relativity and Differential Geometry

SHIING-SHEN CHERN

The title for this paper originally suggested by the organizing committee was "The Interaction of General Relativity and Differential Geometry." It is a strange feeling to speak on a topic of which I do not know half of the title. I will speak as a differential geometer looking at the impressive structure of general relativity. As I understand it, general relativity is physics; its basis is physical experimentation. The aim of geometry should be the study of spaces. But this is a tautological statement. It is perhaps better to say that geometrical investigations are guided by tradition and continuity. Their criterion is mathematical originality, simplicity, and depth, and a good combination and compromise of the latter. Geometry has thus more freedom and can indulge a little bit more in ideal topics. It has been, however, a historical fact that it could be rudely awakened to find its abstract objects close to reality. Such an example is furnished by the relation between differential geometry and general relativity.

General relativity was born in 1915. Differential geometry at its early stage was identical with the infinitesimal calculus, the derivative and the tangent line and the integral and the area being synonymous objects. As an independent discipline differential geometry was born in 1827. It was the year that Gauss published his "General Investigations of Curved Surfaces (Disquisitiones circa superficies curvas)," in which he laid the foundations of a local two-dimensional geometry based on a quadratic differential form. Even Gauss could not have foreseen that its four-dimensional generalization would be the basis of gravitation.

1. Differential Geometry before Einstein

In a historical paper in 1854, "Über die Hypothesen, welche der Geometrie zugrunde liegen," Riemann generalized Gauss's work to high dimensions and laid the foundations of Riemannian geometry. In that paper he first introduced the idea of an n-dimensional manifold whose points are described by n real numbers as coordinates.

Harry Woolf (ed.), Some Strangeness in the Proportion: A Centennial Symposium to Celebrate the Achievements of Albert Einstein ISBN 0-201-09924-1

This was a giant step from Gauss, whose curved surfaces lie in the three-dimensional Euclidean space and are not intrinsic. Being mathematically a purist, Einstein was reluctant to accept such an idea. It took him seven years to pass from his special relativity in 1908 to his general relativity in 1915. He gave the following reason: "Why were another seven years required for the construction of the general theory of relativity? The main reason lies in the fact that it is not so easy to free oneself from the idea that coordinates must have an immediate metrical meaning."[1]

The fundamental problem in Riemannian geometry is the form problem. Given two quadratic differential forms

$$ds^2 = \sum_{i,k} g_{ik}\, dx^i\, dx^k \tag{16.1}$$

$$ds'^2 = \sum_{i,k} g'_{ik}\, dx'^i\, dx'^k, \qquad 1 \leq i, k \leq n, \tag{16.2}$$

in two different systems of coordinates $x^1, \ldots, x^n$ and $x'^1, \ldots, x'^n$, find the conditions that there is a coordinate transformation

$$x'^i = x'^i(x^1, \ldots, x^n), \qquad 1 \leq i \leq n, \tag{16.3}$$

carrying one into the other. This problem was solved by E. B. Christoffel and R. Lipschitz in 1869. Christoffel's solution involves the symbols associated with his name and the notion of covariant differentiation. Based on it, Ricci developed in 1887–96 the tensor analysis that plays a fundamental role in general relativity. An account of the Ricci calculus was given in a historical memoir, "Méthodes de calcul differentiel absolu et leurs applications," published in *Mathematische Annalen* in 1901, by him and his student, T. Levi-Civita. Christoffel taught at the Eidgenossische Technische Hochschule in Zürich (where Einstein was later a student) and thereby had an influence on Italian geometers. It may be of interest to note that this year is the one-hundred-fiftieth anniversary of his birth.

Important as these developments were, the main activities in differential geometry at the turn of the century centered on the geometry in Euclidean space, following the tradition of Euler and Monge. A representative work was Darboux's four-volumed "Théorie des surfaces," which was and still is a classic. It was difficult for geometers to free themselves from an absolute ambient space, which is usually the Euclidean space.

At about the same time as the Christoffel–Lipschitz solution of the form problem, Felix Klein formulated in 1872 his *Erlangen* program. This is to define geometry as the study of a space with a continuous group of automorphisms, such as the Euclidean space with the group of rigid motions, the projective space with the group of projective collineations, and so on. The *Erlangen* program unified geometry under group theory and was the guiding principle of geometry for about a half century. In practice it can be used, as a consequence of the isomorphism of groups, to derive from a given geometrical result new and seemingly unrelated results. A famous example is the line-sphere transformation of Sophus Lie. A more elementary example is based on Study's dual numbers. From the theorem in elementary geometry that the three heights of a triangle meet in a point one can derive the following theorem of Morley–Petersen: Given a simple skew hexagon in space whose adjacent sides are perpendicular, there is a line that meets perpendicularly the three common perpendiculars of the three pairs of opposite sides.[2]

Klein's *Erlangen* program fits perfectly with the special theory of relativity, one of whose principles is the invariance of the field equations under the Lorentz group. As a result Klein, the most influential German mathematician at the turn of the century, was one of the earliest supporters of the special theory of relativity. The Lorentzian structure plays a fundamental role in relativity. It has also geometrical interpretations. Indeed this happens when we study the geometry of spheres in space. All the contact transformations in space carrying spheres to spheres form a 15-parameter Lie group, and those that carry planes to planes a 10-parameter subgroup. The latter is isomorphic to the Lorentzian group in four variables. The resulting geometry is known as the Laguerre sphere geometry.[3]

The great success of Klein's *Erlangen* program naturally led to the study of differential geometry in a Klein space, or a homogeneous space as it is now called. In particular, projective differential geometry was initiated in Halphen's thesis in 1878 and later developed by the American school under E. J. Wilczynski beginning in 1906 and the Italian school under G. Fubini beginning in 1916.

At the beginning of the twentieth century, global differential geometry was at its infancy. The four-vertex theorem was formulated by Mukhopadhyaya in 1909. The topological conclusion drawn from the Gauss–Bonnet formula that the integral of the Gaussian curvature of a closed orientable surface is equal to $2\pi\chi$, where χ is the Euler characteristic of the surface, was deduced by von Dyck in 1888. With typical foresight Hilbert wrote a paper in 1901 on surfaces of constant Gaussian curvature, in which he gave a new proof of Liebmann's theorem that a closed surface of constant Gaussian curvature is a sphere and the theorem (Hilbert's theorem) that a complete surface of constant negative curvature cannot be regular everywhere. Under Hilbert's supervision Zoll found in 1903 closed surfaces of revolution, not the sphere, all of whose geodesics are closed. Motivated by dynamics, Poincaré and G. D. Birkhoff proved the existence of closed geodesics on convex surfaces.

The final aim of differential geometry is global results. However, local differential geometry should not be minimized because every global result must have a local basis. To have a systematic development of global differential geometry its foundations must be laid. This should come from topology. General relativity provided the impetus.

2. The Impact of General Relativity

When Einstein founded general relativity, the mathematical tools were available in the form of Riemannian geometry treated by the Ricci calculus. Einstein introduced the useful summation convention. The effect on differential geometry was electrifying; Riemannian geometry became the central topic. It may be of interest to note that almost all the standard books on Riemannian geometry, by Schouten, Levi–Civita, Elie Cartan, and Eisenhart, appeared in the period 1924–26.

These developments immediately led to generalizations. It soon became clear that in the applications of Riemannian geometry to relativity, the Levi–Civita parallelism, and not the Riemannian metric itself, plays the crucial role. In his famous book *Raum, Zeit, Materie*, published in 1918, Hermann Weyl introduced the notion of an affine connection. It is a structure, not necessarily Riemannian, where parallelism and covariant differentiation are defined, with the desired properties. Weyl's affine connections are symmetric or without torsion.

A definitive treatment of affine connections, together with a generalization to connections with torsion, was given by Elie Cartan in his fundamental paper, "Sur les variétés a connexion affine et la théorie de la relativité généralisé," published in 1923–24. The paper did not receive the attention it deserved for the simple reason it was ahead of its time. For it is more than a theory of affine connections. Its ideas can be easily generalized to give a theory of connections in a fiber bundle with any Lie group, for whose treatment the Ricci calculus is no longer adequate. The paper shows, among other things, how Einstein's theory is a direct generalization of Newton's theory when the latter is expressed in an intrinsic form, that is, in general coordinates. Specifically the following contributions could be singled out: 1) the introduction of the equations of structure and the interpretation of the so-called Bianchi identities as the result of exterior differentiation of the equations of structure; 2) the recognition of curvature as a tensor-valued exterior quadratic differential form.

Geometrically an affine connection is a family of affine spaces, the fibers, parametrized by a space, the base space, such that the family is locally trivial and that there is a law of "development" of the fibers along the curves of the base space, preserving the linear properties. In the same vein we can take Klein spaces as fibers and replace the general linear group by a Lie group acting on the Klein space, with a corresponding law of development. Such a structure was called by Cartan a generalized space ("espace généralisé"). In general, the connection is nonholonomic; that is, the development depends on the curve in the base space. In other words, the space does not return to its original position after being developed along a closed curve; the measure of the deviation is given by the curvature of the connection. Clearly, a Klein space is itself a generalized space, with curvature identically zero.

When Klein formulated his *Erlangen* program, it was observed that Riemannian geometry was not included, because a generic Riemannian space admits no isometry other than the identity. From Cartan's viewpoint it is a generalized space with Euclidean spaces as fibers and provided with the Levi–Civita connection. This settles a foundational issue in differential geometry, as we have now a notion that includes Klein spaces and Riemannian spaces, and generalizations of both.

In practice, geometrical structures are given in a nonintuitive form. Frequently it is either a metric defined by an integral or a family of submanifolds defined by a system of differential equations. The two most familiar examples are the Riemannian metric and the paths defined by a system of ordinary differential equations of the second order. The association of a connection is not an easy problem. In fact, even the definition of the Levi–Civita connection of a Riemannian metric is, from the purely algebraic viewpoint, fairly nontrivial. As to be expected, the geometry of paths (E. Cartan, O. Veblen, T. Y. Thomas) involves a projective connection.

These developments in what is commonly described as non-Riemannian geometry were accompanied by parallel developments in general relativity. With special relativity for the electromagnetic field and general relativity for the gravitational field there was clearly a need for a unified field theory where the two could be combined. The first significant step was taken by Hermann Weyl in 1918 in his gauge theory. Weyl used a generalized space having as group the group of homothetic transformations. It was found to be physically untenable. It is now understood that his gauge group should not be the noncompact group of homothetic transformations but the compact circle group. Recent developments on the gauge theory will be discussed in Sec. 4.

Following Weyl other unified field theories were proposed, among which were those of Kaluza–Klein, Einstein–Mayer (1931), and Veblen (the projective theory of relativity, 1933). A common feature is the introduction of a fifth dimension, to account for electricity and magnetism. (Veblen's theory is four-dimensional, but the tangent projective space has five homogeneous coordinates.) Veblen's projective theory is geometrically simple, in taking as a starting point the paths in the space, which are to be the trajectories of the charged particles.

Einstein himself pursued his search for a unified field theory throughout his last years, often with collaborators. In this respect I wish to insert a personal note. I came to the Institute for Advanced Study in 1943 from Kunming in Southwest China. It was at the height of the Second World War, and Einstein greeted me with great warmth and sympathy. It was my great fortune to be able to discuss with him from time to time different topics, general relativity included. I soon saw the extreme difficulty of his problem and the difference between mathematics and physics. The famous problems in mathematics are usually well formulated, but in physics the formulation is part of the problem.

Einstein, with a strict standard for the final answer, was not content with the aforementioned proposals, and in fact with many others. He tested the geometrical structures that might possibly underlie a unified field theory. Among them are the following:

1. A nonsymmetric g_{ij}
2. Four-dimensional complex space with an Hermitian structure
3. Metric spaces more general than Riemannian

The geometry of general metric spaces was founded and investigated by Karl Menger, and Einstein's friend Kurt Gödel made important contributions to it. Structure 1, the g_{ij}, splits uniquely into a symmetric and an antisymmetric part. If the former is nondegenerate, the structure is equivalent to a pseudo-Riemannian structure with an exterior quadratic differential form. The pseudo-Riemannian structure is Riemannian or Lorentzian if the signature of the symmetric part of g_{ij} is $++++$ or $+++-$. Structure 2 is closely related to the geometry of complex algebraic varieties and to functions of several complex variables, which are areas of mathematics with spectacular developments in the last decades.

3. Positive-Mass Conjecture, Minimal Surfaces, and Manifolds of Positive Scalar Curvature

In the post-Einstein era general relativity made great progress in its emphasis on a global theory (or "large-scale space-time"). The origin was cosmology, in which Einstein himself took an active part, but the influence of the developments in global differential geometry is unmistakable. The universe is identified as a four-dimensional connected Lorentzian manifold, and physics and geometry are intertwined more than ever. However, the purely geometrical problems are usually simpler. Two of the reasons are that geometry is Pythagoriasian or Riemannian and the geometers can idealize their spaces by assuming them to be compact.

It is natural to record the data at a given instant by a data-set, which is a hypersurface $\sum$ having everywhere a timelike normal (so that the induced metric is Riemannian). In this way the theory of hypersurfaces in a four-dimensional manifold, an immediate generalization of classical surface theory, plays a role in general relativity. The local invariants of $\sum$ are given by two quadratic differential forms, its first and second fundamental forms. The trace of the second fundamental form is called the mean curvature. Its vanishing characterizes the maximal hypersurfaces. On the other hand, the induced Riemannian metric on $\sum$ has a scalar curvature. All these quantities are connected by the Gauss–Codazzi equations. The mass density μ and the momentum density J^a are, as a consequence of Einstein's field equations, combinations of the coefficients of the first and second fundamental forms and their covariant derivatives. Since momentum density should not exceed the mass density, we have

$$\mu \geq \left| \sum_{1 \leq a \leq 3} J^a J_a \right|^{1/2} \geq 0. \tag{16.4}$$

On a maximal hypersurface the scalar curvature is non-negative.

A data-set is called asymptotically flat if for some compact C, $\sum - C$ consists of a finite number of components N_i (to be called the ends), such that each N_i is diffeomorphic to the complement of a compact set in R^3 and has asymptotically the metric

$$ds^2 = \left(1 + \frac{M_i}{2r}\right)^4 \left(\sum_{1 \leq a \leq 3} (dx^a)^2\right), \tag{16.5}$$

r being the distance from the origin. The number M_i coincides with the Schwarzschild mass in the case of the Schwarzschild metric and is hence called the total mass of N_i. The positive-mass conjecture states that for an asymptotically flat data set, each end has total mass $M_i \geq 0$, and that if one $M_i = 0$, the data set is flat in the sense that the induced Riemannian metric is flat and the second fundamental form is zero. This conjecture is of fundamental importance in general relativity. For physical reasons Einstein assumed it to be true.

Under the assumption that $\sum$ is a maximal hypersurface this was proved in its full generality in 1978 by R. Schoen and S. T. Yau.[6] The complete story of this work is a perfect example of the fruit of contact and collaboration between relativists and differential geometers. At the Differential Geometry Summer Institute of the American Mathematical Society at Stanford University in 1973, Robert Geroch was invited to give a series of lectures on general relativity. The positive-mass conjecture is clearly one of the open problems. To simplify its statement Geroch formulated some pilot conjectures, one of which says the following: "On the three-dimensional number space R^3 consider a Riemannian metric which is flat outside a compact set. If its scalar curvature is ≥ 0, then the metric is flat." By enclosing the compact in a big box and identifying the opposite faces, J. Kazdan and F. Warner reformulated the conjecture as follows: "A Riemannian metric on the three-dimensional torus with scalar curvature ≥ 0 is flat." Geroch remarked: "It is widely felt that proofs of several of these special cases could be generalized to a proof of the full conjecture."

This Georch conjecture falls in the realm of differential geometry; Schoen and Yau proved it first. The idea of the proof makes use of closed minimal surfaces. In fact, from the formula for the second variation of area it is seen that in a three-dimensional compact orientable Riemannian manifold of positive scalar curvature a closed minimal surface of positive genus is not stable; that is, it will have a smaller area by a perturbation. On the other hand, the three-dimensional torus has a large fundamental group (isomorphic to $Z \oplus Z \oplus Z$), and has second Betti number equal to 3. These topological properties should imply the existence of a closed regular surface of smallest area in a nonzero homotopy class of closed surfaces. Such a result is a generalization of the fact that on a compact Riemannian manifold there is in every nonzero homotopy class of closed curves a shortest closed smooth geodesic. The proof of a corresponding result for minimal surfaces is of course much more subtle. Schoen and Yau went on to a proof of the positive-mass conjecture for a maximal hypersurface. Later in 1978 the result was generalized to higher dimensions.

These developments touch on topics that are dear to differential geometers. They are: minimal surfaces and manifolds of positive scalar curvature.

Early investigations on minimal surfaces were concentrated to the plateau problem: Given a closed curve in R^3, find a surface of smallest area bounded by it. Only in recent years has attention been directed to the study of closed or complete minimal surfaces in a given manifold, such as the n-dimensional Euclidean space R^n *or* the n-dimensional unit sphere S^n. The study generalizes that of closed geodesics, which have been found to play an important role in the geometry and topology of Riemannian manifolds.[8] Closed or complete minimal surfaces, particularly the regular ones, are destined to be a richer and even more interesting object. It is natural to take as a data-set a maximal hypersurface. Recently, J. Sachs and K. Uhlenbeck proved that a compact simply connected Riemannian manifold always has a minimally immersed 2-sphere.

The basic question on manifolds of positive scalar curvature is: What compact manifolds can be given a Riemannian metric of positive scalar curvature? The interest in this question is enhanced by the fact that every compact manifold of dimension ≥ 3 can be given a Riemannian metric of negative scalar curvature. In fact, it can be given a metric whose total scalar curvature (that is, the integral of the scalar curvature) is ≤ 0. The latter can then be conformally deformed into one with constant negative scalar curvature. On the other hand, by studying harmonic spinors, A. Lichnerowicz proved in 1963 that if a compact spin manifold has a Riemannian metric of positive scalar curvature, its Â-genus is zero. Yau and Schoen showed that the three-dimensional torus cannot have a Riemannian metric of positive scalar curvature; the same has since been proved true for the n-dimensional torus (M. Gromov, B. Lawson, R. Schoen, S. T. Yau). Motivated by general relativity, these authors are close to a complete topological description of all compact manifolds that can carry a Riemannian metric of positive scalar curvature.

The same questions can be asked for the Ricci curvature or the sectional curvature. A classical theorem of S. Myers says that a complete Riemannian manifold of positive Ricci curvature must be compact and hence must have a finite fundamental group. The condition for a Riemannian manifold to have positive sectional curvature is even stronger, and such manifolds are expected to be few. The compact symmetric spaces of rank 1 have this property, but there are others of a sporadic nature. A complete topological description of compact Riemannian manifolds of positive sectional curvature seems to be difficult.

4. Gauge Field Theory

Gauge field theory was introduced by Hermann Weyl in his paper "Gravitation und Elektrizität" in 1918. The idea is to use a pair of quadratic and linear differential forms

$$ds^2 = \sum_{i,k} g_{ik}\, dx_i\, dx_k, \tag{16.6}$$

$$\phi = \sum_i \phi_i\, dx_i, \qquad 1 \leq i, k \leq 4, \tag{16.7}$$

defined up to the gauge transformation

$$\begin{aligned} ds^2 &\to \lambda\, ds^2, \\ \phi &\to \phi + d \log \lambda. \end{aligned} \tag{16.8}$$

With ϕ as the electromagnetic potential and its exterior derivative $d\phi$ as the electromagnetic strength or Faraday, this was the first attempt of a unified field theory. Einstein objected to the indeterminacy of ds^2 but expressed admiration of the depth and boldness of Weyl's proposal.

All the objections, then and later, disappear if we interpret Weyl's theory as based on the geometry of a circle bundle over a Lorentzian manifold.[9] Then the form ϕ, subject to the gauge transformation, can be interpreted as defining a connection in the circle bundle and ds^2 remains unaltered, eliminating Einstein's objection.

The mathematical basis of gauge field theory lies in vector bundles and the connections in them. The notion of a fiber bundle or fiber space, being global in character, arose in topology. At first it was an attempt to find new examples of manifolds.[10] Fiber spaces are locally, but not globally, product spaces. The presence of such a distinction is a sophisticated mathematical fact. The development of fiber spaces has to wait until invariants are found to distinguish the fiberings or even to show that globally there are nontrivial ones. The first such invariants are the characteristic classes introduced by H. Whitney and by E. Stiefel in 1935.[11] Topology, however, forgets the algebraic structure, and in applications vector bundles, with the linear structure intact, are more useful. A vector bundle $\pi: E \to M$ over a manifold M is, roughly speaking, a family of vector spaces parametrized by M such that it is locally a product. The vector space $E_x = \pi^{-1}(x)$ corresponding to $x \in M$ is called the fiber at x. Examples are the tangent bundle of M and all tensor bundles associated to it. A more trivial bundle is the product bundle $M \times V$, where V is a fixed vector space and (x, V), $x \in M$, is the fiber at x. A vector bundle is called real or complex according as the fiber is a real or complex vector space. Its dimension is the dimension of the fibers.

It is important that the linear structure on the fibers has a meaning so that the general linear group plays a fundamental role in matching the fibers; it is called the structure group. A real (respectively complex) vector bundle is called Riemannian (respectively Hermitian) if the fibers are provided with inner products. In this case the structure group is reduced to $0(n)$ (respectively $U(n)$), n being the dimension of the fibers; the bundle is then called an $0(n)$ (respectively $U(n)$)-bundle. Similarly, we have the notion of an $SU(n)$-bundle.

A section of the bundle E is an attachment, in a continuous and smooth manner, to every point $x \in M$, a point of the fiber E_x. In other words, it is a continuous mapping $s: M \to E$ such that the composition $\pi \circ s$ is the identity. The notion is a natural generalization of a vector-valued function and of a tangent vector field. In order to differentiate s we need a so-called connection in E. The latter allows the definition of the covariant derivative $D_X s$ (X being a vector field in M), which is a new section of E. Covariant differentiation is generally not commutative; that is, $D_X \circ D_Y \neq D_Y \circ D_X$ for two vector fields X, Y in M. The measure of the noncommutativity gives the curvature of the connection; this is an analytic version of the geometric concept of nonholonomy described in Sec. 2. Following Elie Cartan it is important to regard the curvature as a matrix-valued exterior quadratic differential form. Its trace is a closed 2-form. More generally, the sum of all its principal minors of order k is a closed $2k$-form. It is called a characteristic form (Pontrjagin form or Chern form, according as the bundle is real or complex). By the de Rham theory the characteristic form of degree $2k$ determines a cohomology class of dimension $2k$, to be called a characteristic class. Whereas the characteristic forms depend on the connection, the characteristic classes depend only on the bundle. They are the simplest global invariants of the bundle. It must be an act of nature that the nontriviality of a vector bundle is recognized through the need of a covariant differentiation and that its noncommutativity accounts for the first global invariants. This introduction of the characteristic classes gives emphasis on its local character, and the characteristic forms contain more information than the classes. When M is a compact oriented manifold, a characteristic class of the top dimension (that is, of dimension equal to that of M) gives by integration a characteristic number. When it is an integer, it is called a topological quantum number.

These differential-geometric notions have been found to be the likely mathematical basis of a unified field theory. Weyl's gauge theory deals with a circle bundle or a $U(1)$-bundle, that is, a complex Hermitian bundle of dimension one.

In studying the isotopic spin Yang–Mills used what is essentially a connection in an $SU(2)$-bundle. It is the first instance of a non-Abelian gauge theory. From the connection the "action" can be defined. A connection in an $SU(2)$-bundle at which the action takes the minimum is called an instanton. Its curvature has a simple expression and is called self-dual. An instanton is thus a self-dual solution of the Yang–Mills equation. When the space R^4 is compactified into the four-dimensional sphere S^4, the $SU(2)$-bundles are determined up to an isomorphism by a topological quantum number k, which is an integer. Atiyah, Hitchin, and Singer[12] proved that over S^4 the moduli (or parameter space) for the set of connections with self-dual curvature on the $SU(2)$-bundle with given $k > 0$ is a smooth manifold of dimension $8k - 3$. In physical terms this is the dimension of the space of instantons with topological quantum number $k > 0$.

Atiyah and Ward observed that the self-dual Yang–Mills fields fit well into the Penrose Twistor program. They were able to translate the problem of finding all self-dual solutions into a problem of algebraic geometry: classifying holomorphic vector bundles on the complex projective 3-space. This problem had been looked at closely by K. Barth, G. Horrocks, R. Hartshone, and others. Using their results, one can finally find all self-dual solutions,[13] and in fact, coming back to physics, translate the mathematical results into explicit formulas that physicists find satisfactory.[14]

Instantons can claim a relation to Einstein through the following result. The group $SO(4)$ is locally isomorphic to $SU(2) \times SU(2)$, so that a Riemannian metric on a four-dimensional manifold M gives rise through projection to connections in the $SU(2)$-bundles. M is an Einstein manifold if and only if these connections are self-dual or anti-self-dual.[15]

It remains to be seen whether a satisfactory unified field theory, including even weak and strong interactions, will be achieved through a nonabelian gauge theory. It suffices to say that the geometric notion of bundle and connection, as explained here, is exceedingly simple. I believe Einstein would have liked it.

5. Concluding Remarks

My own limitation restricts the scope of this account, which is conspicuously incomplete. The most obvious omission is the important topic of singularities as culminated in the work of Penrose and Hawking.

Finally, as an amateur I wish to express the hope that general relativity not be restricted to the gravitational field. A global unified field theory, whatever it is, must be closer to Einstein's grand design. More mathematical concepts and tools are now available.

Acknowledgement

This material is based on work supported by the National Science Foundation under Grant MCS 77-23579.

Notes

1. A. Einstein, "Autobiographical Notes," in *Albert Einstein: Philosopher-Scientist* edited by P. A. Schilpp (Open Court, La Salle, Ill., 1949) Vol. 1, p. 67.
2. F. Klein, *Höhere Geometrie* (Springer, 1926), p. 314.
3. W. Blaschke, *Differentialgeometrie*, Vol. III (Springer, 1929).
4. See Appendix II of *The Meaning of Relativity*, 5th ed. (1955).
5. Y. Choquet-Bruhat, A. E. Fischer, and J. E. Marsden, "Maximal Hypersurfaces and Positivity of Mass," *Isolated Gravitating Systems in General Relativity*, Ed. J. Ehlers, Italian Physical Society, 1979, 396–456.
6. R. Schoen and S. T. Yau, "Incompressible Minimal Surfaces, Three-dimensional Manifolds with Nonnegative Scalar Curvature, and the Positive Mass Conjecture in General Relativity," Proc. NAS, USA, **75**, 2567 (1978); "On the Proof of the Positive Mass Conjecture in General Relativity," Comm. Math. Phys., **65**, 45–76 (1979); "Complete Manifolds with Nonnegative Scalar Curvature and the Positive Action Conjecture in General Relativity," Proc. NAS, USA **76**, 1024–25 (1979).

7. Robert Geroch, "General Relativity," in *Proceedings of the Symposium in Pure Math*, 27, Pt 2, (1975), pp. 401–14.
8. W. Klingenberg, *Lectures on Closed Geodesics* (Springer, 1978).
9. S. Chern, "Circle Bundles," Geometry and Topology, III Latin American School of Mathematics, Springer Lecture Notes, No. 597 (1977) pp. 114–31.
10. H. Hotelling, "Three Dimensional Manifolds of States of Motion," Trans. Amer. Math. Soc. **27**, 329–44 (1929), and **28**, 479–90 (1929); H. Seifert, "Topology 3 dimensionaler gefaserter Raume," Acta Math. **60**, 137–238 (1932).
11. E. Stiefel, "Richtungsfelder und Fernpurallelismus in Mannigfaltigkeiten," Comm. Math. Helv. **8**, 3–51 (1936); H. Whitney, "Sphere Spaces," Proc. Math. Acad. Sci. USA **21**, 462–68 (1935).
12. M. F. Atiyah, N. J. Hitchin, and I. M. Singer, "Deformations of Instantons," Proc. NAS, USA, **74**, 2662-63 (1977); "Self-duality in Four-dimensional Riemannian Geometry," Proc. R. Soc. Lond. **A362**, 425–61 (1978).
13. M. F. Atiyah, et al. "Construction of Instantons," Phys. Lett. **65A**, 185 (1978).
14. N. H. Christ, E. J. Weinberg, and N. K. Stanton, "General Self-Dual Yang–Mills Solutions," Phys. Rev. **D18**, 2013–25 (1978).
15. Atiyah, Hitchin, and Singer, op. cit. in n. 12.

The Mathematical Work of H. C. Wang

W. M. BOOTHBY, S. S. CHERN, and S. P. WANG

H. C. Wang was so unassuming and excessively modest about his work and so reluctant to speak of his own achievements that the quality and originality of his research are not as well known as they should be. This is especially true today since much of his best work was done some years ago and has, in the meantime, been incorporated into many books, articles, etc. which are often cited rather than the original papers. This is testimony to their importance, but at the same time it is of considerable interest to look at the papers themselves and to see how they fitted into the mathematics of the time. Although all of his research was in the general area of what might be called differential geometry, he showed breadth and versatility within this very broad category. Indeed, he has written papers in classical differential geometry (i.e. differential invariants), algebraic topology, topological groups and transformation groups, homogeneous spaces of Lie groups, discrete subgroups of Lie groups, and more. In every case he has written one or more papers of which any mathematician would be proud and usually with very original and interesting ideas which influenced others. It was not his style to concentrate in one area for some time and forge general theories, but rather to solve, often in a very elegant and interesting fashion, a special problem of genuine interest to many other people.

This style was manifest already in his earliest papers. His first published paper, [1], which appeared in 1942, was evidently written when he was in his early twenties. It was an announcement of results whose details appeared in [2]. There was at that time great interest in the so-called non-holonomic systems in mechanics and Wang studied a concrete problem in this area. This with [3] and [6] deal with differential invariants, much in the spirit of Cartan. Paper [4] was motivated by the work of J. Douglas on the inverse problem of the calculus of variations. It gave a beautiful geometrical interpretation, in the spirit of Lie and Engel, of the main condition of Douglas. Papers [5], [7], and [10] settled some outsanding problems in the geometry of paths by using ideas and methods in Lie group theory, which at that time was completely novel. In [7] he resolved an open question, posed by Knebelmanto determine the Finsler spaces with completely integrable equations of Killing—by proving the following theorem: *If a Finsler space E of N dimensions admits a group G of motions depending on* $r > \frac{1}{2}N(N-1) + 1$ *essential parameters, then E is a Riemannian space of constant curvature.* In [10] he proved the theorem that if a space of paths satisfies the axiom of the plane, it is projectively flat.

In 1946 he arrived in England where he began work on his doctorate at the University of Manchester. A short but interesting paper which was probably written soon after his arrival was [8], the first of two papers on algebraic topology. In it he compared homology and homotopy groups of $M_1 = S^n \times P^{n+2}(R)$ (the n-sphere and real $(n+2)$-dimensional projective space), with those of $M_2 = S^{n+2} \times P^n(R)$. He showed that, although the homotopy groups are the same in every dimension (even vanishing in dimensions 2 to $n-1$), neither the integral homology groups nor these groups modulo the spherical homology classes are the same. (As pointed out by S. Eilenberg, when n is odd the homotopy groups are even operator isomorphic, with respect to the fundamental groups).

The second paper on algebraic topology alluded to, which was evidently written during his stay at Manchester, is [11]—one of his more important papers. There was much interest at this time in the relations among homology and cohomology groups of base space, fibre and total space of fibre spaces (the powerful spectral sequence methods of Leray were just then being developed). In 1941, Gysin had proved theorems which nowadays can be given as an exact sequence relating the cohomology of the three spaces when the fibre was (cohomologically) a sphere. Wang in this paper gave a sequence relating homology of total space, base space, and fibre when the base space is a sphere. This fitted in very nicely with the work of Leray and was discussed in one of Leray's first fundamental papers on spectral sequences (J. Math. Pures et Appl., 29 [1950], pp. 169–213). The exact sequence of [11] was called the "Wang sequence" by Serre in his thesis (Annals of Math., 54 (1951), pp. 425–505), a name which has remained in use since then—which must have given Wang much personal satisfaction (although he was too modest ever to say as much).

Wang's work in the first few years after 1948 shows diversity and talent. Several of his best papers were written in this period. The next important paper, which is in a very different direction from both his early work and the two papers discussed above, was [13] (results announced in [12]) which takes its starting point from a paper of H. Hopf and H. Samelson in which it is shown that if G is a connected compact Lie group and H a closed subgroup, then the rank of G and H are the same (i.e. H contains a maximal torus of G) if and only if the Euler characteristic $\chi(G/H) > 0$. Wang showed easily using the same reasoning as in Montgomery and Samelson's paper determining groups transitive on spheres, that $G/K \approx G_1/K_1 \times \cdots \times G_r/K_r$ where G_i are the simple constituents of G and $K_i = K \cap G_i$. He then determined for each of the four main classes of compact simple groups all possible K. Additionally, the Euler characteristic and first five homotopy groups are determined in each case. A paper containing essentially the same results was written almost simultaneously (and quite independently) by Borel and de Siebenthal, appearing in the Commentarii Helvetici in the same year. (Their paper includes the exceptional groups.) Oddly enough the authors have noted that many researchers who cite the latter are apparently unaware of Wang's paper. Wang's paper [14] is related—in it he determined the (simply connected) homogeneous spaces of compact Lie groups (effective and transitive action) for which the Euler characteristic is 1 (a single point), 2 (an even dimensional sphere), or an odd prime. In the last case, there are the following possibilities: complex or quaternionic projective spaces of suitable dimension and one exceptional case. [13] was Wang's Ph. D. thesis at the University of Manchester where he received his Ph. D. in July, 1948. His thesis supervisor was M. H. A. Newman and his External Examiner was J. H. C. Whitehead. The following year he came to the United States where he was appointed as a lecturer at Louisiana State University from 1949 until 1951 when he made the first of four visits to the Institute for Advanced Study.

During this period he wrote an interesting series of papers, [15]–[20], which all deal with various aspects of topological groups and topological transformation groups, usually acting as isometries on a metric space. Of them the most important is [18] in which he considered a connected compact metric space E and a compact group G of isometries of E which is locally strongly transitive in the sense that for some $r > 0$ there is an isometry of G carrying a pair (a_1, b_1) of points of E to a pair (a_2, b_2) whenever the distance $d(a_1, b_1) = d(a_2, b_2) < r$. Using a short, clever argument he showed that G does not have arbitrarily small normal subgroups and so must be a Lie group. Hence E is a compact homogeneous space of G which is, then, transitive on directions as well as points. He then determined all such spaces. They are spheres, real and complex projective spaces, quaternion projective spaces and the Cayley projective plane (symmetric spaces of rank one). In [17] partial results were obtained for the non-compact case, but it was necessary to assume further properties of the metric. The noncompact case was further studied by J. Tits, who resolved questions left open by [17]. The net result is a very beautiful classification theory for two point homogeneous metric spaces, proceeding from very few assumptions on the metric space beyond two point homogeneity and local compactness (for example, no differentiability is assumed). This is very much in the spirit of the questions raised by Hilbert's fifth problem, which many mathematicians were trying to solve at the time.

As to the other papers of this group, [15] was an abstract of some of the results of [17] and [18], [16] is a short note which answered a conjecture of P. A. Smith by showing that any separable, locally compact, non-discrete metric group G must contain a proper non-countable subgroup dense in G. In [19], as in [17], a connected non-compact metric space is studied but it is only assumed that it is locally compact and homogeneous with respect to its group of isometries. It is then shown that X is contractible to a point in its one-point compactification. Further, the one-dimensional cohomology group $H_c^1(X)$ with compact supports either vanishes or is isomorphic to the coefficient group (in which case $X \approx R \times C$, C compact).

The final paper dealing with rather general transformation groups is [20]; it answered a question raised by Wallace concerning a Peano continuum. To state the result we define a point e of a connected Hausdorff space X to be an end point if for any neighborhood U of e there is a point x of U such that $X - \{x\}$ separates: $X - \{x\} = X_1 \cup X_2$ with $e \in X_1 \subset U$. The main result asserts that if X is arcwise connected and G is a transformation group of X leaving fixed an end point e, then G has no other fixed point if and only if the G-orbit of every neighborhood of e is all of X. Moreover, if G is compact, then G must have another fixed point.

In 1954 Wang returned to the Institute for Advanced Study for a second time. Just prior to this, in March 1953, he submitted [21] to the American Journal of Mathematics. This is one of his most frequently quoted papers. It was known through the work of Bochner and Montgomery that the group of automorphisms of a compact complex analytic manifold can be given a complex analytic structure so that the action is complex analytic. Moreover, some isolated examples of compact complex coset spaces with more than one complex structure had been given by Hopf, Eckmann and Calabi. In a considerable tourdeforce this paper characterizes and (eventually) enumerates all simply connected compact complex analytic manifolds whose automorphism group is transitive. Many new examples of complex analytic manifolds result, including some which are not Kähler and some which have an uncountable number of inequivalent complex structures. The examples mentioned above plus some obtained concurrently by Samelson are special cases. Briefly, it is shown, using a theorem of Montgomery, that the maximal compact subgroup of the automorphism group acts transitively on the manifold, that the manifold is homeomorphic with the quotient of a compact semi-simple group by a closed connected subgroup whose semi-simple part coincides with the centralizer of a toral subgroup of K. Conversely, given a compact semi-simple group K and a subgroup of this type, the coset space may be given an invariant complex structure. The number of inequivalent complex structures is shown to be infinite only when the Euler characteristic is zero.

In the same year he wrote a short note, [22], in which he showed that a compact complex manifold of dimension n with n complex analytic forms, linearly independent at every point, is necessarily the quotient G/D of a complex Lie group by a discrete subgroup. Moreover, it is Kählerian if and only if it is a torus.

Paper [23] is somewhat reminiscent of [7] in that manifolds with a geometric object—in this case a symmetric affine connection—are studied under the assumptions that the group of automorphisms is "large". In this instance it is assumed that its dimension is greater than $n^2 - n + 5$, n being the dimension of the manifold. All possible such spaces are constructed as coset spaces and their curvature is determined. The situation is complicated by the fact that there are diffeomorphic affine spaces with quite different automorphism groups, even among those with curvature 0. Paper [26] is along the same lines but with a different geometric structure, this time a contact structure, which is invariant under the automorphism group—which itself acts transitively on the manifold. This paper has had some very interesting generalizations in the hands of R. Lutz and his students (e.g. Ann. Inst. Fourier de Grenoble 27 [1977], 1–15 and 29 [1979], 283–306).

The general method, going back to E. Cartan, and the interesting questions concerning the study of spaces with a geometric structure, invariant under some group acting on the space, is beautifully explained in [27] which is the substance of Wang's invited address to the International Congress of Mathematicians, 1958 at Edinburgh. The papers [18], [21], [22], [23], and [26] all come in this category and also to some extent even [25] where he generalized in a significant way earlier work of Nomizu (Amer. J. Math. 76 [1954], 33–65) on invariant affine connections. Here one considers connections on a principal fibre bundle E with group S over a base space B. G is assumed to be a Lie group of bundle transformations which is transitive in the sense that given any two fibres there is a group element carrying one to the other. The G-invariant connections in such bundles are characterized and their holonomy algebras are determined.

Although much preoccupied with the differential geometry of coset spaces in the years 1952–1958, he also wrote his first significant paper, [24], on discrete subgroups of Lie groups (in this case solvable groups). His interest in this subject was first aroused in analyzing complex parallelizable manifolds, [22], where discrete subgroups of Lie groups play a key role. Moreover, during this period (in 1958) he listened to the lectures of André Weil on discrete subgroups and then became even further interested during his third years, 1961–62, at the Institute for Advanced Study in Princeton—this time as a Guggenheim Fellow. Several further papers followed [24] and these exposures to "arithmetic" questions: [31], [32], [34], [35], [38], etc., which will all be described in turn. It is fair to say that such questions became his major mathematical preoccupation from around 1963 onward. We discuss these papers as a group below. First, however, some further comments.

The paper [28] is notable in that it is a venture into an area rather different from those above, namely it is a study of transformation groups of S^n—no longer assumed transitive, but assumed to have an $(n-1)$-dimensional orbit. Thus the methods are quite different from those in the cases where transitivity and invariance of a geometric object are assumed. From a historical point of view this is one of the first attempts to study the global orbit structures under the intransitive group actions. It was a remarkable piece of original work which initiated much of the later works along that direction, an example being the recent important work of H. F, Münzner on isoparametric hypersurfaces (Mathematische Annalen, to appear). The paper, a long one, involves quite a few interesting ideas and formidable computations both in algebraic topology and in Lie group theory. It was unfortunate for him that in the classification he missed some slightly twisted orbit structures modelled after the orthogonal action of the tensor representation of $SO(2) \times SO(n)$ on $R^2 \otimes R^n$. This was pointed out and corrected in the papers of W. C. and W. Y. Hsiang (Annals of Math. 85 [1967], pp. 351–369). In fact, according to Wu-Yi Hsiang, by this study he could have stumbled on the Kervaire exotic spheres.

Paper [30] is a generalization to almost complex structures of the Bochner and Montgomery theorem which asserts that the group of automorphisms of a complex structure on a compact manifold M is necessarily a Lie group, acting differentiably on M. It involves application of the theory of elliptic differential equations to a system of partial differential equations derived from the assumption that an almost complex structure is preserved.

[29] proves the following interesting result: If G is a complex semi-simple Lie group, then every element z is a commutator, i.e. $z = xyx^{-1}y^{-1}$ for some $x, y \in G$. [33] is a short note outlining some rather general facts concerning abelian subgroups of a compact Lie group which are not subgroups of a toral group. This is surely a fragment of a more general program he had in mind.

Turning now to his papers on discrete subgroups of Lie groups, we consider [24]. This may be considered his first paper on the subject, his interest having been aroused by his discovery of the role of such subgroups in determining complex parallelisable manifolds [22], and also, no doubt, by the work of Malcev on discrete subgroups of Lie groups which appeared in translation in 1951. This was one of the first extensions of Malcev's results and is one of his most quoted papers. In it he gave necessary and sufficient conditions for an abstract (discrete) group Γ to be imbeddable as a uniform discrete subgroup of a simply connected solvable group with a finite number of components. In particular it was one of the first papers in which algebraic group techniques—which have since proved extremely useful and fruitful—were used in the study of discrete subgroups.

The general study of discrete subgroups of connected Lie groups often neglects the study of finite subgroups of Lie groups, in part because of lack of general methods rather than lack of interest and importance. The paper [32] treats this subject, generalizing some classical results of Jordan. The methods developed here were later used in [35].

An important sub-theme often found in the study of lattices Γ and of discrete uniform subgroups of a Lie group G is whether the structure is rigid or deformable, i.e. whether Γ permits displacement within G, other than by inner automorphisms. One of the most important results in this area is Weil's study of the space of deformations of a lattice in a connected semi-simple group G. In [31] Wang gave an extension of Weil's results including a description of the space of deformations of a discrete uniform subgroup Γ in a connected Lie group whose semi-simple part contains no compact factor. In [34] he showed further that in a connected semi-simple Lie group without compact factors a discrete subgroup Γ such that G/Γ has finite volume (i.e. Γ is a "lattice") can be contained in only a finite number of discrete subgroups. The treatment of lattices in [34] gives some hope that it can lead to a discussion of maximal discrete subgroups.

In [35] he verified a conjecture of Selberg by showing that if G is a connected semi-simple Lie group without compact factors, then there exists a neighborhood U of the identity, which depends on G only, such that for every lattice Γ of G there is a $g \in G$ with $g\Gamma g^{-1} \cap U = \{e\}$, the identity. This was first shown by D. A. Kazdan and G. A. Margulis, but the proof here is different and is a quantitative study of U, related to the method of [32]. The geometric method given here has been useful in other quantitative discussions of discrete subgroups of Lie groups.

In [36] and [37], written with M. Goto, he studied the rigidity of non-discrete uniform subgroups of semi-simple Lie groups. Theorems are obtained that relate to those of the discrete case. It is shown, for example, as a special case, that if G is connected, simple and non-compact and H a non-discrete closed uniform subgroup of G, then there exists a finite set of subgroups $H_1, \ldots, H_s$ such that any $H^* \subset G$ and isomorphic to H must be conjugate to one of the H_i. Finally, [38] is a beautiful expository paper on discrete subgroups of Lie groups discussing recent results as well as their historical origins. It is highly worth reading for anyone who is interested in the subject, written as it is from the broad perspective of one of the innovators. Just as [27] gives an elegant treatment of the geometry of homogeneous spaces and unifies much of his early work, so this paper does for his later work and that of many others including his students and collaborators.

His last paper, [39], appeared thirty-one years after his first, and is a clear indication that his intellectual powers were as strong as ever. It too might be considered a generalization of some of his earlier as well as recent work. The following question is considered: let $\{X, G\}$ be a topological transformation group with space X and group G, then when is the orbit space X/G compact and Hausdorff? He gave a criterion which shows that this depends on X and G separately and *not* on the particular action of G and X. In so doing he tied together a number of isolated results of this sort, including some of his own. The proofs involve ingenious use of group properties as well as spectral sequence arguments on the bundle X over X/G. The paper is a fitting capston to a distinguished career of research and teaching. H. C. Wang is and will be sorely missed by his many friends and his family, but his talent, his industry, and his devotion to mathematics have left a legacy which will be gratefully studied and used by many mathematicians for years to come.

Publications of Hsien-Chung Wang

1. On the projective linear elements of a non-holonomic surface. Science Record, Academia Sinica, **1** (1942), pp. 84–86. MR 5–15.
2. A projective invariant of a non-holonomic surface. Annals of Mathematics, **44** (1943), pp. 562–571. MR 5–15.
3. On the projective deformation of a family of elements of contact. Journal of Indian Mathematical Society, **7** (1943), pp. 51–57. MR 5–157.
4. On the paths with Monge's equations of the second degree as conditions of intersection. Bulletin of American Mathematical Society, **50**, (1944), pp. 935–942. MR 6–128.
5. Path manifolds in a general space of paths. Journal of London Mathematical Society, **21** (1946), pp. 134–139. MR 8–491.
6. The projective deformation of non-holonomic surfaces. Duke Mathematical Journal, **14** (1947), pp. 159–166. MR 8–601.
7. Finsler spaces whose equations of Killing are completely integrable. Journal of London Mathematical Society, **22** (1947), pp. 5–9. MR 9–206.
8. Some examples concerning the relations between homology and homotopy groups. Koninklijke Nederlandse Akademie van Wetenschappen, Proceedings, **50** (1947), p. 873–875. Koninklijke Nederlandse Akademie van Wetenschappen, Indagationes Mathematicae ex Actis Quibus Titulus, **9** (1947), pp. 384–386. MR 9–197.
9. Differential geometry in symplectic space. I. Science Reports, National Tsing Hua University, **4** (1947), pp. 453–477. (with Chern, Shiing-Shen). MR 10–65.
10. Axiom of the plane in a general space of paths. Annals of Mathematics, **49** (1948), pp. 731–737. MR 10–65.
11. The homology groups of the fibre bundles over a sphere. Duke Mathematical Journal, **16** (1949), pp. 33–38. MR 10–468.
12. Homogeneous spaces with non-vanishing Euler characteristic. Science Record, Academia Sinica, **2** (1949), pp. 215–219. MR 11–79.
13. Homogeneous space with non-vanishing Euler characteristics. Annals of Mathematics, **50** (1949), pp. 925–953. MR 11–326.
14. A new characterization of spheres of even dimension. Koninklijke Nederlandse Akademie van Wetenschappen, Proceedings, **52**, pp. 838–847. MR 11–377. Koninklijke Nederlandse Akademie van Wetenschappen, Indagationes Mathematicae ex Actis Quibus Titulus, **11** (1949), pp. 286–295.
15. Metric space and its group of isometries. Proceedings of the International Congress of Mathematicians, Cambridge, **1** (1950), p. 538.
16. A problem of P. A. Smith. Proceedings of American Mathematical Society, **1** (1950), pp. 18–19. MR 11–416.
17. Two theorems on metric spaces. Pacific Journal of Mathematics, **1** (1951), pp. 473–480. MR 13–673.
18. Two-point homogeneous spaces. Annals of Mathematics, **55** (1952), pp. 177–191. MR 13–863.
19. One-dimensional cohomology group of locally compact metrically homogeneous spaces. Duke Mathematical Journal, **19** (1952), pp. 303–310. MR 13–912.

20. A remark on transformation groups leaving fixed an end point. Proceedings of American Mathematical Society, **3** (1952), pp. 548–549. MR 14–72.
21. Closed manifolds with homogeneous complex structure. American Journal of Mathematics, **76** (1954), pp. 1–32. MR 16–518.
22. Complex parallisable manifolds. Proceedings of American Mathematical Society, **5** (1954), pp. 771–776. MR 17–531.
23. A class of affinely connected spaces. Transactions of American Mathematical Society, **80** (1955), pp. 72–79. MR 17–407.
24. Discrete subgroups of solvable Lie groups. I. Annals of Mathematics, **64** (1956), pp. 1–10. MR 17–1224.
25. On invariant connections over a principal fibre bundle. Nagoya Mathematical Journal, **13** (1958), pp. 1–19. MR 21–6001.
26. On contact manifolds. Annals of Mathematics, **68** (1958), pp. 721–734. MR 22–3105. (with W. M. Boothby)
27. Some geometrical aspects of coset spaces of Lie groups. Proceedings of the International Congress of Mathematicians, Edinburgh, 1958, pp. 500–509. MR 28–3117.
28. Compact transformation groups of S^n with an $(n-1)$-dimensional orbit. American Journal of Mathematics, **82** (1960), pp. 698–748. MR 29–1285.
29. Commutators in a semi-simple Lie group. Proceedings of American Mathematical Society, **13** (1962), pp. 907–913, MR 30–190. (with S. Pasiencier)
30. A note on mappings and automorphisms of almost complex manifolds. Annals of Mathematics, **77** (1963), pp. 329–334. MR 26–4284. (with W. M. Boothby and S. Kobayashi)
31. On the deformations of lattice in a Lie group. American Journal of Mathematics, **85** (1963), pp. 189–212. MR 27–2582.
32. On the finite subgroups of connected Lie groups. Commentarii Mathematici Helvetici, **39** (1965) pp. 281–294. MR 31–4856. (with W. M. Boothby)
33. Root systems and abelian subgroups of compact Lie groups. Proceedings of the United States-Japan Seminar in Differential Geometry, Kyoto, Japan, 1965. MR 36–305.
34. On a maximality property of discrete subgroups with fundamental domain of finite measure. American Journal of Mathematics, **89** (1967), pp. 124–132.
35. On discrete nilpotent subgroups of Lie groups. Journal of Differential Geometry, **3** (1969), pp. 481–492.
36. Uniform subgroups of simple Lie groups (with M. Goto). Studies and Essays presented to Yu-Why Chen on his Sixtieth Birthday, 1970, Academia Sinica, Taipei, MR 43–2159.
37. Non-discrete uniform subgroups of semi-simple Lie groups. Mathematische Annalen, **198** (1972), pp. 259–286. MR 50–7411. (with M. Goto)
38. Topics on totally discontinuous groups. Symmetric Spaces (edited by W. M. Boothby and G. Weiss), Marcel-Dekker, New York (1972).
39. On co-compactness of group of transformation groups. American Journal of Mathematics, **95** (1973), pp. 885–903. MR 50–11304.

Washington University, St. Louis, Missouri
University of California, Berkeley, California
Institute for Advanced Study, Princeton, New Jersey

Bulletin of the
Institute of Mathematics
Academia Sinica
8, 2/3, Part I, (July, 1980)

Permissions

Springer-Verlag would like to thank the original publishers of Chern's papers for granting permissions to reprint specific papers in his collection. The following list contains the credit lines for those articles.

[81] Reprinted from *Abh. Math. Sem. Hamburg* **29,** © 1965 by Universität Hamburg.

[82] Reprinted from *Proc. of U.S. Japan Seminar in Differential Geometry,* © 1965 by Yozo Matshushima.

[83] Reprinted from *Bulletin American Mathematical Society* **72,** © 1966 by American Mathematical Society.

[84] Reprinted from *J. of Math. and Mech.* **16,** © 1966 by Indiana University Mathematics Journal.

[86] Reprinted from *J. de l'Analyse Math.* **19,** © 1967 by Weizmann Science Press of Israel.

[87] Reprinted from *J. Diff. Geometry* **1,** © 1967 by Dr. Chuan Hsiung.

[95] Reprinted from *Actes, Congrés intern. Math.* **1,** © 1970 by Editions Bordas, Dunod, Gauthier-Villars.

[96] Reprinted from *Problems in Analysis,* © 1970 by Princeton University Press.

[97] Reprinted from *Differentialgeometrie im Grossen, Berichte aus dem Forschungsinstitut Oberwolfach,* © 1970 by W. Klingenberg.

[99] Reprinted from *Differential Geometry in Honor of K. Yano,* © 1972 by Kinokuniya Book Store Co., Ltd.

[104] Reprinted from *Proc. of conf. on value distribution theory, Tulane Univ.,* © 1974 by Marcel Dekker, Inc.

[105] Reprinted from *Acta Math.* **133,** © 1974 by Institut Mittag-Leffler.

[106] Reprinted from *Amer. J. of Math.* **97,** © 1975 by The Johns Hopkins University Press.

[107] Reprinted from *Math. Scandanavia* **36,** © 1975 by Matematisk Institut, Aarhus, Denmark.

[111] Reprinted from *Science of Matter, dedicated to Ta-You Wu,* © 1978 by Gordon and Breach.

[112] Reprinted from *Jahresbericht deut. Math. Ver.* **80,** © 1978 by B.G. Teubner-Verlag.

[113] Reprinted from *Annali Sc. Norm. Pisa* **5,** © 1978 by Scuola Normale Superiore.

[114] Reprinted from *Proc. U.S. Japan Seminar on Minimal Submanifolds and Geodesics,* © 1978 by Kaigai Publications.

[116] Reprinted from *Rocky Mountain Math. Journal* **10,** © 1979 by Rocky Mountain Mathematics Consortium.